Basic Biostatistics
Statistics for Public Health Practice
SECOND EDITION

B. Burt Gerstman

Professor
Department of Health Science
San Jose State University
San Jose, California

JONES & BARTLETT
LEARNING

World Headquarters
Jones & Bartlett Learning
5 Wall Street
Burlington, MA 01803
978-443-5000
info@jblearning.com
www.jblearning.com

Jones & Bartlett Learning books and products are available through most bookstores and online booksellers. To contact Jones & Bartlett Learning directly, call 800-832-0034, fax 978-443-8000, or visit our website, www.jblearning.com.

Production Credits
Executive Publisher: William Brottmiller
Publisher: Michael Brown
Associate Editor: Chloe Falivene
Editorial Assistant: Nicholas Alakel
Director of Production: Amy Rose
Production Manager: Tracey McCrea
Senior Marketing Manager: Sophie Fleck Teague
Art Development Editor: Joanna Lundeen
Manufacturing and Inventory Control Supervisor: Amy Bacus
Composition: Laserwords Private Limited, Chennai, India
Cover Design: Theresa Day
Photo Research and Permissions Coordinator: Lauren Miller
Cover Image: (crowd) © Henrik Winther Anderson/ShutterStock, Inc.; (light) © Subbotina Anna/ShutterStock, Inc.; (scatterplot) © Jones & Bartlett Learning
Printing and Binding: Edwards Brothers Malloy
Cover Printing: Edwards Brothers Malloy
To order this product, use ISBN: 978-1-284-03601-5

Library of Congress Cataloging-in-Publication Data
Gerstman, B. Burt, author.
  Basic biostatistics : statistics for public health practice / B. Burt Gerstman. -- Second edition.
     p. ; cm.
  Includes index.
  ISBN 978-1-284-02546-0 (pbk.) -- ISBN 1-284-02546-2 (pbk.)
  I. Title.
  [DNLM: 1. Biostatistics--methods. 2. Public Health Practice. WA 950]
  RA409
  362.1072'7--dc23
                                2013034384
6048
Printed in the United States of America
21 20 19 18 17   10 9 8

Dedication

In memory of my father, Joseph, and my mother, Bernadine.

Contents

Preface

Basic Biostatistics: Statistics for Public Health, Second Edition is an introductory text that presents statistical ideas and techniques for students and workers in public health and biomedical research. The book is designed to be accessible to students with modest mathematical backgrounds. No more than high school algebra is needed to understand this book. With this said, I hope to get past the notion that biostatistics is just an extension of math. Biostatistics is much more than that—it is a combination of mathematics and careful reasoning. Do not let the former interfere with the later.

Biostatistical analysis is more than just number crunching. It considers how research questions are generated, studies are designed, data are collected, and results are interpreted.

> Analysis of data, with a more or less statistical flavor, should play many roles.[1]

Basic Biostatistics pays particular attention to exploratory and descriptive analyses. Whereas many introductory biostatistics texts give this topic intermittent attention, this text gives it ongoing consideration.

> Both exploratory and confirmatory data analysis deserves our attention.[2]

Biostatistics entails formulating research questions and designing processes for exploring and testing theories. I hope students who come to the study of biostatistics asking "What's the right answer?" leave asking "Was that the right question?"

> Far better an approximate answer to the *right* question, which is often vague, than an *exact* answer to the wrong question, which can always be made precise.[3]

---

[1] Tukey, J. W. (1980). We need both exploratory and confirmatory. *American Statistician, 34*(1), 23–25.

[2] Tukey, J. W. (1969). Analyzing data: Sanctification or detective work? *American Psychologist, 24*, 83–91. Quote on p. 83.

[3] Tukey, J. W. (1962). The future of data analysis. *Annals of Mathematical Statistics, 33*(1), 1–67. Quote is on pp. 13–14.

Several additional points bear emphasis:

Point 1: Practice, practice, practice. In studying biostatistics, you are developing a new set of reasoning skills. What is true of developing other skills is true of developing biostatistical skills—the only way to get better is to practice with the proper awareness and attention. To this end, illustrative examples and exercises are incorporated throughout the book. I've tried to make the illustrations and exercises relevant. Many have historical importance. Carefully following the reasoning of illustrations and exercises is an opportunity to learn. ("Never regard study as a duty, but as the enviable opportunity to learn.") Answers to odd-numbered exercises are provided in the back of the book. Instructors may request the answers to even-numbered exercises from the publisher.

Point 2: Structure of book. The structure of this book may differ from that of other texts. Chapters are intentionally brief and limited in scope. This allows for flexibility in the order of coverage. The book is organized into three main parts. Part I (Chapters 1–10) addresses basic concepts and techniques. Students should complete these chapters (or a comparable introductory course) before moving on to Parts II and III.

Part II (Chapters 11–15) covers analytic techniques for quantitative responses. Part III (Chapters 16–19) covers techniques for categorical responses. Chapters in these sections can be covered in different orders, at the discretion of the instructor. One instructor may choose to cover these chapters in sequence, while another may cover Chapter 11 and Chapter 16 simultaneously (as an example), since these chapters both address one-sample problems. (Chapter 11 covers one-sample problems for quantitative response variables; Chapter 16 covers one-sample problems for binary response variables.) As another example, one could cover the chapters on categorical responses (Chapters 16–19) before covering the chapters on quantitative response (Chapter 11–15), if this was the focus of the course.

Point 3: Hand calculations *and* computational support. While I believe there is still benefit in learning to calculate statistics by hand, students are encouraged to use statistical software to supplement hand calculations. Use of software tools can free us from some of the tedium of numerical manipulations, leaving more time to think about practical implications of results.

> The only way humans can do BETTER than computers is to take a chance of doing WORSE. So we have got to take seriously the need for steady progress toward teaching routine procedures to computers rather than to people. That will leave the teachers of people with only things hard to teach, but this is our proper fate.[4]

---

[4]Tukey, J. W. (1980). We need both exploratory and confirmatory. *American Statistician, 34*, 23–25.

The book is not tied to any particular software package,[5] but does make frequent use of these four programs: *StaTable*, *SPSS*, *BrightStat.com*, and *WinPepi*.

- *StaTable*[6] is a freeware program that provides access to twenty-five commonly used statistical distributions. It is runs on Windows®, Palm®, and Web-browser (Java) platforms. This program eliminates the need to look-up probabilities in hardcopy tables. It also allows for more precise interpolations for probabilities, especially for continuous random variables.

- *BrightStat.com* is a free statistical program that runs in the cloud through a browser window on your computer. The program is the creation of Dr. Daniel Stricker of University of Bern (Switzerland), is easy to use, requires no special installation, and has an interface and elementary statistical procedures that are analogous to those of SPSS® (IBM Corporation, Armonk, New York). All the datasets have been uploaded and made public to users on the BrightStat.com server for easy access.

- *SPSS*[7]—SPSS is a commercial software package with versions that run on Windows® and Macintosh® computers. A student version of the program can be purchased at campus bookstores and online at www.journeyed.com. Another economical alternative is to lease SPSS for short-term use through the website www.e-academy.com.

- *WinPepi*[8] stands for WINdows Programs for EPIdemiologists. This is a series of computer programs written by Joe Abramson of the Hebrew University-Hadassah School of Public Health and Community Medicine (Jerusalem, Israel) and Paul Gahlinger (University of Utah, Salt Lake City, Utah). The programs are designed for use in the practice of biostatistics, but are also excellent learning aids. *WinPepi* is free.

---

[5] Other commercial and non-commercial statistical software products and utilities can be used to similar effect.
[6] www.cytel.com/Products/StaTable/, Cytel Inc., 675 Massachusetts Ave., Cambridge, MA 02139.
[7] SPSS, Inc., Chicago, IL.
[8] Abramson, J. H. (2004). *WINPEPI* (PEPI-for-Windows): computer programs for epidemiologists. *Epidemiologic Perspectives & Innovations, 1*(1), 6.

Acknowledgments

I wish to express my appreciation to San Jose State University for affording me the opportunity to work on this book. I would especially like to thank Jane Pham for her assistance in the preparation of this edition. I also greatly appreciate the artistic and technical support of Jean Shiota of the Center for Faculty Development for her work in preparing the illustrations for the text. Thanks, Jean. Finally, I wish to express my thanks to those many students in my classes over the years that have provided me with helpful comments, encouragement, and camaraderie. I am especially grateful to Deborah Danielewicz for her concise and accurate corrections of errors made in the first printing of this book.

While writing this book, I had many constructive discussions with Joe Abramson of the Department of Social Medicine, Hebrew University-Hadassah School of Public Health and Community Medicine. I thank Joe for sharing his insights generously. I also greatly appreciate his careful work in developing WINdows Programs for EPIdemiologists.[9] This is really an exceptional set of programs for public health workers. Along these same lines, Paul Gahlinger (University of Utah) deserves credit for conceiving and creating the progenitor of *WinPepi*, *PEPI* (Programs for EPIdemiologists).[10] I also wish to acknowledge Mads Haahr (University of Dublin, Trinity College, Ireland) for creating his true random number generator at www.random.org, and to John C. Pezzullo (Georgetown University) for his helpful compilation of web pages that perform statistical calculations at www.statpages.org.

Finally, I would like to acknowledge the contributions of my wife, who has been patient, understanding, supportive, and encouraging throughout work on this marathon project. As Ralph Kramden (Jackie Gleason) used to tell his wife Alice (Audrey Meadows), "[Honey], you're the greatest!"

---

[9] Abramson, J. H. (2004). *WINPEPI* (PEPI-for-Windows): computer programs for epidemiologists. *Epidemiologic Perspectives & Innovations, 1*(1), 6.
[10] Abramson, J. H., & Gahlinger, P. M. (2001). *Computer Programs for Epidemiologic Analyses: PEPI v.4.0.* Salt Lake City, UT: Sagebrush Press.

About the Author

Dr. Gerstman did his undergraduate work at Harpur College (State University of New York, Binghamtom). He later received a doctor of veterinary medicine (Cornell University), a masters of public health (University of California at Berkeley), and a doctor of philosophy degree (University of California, Davis). He has been a U.S. Public Health Service Epidemiology Fellow and epidemiologist at the U.S. Food and Drug Administration and was an instructor at the National Institutes of Health Foundation Graduate of Health Science at San Jose State University where he teaches epidemiology, biostatistics, and general education courses. Dr. Gerstman's research interests are in the areas of epidemiologic methods, the history of public health, drug safety, and medical and public health record linkage.

Part I

General Concept and Techniques

1 Measurement

■ 1.1 What Is Biostatistics?

Biostatistics is the discipline concerned with the treatment and analysis of numerical data derived from biological, biomedical, and health-related studies. The discipline encompasses a broad range of activities, including the design of research, collection and organization of data, summarization of results, and interpretation of findings. In all its functions, biostatistics is a *servant of the sciences.*[a]

Biostatistics is *more* than just a compilation of computational techniques. It is *not* merely pushing numbers through formulas or computers, but rather it is a way to *detect* patterns and *judge* responses. The statistician is both a data detective and judge.[b] The data detective uncovers patterns and clues, while the data judge decides whether the evidence can be trusted. Goals of biostatistics include[c]:

- Improvement of the intellectual content of the data
- Organization of data into understandable forms
- Reliance on tests of experience as a standard of validity

■ 1.2 Organization of Data

Observations, Variables, Values

Measurement is how we get our data. More formally, *measurement* is "the assigning of numbers or codes according to prior-set rules."[d] Measurement may entail either positioning observations along a numerical continuum (e.g., determining a

---

[a] Neyman, J. (1955). Statistics—servant of all sciences. *Science, 122,* 401–406.
[b] Tukey, J. W. (1969). Analyzing data: sanctification or detective work? *American Psychologist, 24,* 83–91.
[c] Tukey, J. W. (1962). The future of data analysis. *Annals of Mathematical Statistics, 33*(1), 1–67, esp. p. 5.
[d] Stevens, S. S. (1946). On the theory of scales of measurement. *Science, 103,* 677–680.

person's age) or classifying observations into categories (e.g., determining whether an individual is seropositive or seronegative for HIV antibodies).

The term **observation** refers to the unit upon which measurements are made. Observations may correspond to individual people or specimens. They may also correspond to aggregates upon which measurements are made. For example, we can measure the smoking habits of an individual (in terms of "pack-years" for instance) or we can measure the smoking habits of a region (e.g., per capita cigarette consumption). In the former case, the *unit of observation* is a person; in the latter, the *unit of observation* is a region.

Data are often collected with the aid of a **data collection form**, with data on individual data forms usually corresponding to observations. Figure 1.1 depicts four such observations. Each field on the form corresponds to a **variable**. We enter **values** into these fields. For example, the *value* of the fourth *variable* of the first *observation* in Figure 1.1 is "45."

Do not confuse *variables* with *values*. The *variable* is the generic thing being measured. The value is a number or code that has been realized.

> **Observations** are the units upon which measurements are made.
> **Variables** are the characteristics being measured.
> **Values** are realized measurements.

Data Table

Once data are collected, they are organized to form a **data table**. Typically, each row in a data table contains an observation, each column contains a variable, and each cell contains a value.

> **Data table**
> **Observations** → rows
> **Variables** → columns
> **Values** → table cells

Table 1.1 corresponds to data collected with the form depicted in Figure 1.1. This data table has 4 *observations*, 5 *variables*, and 20 *values*. For example, the value of VAR1 for the first observation is "John." As another example, the value of VAR4 for the second observation is "75."

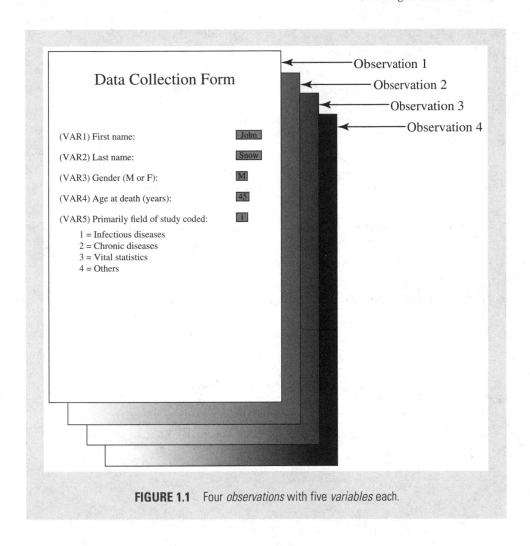

FIGURE 1.1   Four *observations* with five *variables* each.

TABLE 1.1   Data table for data collected with the forms in Figure 1.1.

VAR1	VAR2	VAR3	VAR4	VAR5
John	Snow	M	45	1
William	Farr	M	75	3
Joseph	Goldberger	M	54	2
Janet	Lane-Claypon	F	90	2

Table 1.2 is a data table composed of three variables: country of origin (COUNTRY), per capita cigarette consumption (CIG1930), and lung cancer mortality (LUNGCA). The unit of observation in this data set is a country, not an individual person. Data of this type are said to be *ecological*.[e] This data table has 11 *observations,* 3 *variables,* and 33 *values.*

Exercises

1.1   *Value, variable, observation.* In Table 1.2, what is the value of the LUNGCA variable for the 7th observation? What is the value of the COUNTRY variable for the 11th observation?

1.2   *Value, variable, observation (cont.).* What is the value of the CIG1930 variable for observation 3 in Table 1.2?

1.3   *Value, variable, observation (cont.).* In the form depicted in Figure 1.1, what does VAR3 measure?

1.4   *Value, variable, observation (cont.).* In Table 1.1, what is the value of VAR4 for observation 3?

TABLE 1.2   Per capita cigarette consumption in 1930 (CIG1930) and lung cancer cases per 100,000 in 1950 (LUNGCA) in 11 countries.

COUNTRY	CIG1930	LUNGCA
USA	1300	20
Great Britain	1100	46
Finland	1100	35
Switzerland	510	25
Canada	500	15
Holland	490	24
Australia	480	18
Denmark	380	17
Sweden	300	11
Norway	250	9
Iceland	230	6

Data from U.S. Department of Health, Education, and Welfare. (1964). Smoking and Health. Report of the Advisory Committee to the Surgeon General of the Public Health Service, Page 176. Retrieved April 21, 2003, from http://sgreports.nlm.nih.gov/NN/B/B/M/Q/ segments.html. Original data by Doll, R. (1955). Etiology of lung cancer. *Advances Cancer Research, 3,* 1–50.

[e]The term *ecological* in this context should not be confused with its biological use.

■ 1.3 Types of Measurement Scales

There are different ways to classify variables and measurements. We consider three types of measurement scales: categorical, ordinal, and quantitative.[f] As we go from categorical to ordinal to quantitative, each scale will take on the assumptions of the prior type and adds a further restriction.

- Categorical measurements place observations into unordered categories.
- Ordinal measurements place observations into categories that can be put into rank order.
- Quantitative measurements represent numerical values for which arithmetic operations make sense.

Additional explanation follows.

Categorical measurements place observations into classes or groups. Examples of categorical variables are SEX (male or female), BLOOD_TYPE (A, B, AB, or O), and DISEASE_STATUS (case or noncase). Categorical measurements may occur naturally (e.g., diseased/not diseased) or can be created by grouping quantitative measurements into classes (e.g., classifying blood pressure as normotensive or hypertensive). Categorical variables are also called *nominal variables* (nominal means "named"), *attribute variables,* and *qualitative variables.*

Ordinal measurements assign observations into categories that can be put into rank order. An example of an ordinal variable is STAGE_OF_CANCER classified as stage I, stage II, or stage III. Another example is OPINION ranked on a 5-point scale (e.g., 5 = "strongly agree," 4 = "agree," and so on). Although ordinal scales place observation into order, the "distance" (difference) between ranks is not uniform. For example, the difference between stage I cancer and stage II cancer is not necessarily the same as the difference between stage II and stage III. Ordinal variables serve merely as a ranking and do not truly quantify differences.

Quantitative measurements position observations along a meaningful numeric scale. Examples of quantitative measures are chronological AGE (years), body WEIGHT (pounds), systolic BLOOD_PRESSURE (mmHg), and serum GLUCOSE (mmol/L). Some statistical sources use terms such as *ratio/interval measurement, numeric variable, scale variable,* and *continuous variable* to refer to quantitative measurements.

---

[f]Distinctions between measurement scales often get blurred in practice because the scale type is partially determined by the questions we ask of the data and the purpose for which it is intended. See Velleman, P. F. & Wilkinson, L. (1993). Nominal, ordinal, interval, and ratio typologies are misleading. *American Statistician, 47,* 65–72.

ILLUSTRATIVE EXAMPLE

Weight change and coronary heart disease.[g] A group of 115,818 women between 30 and 55 years of age were recruited to be in a study. Individuals were free of coronary heart disease at the time of recruitment. Body weight of subjects was determined as of 1976. Let us call this variable WT_1976. Weight was also determined as of age 18. Let's call this variable WT_18. From these variables, the investigators calculated weight change for individuals (WT_CHNG = WT_1976 − WT_18). Adult height in meters was determined (HT) and was used to calculate body mass index according to the formula: BMI = weight in kilograms ÷ (height in meters)^2. BMI was determined as of age 18 (BMI_18) and at the time of recruitment in 1976 (BMI_1976). All of these variables are quantitative.

BMI was classified into *quintiles*. This procedure divides a quantitative measurement into five ordered categories with an equal number of individuals in each group. The lowest 20% of the values are put into the first quintile, the next 20% are put into the next quintile, and so on. The quintile cutoff points for BMI at age 18 were <19.1, 19.1–20.3, 20.4–21.5, 21.6–23.2, and ≥23.3. Let us put this information into a new variable called BMI_18_GRP encoded 1, 2, 3, 4, 5 for each of the quintiles. This is an ordinal variable.

The study followed individuals over time and monitored whether they experienced adverse coronary events. A new variable (let us call it CORONARY) would then be used to record this new information. CORONARY is a categorical variable with two possible values: either the person did or did not experience an adverse coronary event. During the first 14 years of follow-up, there were 1292 such events.

Exercises

1.5 *Measurement scale.* Classify each variable depicted in Figure 1.1 as either quantitative, ordinal, or categorical.

1.6 *Measurement scale (cont.).* Classify each variable in Table 1.2 as quantitative, ordinal, or categorical.

---

[g]Willett, W. C., Manson, J. E., Stampfer, M. J., Colditz, G. A., Rosner, B., Speizer, F. E., et al. (1995). Weight, weight change, and coronary heart disease in women. Risk within the "normal" weight range. *JAMA, 273,* 461–465.

■ 1.4 Data Quality

Meaningful Measurements

How reliable is a single blood pressure measurement? What does an opinion score *really* signify? How is cause of death determined on death certificates? Responsible statisticians familiarize themselves with the measurements they use in their research. This requires a critical mind and, often, consultation with a subject matter specialist. We must always do our best to understand the variables we are analyzing.

In our good intentions to be statistical, we might be tempted to collect data that is several steps removed from what we really want to know. This is often a bad idea.

> A drunken individual is searching for his keys under a street lamp at night. A passerby asks the drunk what he is doing. The drunken man slurs that he is looking for his keys. After helping the man unsuccessfully search for the keys under the streetlamp, the Good Samaritan inquires whether the drunk is sure the keys were lost under the street lamp. "No," replies the drunk, "I lost them over there." "Then why are you looking here?" asks the helpful Samaritan. "Because the light is here," says the drunk.

Beware of looking for statistical relationships in data that are far from the information that is actually required.

Here is a story you may be less familiar with. This story comes from the unorthodox scientist Richard Feynman. Feynman calls pseudoscientific work **Cargo Cult science**.[h] This story is based on an actual occurrence in a South Seas island during World War II. During the war, the inhabitants of the island saw airplanes land with goods and materials. With the end of the war, the cargo airplanes ceased and so did deliveries. Since the inhabitants wanted the deliveries to continue, they arranged to imitate things they saw when cargo arrived. Runway lights were constructed (in the form of fires), a wooden hut with bamboo sticks to imitate antennas was built for a "controller" who wore two wooden pieces on his head to emulate headphones, and so on. With the Cargo Cult in place, the island inhabitants awaited airplanes to land. The form was right on the surface, but of course things no longer functioned as they had hoped. Airplanes full of cargo failed to bring goods and services to the island inhabitants. "Cargo Cult" has come to mean a pseudoscientific method that follows precepts and forms, but it is missing in the honest, self-critical assessments that are essential to scientific investigation.

---

[h] Feynman, R. P. (1999). Cargo Cult science: Some remarks on science, pseudoscience, and learning how not to fool yourself. In *The Pleasure of Finding Things Out* (pp. 205–216). Cambridge, MA: Perseus.

These two stories are meant to remind us that sophisticated numerical analyses cannot compensate for poor-quality data. Statisticians have a saying for this: "**Garbage in, garbage out,**" or **GIGO**, for short.

> **GIGO stands for "garbage in, garbage out."**

When nonsense is input into a public health statistical analysis, nonsense comes out. The resulting nonsensical output will look just as "scientific" and "objective" as a useful statistical analysis, but it will be worse than useless—it will be counterproductive and could ultimately have detrimental effects on human health.

Objectivity (the intent to measure things as they are without shaping them to conform to a preconceived worldview) is an important part of measurement accuracy. Objectivity requires a suspension of judgment; it requires us to look at *all* the facts, not just the facts that please us.

Consider how subtle word choices may influence responses. Suppose I ask you to remember the word "jam." I can influence the way you interpret the word by preceding it with the word "traffic" or "strawberry." If I influence your interpretation in the direction of traffic jam, you are less likely to recognize the word subsequently if it is accompanied by the word "grape."[i] This effect will occur even when you are warned not to contextualize the word. The point is that we do not interpret words in a vacuum. When collecting information, nothing should be taken for granted.

Two Types of Measurement Inaccuracies: Imprecision and Bias

We consider **two forms of measurement errors**: imprecision and bias. **Imprecision** expresses itself in a measurement as the inability to get the same result upon repetition. **Bias**, on the other hand, expresses itself as a tendency to overestimate or underestimate the true value of an object. The extent to which something is imprecise can often be quantified using the laws of probability. In contrast, bias is often difficult to quantify in practice. When something is *un*biased, it is said to be *valid*.

Figure 1.2 depicts how imprecision and bias may play out in practice. This figure considers repeated glucose measurement in a single serum sample. The true glucose level in the sample is 100 mg/dl. Measurements have been taken with four different instruments.

---

[i] Example based on Baddeley cited in Gourevitch, P. (1999, 14 June). The memory thief. *The New Yorker*, 48–68.

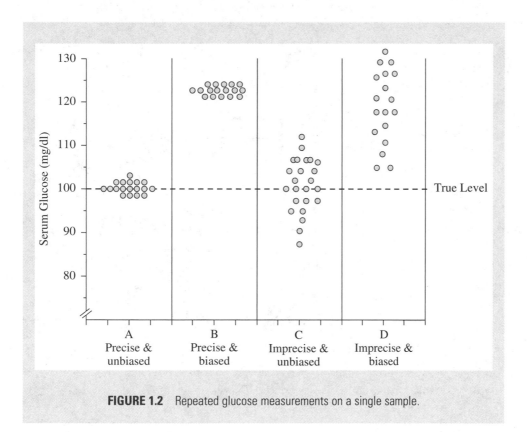

FIGURE 1.2 Repeated glucose measurements on a single sample.

- Instrument A is precise and unbiased.
- Instrument B is precise and has a positive bias.
- Instrument C is imprecise and unbiased.
- Instrument D is imprecise and has a positive bias.

In practice, it is easier to quantify imprecision than bias. This fact can be made clear by an analogy. Imagine an archer shooting at a target. A brave investigator is sitting behind the target at a safe distance. Because the investigator is behind the target, he cannot see the location of the actual bull's-eye. He can, however, see where the arrow pokes out of the back of the target (Figure 1.3). This is analogous to looking at the results of a study—we see where the arrows stick out but do not actually know the location of the bull's-eye.

Figure 1.4 shows exit sites of arrows from two different archers. From this we can tell that Archer B is more **precise** than Archer A (values spaced tightly). We cannot, however, determine which Archer's aim centers in on the bull's-eye. Characterization

FIGURE 1.3   A brave investigator sits behind the target to see what he can see.

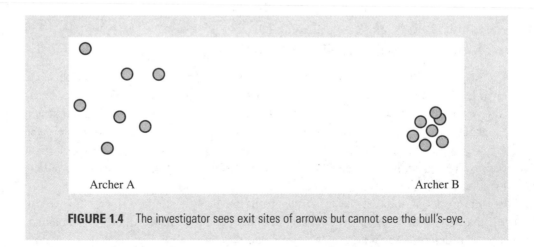

FIGURE 1.4   The investigator sees exit sites of arrows but cannot see the bull's-eye.

of precision is straightforward—it measures the scatter in the results. Characterization of validity, however, requires additional information.

Summary Points (Measurement)

1. **Biostatistics** involves a broad range of activities that help us improve the intellectual content of data from biological, biomedical, and public health–related studies; it is *more* than just a compilation of computational methods.

2. **Measurement** is the assigning of numbers or codes according to prior-set rules.

3. The three basic **measurement scales** are as follows:

   (a) **Categorical** (nominal), which represent unordered categories.

   (b) **Ordinal**, which represent categories that can be put into rank order.

   (c) **Quantitative** (scale, continuous, interval, and ratio), which represent meaningful numerical values for which arithmetic operations such as addition and multiplication make sense.

4. An **observation** is a unit upon which measurements are made (e.g., individuals). Data from observations are stored in rows of data tables.

5. A **variable** is a characteristic that is measured, such as age, gender, or disease status. Data from variables form columns of data tables.

6. **Values** are realized measurements. For example, the value for the variable AGE for observation #1 is, say, "32." Values are stored in table cells.

7. The utility of a study depends on the **quality of its measurements**. When nonsense is input into a biostatistical analysis, nonsense comes out ("garbage in, garbage out").

8. Measurements vary in their **precision** (ability to be replicated) and **validity** (ability to objectively identify the true nature of the observation).

Vocabulary

Bias	Observation
Cargo Cult science	Ordinal measurements
Categorical measurements	Precise
Data table	Quantitative measurements
Garbage in, garbage out (GIGO)	Valid
Imprecision	Values
Measurement	Variable
Objectivity	

Review Questions

1.1   What types of activities other than "calculations" and "math" are associated with the practice of statistics?

1.2   Define the term *measurement*.

1.3   Select the best response: Data in a column in a data table corresponds to a(n):

(a) observation

(b) variable

(c) value

1.4   Select the best response: Data in a row in a data table corresponds to a(n):

(a) observation

(b) variable

(c) value

1.5   List the three main measurement scales addressed in this chapter.

1.6   What type of measurement assigns a name to each observation?

1.7   What type of measurement is based on categories that can be put in rank order?

1.8   What type of measurement assigns a numerical value that permits for meaningful mathematical operations for each observation?

1.9   What does GIGO stand for?

1.10 Provide synonyms for *categorical data*.

1.11 Provide synonyms for *quantitative data*.

1.12 Differentiate between imprecision and bias.

1.13 How is imprecision quantified?

Exercises

1.7   *Duration of hospitalization.* Table 1.3 contains data from an investigation that studied antibiotic use in hospitals.

(a) Classify each variable as quantitative, ordinal, or categorical.

(b) What is the value of the DUR variable for observation 4?

(c) What is the value of the AGE variable for observation 24?

1.8   *Clustering of adverse events.* An investigation was prompted when the U.S. Food and Drug Administration received a report of an increased frequency of an adverse drug-related event after a hospital switched from the innovator company's product to a generic product. To address this issue, a team of investigators

TABLE 1.3 Twenty-five observations derived from hospital discharge summaries.

ID	DUR	AGE	SEX	TEMP	WBC	AB	CULT	SERV
1	5	30	2	99.0	8	2	2	1
2	10	73	2	98.0	5	2	1	1
3	6	40	2	99.0	12	2	2	2
4	11	47	2	98.2	4	2	2	2
5	5	25	2	98.5	11	2	2	2
6	14	82	1	96.8	6	1	2	2
7	30	60	1	99.5	8	1	1	1
8	11	56	2	98.6	7	2	2	1
9	17	43	2	98.0	7	2	2	1
10	3	50	1	98.0	12	2	1	2
11	9	59	2	97.6	7	2	1	1
12	3	4	1	97.8	3	2	2	2
13	8	22	2	99.5	11	1	2	2
14	8	33	2	98.4	14	1	1	2
15	5	20	2	98.4	11	2	1	2
16	5	32	1	99.0	9	2	2	2
17	7	36	1	99.2	6	1	2	2
18	4	69	1	98.0	6	2	2	2
19	3	47	1	97.0	5	1	2	1
20	7	22	1	98.2	6	2	2	2
21	9	11	1	98.2	10	2	2	2
22	11	19	1	98.6	14	1	2	2
23	11	67	2	97.6	4	2	2	1
24	9	43	2	98.6	5	2	2	2
25	4	41	2	98.0	5	2	2	1

Here's a codebook for the data:

Variable	Description
DUR	Duration of hospitalization (days)
AGE	Age (years)
SEX	1 = male, 2 = female
TEMP	Body temperature (degrees Fahrenheit)
WBC	White blood cells per 100 ml
AB	Antibiotic use: 1 = yes, 2 = no
CULT	Blood culture taken: 1 = yes, 2 = no
SERV	Service: 1 = medical, 2 = surgical

Data from Townsend, T. R., Shapiro, M., Rosner, B., & Kass, E. H. (1979). Use of antimicrobial drugs in general hospitals. I. Description of population and definition of methods. *Journal of Infectious Disease*, *139*(6), 688–697 and Rosner, B. (1990). *Fundamentals of Biostatistics* (3rd ed.). Belmont, CA: Duxbury Press, p. 36.

completed chart reviews of patients who had received the drugs in question. Table 1.4 lists data for the first 25 patients in the study.

(a) Classify each variable in the table as either quantitative, ordinal, or categorical.

(b) What is the value of the AGE variable for observation 4?

(c) What is the value of the DIAG (diagnosis) variable for observation 2?

1.9 *Dietary histories.* Prospective studies on nutrition often require subjects to keep detailed daily dietary logs. In contrast, retrospective studies often rely on recall. Which method—dietary logs or retrospective recall—do you believe is more likely to achieve accurate results? Explain your response.

TABLE 1.4   First 25 observations from a study of cerebellar toxicity.

I	AGE	SEX	MANUF	DIAG	STAGE	TOX	DOSE	SCR	WEIGHT	GENERIC
1	50	1	J	1	1	1	36.0	0.8	66	1
2	21	1	J	1	2	2	29.0	1.1	68	1
3	35	1	J	2	2	2	16.2	0.7	97	1
4	49	2	S	1	1	2	29.0	0.8	83	2
5	38	1	J	2	2	1	16.2	1.4	97	1
6	42	1	S	2	2	2	18.0	1.0	82	2
7	17	1	J	1	2	2	17.4	1.0	64	1
8	20	1	S	2	2	2	17.4	1.0	73	2
9	49	2	J	1	1	2	37.2	0.7	103	1
10	41	2	J	1	2	2	18.6	0.9	58	1
11	20	1	S	2	2	2	18.0	1.1	113	2
12	55	1	S	1	1	2	36.0	0.8	87	2
13	44	2	J	1	1	1	22.4	1.2	59	1
14	23	1	S	2	2	2	39.6	0.8	83	2
15	64	2	S	1	1	2	30.0	0.9	69	2
16	65	1	S	1	1	1	23.2	1.7	106	2
17	23	2	S	1	2	2	16.8	0.9	66	2
18	44	1	S	1	2	2	17.4	1.0	84	2
19	29	2	S	2	1	2	18.0	0.7	56	2
20	32	1	S	1	2	2	18.0	1.0	84	2
21	18	1	S	2	2	2	17.4	0.9	70	2
22	22	1	S	1	1	1	26.1	1.7	69	2
23	43	2	J	2	2	2	18.0	0.8	63	1
24	39	2	S	1	2	2	18.0	0.9	55	2
25	38	2	J	1	1	1	16.0	1.0	112	1

Here's a codebook for the data:

Variable	Description
AGE	Age (years)
SEX	1 = male; 2 = female
MANUF	Manufacturer of the drug: Smith or Jones
DIAG	Underling diagnosis: 1 = leukemia; 2 = lymphoma
STAGE	Stage of disease: 1 = relapse; 2 = remission
TOX	Did cerebellar toxicity occur?: 1 = yes; 2 = no
DOSE	Dose of drug (g/m^2)
SCR	Serum creatinine (mg/dl)
WEIGHT	Body weight (kg)
GENERIC	Generic drug: 1 = yes; 2 = no

Data from Jolson, H. M., Bosco, L., Bufton, M. G., Gerstman, B. B., Rinsler, S. S., Williams, E., et al. (1992). Clustering of adverse drug events: analysis of risk factors for cerebellar toxicity with high-dose cytarabine. *JNCI, 84,* 500–505.

1.10 *Variable types.* Classify each of the measurements listed here as quantitative, ordinal, or categorical.

(a) Response to treatment coded as 1= no response, 2 = minor improvement, 3 = major improvement, 4 = complete recovery

(b) Annual income (pretax dollars)

(c) Body temperature (degrees Celsius)

(d) Area of a parcel of land (acres)

(e) Population density (people per acre)

(f) Political party affiliation coded 1 = Democrat, 2 = Republican, 3 = Independent, 4 = Other

1.11 *Variable types 2.* Here is more practice in classifying variables as quantitative, ordinal, or categorical.

(a) White blood cells per deciliter of whole blood

(b) Leukemia rates in geographic regions (cases per 100,000 people)

(c) Presence of type II diabetes mellitus (yes or no)

(d) Body weight (kg)

(e) Low-density lipoprotein level (mg/dl)

(f) Grade in a course coded: A, B, C, D, or F

(g) Religious identity coded 1 = Protestant, 2 = Catholic, 3 = Muslim, 4 = Jewish, 5 = Atheist, 6 = Buddhist, 7 = Hindu, 8 = Other

(h) Blood cholesterol level classified as either 1 = hypercholesterolemic, 2 = borderline hypercholesterolemic, 3 = normocholesterolemic

(i) Course credit (pass or fail)

(j) Ambient temperature (degrees Fahrenheit)

(k) Type of life insurance policy: 1 = none, 2 = term, 3 = endowment, 4 = straight life, 5 = other

(l) Satisfaction: 1 = very satisfied, 2 = satisfied, 3 = neutral, 4 = unsatisfied, 5 = very unsatisfied

(m) Movie review rating: 1 star, 1½ stars, 2 stars, 2½ stars, 3 stars, 3½ stars, 4 stars

(n) Treatment group: 1 = active treatment, 2 = placebo

1.12 *Rating hospital services.* A source ranks hospitals based on each of the following items. (The unit of observation in this study is "hospital.") Identify the measurement scale of each item as quantitative, ordinal, or categorical.

(a) Percentage of patients who survive a given surgical procedure.

(b) Type of hospital: general, district, specialized, or teaching.

(c) Average income of patients that are admitted to the hospital.

(d) Mean salary of physicians working at the hospital.

1.13 *Age recorded on different measurement scales.* We often have a choice of whether to record a given variable on either a quantitative or a categorical scale. How does one measure age quantitatively? Provide an example by which age can be measured categorically.

1.14 *Physical activity in elementary school children.* You are preparing to study physical activity levels in elementary school students. Describe two quantitative variables and two categorical variables that you might wish to measure.

1.15 *Binge drinking.* "Binge alcohol use" is often defined as drinking five or more alcoholic drinks on the same occasion at least one time in the past 30 days. The following table lists data from the National Survey on Drug Use and Health based on a representative sample of the U.S. population of age 12 years and older. Data represent estimated percentages reporting binge drinking in 2003 and 2008. There were about 68,000 respondents in each time period.

AGEGRP Age group (years)	BINGE2003 % Binge drinkers, 2003	BINGE2008 % Binge drinkers, 2008
12–17	10.6	8.8
18–25	41.6	41.8
26 and above	21.0	22.1

Data from *National Data Book, 2012 Statistical Abstract*, United States Census Bureau, Table 207. Retrieved from www.census.gov/compendia/statab/2012edition.html. Accessed February 18, 2013.

Classify the measurement scale of each of the variables in this data table as categorical, ordinal, or quantitative.

1.16 *Assessing two sets of measurements*. Two sets of measurements are given in the following list. Which set of measurements is more precise? Can you determine which is less biased? Explain your reasoning.

Set A:	12.1	13.2	14.3
Set B:	12.1	12.2	12.3

2 | Types of Studies

This chapter considers how data are generated for studies. Two general classes of studies are considered: **surveys** and **comparative studies**. Comparative studies come in experimental and nonexperimental forms. Therefore, our study of taxonomy consists of:

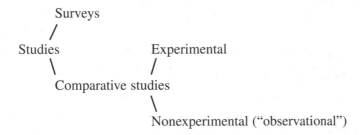

The general purpose of a survey is to quantify population characteristics. Comparative studies are done to quantify relationships between variables. Because these are distinct goals, we consider surveys and comparative studies separately, starting first with surveys.

■ 2.1 Surveys

Surveys are used to quantify population characteristics. The population consists of all entities worthy of study. A survey that attempts to collect information on all individuals in the population is a **census**. Because a true census is seldom feasible, most surveys collect data on only a portion or **sample** of the population. Data in the sample are then used to **infer** population characteristics. Sampling has many advantages. It saves time and money. It can also be more accurate, because when fewer individuals are studied, a larger percentage of resources can be devoted to collecting high-quality information on a broad scope of variables. Sampling is the "rule" in statistics; rarely are data collected for the entire population.

ILLUSTRATIVE EXAMPLE

Youth Risk Behavior Surveillance (YRBS). The Youth Risk Behavior Surveillance System monitors health behaviors in youth and young adults in the United States. Six categories of health-risk behaviors are monitored. These include: (1) behaviors that contribute to unintentional injuries and violence; (2) tobacco use; (3) alcohol and drug use; (4) sexual behaviors; (5) unhealthy dietary behaviors; and (6) physical activity levels and body weight. The 2003 report used information from 15,240 questionnaires completed at 158 schools to infer health-risk behaviors for the public and private school student populations of the United States and District of Columbia.[a] The 15,240 students who completed the questionnaires comprise the sample. This information is used to infer the characteristics of the several million public and private school students in the United States for the period in question.

Simple Random Samples

Samples must be collected in such a way to allow for generalizations to be made to the entire population. To accomplish this goal, the sample must entail an element of chance; a random sample must be used. The most fundamental type of random sample is a **simple random sample (SRS)**. The idea of simple random sampling is to collect data from the population so (1) each population member has the same probability of being selected into the sample and (2) the selection of any individual into the sample does not influence the likelihood of selecting any other individual. These characteristics are referred to as **sampling independence**.

Here is a more formal definition of what constitutes a simple random sample:

> A simple random sample (SRS) of size *n* is selected so that all possible combinations of *n* individuals in the population are equally likely to comprise the sample.

An SRS can be achieved by placing identifiers for population members in a hat, thoroughly mixing up the identifiers, and then blindly drawing entries. In practice, a table of random digits or a software program is used to aid in the selection process.

---

[a] Grunbaum, J. A., Kann, L., Kinchen, S., Ross, J., Hawkins, J., Lowry, R., et al. (2004). Youth risk behavior surveillance—United States, 2003. *MMWR Surveillance Summary, 53*(2), 1–96. Available: www.cdc.gov/mmwr/preview/mmwrhtml/ss5302a1.htm.

Tables of Random Digits

Appendix A Table A lists 2000 random digits. Each digit 0 through 9 is equally likely to occur at any position in the table. This makes selection of any sequence of digits equally likely no matter where you enter the table.

Numbers are listed in groups of five to make it easier to follow sequences along the table.[b] Here is the first line of digits in the table:

79587   19407   49825   58687   99639   82670   73457   53546   30292   75741

To select an SRS of size n from a population with N individuals:

1. Number each population member with a unique identifier number 1 through N.
2. Pick an arbitrary spot to enter Table A. Change the spot of entry each time you make a new selection.
3. Go down the rows (or columns) of the table, selecting appropriate tuples of numbers. For example, select digit-pairs if N is no more than 99, select triplets if N is no more than 999, and so on.
4. Continue selecting tuples until you have n relevant entries.

ILLUSTRATIVE EXAMPLE

Selecting a simple random sample. Suppose a high school population has 600 students and you want to choose three students at random from this population. To select an SRS of $n = 3$:

1. Get a roster of the school. Assign each student a unique identifier 1 through 600.
2. Enter Table A at (say) line 15. Line 15 starts with these digits:

    76931   95289   55809   19381   56686

3. The first six triplets of numbers in this line are 769, 319, 528, 955, 809, and 193.
4. The first triplet (769) is excluded because there is no individual with that number in the population. The next two triplets (319 and 528) identify the first two students to enter the sample. The next two triplets (955 and 809) are not relevant. The last student to enter the sample is student 193.

The final sample is composed of students with the IDs 319, 528, and 193.

---

[b] Number groupings have no special meaning.

Software programs that generate random numbers are often used in practice.[c]

Notes

1. **Sampling fraction.** The ratio of the size of sample (n) to the population size (N) is the sampling fraction. For example, in selecting $n = 6$ individuals from a population of $N = 600$, the sampling fraction $= \dfrac{6}{600} = 0.01$ or 1%.[d]

2. **Sampling with replacement and sampling without replacement.** Sampling a finite population can be done with replacement or without replacement. Sampling with replacement is accomplished by "tossing" selected members back into the mix after they have been selected. In this way, any given unit can appear more than once in a sample. This may seem odd, but all N members of the population still have a $\dfrac{n}{N}$ chance of being selected at each draw and random sampling is still achieved. **Sampling without replacement** is done, so that once a population member has been selected, the selected unit is removed from possible future reselection. This too is a legitimate way to select a simple random sample. The distinction between sampling with replacement and without replacement is of consequence only when more complex sampling designs are used.

3. **Cautions.** Samples that tend to overrepresent or underrepresent certain segments of the population can **bias**[e] the results of a survey in favor of a certain outcome. Here are examples of selection biases we wish to avoid:

   - **Undercoverage** occurs when some groups in the source population are left out of or are underrepresented on the population listing. Sampling from such a list will undermine the goal of achieving equal probabilities of selection.

   - **Volunteer bias** can occur because self-selected participants of a survey tend to be atypical of the population. For example, web surveys may produce flawed results because people who are aware of a website or its sponsor are often interested in a particular point of view. Individuals who take the time to respond to a web survey are further self-selected.

---

[c] A true random number generator is available at www.random.org. This site is owned and maintained by Mads Haahr, Distributed Systems Group, Department of Computer Science, University of Dublin, Trinity College, Ireland.

[d] If the sampling fraction is above 5%, a finite population correction factor must be incorporated into some statistical formulas. In this introductory text, we consider only sampling fractions that are smaller than 5%. For addressing samples that use large sampling fractions see Cochran, W. G. (1977). *Sampling Techniques* (3rd ed.). New York: Wiley.

[e] The term *bias* in statistics refers to a systematic error.

- **Nonresponse bias** occurs when a large percentage of individuals refuse to participate in a survey or cannot otherwise be contacted. Nonresponders often differ systematically from responders, skewing results of the survey.

4. **When random sampling is not possible.** Surveys should use random sampling techniques to collect data whenever possible. However, practical limitations often prevent this from happening. When a sample is not random, it is necessary to ask if the data in the survey can reasonably be viewed as if it were generated by a random sampling technique.

Other Types of Probability Samples[f]

A **probability sample** is a sample in which each member of the population has a known probability of being selected into the sample. SRS is the most basic type of probability sample, serving as the building block for more complex sampling designs. In practice, more complex probability samples are often required. Three such schemes are stratified random sampling, cluster sampling, and multistage sampling.

A **stratified random sample** is a sample that draws independent SRSs from within relatively homogeneous groups or "strata." For example, the population can be divided into 5-year age groups (0–4, 5–9, ...) with simple random samples of varying sizes drawn from each age-strata. Results can then be combined based on the sampling fractions used within each stratum.

Cluster samples randomly select large units (clusters) consisting of smaller subunits. This allows for random sampling when there is no reliable list of subunits but there is a reliable list of the clusters. For example, we may have a list of household addresses but not of individuals within the households. Households (clusters) are selected at random, and all individuals are studied within the clusters.

Large-scale surveys often use **multistage sampling** techniques in which large-scale units are selected at random. Then, subunits are sampled in successive stages. The Youth Behavior Survey (introduced earlier) used a multistage sample.

ILLUSTRATIVE EXAMPLE

Youth Risk Behavior Survey (YRBS). The population ("sampling frame") for the YRBS consisted of all public and private schools with students in grades 9–12 in the 50 United States plus the District of Columbia. This sampling frame

continues

---

[f]Your instructor may choose to skip this section without hindering further study.

was divided into 1262 primary sampling units, which consisted of counties, areas of large counties, or groups of small counties. From these 1262 primary units, 57 were selected at random with probability proportional to overall school enrollments. The next stage of sampling selected 195 schools at random with probability proportional to the school enrollment size. The third sampling stage consisted of randomly selecting one or two intact classes from each grade from within the chosen school. All students in the selected classes were then eligible to participate in the survey. Here's a schematic representation of the three stages of sampling used by the YRBS:

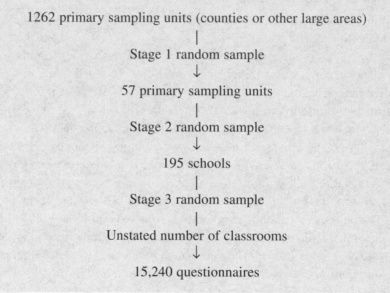

1262 primary sampling units (counties or other large areas)
|
Stage 1 random sample
↓
57 primary sampling units
|
Stage 2 random sample
↓
195 schools
|
Stage 3 random sample
|
Unstated number of classrooms
↓
15,240 questionnaires

Exercises

2.1   *Sample and population.* For the scenarios presented here, identify the source population and sample as specifically as possible. If information is insufficient, do your best to provide a reasonable description of the population and sample and then suggest additional "person, place, and time characteristics" that are needed to better define the population.

(a) A study that reviewed 125 discharge summaries from a large university hospital in metropolitan Detroit found that 35% of the individuals in the hospital received antibiotics during their stay.

(b) A study of eighteen 35- to 44-year-old diabetic men found a mean body mass index that was 13% above what is considered to be normal.

2.2 *A survey of rheumatoid arthritis patients.* A survey mailed questionnaires to 486 patients with rheumatoid arthritis. Responses were received from 334 patients, corresponding to a response rate of 69%. Nonresponders were traced and approached for a telephone interview. Two percent of the responders reported "never having pain" during the study interval, while 6% of eligible participants reported "no pain." Two percent of responders reported "no contact" with healthcare services during the follow-up interval, compared with 4% of eligible responders.[g] Would you trust the results of the survey based solely on the information from initial responders? Explain your response.

2.3 *California counties.* As a pilot investigation for a survey of California county health departments, you want to select four counties at random from the list of counties in Table 2.1. Use the table of random digits (Table A) starting in row 33 to select your simple random sample.

TABLE 2.1 California counties.

01 Alameda	21 Marin	41 San Mateo
02 Alpine	22 Mariposa	42 Santa Barbara
03 Amador	23 Mendocino	43 Santa Clara
04 Butte	24 Merced	44 Santa Cruz
05 Calaveras	25 Modoc	45 Shasta
06 Colusa	26 Mono	46 Sierra
07 Contra Costa	27 Monterey	47 Siskiyou
08 Del Norte	28 Napa	48 Solano
09 El Dorado	29 Nevada	49 Sonoma
10 Fresno	30 Orange	50 Stanislaus
11 Glenn	31 Placer	51 Sutter
12 Humboldt	32 Plumas	52 Tehama
13 Imperial	33 Riverside	53 Trinity
14 Inyo	34 Sacramento	54 Tulare
15 Kern	35 San Benito	55 Tuolumne
16 Kings	36 San Bernardino	56 Ventura
17 Lake	37 San Diego	57 Yolo
18 Lassen	38 San Francisco	58 Yuba
19 Los Angeles	39 San Joaquin	
20 Madera	40 San Luis Obispo	

[g] Rupp, I., Triemstra, M., Boshuizen, H. C., Jacobi, C. E., Dinant, H. J., & van den Bos, G. A. (2002). Selection bias due to non-response in a health survey among patients with rheumatoid arthritis. *European Journal of Public Health, 12*(2), 131–135.

■ 2.2 Comparative Studies

The Basics

The general objective of comparative studies is to learn about the relationship between an explanatory variable and response variable. While being able to infer population characteristics based on a sample was the leading concern in surveys, "comparability" is the leading concern in comparative studies; that is, when we compare groups, they should be similar in as many ways possible except for that of the explanatory factor. This is because the effects of a factor can be judged only in relation to what would happen in its absence; "like-to-like" comparisons are necessary for valid results.

In their most basic form, comparative studies compare responses from one group that is exposed to an explanatory factor to those of a group that is not exposed. There are two general ways to accomplish this; either experimentally or nonexperimentally. In **experimental studies,** the investigator assigns the exposure to one group while leaving the other nonexposed. In **nonexperimental studies,** the investigator merely classifies individuals as exposed or nonexposed without intervention. Nonexperimental studies are also called **observational studies.**[h] Figure 2.1 depicts schematics of experimental and nonexperimental study designs.

Various types of control groups may be used in experimental studies. One type is the **placebo** control. A placebo is an inert or innocuous intervention. Even though the placebo is inert, we expect changes to occur after its administration. Some of this change is random, some is due to the natural history of events, and some is due to "the placebo effect."[i] Another type of control group is the active control group. Active controls receive an alternative physiologically active treatment or intervention. In using either a placebo control or active control, the control group is handled and observed in the same manner as the treatment group. This helps sort out changes due to the treatment and those due to other factors.

Explanatory Variable and Response Variable

Comparative studies have explanatory variables and response variables. The **explanatory variable** is the treatment or exposure that explains or predicts changes in the response variable. The **response variable** is the outcome or response being

---

[h] "Nonexperimental study" is preferable because both experimental studies and nonexperimental studies require observation.

[i] The placebo effect is the perceived improvement following treatment with a placebo. Why the placebo effect occurs is a bit of a mystery but may be related to the fact that subjects know they are being observed and cared for.

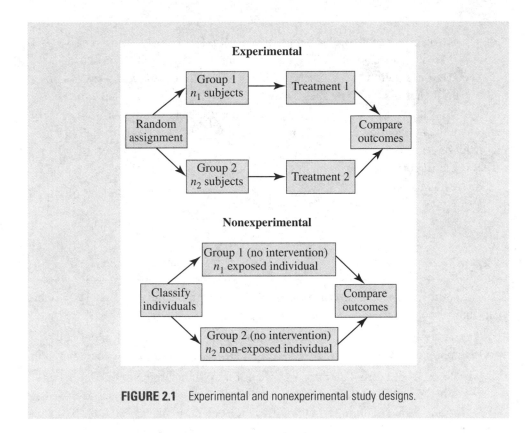

Experimental

Random assignment → Group 1 n_1 subjects → Treatment 1 → Compare outcomes

Random assignment → Group 2 n_2 subjects → Treatment 2 → Compare outcomes

Nonexperimental

Classify individuals → Group 1 (no intervention) n_1 exposed individual → Compare outcomes

Classify individuals → Group 2 (no intervention) n_2 non-exposed individual → Compare outcomes

FIGURE 2.1 Experimental and nonexperimental study designs.

investigated. In the MRFIT illustrative example, the explanatory variable is enrollment in the special intervention program. The response variable is whether a cardiovascular incident occurred during the observation period.

ILLUSTRATIVE EXAMPLE

Multiple Risk Factor Intervention Trail (MRFIT). The Multiple Risk Factor Intervention Trial (MRFIT is pronounced *Mister Fit*) was conducted in the late 1970s and early 1980s to test the effectiveness of coronary disease prevention programs.[j] Study subjects were randomly assigned to either a treatment group that received special interventions intended to reduce the risk of heart disease

continues

---

[j] Multiple Risk Factor Intervention Trial Research Group. (1982). Risk factor changes and mortality results. *JAMA, 248*(12), 1465–1477.

or a control group that received their usual sources of health care. The study found that heart disease mortality declined significantly in both groups. This is because observations took place at a time when the population was learning about the benefits of low-fat diets, exercise, smoking cessation, and other cardiovascular disease-prevention strategies. Had there been no concurrent control group, the interventions would have mistakenly been viewed as effective and considerable resources might have been wasted on ineffective programs. This shows the importance of using a concurrent control group to establish the effect of a treatment.

Exercises

2.4   *Explanatory variable and response variable.* Identify the explanatory variable and response variable in each of the studies described here.

    (a) A study of cell phone use and primary brain cancer suggested that cell phone use was not associated with an elevated risk of brain cancer.[k]

    (b) Records of more than three-quarters of a million surgical procedures conducted at 34 different hospitals were monitored for anesthetics safety. The study found a mortality rate of 3.4% for one particular anesthetic. No other major anesthetics was associated with mortality greater than 1.9%.[l]

    (c) In a landmark study involving more than three-quarters of a million individuals in the United States, Canada, and Finland, subjects were randomly given either the Salk polio vaccine or a saline (placebo) injection. The vaccinated group experienced a polio rate of 28 per 100,000 while the placebo group had a rate of 69 per 100,000. A third group that refused to participate had a polio rate of 46 per 100,000.[m]

2.5   *Experimental or nonexperimental?* Determine whether each of the studies described in Exercise 2.4 are experimental or nonexperimental. Explain your reasoning in each instance.

---

[k] Muscat, J. E., Malkin, M. G., Thompson, S., Shore, R. E., Stellman, S. D., McRee, D., et al. (2000). Handheld cellular telephone use and risk of brain cancer. *JAMA*, *284*(23), 3001–3007.

[l] Moses, L. E., & Mosteller, F. (1972). Safety of anesthetics. In J. M. Tanur, F. Mosteller, W. Kruskal, R. F. Link, R. S. Pieters & G. R. Rising (Eds.). *Statistics: A Guide to the Unknown* (pp. 14–22). San Francisco: Holden-Day.

[m] Francis, T. (1957). Symposium on controlled vaccine field trials: Poliomyelitis. *American Journal of Public Health*, *47*, 283–287.

Confounding

Studies (a) and (b) in Exercise 2.4 are nonexperimental because the explanatory factors (cell phone use and anesthetic type, respectively) were *not* interfered with by the study protocol. Instead, study subjects were merely classified into exposed and nonexposed groups based on what had already occurred. Because of this, nonexperimental studies like these present special challenges in interpretation.

It turns out that discrepancies in experimental and nonexperimental study results are not uncommon. In the example just mentioned, the nonexperimental studies did not interfere with hormones use in subjects, so exposure was self-selected. It was later determined that hormone users and nonusers differed in ways other than hormone use. Hormone users were generally healthier, wealthier, and better educated than nonusers. These **lurking variables** got mixed up with the effects of hormone use, **confounding** the results of the nonexperimental studies.

> Confounding is a distortion in an association between an
> explanatory variable and response variable brought about
> by the influence of extraneous factors.

ILLUSTRATIVE EXAMPLE

WHI postmenopausal hormone use trial. Until 2002, postmenopausal hormone use in women was routine in the United States. Nonexperimental studies had suggested that estrogen use reduced heart attack risk by about 50%. Because other risks associated with postmenopausal hormone use were small by comparison, there was a general consensus that postmenopausal hormones had an overall health benefit.

Things changed following publication of an experimental trial done as part of an NIH-sponsored *Women's Health Initiative* (WHI) investigation. Results from the WHI trial revealed that hormone use did *not* reduce the risk of heart attacks. In fact, overall cardiovascular disease mortality was significantly increased in the women who were assigned hormones.[n] How can we explain the fact that the experimental studies contradicted the nonexperimental studies and the then-current state of knowledge?

---

[n] Writing Group for the Women's Health Initiative Investigators. (2002). Risks and benefits of estrogen plus progestin in healthy postmenopausal women: Principal results from the Women's Health Initiative randomized controlled trial. *JAMA, 288*(3), 321–333.

Confounding occurs when the effects of the explanatory variable get mixed up with those of extraneous variables. The lower risk of coronary disease in the hormone users in the nonexperimental studies of postmenopausal hormone use reflected the effect of these extraneous variables, not of hormone use. Because the experimental WHI study assigned hormone use according to the "toss of a coin," differences at the end of the study would have to be attributable to either the hormone treatment or to chance differences in the groups. It is for this reason that experimental studies provide an important point of reference for nonexperimental studies.[o] Therefore, we will initially focus on experimental studies, later applying some of these principles to nonexperimental studies.

Factors and Treatments

The term **subject** refers to an individual who participates in a study. Explanatory variables in experiments will be referred to as **factors**. A **treatment** is a specific set of factors applied to subjects.

Figure 2.2 displays the WHI study design in schematic form. Schematics such as this will be used to depict the number of study subjects, randomization scheme, and study outcome.

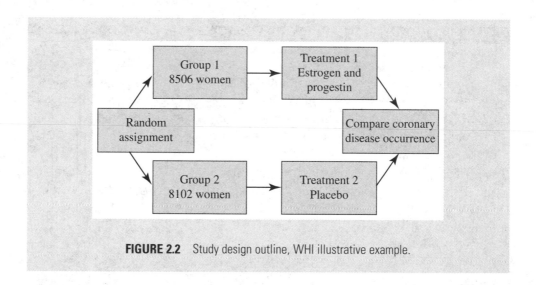

FIGURE 2.2   Study design outline, WHI illustrative example.

---

[o]Miettinen, O. S. (1989). The clinical trial as a paradigm for epidemiologic research. *Journal of Clinical Epidemiology, 42*(6), 491–496.

ILLUSTRATIVE EXAMPLE

WHI postmenopausal hormone use trial. In the WHI trial, the primary study *factor* was postmenopausal hormone use. The study was initially designed to address three treatments: unopposed estrogen, estrogen plus progestin, and an inert placebo. It was later discovered that unopposed estrogen was contraindicated in postmenopausal women with intact uteri, so the protocol was altered to include only two treatments: estrogen with progestin and a placebo. Treatments were then randomized to subjects in approximately equal proportions.

Exercises

2.6 *MRFIT.* The MRFIT study discussed in an earlier illustrative example studied 12,866 high-risk men between 35 and 57 years of age. Approximately half the study subjects were randomly assigned to a special care group; the other half received their usual source of care. Death from coronary disease was monitored over the next seven or so years. Outline this study's design in schematic form.

2.7 *Five-City Project.* The Stanford Five-City Project is a comprehensive community health education study of five moderately sized Northern California towns. Multiple-risk factor intervention strategies were randomly applied to two of the communities. The other three cities served as controls.[p] Outline the design of this study in schematic form.

By applying factors in combination, experiments can study more than one factor at a time.

ILLUSTRATIVE EXAMPLE

Hypertension trial. A trial looked at two explanatory factors in the treatment of hypertension. Factor A was a health-education program aimed at increasing physical activity, improving diet, and lowering body weight. This factor had two levels: active treatment or passive treatment. Factor B was pharmaceutical treatments at three levels: medication A, medication B, and placebo. Because there

continues

[p] Fortmann, S. P. & Varady, A. N. (2000). Effects of a community-wide health-education program on cardiovascular disease morbidity and mortality: The Stanford Five-City Project. *American Journal of Epidemiology, 152*(4), 316–323.

were two levels of the health-education variable and three levels of pharmacological variable, the experiment evaluated six treatments, as shown in Table 2.2.

The response variable was "change in systolic blood pressure" after six months. One hundred and twenty subjects were studied in total, with equal numbers assigned to each group. Figure 2.3 is a schematic of the study design.

TABLE 2.2   Hypertension treatment trial with two factors and six treatments.

		Factor B		
		Medication A	Medication B	Placebo
Factor A	Active	1	2	3
Health education	Passive	4	5	6

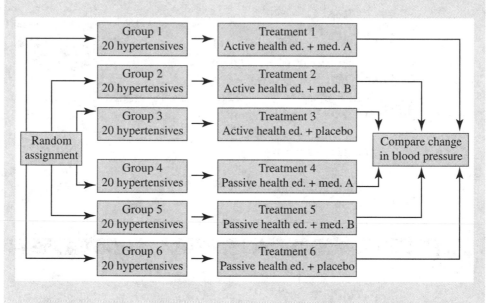

FIGURE 2.3   Study design outline, hypertensive treatment trial illustrative example.

Studies with multiple factors allow us to learn about the individual and combined effects of factors. For example, in the hypertension trial illustration, one of the medications might be particularly effective when combined with a lifestyle change. When two factors produce an effect that would otherwise not be predicted by studying them separately, an **interaction** is said to be present.

> **An interaction occurs when factors in combination produce an effect that could not be predicted by looking at the effect of each factor separately.**

Random Assignment of Treatments

Experiments involving human subjects are called **trials**. Trials with one or more control groups are **controlled trials**. When the assignment of the treatment is based on chance, this is a **randomized controlled trial**. As noted earlier, randomization helps us sort out the effects of the treatment from those of lurking variables.

To randomly assign a treatment:

- Label each subject who will be participating in the study with numbers 1 through N.

- Let n_1 represent the number of individuals you will have in the first treatment group.

- Enter Table A at an arbitrary location. Change the location each time you use the table.

- Read across (or down) lines of the random number table to identify the first n_1 random numbers between 1 and N. Assign these individuals to treatment group 1.

ILLUSTRATIVE EXAMPLE

Poliomyelitis trial. In the 1954 poliomyelitis field trial mentioned in Exercise 2.4(c), subjects were randomly assigned to either the Salk polio vaccine group or a saline placebo group. When the trial was initially proposed, there was a suggestion that everyone who agreed to participate in the study be given the vaccine while those who refused to participate serve as the control group. Fortunately, this did not occur and a placebo control group was included in the study. It was later revealed that "refusers" had atypically low polio rates. Had the refusers been used as the control group, the benefits of the vaccine would have been greatly underestimated.[q]

---

[q] Francis, T., Jr., Napier, J. A., Voight, R. B., Hemphill, F. M., Wenner, H. A., Korns, R. F., et al. (1957). *Evaluation of the 1954 Field Trial of Poliomyelitis Vaccine*. Ann Arbor: University of Michigan.

ILLUSTRATIVE EXAMPLE

HCV trial. Chronic infection with the hepatitis C virus (HCV) can lead to liver cirrhosis, hepatocellular cancer, and the need for liver transplantation. More than 4 million people in the United States are infected with the hepatitis C virus.[r] Suppose a pharmaceutical firm develops a new treatment for HCV infection and wants to test it in 24 individuals. The new treatment will be assigned at random to half the available subjects. The other subjects will receive the current standard treatment. Decreases in viral titer will be measured after three months on both treatments. Figure 2.4 shows a schematic of this study's design.

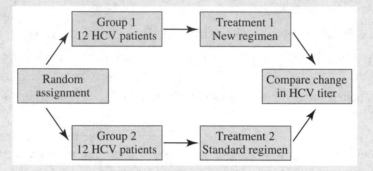

FIGURE 2.4   Study design outline, HCV treatment trial illustration.

To randomize the treatment:
- Label subjects 1 through 24.
- Twelve individuals will be assigned to each group.
- Let us enter Table A at row 34. Here are the digits in lines 34 and 35 of Table A:

  34   **06|16|2   4|02|**56   69|68|8   9|89|**04**   82|39|**1**
        8|29|20| **13|21|**4   2|57|43   31|80|5   **8|24|01**
  35   63|71|6   6|43|**11**   etc.

- The first 12 pairs of digits between 01 and 24 are 06, 16, 24, 02, 04, 18, 20, 13, 21, 24, 01, and 11 (**highlighted**). Individuals with these identification numbers are assigned to the treatment group. The remaining subjects are assigned to the control group.

---

[r]CDC. (2005). *Viral hepatitis* (*NHANES Data Brief*). Available: www.cdc.gov/nchs/data/nhanes /databriefs/viralhep.pdf.

- If there are two groups, the remaining n_2 study subjects are assigned to the control group. If there are more than two treatment groups, continue to select numbers from Table A to identify subjects for the next treatment group until only one group remains.

Exercises

2.8 **_MRFIT._** The MRFIT field trial discussed as an illustrative example studied 12,866 high-risk men between 35 and 57 years of age. Use Table A starting in row 03 to identify the first two members of the treatment group.

2.9 **_Five-City Project._** The Stanford Five-City Project (Exercise 2.7) randomized cities to either a treatment or a control group. Number the cities 1 through 5. Use Table A starting in line 17 to randomly select the two treatment cities.

Blinding

Blinding is an experimental technique in which individuals involved in the study are kept unaware of the treatment assignments. In a **single-blinded study**, the subjects are kept in the dark about the specific type of treatment they are receiving. In a **double-blinded study**, the study subjects and the investigators making measurements are kept in the dark. In **triple-blinded** studies, the study subjects, investigators, and statistician analyzing the data are kept in the dark.

ILLUSTRATIVE EXAMPLE

**Ginkgo and memory enhancement.** Ginkgo biloba is a commonly used herb that claims to improve memory and cognitive function. A randomized, double-blinded study in an elderly population was conducted to evaluate whether this claim is true. The treatment group consisted of subjects who took the active product according to the manufacturer's recommendation. The control group received lactose gelatin capsules that looked and tasted like the ginkgo pill. The study was _double blinded_, with subjects and evaluators administering cognitive tests unaware of whether subjects were receiving the ginkgo or placebo. Analysis revealed no difference in any of the cognitive functions that were measured.[s]

---

[s] Solomon, P. R., Adams, F., Silver, A., Zimmer, J., & DeVeaux, R. (2002). Ginkgo for memory enhancement: a randomized controlled trial. _JAMA_, _288_(7), 835–840.

Double blinding helps avoid biases associated with the reporting and recording of the study outcomes. When errors in measurement do occur, they are likely to occur equally in the groups being compared, mitigating more serious forms of bias.

Exercise

2.10 *Controlled-release morphine in patients with chronic cancer pain.* Warfield reviewed 10 studies comparing the effectiveness of controlled-release and immediate-release morphine in cancer patients with chronic pain.[t] The studies that were reviewed were double blinded. How would you double blind such studies?

Ethics

Because experiments require that the explanatory variable be assigned by the study protocol and not by subjects or their agents, they present special ethical concerns. In general, an experiment can neither withhold an effective treatment nor assign a potentially harmful one. Therefore, a genuine void in knowledge must exist before beginning a trial. Doubt about benefits and risk must be in perfect balance before the trial is begun. Balanced doubt of this type is called **equipoise.**

Of course there are other ethical concerns when conducting studies in human subjects. These include respect for individuals, informed consent (subjects must comprehend the consequences of participating in the study and have freely given consent to participate), beneficence (direct and indirect physical, psychological, and socioeconomic risks and benefits of doing the study have been assessed and are on balance beneficial), and justice (selection of study subjects is equitable).[u] **Institutional review boards (IRBs)** independently review study protocols and results to oversee that ethical guidelines are met.

Summary Points (Types of Studies)

1. The two types of statistical studies are surveys and comparative studies. **Surveys** aim to quantify population characteristics. **Comparative studies** aim to quantify relationships between variables.

2. A statistical **sample** is a subset of a population.

3. Scientific samples incorporate an element of chance. A **probability sample** is a sample in which each member of the population has a known probability of

---

[t]Warfield, C. A. (1998). Controlled-release morphine tablets in patients with chronic cancer pain. *Cancer, 82*(12), 2299–2306.
[u]Belmont Report. (1979). *Ethical principles and guidelines for the protection of human subjects of research.* Available: http://ohrp.osophs.dhhs.gov/humansubjects/guidance/belmont.htm.

entering the sample. A **simple random sample (SRS)** is a probability sample in which each population member has the same chance of entering the sample. Some modern surveys often use **complex sampling designs**, such as stratified samples, cluster samples, and multistage samples.

4. **Comparative studies** can be either experimental or observational. **Experimental studies** permit the investigator to assign study exposures to study subjects. (Often the assignment of exposures is randomized.) **Observational studies** do not permit the assignment of study exposures to study subjects.

5. The **explanatory variable** in a comparative study is the exposure being investigated. The **response variable** is the study outcome.

6. Explanatory variables in experiments are referred to as **factors**. A **treatment** is a specific set of factors applied to study subjects.

7. **Blinding** is a study technique in which individuals involved in the study are kept in the dark about the exposure status of study subjects. Double-blinded studies mask the study subjects and investigators as to the treatment type being received by study subjects.

8. **Confounding** occurs when the effects of a lurking variable become mixed up with the effects of the explanatory variable.

9. Experiments in humans present many **ethical challenges** and are restricted to situations in which equipoise (balanced doubt) is present. Ethical requirements of experiments in humans include respect for individuals, beneficence, justice, and institutional oversight.

Vocabulary

Bias	Lurking variables
Blinding	Multistage sampling
Census	Nonexperimental studies
Cluster samples	Nonresponse bias
Comparative study	Placebo
Confounding	Population
Controlled trial	Probability sample
Double-blinded	Randomized, controlled trial
Equipoise	Sample
Experimental studies	Sampling frame
Explanatory variable	Sampling independence
Factors	Simple random sample (SRS)
Infer	Single-blinded
Interaction	Stratified random sample

Subjects   Triple-blinded
Survey   Undercoverage
Treatment   Volunteer bias
Trials

Review Questions

2.1   What is the general goal of a statistical survey?

2.2   What is the general goal of a comparative statistical study?

2.3   List the advantages of sampling.

2.4   Define the term *simple random sample* (SRS).

2.5   What is *sampling independence*?

2.6   What is a *sampling fraction*?

2.7   What is *sampling bias*?

2.8   Provide examples of *sampling bias*.

2.9   What is a *probability sample*?

2.10 What is *multistage sampling*?

2.11 Why are "like-to-like comparisons" important in comparative studies?

2.12 What is the key distinction between experimental studies and observational studies?

2.13 What is a placebo?

2.14 What is the placebo effect?

2.15 A study seeks to determine the effect of postmenopausal hormone use on mortality. What is the explanatory variable in this study? What is the response variable?

2.16 Select the best response: This is the mixing-up of the effects of the explanatory factor with that of extraneous "lurking" variables.

   (a) the placebo effect

   (b) blinding

   (c) confounding

2.17 Select the best response: An individual who participates in a statistical study is often referred to as a

   (a) study subject

   (b) study factor

   (c) study treatment

2.18 Select the best response: A particular set of explanatory factors applied in an experiment is called a

   (a) subject

   (b) factor

   (c) treatment

2.19 What do we call a trial in which the assignment of the treatment type is based on chance?

2.20 How does randomization produce comparability?

2.21 What does the term *blinding* refer to in statistical studies?

2.22 Who is usually "kept in the dark" in double-blinded studies?

2.23 What does *equipoise* mean?

2.24 What does IRB stand for?

Exercises

2.11 *Campus survey.* A researcher conducts a survey to learn about the sexual behavior of college students on a particular campus. A list of the undergraduates at the university is used to select participants. The investigator sends out 500 surveys but only 136 are returned.

   (a) Consider how the low response rate could bias the results of this study.

   (b) Speculate on potential limitations in the quality of information the researcher will receive on questionnaires that are returned.

2.12 *Sampling nurses.* You want to survey nurses who work at a particular hospital. Of the 90 nurses who work at this hospital, 40 work in the maternity ward, 20 work in the oncology ward, and 30 work in the surgical ward. You decide to study 10% of the nurse population so you choose nine nurses as follows: four nurses are chosen at random from the 40 maternity nurses, two are chosen at random from the 20 oncology nurses, and three are chosen at random from the 30 surgical nurses. Is this a simple random sample? Explain your response.

2.13 *Telephone directory sampling frame.* Telephone surveys may use a telephone directory to identify individuals for study. Speculate on the type of household that would be undercovered by using this sampling frame.

2.14 *Random-digit dialing.* Random-digit dialing is often used to select telephone survey samples. This technique randomly selects the last four digits of a telephone number from a telephone exchange. (An exchange is the first three telephone numbers after the area code.) This type of sample gets around the problem of using a telephone book as a sampling frame; however, other types of problems may still be encountered. What types of selection biases may ensue from this method?

2.15 *Four-naughts.* Could the number "0000" appear in a table of random digits? If so, how likely is this?

2.16 *Class survey.* A simple random sample of students is selected from students attending a class. Identify a problem with this sampling method.

2.17 *Employee counseling.* An employer offers its employees a program that will provide up to four free psychological counseling sessions per calendar year. To evaluate satisfaction with this service, the counseling office mails questionnaires to every 10th employee who used the benefit in the prior year. There were 1000 employees who used the benefit. Therefore, 100 surveys were sent out. However, only 25 of the potential respondents completed and returned their questionnaire.

(a) Describe the population for the study.

(b) Describe the sample.

(c) What concern is raised by the fact that only 25 of the 100 questionnaires were completed and returned?

(d) Suppose all 100 questionnaires were completed and returned. Would this then represent an SRS? What type of sample would it be?

2.18 *How much do Master of Public Health (MPH) students earn?* We are all very concerned with the rising cost of higher education and the amount of money that many students must borrow to complete their studies. A university official wants to know how much MPH students earn from employment during the academic year and during the summer. The student population at the official's school consists of 378 MPH students who have completed at least one year of MPH study at three different campuses. A questionnaire will be sent to an SRS of 75 of these students.

(a) You have a list of the current email addresses and telephone numbers of all the 378 students. Describe how you would derive an SRS of $n = 30$ from this population.

(b) Use Table A starting in line 13 to identify the first 3 students in your sample.

3 Frequency Distributions

Frequency distributions tell us how often we see the various values in a batch of numbers. We begin each data analysis by visually exploring the shape, location, and spread of each variable's distribution.

■ 3.1 Stemplots

The **stem-and-leaf plot (stemplot)** is a graphical technique that organizes data into a histogram-like display. It is an excellent way to begin an analysis and is a good way to learn several important statistical principles.

To construct a stemplot, begin by dividing each data point into a stem component and a leaf component. Considering this small sample of $n = 10$:

$$21 \quad 42 \quad 05 \quad 11 \quad 30 \quad 50 \quad 28 \quad 27 \quad 24 \quad 52$$

For these data points, the "tens place" will become stem values and the "ones place" will become leaf values. For example, the data point 21 has a stem value of 2 and leaf value of 1.

A stem-like axis is drawn. Because data range from 5 to 52, stem values will range from 0 to 5. Stem values are listed in ascending (or descending) order at regularly spaced intervals to form a number line. A vertical line may be drawn next to the stem to separate it from where the leaves will be placed.

```
0 |
1 |
2 |
3 |
4 |
5 |
× 10
```

An **axis multiplier** ($\times 10$) is included below the stem to show that a stem value of 5 represents 50 (and not say 5 or 500). Leaf values are placed adjacent to their associated stem values. For example, "21" is:

```
0 |
1 |
2 | 1
3 |
4 |
5 |
✕ 10
```

The remaining leaves are plotted:

```
0 | 5
1 | 1
2 | 1874
3 | 0
4 | 2
5 | 02
✕ 10
```

Leaves are then rearranged to appear in rank order:

```
0 | 5
1 | 1
2 | 1478
3 | 0
4 | 2
5 | 02
✕ 10
```

The stemplot now resembles a histogram on its side. Rotate the plot 90 degrees (in your imagination) to display the distribution in the more familiar horizontal orientation.

```
        8
        7
        4       2
5  1  1  0  2  0
- - - - - - - - - -
0  1  2  3  4  5  (✕10)
- - - - - - - - - -
```

Three aspects of the distribution are now visible. These are its:

1. **Shape**
2. **Location**
3. **Spread**

Shape

Shape refers to the configuration of data points as they appear on the graph. This is seen as a "skyline silhouette":

```
    X
    X
    X    X
X  X  X  X  X
----------
0  1  2  3  4  5  (×10)
----------
```

It is difficult to make statements about shape when the data set is this small (a few more data points landing just so can entirely change our impression of its shape), so let us look at a larger data set.

Figure 3.1 is a histogram of about a thousand intelligence quotient scores. Overlaying the histogram is a fitted curve. Although the fit of the curve is imperfect, the curve still provides a convenient way to discuss the shape of the distribution.

A distribution's shape can be discussed in terms of its symmetry, modality, and kurtosis.

- **Symmetry** refers to the degree to which the shape reflects a mirror image of itself around its center.
- **Modality** refers to the number of peaks on the distribution.
- **Kurtosis** refers to the steepness of the mound.

Figure 3.2 illustrates these characteristics.

- Distributions (a)–(c) are *symmetrical.*
- Distributions (d)–(f) are *asymmetrical.*
- Distribution (d) is **bimodal**; the rest of the distributions are **unimodal**.
- Distribution (b) is flat with broad tails. This is a **platykuric** distribution (like a platypus). A tall curve with long skinny tails (not shown) is said to be **leptokurtic**. A curve with medium kurtosis is *mesokurtic.*[a]
- Distributions (e) and (f) are **skewed**. Figure (e) has a **positive skew** (tail toward larger numbers on a number line). Figure (f) has a **negative skew** (tail toward smaller numbers).

---

[a]Be aware that it is often difficult to assess the degree of kurtosis visually in applied situations.

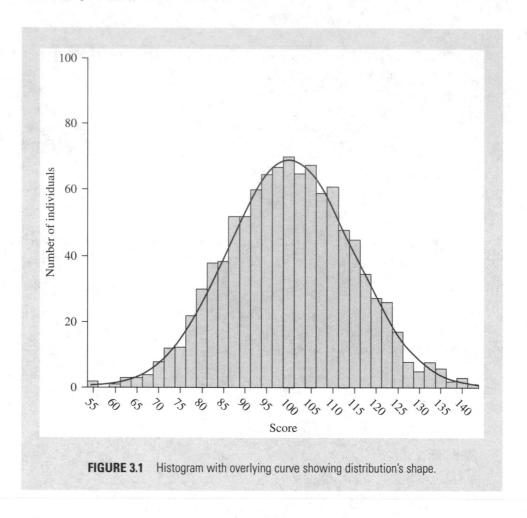

FIGURE 3.1   Histogram with overlying curve showing distribution's shape.

An **outlier** is a striking deviation from the overall pattern or shape of the distribution. As an example, the value of 50 on this stemplot is an outlier:

```
0|689
1|0124667
2|
3|
4|
5|0
×10
```

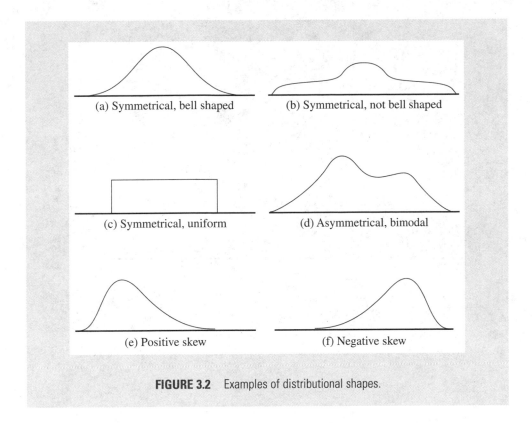

FIGURE 3.2   Examples of distributional shapes.

Location

We summarize the **location** of a distribution in terms of its center. Figure 3.3 shows distributions with different locations. Although the two distributions overlap, distribution 2 has higher values on average, as portrayed by its shift toward the right.

The term **average** refers to the center of a distribution.[b] There are different ways to identify a distribution's average, the two most common being the arithmetic average and the median.

The **arithmetic average** is a distribution's gravitational center. This is where the distribution would balance if placed on a scale. The balancing point for the stemplot here is somewhere between 20 and 30:

---

[b]Sometimes the term *average* is used restrictively to refer only to the arithmetic mean of a data set.

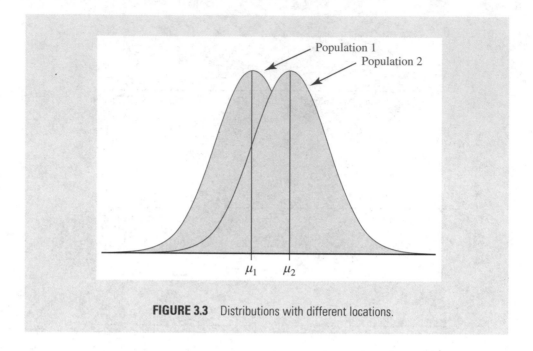

FIGURE 3.3 Distributions with different locations.

```
    8
    7
    4       2
5 1 1 0 2 0
- - - - - - - - - - -
0 1 2 3 4 5  (×10)
- - - - - - - - - - -
        ^
```

Gravitational
   Center

Of course this is only the *approximate* balancing point. The exact balancing point (arithmetic average) is determined by adding up all the values in the data set and dividing by the sample size. In this case, the arithmetic average = (21 + 42 + 5 + 11 + 30 + 50 + 28 + 27 + 24 + 52) ÷ 10 = 29. Our "eyeball estimate" was pretty good.

The **median** is the point that divides the data set into a top half and bottom half; it is halfway up (or down) the ordered list.

> **The depth of a data point corresponds to its rank from either the top or bottom of the ordered list of values.**

It is a little easier to determine the median if we stretch out the data to form an **ordered array**. The ordered array and median for the current data is:

```
05 11 21 24 27 28 30 42 50 52
                ^
              median
```

More formally, the median has a depth of $\dfrac{n + 1}{2}$, where n is the sample size. When n is even, the median will fall between two values, in which case you simply average these values to get the median. For example, the median in our illustrative data set has a depth of $\dfrac{10 + 1}{2} = 5.5$, placing it between the fifth (27) and sixth (28) ordered value. The average of these two points (27.5) is the median.

> **The arithmetic average is a distribution's balancing point.
> The median is its "middle value."**

Spread

The term **spread** is an informal way to refer to the dispersion or *variability* of data points. Figure 3.4 shows distributions with different variability. Populations 1 and 2 have the same central locations, but population 2 has greater spread (variability).

It is best to quantify spread with a statistic based on typical distance around the center of the distribution.

There are several ways to measure spread, and we will learn several of these methods in Chapter 4. For now, let us simply describe spread in terms of the range of values, lowest to highest.

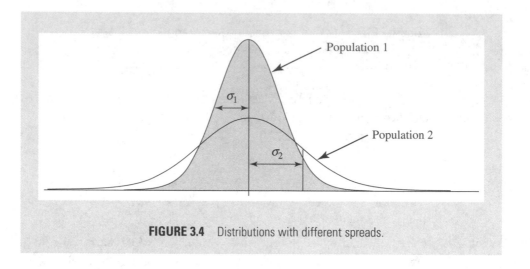

FIGURE 3.4   Distributions with different spreads.

Additional Illustrations of Stemplots

The next couple of illustrations show how to draw a stemplot for data that might not immediately lend itself to plotting.

ILLUSTRATIVE EXAMPLE

Truncating leaf values. Consider these eight data points:

$$1.47 \quad 2.06 \quad 2.36 \quad 3.43 \quad 3.74 \quad 3.78 \quad 3.94 \quad 4.42$$

Data have three significant digits, although only two are needed for plotting. Our rule will be to "prune" the leaves by **truncating** extra digits before plotting. For example, the value 1.47 is truncated to 1.4, the value 2.06 is truncated to 2.0, and so on. We also drop the decimal point before plotting. For example, "1.47" appears as "1 | 4".

Here is the stemplot:

```
1|4
2|03
3|4779
4|4
×1
```

How do we interpret this plot? As always, consider its shape, location, and spread.

- Shape: A mound shape with no apparent outliers. (With such a small data set, little else can be said about shape.)

- Location: Because there are $n = 8$ data points, the median has a depth of $\frac{8 + 1}{2} = 4.5$. Count to a depth of 4½ to see that the median falls between 3.4 and 3.7 (underlined in the stemplot). Average these values; the median is about 3.55.

- Spread: Data spread from about 1.4 to 4.4.

ILLUSTRATIVE EXAMPLE

Irish healthcare websites. The Irish Department of Health recommends a reading level of 12 to 14 years of age for health information leaflets aimed at the public. Table 3.1 lists reading levels for $n = 46$ Irish healthcare websites.

TABLE 3.1   Reading levels for Irish healthcare websites ($n = 46$).

08	10	11	11	12	13	13	13	13	14
14	15	15	15	15	15	15	15	16	16
16	14	17	17	17	17	17	17	17	17
17	17	17	17	17	17	17	17	17	17
17	17	17	17	17	17				

Data from O'Mahoney, B. (1999). Irish healthcare websites: A Review. *Irish Medical Journal,* 92(4), 334–336. Data are stored online in the file IRISHWEB.*.

If we imagine that each data point has an invisible ".0", plotting these zeros makes this revealing plot:

```
08|0
09|
10|0
11|00
12|0
13|0000
14|000
15|0000000
16|000
17|000000000000000000000000000
×1
```

continues

This distribution has a negative skew and a low outlier of 8.0 (shape). The median is 17 (location).[c] Data range from 8 to 17 (spread).

Splitting Stem Values

Sometimes a stemplot will be too squished to reveal its shape. In such circumstances, we can use **split stem values** to stretch out the stem. As an example, consider this plot.

```
1|4789
2|223466789
3|000123445678
×1
```

This plot is too squashed to reveal its shape, so we will split each stem value in to two, listing two "1s" where there had been one, two "2s" and so on. Think of each stem value as a "bin." The first "1" will be a bin to hold values between 1.0 and 1.4. The second "1" will hold values between 1.5 and 1.9 (and so on). Here's the plot with split stem values:

```
1|4
1|789
2|2234
2|66789
3|00012344
3|5678
×1
```

This plot does a better job showing the shape of the distribution, revealing its negative skew.

When needed, we can also split stem values into five subunits. The following codes can be used to tag stem values:

*	for leaves of zero and one
T	for leaves of two and three
F	for leaves of four and five
S	for leaves of six and seven
.	for leaves of eight and nine

---

[c]The median has a depth of $(n + 1)/2 = (46 + 1)/2 = 23.5$. This position is underlined revealing a median of 17.

Consider these nine values:

$$3.5 \quad 8.1 \quad 7.4 \quad 4.0 \quad 0.7 \quad 4.9 \quad 8.4 \quad 7.0 \quad 5.5$$

A stemplot with quintuple-split stem values makes a nice picture:

```
0* | 0
 T | 3
 F | 445
 S | 77
 . | 88
 X   10
```

How Many Stem Values?

When creating stemplots, you must choose how to scale the stem. Again, think of stem values as "bins" for collecting leaves. You can start with between 3 and 12 "bins" and make adjustments from there. If the plot is too squished, split the stem values. If it is too spread out, use a larger stem multiplier. Finding the most revealing plot may entail trial and error.

ILLUSTRATIVE EXAMPLE

Health insurance coverage. A U.S. Census Bureau report looked at health insurance coverage in the United States for the period 2002 to 2004. Table 3.2 lists the average percentage of people without health insurance coverage by state for this period.

TABLE 3.2   Percentage of residents without health insurance by state, United States 2004, $n = 51$.

State	%	State	%	State	%
Alabama	13.5	Kentucky	13.9	N. Dakota	11.0
Alaska	18.2	Louisiana	18.8	Ohio	11.8
Arizona	17.0	Maine	10.6	Oklahoma	19.2
Arkansas	16.7	Maryland	14.0	Oregon	16.1
California	18.4	Massachusetts	10.8	Pennsylvania	11.5
Colorado	16.8	Michigan	11.4	Rhode Is.	10.5
Connecticut	10.9	Minnesota	08.5	S. Carolina	13.8
Delaware	11.8	Mississippi	17.2	S. Dakota	11.9
Dist. Col.	13.5	Missouri	11.7	Tennessee	12.7
Florida	18.5	Montana	17.9	Texas	25.1

continues

State	%	State	%	State	%
Georgia	16.6	Nebraska	11.0	Utah	13.4
Hawaii	09.9	Nevada	19.1	Vermont	10.5
Idaho	17.3	N. Hamp.	10.6	Virginia	13.6
Illinois	14.2	N. Jersey	14.4	Washington	14.2
Indiana	13.7	N. Mexico	21.4	W. Virginia	15.9
Iowa	10.1	New York	15.0	Wisconsin	10.4
Kansas	10.8	N. Carolina	16.6	Wyoming	15.9

Data from DeNavas-Walt, C., Proctor, B. D., & Lee, C. H. (2005). *Income, Poverty, and Health Insurance Coverage in the United States: 2004* (No. P60-229). Washington, D.C.: U.S. Government Printing Office. Data are stored online in the file INC-POV-HLTHINS.* as the variable NOINS.

The stemplot with single stem values looks like this:

```
0 | 89
1 | 00000000001111111123333333444455566667777888899
2 | 15
×10
```

This plot is too squished, so we split the stem values to come up with the following plot:

```
0 | 89
1 | 0000000000111111123333333444
1 | 555666667777888899
2 | 1
2 | 5
×10
```

This plot is improved, but still seems too compressed. Let's try a quintuple split of stem values:

```
0. | 89
1* | 0000000000111111111
 T | 23333333
 F | 4444555
 S | 666667777
 . | 888899
2* | 1
 T |
 F | 5
×  10
```

This plot reveals a positive skew and high outlier.

ILLUSTRATIVE EXAMPLE

Student weights. Table 3.3 lists body weights of 53 students.

TABLE 3.3 Body weight (pounds) of students in a class, $n = 53$.

192	110	195	180	170	215
152	120	170	130	130	125
135	185	120	155	101	194
110	165	185	220	180	
128	212	175	140	187	
180	119	203	157	148	
260	165	185	150	106	
170	210	123	172	180	
165	186	139	175	127	
150	100	106	133	124	

Data are stored online in the file BODY-WEIGHT.*.

How would we plot this data in a way that is most revealing? First notice that values range from 100 to 260 pounds. Using a multiplier of ×100 would result in only two stem values (100–199 and 200–299). Splitting stem values with the ×100 in two would help, but would still result in only four stem values: 100–149, 150–199, 200–249, and 250–299. Using quintuple-split stem values produces this nice plot:

```
1*|0000111
1T|222222233333
1F|4455555
1S|666777777
1.|888888888999
2*|0111
2T|2
2F|
2S|6
×100
```

This plot has a positive skew and high outlier. The location of its median is underlined (median = 160), and data spread from 100 to 260.

Note that each illustration concludes with a brief narrative summary. "A picture is worth a 1,000 words, but to be so it may have to conclude 100 words."[d]

Back-to-Back Stemplots

We can compare two distributions with **back-to-back stemplots**. To create this type of plot, draw a stem in a central gutter and place leaves from groups on either side of this central stem. Back-to-back plots make it easy to compare group shapes, locations, and spreads.

ILLUSTRATIVE EXAMPLE

Back-to-back stemplots. Table 3.4 lists fasting cholesterol values (mg/dL) for two groups of men.

TABLE 3.4   Fasting cholesterol values (mg/dL) in two groups of men. Group 1 men were classified as type A personalities.

Group 1

233	291	312	250	246	197	268	224	239	239
254	276	234	181	248	252	202	218	212	325

Group 2

344	185	263	246	224	212	188	250	148	169
226	175	242	252	153	183	137	202	194	213

Data stored online in the file WCGS.*.

Here is the back-to-back stemplot of the data using quintuple-split stem values:

```
Group 1|   |Group 2
------------------
         |1T|3
         |1F|45
         |1S|67
      98|1.|8889
     110|2*|011
   33332|2T|22
   55544|2F|4455
      76|2S|6
       9|2.|
```

```
1|3*|
2|3T|
 |3F|4
 (×100)
```

Notice that the distribution of group 1 is shifted down the axis toward the higher values on the stem showing it to have higher values on average.

Exercises

3.1 *Poverty in eastern states, 2000.* Table 3.5 lists the percentage of people living below the poverty line in the 26 states east of the Mississippi River for the year 2000. Make a stemplot of these values. After creating the plot, describe the distribution's shape, location, and spread. Are there any outliers? Which states straddle the median?

3.2 *Hospitalization.* Table 3.6 lists lengths of stays (days) for a sample 25 patients.

    (a) Create a stemplot with single stem values for these data. (Use an axis multiplier of ×10).

    (b) Create a stemplot with split stem values.

    (c) Which of the stemplots do you prefer?

    (d) Describe in plain language the distribution's shape, location, and spread.

TABLE 3.5 Percentage of people living below the poverty line in each of the 26 states east of the Mississippi River for the year 2000.

Alabama	14.6	Maryland	07.3	Pennsylvania	09.9
Connecticut	07.6	Massachusetts	10.2	Rhode Is.	10.0
Delaware	09.8	Michigan	10.2	S. Carolina	11.9
Florida	12.1	Mississippi	15.5	Tennessee	13.3
Georgia	12.6	New Hamp.	07.4	Vermont	10.1
Illinois	10.5	New Jersey	08.1	Virginia	08.1
Indiana	08.2	New York	14.7	West Virginia	15.8
Kentucky	12.5	N. Carolina	13.2	Wisconsin	08.8
Maine	09.8	Ohio	11.1		

Data from Delaker, J. (2001). *Poverty in the United States, 2000* (No. P60-214). Washington, DC: U.S. Census Bureau. Table D, p. 11. Data are stored in the file POV-EAST-2000.*.

TABLE 3.6   Duration of hospitalization (days), $n = 25$.

5	10	6	11	5	14	30	11	17	3
9	3	8	8	5	5	7	4	3	7
9	11	11	9	4					

Data are stored online in the file HDUR.* as the variable DUR.

3.3   *Leaves on a common stem.* For each of the following comparisons, plot the data as back-to-back stemplots on a common stem. Then, compare group locations and spreads.

(a) Comparison A

Group 1:	90	70	50	30	10
Group 2:	70	60	50	40	30

(b) Comparison B

Group 1:	90	80	70	60	50
Group 2:	70	60	50	40	30

(c) Comparison C

Group 1:	90	70	50	30	10
Group 2:	90	80	70	60	50

3.4   *Cholesterol comparison.* Table 3.7 lists plasma cholesterol levels (mmol/m^3) in two independent groups. Plot these data on a common stem. Then, compare group locations and spreads.

TABLE 3.7   Plasma cholesterol levels (mmol/m^3) in two independent groups.

Group 1 (mildly hypercholesterolemic)											
6.0	6.4	7.0	5.8	6.0	5.8	5.9	6.7	6.1	6.5	6.3	5.8
Group 2 (controls)											
6.4	5.4	5.6	5.0	4.0	4.5	6.0					

Source: Data for group 1 are from Rassias, G., Kestin, M., & Nestel, P. J. (1991). Linoleic acid lowers LDL cholesterol without a proportionate displacement of saturated fatty acid. *European Journal of Clinical Nutrition, 45*(6), 315–320. Data for group 2 were generated with a Normal random variable number generator. Data for both groups are stored online in the file CHOLESTEROL.SAV.

■ 3.2 Frequency Tables

Frequency Counts from Stemplots

Frequency means the number of times something occurs. Figure 3.5 is a stemplot with frequency counts shown to the left of the stem. Figure 3.6 displays a similar technique for a larger data set. Below the second plot, it states: Each leaf: 2 case(s):[e] This plot uses 327 leaves to represent the 654 observations. Listing counts next to the stem can be very convenient.

Frequency Tables

Frequency tables are a traditional way to describe the distribution of counts in a data set. Table 3.8 shows a typical frequency table. Three types of frequencies are listed.

- The **frequency** column contains counts.
- The **relative frequency (%)** column contains frequency counts divided by the total with values expressed as a percentage.[f] For example:

$$2 \div 654 = .003, \text{ or } 0.3\% \text{ are 3 years of age}$$
$$9 \div 654 = .014, \text{ or } 1.4\% \text{ are 4 years of age}$$
$$28 \div 654 = .043, \text{ or } 4.3\% \text{ are 5 years of age}$$
And so on.

Frequency	Stem and Leaf
1.00	0 . 5
1.00	1 . 1
4.00	2 . 1478
1.00	3 . 0
1.00	4 . 2
2.00	5 . 02
Stem width: 10	
Each leaf: 1 case(s)	

FIGURE 3.5   Frequency counts from stemplot, SPSS output.

---

[e] SPSS uses the term "cases" to refer to "observations."
[f] "Per-cent" literally means "per hundred." To turn a proportion into a percentage, simply multiply by 100%. This does *not* change the value, since multiplying 100% is the same as multiplying by 1.

Frequency	Stem and Leaf
2.00	3. 0
9.00	4. 0000
28.00	5. 000000000000000
37.00	6. 0000000000000000000
54.00	7. 000000000000000000000000000
85.00	8. 000000000000000000000000000000000000000000
94.00	9. 00000000000000000000000000000000000000000000000
81.00	10. 0000000000000000000000000000000000000000
90.00	11. 000000000000000000000000000000000000000000000
57.00	12. 0000000000000000000000000000
43.00	13. 000000000000000000000
25.00	14. 000000000000
19.00	15. 000000000
13.00	16. 000000
8.00	17. 0000
9.00	Extremes (>=18)

Stem width: 1
Each leaf: 2 case(s)

FIGURE 3.6 Ages of participants in a childhood and adolescent health survey. Plot produced with SPSS for Windows, Rel. 11.0.1. 2001. Chicago: SPSS Inc.

- The **cumulative frequency (%)** column contains percents that fall within or below a given level. To determine cumulative percents, add the current relative frequency to the prior cumulative relative frequency. Start with the fact that 0.3% of individuals are 3 years old. Then determine

$$0.3\% + 1.4\% = 1.7\% \text{ are 4 years of age or younger}$$
$$1.7\% + 4.3\% = 6.0\% \text{ are 5 years of age or younger}$$
And so on.

Frequency ≡ count
Relative frequency ≡ proportion
Cumulative relative frequency ≡ proportion that falls in or below a certain level

TABLE 3.8 Frequency table, ages of subjects in a respiratory health study.

Age (years)	Frequency	Relative frequency (%)	Cumulative frequency (%)
3	2	0.3	0.3
4	9	1.4	1.7
5	28	4.3	6.0
6	37	5.6	11.6
7	54	8.3	19.9
8	85	13.0	32.9
9	94	14.3	47.2
10	81	12.4	59.6
11	90	13.8	73.4
12	57	8.7	82.1
13	43	6.6	88.7
14	25	3.8	92.5
15	19	2.9	95.4
16	13	2.0	97.4
17	8	1.2	98.6
18	6	0.9	99.5
19	3	0.5	100.0
Total	654	100.0	—

Class-Interval Frequency Tables (Grouped Data)

It is often necessary to group data into **class intervals** before tallying frequencies. This process is analogous to deciding on the number of stem values to use when creating a stemplot. Again, begin by grouping data into 3 to 12 class intervals. Then, by trial and error, find a grouping that best suits your needs.

Class intervals can be set up with equal spacing or unequal spacing; either way, **endpoint conventions** are needed to ensure that each observation falls into only one interval. Class intervals may either: (a) include the left boundary and exclude the right boundary or (b) include the right boundary and exclude the left boundary. Here are 10-year age intervals that include the left boundary and exclude the right boundary:

$$0.0 \leq \text{years} < 10$$
$$10.0 \leq \text{years} < 20$$
$$\cdot$$
$$\cdot$$
$$\cdot$$
$$50.0 \leq \text{years} < 60$$

To construct a frequency table with class intervals:

1. List class intervals in ascending or descending order.
2. Tally frequencies that fall within each interval.
3. Sum frequencies to determine the total sample size: $n = \Sigma f_i$, where f_i represents the frequency in class i.
4. Determine relative frequencies (proportions) that fall within each interval: $p_i = f_i \div n$, where p_i represents the proportion in class interval i. (It is often useful to report proportions as percents by moving the decimal point over two places to the right and adding a "%" sign.)
5. Determine the cumulative proportions by summing proportions from prior levels to the current level: $c_i = p_i + c_{i-1}$, where c_i represents the cumulative relative frequency in class i.

Here is a listing of the 10 data points that began the chapter:

$$5 \quad 11 \quad 21 \quad 24 \quad 27 \quad 28 \quad 30 \quad 42 \quad 50 \quad 52$$

Here are the data as a frequency table with uniform 15-year class intervals:

Class interval (i)	Age range (years)	Tally	Frequency	Age range frequency (%)	Cum. relative frequency (%)
1	0–14	//	2	20	20
2	15–29	////	4	40	60
3	30–44	//	2	20	80
4	45–59	//	2	20	100
Totals	All	—	10	100	—

Some situations call for **nonuniform class intervals.** Here are the data from Figure 3.6 with data grouped in classes corresponding to school-age populations.

Age range	Frequency	Relative frequency	Cum. relative frequency
Pre-school (3–4 years)	11	1.7	1.7
Elementary (5–11 years)	469	71.7	73.4
Junior High (12–13 years)	100	15.3	88.7
Senior High (14–19 years)	74	11.3	100.0
Total	654	100.0	—

■ 3.3 Additional Frequency Charts

Bar charts display frequencies with bars that correspond in height to frequencies or relative frequencies. **Histograms** are bar charts with contiguous bars and are therefore reserved for displaying frequencies associated with quantitative variables. Figure 3.7 is a histogram of the data from Table 3.8.

Frequency polygons replace histogram bars with a line connecting frequency levels. Figure 3.8 is a frequency polygon of the same data plotted in Figure 3.7.

Histograms and frequency polygons are reserved for use with quantitative variables. Frequencies for categorical variables should be displayed via **bar charts** with noncontiguous bars (e.g., Figure 3.9) or **pie charts** (e.g., Figure 3.10).

For a remarkably interesting book on good graphical practices in statistics, see Tufte, E. R. (1983). *The Visual Display of Quantitative Information.* Cheshire, CT: Graphic Press.

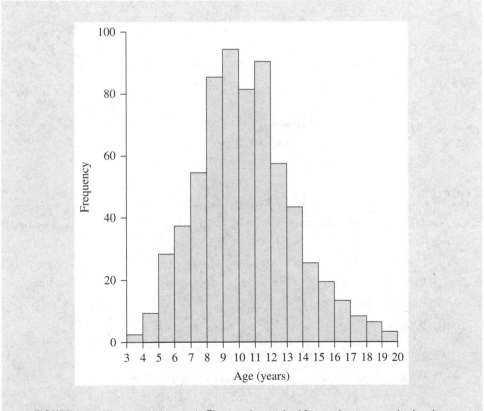

FIGURE 3.7 Histogram bars touch. They are best suited for continuous quantitative data.

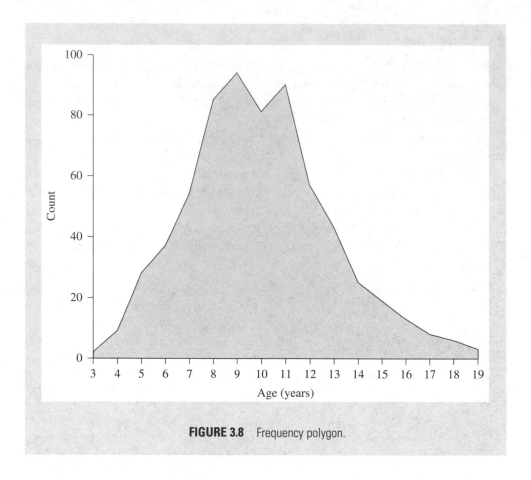

FIGURE 3.8   Frequency polygon.

Exercise

3.5   *Hospital stay duration.* In Exercise 3.2, you created a stemplot of lengths of hospital stays for 25 patients. Table 3.6 lists the data.

   (a) Construct a frequency table for these data using 5-day class intervals. Include columns for the frequency counts, relative frequencies, and cumulative frequencies.

   (b) What percentage of hospital stays were less than 5 days?

   (c) What percentage were less than 15 days?

   (d) What percentage of hospital stays were at least 15 days in length?

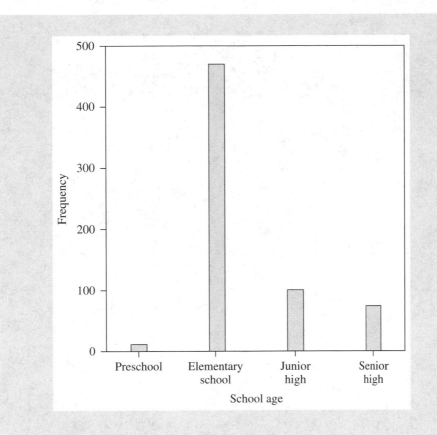

FIGURE 3.9 Bar charts with noncontinuous bars are better suited for nonuniform class intervals and categorical data.

Summary Points (Frequency Distributions)

1. **Frequency distributions** tell us how often various values occur in a batch of numbers. We explore a distribution's shape, location, and spread to begin each data analysis.

2. **Distributional shape** is described in terms of symmetry, direction of skew if asymmetric, modality (number of peaks), and kurtosis (steepness of peaks). Note: Descriptors of shape are unreliable when data sets are small and moderate in size.

3. The **location** of a distribution is summarized by its center. The two measures of central location addressed in this chapter are the mean (arithmetic average) and median (middle value in an ordered array).

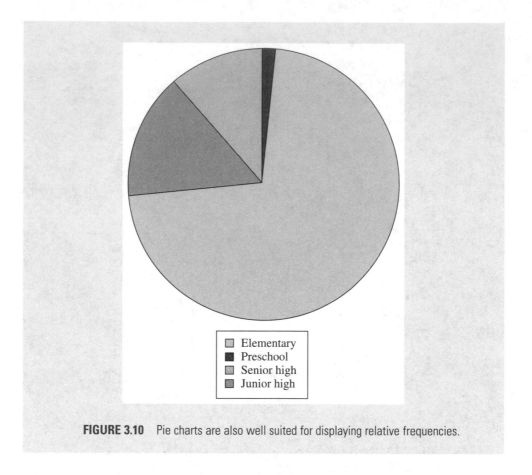

FIGURE 3.10   Pie charts are also well suited for displaying relative frequencies.

4. The **spread** of a distribution refers to the extent to which values are dispersed. This chapter merely reports the minimum and maximum as a crude descriptor of spread. (The next chapter introduces more advance measures of spread.)

5. The **stem-and-leaf plot ("stemplot")** is an excellent way to explore the shape, location, and spread of a distribution. Briefly, the stem on a stemplot represents a number line with "bins." Leaves represent the following significant digit value.

6. **Additional graphical techniques** for exploring "shape, location, and spread" include histograms and frequency polygons.

7. The distribution of **categorical variables** may be displayed in the form of a bar chart or pie chart.

8. **Frequency tables** list frequencies (counts), relative frequencies (proportions), and cumulative frequencies (proportion of values up to and including the current value). Quantitative data may first need to be grouped into class intervals before tallying frequencies.

Vocabulary

Arithmetic average	Median
Average	Modality
Axis multiplier	Negative skew
Back-to-back stemplot	Nonuniform class intervals
Bar charts	Ordered array
Bimodal	Outlier
Class intervals	Pie charts
Cumulative frequency	Platykurtic
Depth	Positive skew
Endpoint conventions	Relative frequency
Frequency	Shape
Frequency polygons	Split stem values
Histograms	Spread
Kurtosis	Stem-and-leaf plot (stemplot)
Leaf	Symmetry
Leptokurtic	Truncate
Location	

Review Questions

3.1 Name three general characteristics of distributions.

3.2 Select the best response: An asymmetrical distribution with a long right tail (toward the higher numbers) is said to have _____ skew.

(a) a positive

(b) a negative

(c) no

3.3 Select the best answer: An asymmetrical distribution with a long left tail is said to have _____ skew.

(a) a positive

(b) a negative

(c) no

3.4 What term refers to the number of peaks in a distribution?

3.5 What is a deviation from the overall pattern of a distribution?

3.6 Select the best response: This refers to a flat curve with thick tails.

(a) platykurtotic

(b) mesokurtotic

(c) leptokurtotic

3.7 Select the best response: This refers to the "peak" of a distribution.

(a) mean

(b) median

(c) mode

3.8 What refers to a distribution with two peaks?

(a) nonmodal

(b) unimodal

(c) bimodal

3.9 The two most common measures of central location are the _____ and _____.

3.10 What statistic identifies the gravitational center of a distribution?

3.11 What statistic identifies the value that is greater than or equal to 50% of the other data points in a distribution?

3.12 Fill in the blank: The median is located at a depth of _____ in the ordered array (where n is the sample size).

3.13 How many leaves are there on a stemplot?

3.14 What is the purpose of a stem (axis) multiplier?

3.15 What is another technical term for a count?

3.16 What is another technical term for a proportion?

3.17 What is the proportion of values at or below a specified value?

3.18 Why are end-point conventions needed when constructing frequency tables with class intervals?

3.19 True or false? Histograms are used to display frequencies for categorical data.

Exercises

3.6 *Outpatient wait time.* Waiting times (minutes) for 25 patients at a public health clinic are[g]:

$$
\begin{array}{cccccccccc}
35 & 22 & 63 & 6 & 49 & 19 & 16 & 31 & 24 & 29 \\
23 & 32 & 72 & 13 & 51 & 45 & 77 & 16 & 33 & 55 \\
10 & 42 & 28 & 72 & 13 & & & & &
\end{array}
$$

(a) Create a stemplot of these data. Describe the distribution's shape, location, and spread.

---

[g]Data are fictitious but realistic. Stored online in the file WAITTIME.*.

(b) From your stemplot, create a frequency table with counts, relative frequencies, and cumulative relative frequencies. What percentage of wait times were less than 20 minutes? What percentage were at least 20 minutes?

3.7 *Body weight expressed as a percentage of ideal.* Body weights of 18 diabetics expressed as a percentage of ideal (defined as body weight ÷ ideal body weight × 100) are shown here.[h] Construct a stem-and-leaf plot of these data and interpret your findings.

107	119	99	114	120	104	88	114	124	116
101	121	152	100	125	114	95	117		

3.8 *Docs' kids.* The numbers of children of 24 physicians who work at a particular clinic are shown here.[i] Create a stemplot with these data. Consider its shape, location, and spread. What percentage of physicians at this clinic have less than three children?

```
3 2 0 1 4 7 3 2 4 1
0 2 5 6 2 1 2 1 0 0
3 6 2 1
```

3.9 *Seizures following bacterial meningitis.* A study examined the induction time between bacterial meningitis and the onset of seizures in 13 cases (months). Data are shown here.[j] Construct a stemplot of these data and describe what you see. [*Suggestion*: Use a stem multiplier of ×10 so that the value 0.10 is truncated to 00 and is plotted as "0|0".]

```
0.10  0.25  0.5  4  12  12  24  24  31  36  42  55  96
```

3.10 *Surgical times.* Durations of surgeries (hours) for 14 patients receiving artificial hearts are shown here.[k] Create a stem plot of these data. Describe the distribution. Are there any outliers?

```
7.0  6.5  3.5  3.1  2.8  2.5  2.6  2.4  2.1  1.8  2.3  3.1  3.0  2.5
```

3.11 *U.S. Hispanic population.* Table 3.9 lists the percent of residents in the 50 states who identified themselves in the 2000 census as Spanish, Hispanic, or Latino. Create a stemplot of these data using single stem values and an axis multiplier of ×10. Then create a stemplot using double-split stem values. Which plot do you prefer?

---

[h] Saudek, C. D., Selam, J. L., Pitt, H. A., Waxman, K., Rubio, M., Jeandidier, N., et al. (1989). A preliminary trial of the programmable implantable medication system for insulin delivery. *New England Journal of Medicine, 321*(9), 574–579. Data are stored online in the file %IDEAL.[*].
[i] Data are fictitious and are stored online in the file DOCKIDS.[*].
[j] Source: Unknown. Data stored online in the file SEIZURE.[*].
[k] Source: Unknown. Data stored online in the file SURG-TIME.[*].

TABLE 3.9 Data for Exercise 3.11. Percent of residents who identified themselves as Spanish, Hispanic, or Latino by state, United States, 2000.

State	Percent	State	Percent	State	Percent
Alabama	1.5	Louisiana	2.4	Ohio	1.9
Alaska	4.1	Maine	0.7	Oklahoma	5.2
Arizona	25.3	Maryland	4.3	Oregon	8.0
Arkansas	2.8	Massachusetts	6.8	Pennsylvania	3.2
California	32.4	Michigan	3.3	Rhode Island	8.7
Colorado	17.1	Minnesota	2.9	South Carolina	2.4
Connecticut	9.4	Mississippi	1.3	South Dakota	1.4
Delaware	4.8	Missouri	2.1	Tennessee	2.0
Florida	16.8	Montana	2.0	Texas	32.0
Georgia	5.3	Nebraska	5.5	Utah	9.0
Hawaii	7.2	Nevada	19.7	Vermont	0.9
Idaho	7.9	New Hampshire	1.7	Virginia	4.7
Illinois	10.7	New Jersey	13.3	Washington	7.2
Indiana	3.5	New Mexico	42.1	West Virginia	0.7
Iowa	2.8	New York	15.1	Wisconsin	3.6
Kansas	7.0	North Carolina	4.7	Wyoming	6.4
Kentucky	1.5	North Dakota	1.2		

Data from the U.S. Census Bureau and Guzman (2001). The Hispanic Population: Census 2000 brief. http://www.census.gov/prod/2001/pubs/c2kbr01-3.pdf. Accessed July 22, 2013. Data stored online in the file PER-HISP.*.

3.12 Low birth weight rates. A birth weight of less than 2500 grams (about 5.5 pounds) qualifies as "low birth weight" according to international standards. Table 3.10 lists low birth weight rates (per 100 births) by country for the year 1991 in 109 countries.

(a) Explore these data with a stemplot. Use an axis multiplier of ×10 and quintuple split stem values to draw your plot. After creating the plot, provide a narrative description of its shape, location, and spread.

(b) Observations in this data set can be **ranked** from 1 to 109. When ranks are tied, the average rank is assigned to each item in the tied set. For example, Finland, Ireland, Norway, and Sweden are tied for ranks 2 through 5. Therefore, these four countries share a rank of $\dfrac{2+5}{2} = 3\frac{1}{2}$. We then resume ranking where we left off: Belgium is initially ranked as sixth but is tied with nine other nations with a value of 5. Therefore, positions 6–15 are tied in the ordered array sharing a rank of $\dfrac{6+15}{2} = 10\frac{1}{2}$. Where does the United States rank in this listing?

TABLE 3.10 Data for Exercise 3.12. Low birth weights per 100 births for 109 countries, 1991.

Spain	1	UK	7	Cote d'Ivoire	14
Finland	4	USA	7	Guatemala	14
Ireland	4	Yugoslavia	7	Indonesia	14
Norway	4	Benin	8	Tanzania	14
Sweden	4	Botswana	8	Zambia	14
Belgium	5	Brazil	8	Central African Rep.	15
Egypt	5	Colombia	8	El Salvador	15
France	5	Cuba	8	Kenya	15
Hong Kong	5	Jamaica	8	Mexico	15
Iran	5	Panama	8	Nicaragua	15
Japan	5	Poland	8	Niger	15
Jordan	5	Tunisia	8	Zimbabwe	15
New Zealand	5	Uruguay	8	Congo	16
Portugal	5	Algeria	9	Dominican Rep.	16
Switzerland	5	Burundi	9	Myanmar	16
Australia	6	China	9	Angola	17
Austria	6	Iraq	9	Ghana	17
Bulgaria	6	Korea, Rep.	9	Haiti	17
Canada	6	Mauritius	9	Mali	17
Czechoslovakia	6	Peru	9	Rwanda	17
Denmark	6	Venezuela	9	Sierra Leone	17
Germany, Fed.	6	Costa Rica	10	Philippines	18
Germany, Rep.	6	Hungary	10	Viet Nam	18
Greece	6	Lebanon	10	Afghanistan	20
Romania	6	Madagascar	10	Honduras	20
Saudi Arabia	6	Malaysia	10	Malawi	20
USSR	6	Mongolia	10	Mozambique	20
Albania	7	Ecuador	11	Nigeria	20
Chile	7	Lesotho	11	Togo	20
Israel	7	Mauritania	11	Pakistan	25
Italy	7	Senegal	11	Papua New Guinea	25
Kuwait	7	Syria	11	Bangladesh	28
Oman	7	Bolivia	12	Sri Lanka	28
Paraguay	7	South Africa	12	India	30
Singapore	7	Thailand	12	Laos	39
Turkey	7	Cameroon	13		
United Arab Emirates	7	Zaire	13		

Data from Grant, J. P. (1992). *The State of the World's Children; Vol. 1991 (United Nations Children's Fund)*. New York: Oxford. Data stored online in the file UNICEF.[*].

3.13 *Air samples.* An environmental study looked at suspended particulate matter in air samples ($\mu g/m^3$) at two different sites. Data are listed here.[1] Construct side-by-side stemplots to compare the two sites.

> Site 1: 68 22 36 32 42 24 28 38
> Site 2: 36 38 39 40 36 34 33 32

3.14 *Low birth weight rates.* Create a frequency table for the low birth weight data in Table 3.10. Use two-unit class intervals to construct your table, starting with the interval 0–1. Include frequency counts, relative frequencies, and cumulative frequencies in your table.

3.15 *Practicing docs.* The *Health United States* series published by the National Center for Health Statistics each year tracks trends in health and health care in the United States. Table 3.11 is derived from a part of Table 104 in *Health United States 2006*. This table lists the number of practicing medical doctors per 10,000 residents for each of the 50 states and the District of Columbia for the years 1975, 1985, 1995, 2002, 2003, and 2004.

TABLE 3.11   Active physicians and doctors of medicine in patient care, by state, United States, for selected years between 1975 and 2004.

State	1975	1985	1995	2002	2003	2004
Connecticut	17.7	24.3	29.5	30.9	31.2	31.3
Maine	10.7	15.6	18.2	22.6	24.1	24.0
Massachusetts	18.3	25.4	33.2	35.1	36.5	37.4
New Hampshire	13.1	16.7	19.8	23.0	24.0	23.6
Rhode Island	16.1	20.2	26.7	29.7	30.7	30.7
Vermont	15.5	20.3	24.2	30.6	32.1	31.9
New Jersey	14.0	19.8	24.9	26.8	27.3	26.8
New York	20.2	25.2	31.6	32.6	33.1	33.0
Pennsylvania	13.9	19.2	24.6	25.5	25.7	25.6
Illinois	13.1	18.2	22.1	23.1	23.9	24.1
Indiana	9.6	13.2	16.6	18.9	19.7	19.6
Michigan	12.0	16.0	19.0	20.1	20.9	21.3
Ohio	12.2	16.8	20.0	22.0	22.6	23.0
Wisconsin	11.4	15.9	19.6	22.0	22.8	23.1
Iowa	9.4	12.4	15.1	15.7	16.6	16.6
Kansas	11.2	15.1	18.0	18.8	19.8	20.1
Minnesota	13.7	18.5	21.5	23.3	24.8	25.3
Missouri	11.6	16.3	19.7	20.6	21.3	21.4
Nebraska	10.9	14.4	18.3	20.8	21.9	21.6

[1]Source: Unknown. Data stored online in the file AIRSAMPLES.*.

State	1975	1985	1995	2002	2003	2004
North Dakota	9.2	14.9	18.9	20.8	21.8	22.1
South Dakota	7.7	12.3	15.7	18.6	19.9	20.3
Delaware	12.7	17.1	19.7	21.5	22.6	22.2
District of Columbia	34.6	45.6	53.6	53.9	60.2	64.2
Florida	13.4	17.8	20.3	21.4	22.4	22.2
Georgia	10.6	14.7	18.0	18.8	19.8	20.0
Maryland	16.5	24.9	29.9	31.2	33.6	33.8
North Carolina	10.6	15.0	19.4	21.4	22.6	22.6
South Carolina	9.3	13.6	17.6	19.9	20.9	21.0
Virginia	11.9	17.8	20.8	22.5	24.4	24.2
West Virginia	10.0	14.6	17.9	19.8	20.5	20.4
Alabama	8.6	13.1	17.0	18.3	19.3	19.5
Kentucky	10.1	13.9	18.0	19.8	20.6	21.0
Mississippi	8.0	11.1	13.0	15.6	16.7	16.8
Tennessee	11.3	16.2	20.8	22.5	23.3	23.4
Arkansas	8.5	12.8	16.0	17.8	18.6	18.9
Louisiana	10.5	16.1	20.3	23.0	23.7	23.9
Oklahoma	9.4	12.9	14.7	14.8	15.5	15.6
Texas	11.0	14.7	17.3	18.1	18.9	18.9
Arizona	14.1	17.1	18.2	17.6	18.9	18.9
Colorado	15.0	17.7	20.6	21.2	22.8	23.3
Idaho	8.9	11.4	13.1	15.2	16.1	16.1
Montana	10.1	13.2	17.1	20.3	21.4	20.9
Nevada	10.9	14.5	14.6	16.1	17.2	17.2
New Mexico	10.1	14.7	18.0	19.0	21.1	21.2
Utah	13.0	15.5	17.6	17.9	19.1	19.0
Wyoming	8.9	12.0	13.9	16.6	18.1	17.7
Alaska	7.8	12.1	14.2	17.7	20.4	20.5
California	17.3	21.5	21.7	21.8	22.8	22.9
Hawaii	14.7	19.8	22.8	25.2	28.1	28.1
Oregon	13.8	17.6	19.5	21.7	23.5	23.6
Washington	13.6	17.9	20.2	22.3	23.7	23.6

Reproduced from the National Center for Health Statistics (2006). *Health United States 2006: Chartbook on Trends in the Health of Americans*, Table 104. Retrieved from http://www.cdc.gov/nchs/data/hus/hus06.pdf on Nov 19, 2010.

(a) Create a stemplot of the data for 2004. Use quintuple split stem values and an axis multiplier of ×10 to create your plot. Describe the shape, central location, and spread of the data.

Students are encouraged to use statistical software to aid in manipulating and analyzing the data. Data are provided on the companion website in SPSS (HealthUnitedStates2006Table104.sav) and Excel (HealthUnitedStates2006Table104.xls) file formats.

(b) Explore the data for 1975 by using a plot similar to the one you produce in part (a) of this exercise.

(c) Put your stemplots back to back or side by side with their axes (stems) aligned. Then, compare the results. Describe the comparison with a brief narrative.

3.16 *Health insurance.* Table 3.12 lists the percentage of people without health insurance in the 50 states and the District of Columbia for the year 2004. In addition to being listed in Table 3.12, data are also stored on the companion website in file INC-POV-HLTHINS.* as the variable NOINS (no insurance). Use a stem multiplier of ×10 and split stem values to create your plot. Interpret what you see.

TABLE 3.12   Percent of residents without health insurance by state, United States, 2004, $n = 51$, presented in rank order.

State	% w/o insurance	Depth	State	% w/o insurance	Depth
Minnesota	8.5	1	Kentucky	13.9	27
Hawaii	9.9	2	Maryland	14.0	28
Iowa	10.1	3	Illinois	14.2	29
Wisconsin	10.4	4	Washington	14.2	30
Rhode Island	10.5	5	New Jersey	14.4	31
Vermont	10.5	6	New York	15.0	32
Maine	10.6	7	West Virginia	15.9	33
New Hampshire	10.6	8	Wyoming	15.9	34
Kansas	10.8	9	Oregon	16.1	35
Massachusetts	10.8	10	Georgia	16.6	36
Connecticut	10.9	11	North Carolina	16.6	37
Nebraska	11.0	12	Arkansas	16.7	38
North Dakota	11.0	13	Colorado	16.8	39
Michigan	11.4	14	Arizona	17.0	40
Pennsylvania	11.5	15	Mississippi	17.2	41
Missouri	11.7	16	Idaho	17.3	42
Delaware	11.8	17	Montana	17.9	43
Ohio	11.8	18	Alaska	18.2	44
South Dakota	11.9	19	California	18.4	45
Tennessee	12.7	20	Florida	18.5	46

State	% w/o insurance	Depth	State	% w/o insurance	Depth
Utah	13.4	21	Louisiana	18.8	47
Alabama	13.5	22	Nevada	19.1	48
Dist. of Columbia	13.5	23	Oklahoma	19.2	49
Virginia	13.6	24	New Mexico	21.4	50
Indiana	13.7	25	Texas	25.1	51
South Carolina	13.8	26			

Source: DeNavas-Walt, C., Proctor, B. D., & Lee, C. H. (2005). *Income, Poverty, and Health Insurance Coverage in the United States: 2004* (No. P60-229). Washington, DC: U.S. Government Printing Office.

Data are stored in INC-POV-HLTHINS.SAV as the variable NOINS (no insurance).

3.17 *Cancer treatment.* A new cancer treatment uses genetically engineered white blood cells to recognize and destroy cancer cells. Counts of activated immune cells in 11 patients are as follows: {27, 7, 0, 215, 20, 700, 13, 510, 34, 86, 108} (data are fictitious). Make a stemplot of the data. Describe the distribution's shape, central location, and spread.

4 | Summary Statistics

The prior chapter used stemplots, frequency tables, and frequency charts to help describe the shape, location, and spread of a distribution. This chapter introduces numerical summaries that are used for similar purposes. Numerical summaries of location and spread are covered. Numerical summaries of shape are *not* covered because they are seldom used in practice.

■ 4.1 Central Location: Mean

When used without specification, **mean** refers to the arithmetic average of a data set. This is the most common measure of central location.

To calculate the mean, add up all the values in the data set and then divide by the number of observations. Although this is a simple procedure, it will help establish **notation** for future use. Consider the following 10 age values:

$$21 \quad 42 \quad 5 \quad 11 \quad 30 \quad 50 \quad 28 \quad 27 \quad 24 \quad 52$$

Let:

- n represent the sample size ($n = 10$)
- x_i denotes the value of the *i*th observation in the data set (e.g., $x_1 = 21$, $x_2 = 42$)
- $\sum_{i=1}^{n} x_i$ tells you to add all the values from x_1 to x_n. We often drop the subscripts, so $\sum x$ tells you the same thing. For this data, $\sum x = 21 + 42 + 5 + 11 + 30 + 50 + 28 + 27 + 24 + 52 = 290$.

The symbol $\bar{x}$ ("x-bar") represents the sample mean:

$$\bar{x} = \frac{x_1 + x_2 + \ldots + x_n}{n}$$

We can write this in succinct form as:

$$\bar{x} = \frac{1}{n}\sum x_i$$

For the data listed, $\bar{x} = \dfrac{1}{10} \times 290 = 29.0$.

Notes

1. **Gravitational center.** The mean $\bar{x}$ is the gravitational center of the distribution; it is where the data would balance if placed on a seesaw. Figure 4.1 depicts this graphically.

2. **Susceptibility to skews.** Because the mean is a balancing point, a small weight far from the center will counterbalance a large weight near the center. This makes the mean highly susceptible to the influence of outliers and skews. Figure 4.2 depicts this graphically.

3. **Three functions of the mean.** The sample mean tells you three things you might want to know: (1) it can be used to predict an individual value drawn at random from the sample, (2) it can be used to predict a value drawn at random from the population, and (3) it can be used to predict the population mean. These three different uses often get confused.

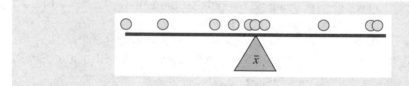

FIGURE 4.1   The mean is the balancing point of a distribution.

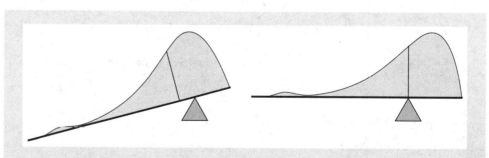

FIGURE 4.2   A skew tips the distribution causing the mean to shift.

4. **The mean from a frequency table.** The mean can be derived from a frequency table by calculating the weighted average of values with weights provided by the relative frequency (proportions) of each value using this formula:

$$\bar{x} = \Sigma p_i x_i$$

where p_i is the proportion of points with value x_i. Table 4.1 illustrates how this is done.

Exercises

4.1 *Gravitational center.* This exercise will demonstrate how the mean is the balancing point of a set of numbers.

(a) **Distribution A.** The values 1 and 5 are marked as "Xs" on the following number line. Calculate the mean and mark its location on the number line.

```
X                       X
1----2----3----4----5
```

(b) **Distribution B.** The values 1, 5, and 5 are shown on the number line. Calculate the mean and show its location on the number line. Notice how the extra 5 pulls the mean to the right.

```
                    X
X                   X
1----2----3----4----5
```

TABLE 4.1 The mean calculated from frequencies. Data are frequencies of fatal horse kicks in the Prussian Army; a classical example from Bortkiewicz (1898).

Interval i	No. of fatalities x_i	Frequency f_i	Proportion p_i
1	0	109	0.545
2	1	65	0.325
3	2	22	0.110
4	3	3	0.015
5	4	1	0.005
6	5	0	0.000

$$\bar{x} = \Sigma p_i x_i = (0.545 \cdot 0) + (0.325 \cdot 1) + (0.110 \cdot 2) + (0.015 \cdot 3) + (0.005 \cdot 4) = 0.61$$

Data from Bortkiewicz, L. V. (1898). *Das Gesetz Der Kleinen Zahlen.* Liepzig: Tuebner.

(c) **Distribution C.** Calculate and show the location of the mean for the data points 2.75, 3.00, and 3.25 on the following number line:

```
      XXX
1----2----3----4----5
```

(d) **Distribution D.** Calculate and show the mean of these three points:

```
      X
      X          X
1----2----3----4----5
```

Exercise 3.1 should convince you that the mean tells you nothing about the spread or shape of a distribution. All four distributions are different, yet distributions A and C have the same means ($\bar{x} = 3$), as do distributions B and D ($\bar{x} = 3.67$). Reliance solely on the mean would have missed the full picture. Consider:

• Describing the central location of a pendulum tells you little of its motion.

• If you have your head in the freezer and your feet in the oven, your average body temperature can still be normal.

• You can drown in a deep area of a lake that has an average depth of just a few inches.

Sole reliance on a mean often misses the true picture.

4.2 *Visualizing the mean.* Consider these eight data points:

```
    1.47  2.06  2.36  3.43  3.74  3.78  3.94  4.42
```

A stemplot of the data looks like this:
```
1 | 4
2 | 03
3 | 4779
4 | 4
×1
```

Visually estimate ("eyeball") the balancing point of the distribution; then calculate the distribution's mean. How well did you do with your "eyeball" estimate?

4.3 *More visualization.* Figure 4.3 contains three stemplots. Stemplot A represents ages (years) of participants in a childhood health survey (*n* = 322). Stemplot B represents body weights of students in a class (*n* = 53). Stemplot C represents

---

Stemplot A

```
 3 | 0
 4 | 0000
 5 | 00000000000000
 6 | 000000000000000000
 7 | 000000000000000000000000000
 8 | 0000000000000000000000000000000000000000000
 9 | 00000000000000000000000000000000000000000000000000
10 | 000000000000000000000000000000000000000
11 | 00000000000000000000000000000000000000000000000
12 | 0000000000000000000000000000000
13 | 000000000000000000000
14 | 000000000000
15 | 000000000
16 | 000000
17 | 0000
18 | 000
19 | 0
x1
```

Stemplot B	**Stemplot C**

<pre>10|0166
11|009
12|0034578
13|00359
14|08
15|00257
16|555
17|000255
18|000055567
19|245
20|3
21|025
22|0
23|
24|
25|
26|0
x10</pre>
	<pre>1|4
	1|789
	2|2234
	2|66789
	3|00012344
	3|5678
	x1</pre>

FIGURE 4.3 Stemplots for Exercise 4.3; visualize the locations of the means.

coliform levels in water samples ($n = 25$). Look at each of these plots and visually estimate each mean. It may help to tip your head to the right to help imagine where the stemplots would balance. The calculated means are listed in the answer key in the back of the book. *After* you have completed your visual estimates, compare them to the actual means.

■ **4.2 Central Location: Median**

The **median** is a different kind of average. It is the midpoint of a distribution—the point that is greater than or equal to half of the values in the data set. To find the median, arrange the data in rank order (i.e., create an ordered array of the data) and then count in from either end of the array to a **depth** of $\dfrac{n + 1}{2}$. When n is odd, you will land on the median. When n is even, you will land between two values. Average these values to determine the actual median. For example, the median of this small data set with 10 observations has a depth $\dfrac{10 + 1}{2} = 5.5$, placing it between 27 and 28.

$$5 \quad 11 \quad 21 \quad 24 \quad 27 \quad 28 \quad 30 \quad 42 \quad 50 \quad 52$$
$$\uparrow$$
$$\text{Median}$$

Average these values to find the median $= \dfrac{27 + 28}{2} = 27.5$.

The median is more **resistant** to outliers than the mean. Consider these five values:

$$4 \quad 7 \quad 8 \quad 9 \quad 12$$
$$\uparrow$$

The median is 8. Now *suppose* there was a data entry error, so that the value of 12 was mistakenly recorded as 120:

$$4 \quad 7 \quad 8 \quad 9 \quad 120$$
$$\uparrow$$

The median is still 8, but the mean goes from 8 to 29.6. The median stayed put, while the mean more than tripled.

The median is relatively resistant to outliers and skew.

ILLUSTRATIVE EXAMPLE

Comparison of mean, median, and mode. The duration of 15 complex surgical procedures are shown in this stemplot:

```
1 | 8
2 | 1345568
3 | 01158
4 |
5 |
6 | 5
7 | 0
×1
```

The median of this distribution is 2.8 (underlined). The mean is about 3.3.[a] The mean has been "pulled up" by the two high outliers (6.5 and 7.0).

■ 4.3 Central Location: Mode

The **mode** is the most frequently occurring value in the data set. For example, the following data set has a mode of 7:

$$4 \quad 7 \quad 7 \quad 7 \quad 8 \quad 8 \quad 9$$

With small data sets, it is often preferable to report a class interval as the mode (as opposed to using a single value). This will be where the stemplot shows a peak. As examples,

- Stemplot A in Figure 4.3 has a mode at 9.
- Stemplot B in Figure 4.3 seems to be **bimodal**, with peaks in the 120s and 180s, probably reflecting different weight distributions for female and male students.
- Stemplot C in Figure 4.3 has a mode in the interval 3.0 to 3.4.

■ 4.4 Comparison of the Mean, Median, and Mode

The mean, median, and mode will coincide in **unimodal** distributions that are symmetrical. With asymmetry, the mean will be pulled toward the skew more so than the median. Figure 4.4 depicts this fact. Because of this, you can predict the shape of a distribution by comparing its mean and median.

---

[a] $\bar{x} = \dfrac{1}{15}(1.8 + 2.1 + \ldots + 7.0) = \dfrac{1}{15} \times 49.0 \cong 3.3$

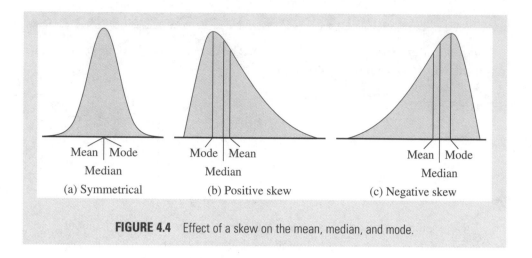

FIGURE 4.4   Effect of a skew on the mean, median, and mode.

When mean = median → distribution is symmetrical
When mean > median → there is a positive skew
When mean < median → there is a negative skew

Exercise

4.4   *Seizures following meningitis.* Exercise 3.9 considered induction times (months) for seizures in 13 cases. Data were {0.10, 0.25, 0.50, 4, 12, 12, 24, 24, 31, 36, 42, 55, 96}.

(a) Calculate the mean and median of this data set.

(b) Compare the mean and median. What does this tell you about the distribution's shape?

(c) Which measure of central location would you use to describe this distribution's center? Explain your response.

■ 4.5 Spread: Quartiles

As noted earlier, sole reliance on a measure of central location to describe a distribution can often miss the point. Accompanying the measure of central location with a measure of spread (**variability**) will help fill out the picture. Here is an example to illustrate this fact:

ILLUSTRATIVE EXAMPLE

Importance of spread (Air samples). Air samples were collected on eight successive days at two different sites. Particulate matter was measured in these samples ($\mu g/m^3$) as an index of environmental quality. Table 4.2 lists the data for the two sites.

TABLE 4.2 Data for "Air samples" illustrative example. Particulates in air samples on successive days at two sites ($\mu g/m^3$).

Site 1:	68	22	36	32	42	24	28	38
Site 2:	36	38	39	40	36	34	33	32

Data are fictitious but realistic. The data file is stored online in the file AIRSAMPLES.[*]

To the nearest unit, both sites have means of 36 $\mu g/m^3$. However, side-by-side stemplots reveal different pictures:

```
Site 1| |Site 2
--------------
  42|2|
   8|2|
   2|3|234
  86|3|6689
   2|4|0
    |4|
    |5|
    |5|
    |6|
   8|6|
   ×10
```

Notice how there is much greater variability of readings at site 1. One particular reading at site 1 was very high (68 $\mu g/m^3$), indicating a very "dirty" day. The occurrence of this very high level of pollution can be hazardous, especially for individuals with compromised cardiorespiratory function.

Range

There are several ways to measure the spread of a distribution. The simplest measure of spread is the **range**, which is merely:

$$\text{Range} = \text{Maximum} - \text{Minimum}$$

The range of the particulate matter in air samples for site 1 in the prior illustrative example is equal to $68 - 22 = 46$. The range at site 2 is equal to $40 - 32 = 8$. Variability is much greater at site 1.

Although suited for some purposes, the range is *not* a very good general measure of spread. It accounts only for two values in the data set (the maximum and minimum), making it a relatively *inefficient* statistic. In addition, it is unlikely to capture both the maximum and minimum in the population, so it has a tendency to underestimate the population's range, making it a *biased* statistic. Therefore, when used, the range should be supplemented with at least one of the other measures of spread described in this section.

Quartiles, Five-Point Summary, Interquartile Range

Quartiles provide an inefficient and intuitive way to describe variability. The idea is to divide the data set into four equal segments.[b] **Quartile 1 (Q1)** cuts off the bottom quarter of data points. **Quartile 3 (Q3)** cuts off the top quarter. We then describe the distance between the quartiles as a measure of spread.

Because data do not always divide neatly into quarters, we need rules for interpolating quartiles. Several interpolation rule systems exist. The rule system we use derives from Tukey's hinges.[c] **Hinges** are where the ordered array "folds" upon itself. To determine quartiles by the hinge method:

1. Put the data in rank order; then locate the median.
2. Divide the data set into a "low group" and a "high group" where they split at the median. When *n* is odd, put the median in both groups.
3. Find the middle value ("median") of the low group. This is **Q1**.
4. Find the middle value ("median") of the high group. This is **Q3**.

ILLUSTRATIVE EXAMPLE

Quartiles. Consider this small ordered array ($n = 10$):

```
5  11  21  24  27  28  30  42  50  52
        ↑           ↑           ↑
       Q1           M          Q3
```

The low group is {5, 11, 21, 24, 27}, so Q1 = 21. The high group is {28, 30, 42, 50, 52}, so Q3 = 42.

---

[b] The idea of dividing a data set into groups of equal size was initially described by Francis Galton in the 19th century. He was also first to use the term "quartile." See: Galton, F. (1879). On a proposed statistical scale. *Nature, 9,* 342–343.

[c] Be aware that there are different rule systems to determine quartiles. If you are using a software program or calculator to determine quartiles, results may differ from the "hinge method" described here.

We can summarize a distribution with the points that define its quartiles. This is known as the **five-point summary,** which consists of the distribution's

- **Q0** (the minimum)
- **Q1** (the lower hinge)
- **Q2** (the median)
- **Q3** (the upper hinge)
- **Q4** (the maximum)

The five-point summary for the data set in the previous illustrative example is 5, 21, 27.5, 42, 52.

The **interquartile range (IQR)** is a summary measure of spread consisting of:

$$IQR = Q3 - Q1$$

This range captures the middle 50% of data points in a set. The IQR for the current illustrative data $= 42 - 21 = 21$.

ILLUSTRATIVE EXAMPLE

IQR (n = 7). What is the five-point summary and IQR of this small data set?

    1.47 2.06 2.36 3.43 3.74 3.78 3.94

- The median is 3.43.
- The low group consisting of {1.47, 2.06, 2.36, 3.43} has a middle value between 2.06 and 2.36. We average these to calculate

$$Q1 = \frac{2.06 + 2.36}{2} = 2.21.$$

- The high group consisting of {3.43, 3.74, 3.78, 3.94} has a middle value between 3.74 and 3.78, so $Q3 = \frac{3.74 + 3.78}{2} = 3.76.$
- The five-point summary is 1.47, 2.21, 3.43, 3.76, 3.94.
- The IQR $= 3.76 - 2.21 = 1.55$.

ILLUSTRATIVE EXAMPLE

IQRs (Air samples data). Recall the "air samples" illustrative data presented earlier in this chapter. Table 4.2 lists data for particulate matter at two sampling sites.

continues

For site 1, the ordered array or data is:

$$22 \quad 24 \quad 28 \quad 32 \quad 36 \quad 38 \quad 42 \quad 68$$

$$\uparrow \qquad\qquad \uparrow \qquad\qquad \uparrow$$

$$Q1 \qquad\qquad M \qquad\qquad Q3$$

Thus, Q1 = the average of 24 and 28 (which is 26) and Q3 = the average of 38 and 42 (which is 40). The IQR = 40 − 26 = 14.

For site 2, the ordered array is:

$$32 \quad 33 \quad 34 \quad 36 \quad 36 \quad 38 \quad 39 \quad 40$$

$$\uparrow \qquad\qquad \uparrow \qquad\qquad \uparrow$$

$$Q1 \qquad\qquad M \qquad\qquad Q3$$

Q1 = 33.5 and Q3 = 38.5. Therefore, IQR = 38.5 − 33.5 = 5.
Data are less variable at site 2 (IQR: 14 vs. 5).

■ 4.6 Boxplots

Box-and-whiskers plots (boxplots) display five-point summaries and "potential outliers" in graphical form. The box of the boxplot spans the IQR of a data set. The median is indicated as a line within the box. Extreme values (potential outliers) are plotted as separate points beyond data set "whiskers." To construct a boxplot:

1. Determine the five-point summary for the data. Draw a **box** extending from hinge to hinge (i.e., from Q1 to Q3).

2. Calculate the IQR and use this to determine **fences** as follows:

$$\text{Fence}_{\text{Lower}} = Q1 - (1.5)(\text{IQR})$$
$$\text{Fence}_{\text{Upper}} = Q3 + (1.5)(\text{IQR})$$

Do *not* plot the fences.

3. Determine whether the data set contains **outside values**. Values below the lower fence are **lower outside values**. Values above the upper fence are **upper outside values**. Plot these as separate points on the graph.

4. The smallest value inside the lower fence is the **lower inside value**. The largest value inside the upper fence is the **upper inside value**. Draw **whiskers** from respective quartiles (hinges) to inside values.

Boxplots provide insight into the distribution's location, spread, and shape:

- **Location:** The location of the distribution is visible. The box cradles the middle 50% of values. The line in the box locates the median.

- **Spread:** The IQR (**hinge spread**) is visible as the height of the box. The **whisker spread** (from whisker to whisker) and range also provide visual clues of "spread".

- **Shape:** Shape is difficult to judge when the sample is small. With large data sets, you can judge symmetry and get a sense of whether the distribution has long or short tails.

ILLUSTRATIVE EXAMPLE

Boxplots (Air samples data). We propose to draw boxplots for the air sample data presented in prior illustrations (see Table 4.2).

Data for site 1:   22   24   28   32   36   38   42   68

1. The five-point summary (determined in a prior illustration) is: 22, 26, 34, 40, and 68. Therefore, the box extends from 26 to 40. A line for the median is drawn in the box at 34.

2. IQR = 40 − 26 = 14. Therefore, $Fence_{Lower}$ = 26 − (1.5)(14) = 5 and $Fence_{Upper}$ = 40 + (1.5)(14) = 61.

3. There are no values below the lower fence. There is one value (68) above the upper fence. The *upper outside value* of 68 is plotted as a separate point above the box.

4. The smallest value still inside the fence (lower inside value) is 22. A whisker is drawn from Q1 to this inside value. The upper inside value is 42. A whisker is drawn from Q3 to this inside value.

Figure 4.5 shows the boxplot for site 1 on the left.

Data for site 2:   32   33   34   36   36   38   39   40

1. The five-point summary is 32, 33.5, 36, 38.5, and 40. The box extends from 33.5 to 38.5. A line for the median is drawn inside the box at 36.
2. IQR = 38.5 − 33.5 = 5. $Fence_{Lower}$ = 33.5 − (1.5)(5) = 26. $Fence_{Upper}$ = 38.5 + (1.5)(5) = 46.
3. There are no outside values in this group.
4. The lower inside value is 32. A whisker is drawn from Q1 to this lower inside value. The upper inside value is 40. A whisker is drawn from Q3 to this upper inside value.

continues

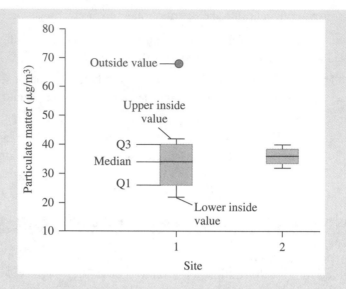

FIGURE 4.5 Side-by-side boxplots, air samples illustrative data.

Figure 4.5 shows the boxplot for site 2 on the right. The side-by-side box-plots in this figure show sites 1 and 2 have similar central locations but different spreads.

Be aware that there are several **different ways to determine quartiles**. We use **Tukey's hinge method** in which the median is included in both the lower half and upper half of the data when determining the locations of Q1 and Q3. (This is only when *n* is odd.) An equally valid method for determining locations of quartiles is to exclude the median when splitting the data set in half (when *n* is odd). Many calculators and computer programs default to this second method. Therefore, one should not expect all calculator- and computer-generated output to match our results. Check your computer program's documentation to determine the method it uses to determine quartiles.

Comment: **John Wilder Tukey** (1915–2000) was an American statistician and math-ematician known for his wisdom and insight into mathematical and natural phenomena. Many of Tukey's innovations in the field of data analysis are documented in his 1977 landmark text *Exploratory Data Analysis* (Addison-Wesley, Reading, MA).

Exercises

4.5 *Outside?* The following stemplot has 18 observations. Prove that the value 152 in this data set is an outside value. Then draw the boxplot for the data.

```
08 | 8
09 | 59
10 | 0147
11 | 444679
12 | 0145
13 |
14 |
15 | 2
×1 0
```

4.6 *Seizures following bacterial meningitis.* In Exercise 4.4, you calculated the mean and median of induction times in 13 seizures cases. Data were {0.10, 0.25, 0.50, 4, 12, 12, 24, 24, 31, 36, 42, 55, 96} months. Now construct a boxplot for these data. Are there any outside values in this data set? Does the boxplot show evidence of asymmetry?

■ 4.7 Spread: Variance and Standard Deviation

The **variance** and **standard deviation** are the most common measures of spread. These statistics are based on the average squared distances of values around the data set's mean. Here is how they are calculated:

- Determine the **deviation** of each data point. A deviation is the data point minus its mean: $x_i - \bar{x}$. Figure 4.6 shows deviations for two of the values in the air samples data (Table 4.2) for site 2.

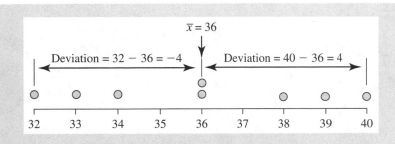

FIGURE 4.6 Deviations of two observations, site 2, air samples illustrative data, Table 4.2.

- **Square each deviation.** This is denoted $(x_i - \bar{x})^2$.
- **Sum the squared deviations.** This statistic is the **sum of squares (SS)**:

$$SS = (x_1 - \bar{x})^2 + (x_2 - \bar{x})^2 + \ldots + (x_n - \bar{x})^2$$

- Divide the sum of squares by n minus 1. This is the **variance**, denoted s^2:

$$s^2 = \frac{SS}{n - 1}$$

An equivalent formula is $s^2 = \dfrac{1}{n-1}\Sigma(x_i - \bar{x})^2$

- Take the square root of the variance. This is the **standard deviation** (s):

$$s = \sqrt{s^2}$$

Equivalently,

$$s = \sqrt{\frac{1}{n-1}\sum(x_i - \bar{x})^2}$$

In practice, we often use software to compute the variance and standard deviation. However, it is instructive to complete these calculations by hand, so let's practice a few.

ILLUSTRATIVE EXAMPLE

Standard deviation (Air samples from site 2). We propose to calculate the variance and standard deviation for the particulate matter in the air samples from site 2 for the air samples data (**Table 4.2** and **Figure 4.6**). At site 2, $n = 8$ and

$$\bar{x} = \frac{36 + 38 + 39 + 40 + 36 + 34 + 33 + 32}{8} = 36 \ (\mu g/m^3).$$

Deviations and squared deviations for each point are:

Data x_i	Deviations $x_i - \bar{x}$	Squared deviations $(x_i - \bar{x})^2$
36	$36 - 36 = 0$	$0^2 = 0$
38	$38 - 36 = 2$	$2^2 = 4$
39	$39 - 36 = 3$	$3^2 = 9$
40	$40 - 36 = 4$	$4^2 = 16$
36	$36 - 36 = 0$	$0^2 = 0$
34	$34 - 36 = -2$	$-2^2 = 4$
33	$33 - 36 = -3$	$-3^2 = 9$
32	$32 - 36 = -4$	$-4^2 = 16$
$\Sigma x_i = 288$	$\Sigma(x_i - \bar{x}) = 0$	$\Sigma(x_i - \bar{x})^2 = 58$

The sum of squares (last column) $\text{SS} = \Sigma(x_i - \bar{x})^2 = 0 + 4 + 9 + 16 + 0 + 4 + 9 + 16 = 58$.

The variance is $s^2 = \dfrac{\text{SS}}{n-1} = \dfrac{58}{8-1} = 8.286 \ (\mu g/m^3)^2$.

The standard deviation is the square root of the variance:

$s = \sqrt{8.286 \ (\mu g/m^3)^2} = 2.88 \ \mu g/m^3$.

Facts About the Standard Deviation

1. **Deviations.** The deviations of a data set always sum to 0. (Data points balance perfectly around the mean in positive and negative directions.) Squaring deviations makes their signs unimportant, so the sum of the squared deviations will always be a positive value.

2. **Units.** The variance carries "units squared." The standard deviation carries the same units as the data, making it a better choice for descriptive purposes.

3. **No variability → standard deviation = 0.** When all values in the data set are the same, there is no spread and the variance and standard deviation equal 0. In all other instances, these statistics are positive values.

4. **Degrees of freedom.** The variance is the average of the sum of squares, with the sum of squares divided by $n - 1$ instead of n. The number $n - 1$ is the **degrees of freedom** of the variance. You lose one **degree of freedom** because knowing $n - 1$ of the deviations determines the last deviation.

5. **The standard deviation is sensitive to outliers and skews.** Like the mean, the standard deviation is not resistant to outliers or strong skews. Consider the data set {4, 7, 8, 11, 12}. These data have mean $\bar{x} = 8.4$ and standard deviation $s = 3.2$. Had we made a data entry problem and entered data as {4, 7, 8, 11, 120}, $\bar{x}$ would go to 30.0 and s would go to 50.4. In contrast, the median and IQR would be unaffected.

6. **Standard deviations are useful when making comparisons.** The greater the variability within a group, the larger its standard deviation. For example, if the standard deviation of an age variable in one group is 15 years and that of another group is 7 years, the first group has much greater age variability than the second.

7. **Percentage of data points falling in a range.** The standard deviation can be used to describe the percentage of the data points that will fall in certain ranges. Two rules are applied: One rule is **Chebychev's rule**; the other rule is the **Normal rule**.

   Chebychev's rule applies to all data sets, regardless of their shape. It says that *at least* three-fourths of the data points will lie within two standard deviations of the mean. These boundaries are $\bar{x} \pm 2s$. For example, if a data set has a mean age of 30 years and a standard deviation of 10 years, then *at least* three-quarters of the values lie in the range $30 \pm (2)(10) = 10$ to 50. The key phrase here is *at least*; it is possible that more than three-fourths of the individuals fall in the range (maybe even 100%).

   The **Normal rule** (also known as the **68-95-99.7 rule**) applies only to distributions with a particular **Normal** shape. We will cover the Normal distribution in Chapter 7, but for now it is important not to confuse the term *N*ormal (meaning a particular bell shape) with the common term *n*ormal (meaning "typical"). Many natural distributions are *not* Normal.[d] However, when a distribution is Normal

- 68% of the data points will lie within one standard deviation of the mean ($\bar{x} \pm s$).
- 95% of data points lie within two standard deviations of the mean ($\bar{x} \pm 2s$).
- 99.7% of the data points lie within three standard deviations of the mean ($\bar{x} \pm 3s$).

For example, a Normal distribution with a mean of 30 and standard deviation of 10 has 68% of its values in the range $30 \pm (1)(10) = 20$ to 40, 95% of its values in the range $30 \pm (2)(10) = 10$ to 50, and 99.7% of its values in the range $30 \pm (3)(10) = 0$ to 60.

By putting Chebychev and the Normal rule together, we can say that between 75% and 95% of the data points usually fall within two standard deviations of the mean.

Exercises

4.7 *Spread.* Each of the following batches of numbers has a mean of 100. Which has the most variability? (Arithmetic *not* required.)

```
Batch A:    0  50  100  150  200
Batch B:   50  75  100  125  150
Batch C:   75  88  100  113  125
```

---

[d]Elveback, L. R., Guillier, C. L., & Keating, F. R., Jr. (1970). Health, normality, and the ghost of Gauss. *JAMA, 211*(1), 69–75.

4.8 *Standard deviation for site 1.* Use a step-by-step approach to calculate the standard deviation of the data for site 1 listed in **Table 4.2**. Compare this standard deviation to that of the data from site 2 ($s_2 = 2.88$ μg/m³). How does this numerical comparison relate to what you see in **Figure 4.5**?

4.9 *Standard deviation via technology.* In practice, we normally use statistical calculators or computers to calculate standard deviations. Using your statistical calculator or computer, calculate the standard deviations and variances for the air samples data originally presented in **Table 4.2.** Make certain results agree with prior hand calculations.

Notes

1. **Sample standard deviation and population standard deviation.** Some calculators and programs report two different standard deviations formulas. One standard deviation is the *sample* standard deviation (s). The other is for the population standard deviation (σ).[e] In Excel (Microsoft Corp.), for instance, the sample standard deviation corresponds to the =STDEV function and the *population* standard deviation formula corresponds to =STDEVP. Hand calculators often have two different keys, one labeled s and one labeled σ. In practice, you should always use the *sample* standard deviation (s) to calculate the standard deviation of a data set.

2. **Calculators and applets.** If you do not own a statistical calculator or statistical software package, consider using one of the many free statistical applets on the Web. The website http://statpages.org/ lists about a dozen free online applets that calculate summary statistics.[f] The program *WinPepi* > Describe.exe can also be used for this purpose.[g] There is a link to download *WinPepi* on the website for this book.

4.10 *Heart rate.* An individual with an irregular heartbeat is given a medication to stabilize his condition. Heart rates (beats per minute) before and after treatment are shown here.[h] Determine the means and standard deviations before and after treatment. Did the drug work?

Before: 65   85   90   65   55   60
After: 68   70   69   70   71   72

---

[e] Population standard deviation. $\sigma = \sqrt{\frac{1}{n}\Sigma(x_i - \mu)^2}$. There is no loss of one degree of freedom when calculating the population standard deviation because the mean is not derived from the data. When n is large, $(n - 1) \leqslant n$ and $s \leqslant \sigma$.

[f] Pezzullo, J. C. (2006). *StatPages.net: web pages that perform statistical calculations.* Available: http://statpages.org/.

[g] Abramson, J. H. (2001). *WINPEPI* (updated): computer programs for epidemiologists and their teaching potential. *Epidemiologic Perspectives & Innovations, 8*(1), 1.

[h] Data are fictitious and are stored online in the file HEARTRATE.*.

4.11 *Units of measure change numeric values of a standard deviation.* Calculate the standard deviations of the batches of numbers here. Which batch has the greatest variability?

Batch A:	0 years	1 year	2 years
Batch B:	0 months	12 months	24 months
Batch C:	0 days	365 days	730 days

This exercise demonstrates a potential problem when comparing standard deviations for variables using different units of measurement. The three batches of numbers describe identical information, but their standard deviations differ. Reporting units of measure mitigates the problem: Batch A has a standard deviation of 1 year, batch 2 has a standard deviation of 12 months, and batch 3 has a standard deviation of 365 days.

When it is necessary to compare standard deviations of measurements on different scales, convert each standard deviation to a **coefficient of variation (CV)**, defined as

$$CV = \frac{s}{\bar{x}}$$

The CV removes the units of measure and expresses the standard deviation in relation to the size of the mean. This allows for direct comparison of numerical results. In Exercise 4.11, batch A has CV = 1 year/1 year = 1, batch 2 has CV = 12 months/12 months = 1, and batch 3 has CV = 365 days/365 days = 1. All three CVs are equal to 1.

4.12 *Test scores with mean 100 and standard deviation 10.* Test scores have a Normal distribution with a mean of 100 and standard deviation of 10. What percentage of scores falls in the range 80 to 120? Explain.

■ 4.8 Selecting Summary Statistics

How do you choose which summary statistics to use in a given situation? Each situation differs, and there is no "one size fits all" solution, but certain general rules apply.

1. Always report the sample size (n), a measure of central location, and a measure of spread. The mean should be accompanied by the standard deviation. The median should be accompanied by the IQR or quartiles.

2. Use the mean/standard deviation when the distribution is symmetrical. You should use the median/quartiles when the distribution is asymmetrical or has outliers.

3. Accompany summary statistics with plots, when possible. Graphs can provide information that escapes numerical summaries: "You can observe a lot by watching."[i]

Exercises

4.13 *Which statistics?* Which measures of central location and spread would you use to describe each of the data sets depicted in **Figure 4.3**?

4.14 *Effect of removing an outlier.*[j] Exercise 4.6 looked at months between bacterial meningitis and the onset of seizures in 13 cases. Data were {0.10, 0.25, 0.50, 4, 12, 12, 24, 24, 31, 36, 42, 55, 96}.

(a) Calculate the mean and standard deviation for these data.

(b) Calculate the median and IQR.

(c) Remove the outlier (96) and recalculate the mean and standard deviation. Also, recalculate the median and IQR. What effect did removing the outlier have on the mean and standard deviation? What effect did it have on the median and IQR?

Summary Points (Summary Statistics)

1. The three main **measures of central location** are the mean, median, and mode. The **mean** ($\bar{x}$) is the gravitational center of the distribution. The **median** is the central value in the order array of the data. The **mode** is the most frequently occurring value in the data. (The mode is used only in large data sets with many repeating values.)

2. The mean is more efficient than the median in describing a distribution's center because it makes use of the quantitative information. However, the median is more robust than the mean because it is less easily influenced by outliers and skews.

3. If the mean and median are about equal, the distribution is more or less **symmetrical**. If the mean is larger than the median, the distribution has a **positive skew**. If the mean is smaller than the median, the distribution has a **negative skew**.

4. The **5-point summary** lists the minimum (quartile 0), 25th percentile (quartile 1, Q1), 50th percentile (median, quartile 2), 75th percentile (quartile 3, Q3), and maximum (quartile 4) of the distribution.

---

[i] Lawrence Peter (Yogi) Berra.
[j] You may wish to use an applet or statistical calculator to complete calculations for this problem.

5. There are several ways to identify quartiles. We rely on **Tukey's hinge method** for this purpose. Check your calculator's documentation to see which method it uses.

6. The main measures of **spread (variability)** are the **inter-quartile range (IQR)** and **standard deviation (s)**.

7. The standard deviation is a more efficient measure of spread than the IQR. However, it is less **robust** than the IQR.

8. Always report a measure of spread along with a measure of central location. The mean and standard deviation are the preferred measures of location and spread when the distribution is symmetrical. The median and IQR (or 5-point summary) are preferred when the distribution is asymmetrical.

9. The **box-and-whiskers plot ("boxplot")** displays quartiles and outside values in a graphical form. These plots are particularly well-suited for comparing distributions in a side-by-side manner.

Vocabulary

Bimodal	Normal rule (68-95-99.7 rule)
Box-and-whiskers plot (boxplots)	Outside values
Chebychev's rule	Q0 (quartile 0; minimum)
Coefficient of variation (CV)	Q1 (quartile 1; 25th percentile)
Degrees of freedom	Q2 (quartile 2; median)
Depth	Q3 (quartile 3; 75th percentile)
Deviation	Q4 (quartile 4; maximum)
Fence	Quartiles
Five-point summary	Range
Hinges	Resistant
Hinge spread	Standard deviation
Inside value	Sum of squares
Interquartile range (IQR)	Unimodal
Mean	Variability
Median	Variance
Mode	Whisker spread

Review Questions for Chapter 4

4.1 In statistical notation, how does x differ from X?

4.2 Name three measures of central location.

4.3 Name four measures of spread.

4.4 A sample mean is used to "predict" three things. List these things.

4.5 What does it mean when we say that a statistic is robust?

4.6 Which is more robust: the mean or the median?

4.7 Which is more robust: the standard deviation or the inter-quartile range?

4.8 Where is the median of a data set located?

4.9 Why is the sample range a poor measure of spread?

4.10 Provide two synonyms for Q1 (quartile 1).

4.11 Provide two synonyms for Q3 (quartile 3).

4.12 What is a *hinge spread*?

4.13 List the elements of a 5-point summary.

4.14 When determining the quartiles by Tukey's hinge method and n is odd, do you include the median in the low- and high-groups when splitting the data set?

4.15 True or false: There is a universally accepted way to interpolate quartiles in datasets.

4.16 How do you determine whether a data point is an outside value?

4.17 What is an upper inside value? What is a lower inside value?

4.18 What is a *deviation*?

4.19 What statistic is the sum of squared deviations divided by $n - 1$?

4.20 What statistic is the square root of a variance?

4.21 What is the standard deviation of this data set: $\{5, 5, 5, 5\}$? [Calculations not required.]

4.22 What rule says "at least 75% of the data points will always fall within two standard deviation of the mean"?

4.23 Select the best response: When do 95% of the values fall within 2 standard deviations of the mean?

   (a) always
   (b) when the distribution is symmetrical
   (c) only when the distribution is normal

4.24 Why do we lose one degree of freedom when calculating the sample standard deviation?

Exercises

4.15 *Leaves on stems.* Calculate the mean and standard deviation of each group depicted in each of the side-by-side stemplots in this problem. Discuss how these statistics relate to what you see.

(a) Comparison A

```
Group 1|  |Group 2
- - - - - - - - - - - - - - - - -
       0 |1|
         |2|
       0 |3| 0
         |4| 0
       0 |5| 0
         |6| 0
       0 |7| 0
         |8|
       0 |9|
          ×10
```

(b) Comparison B

```
  Group 1|  |Group 2
  - - - - - - - - - - - - - - - - -
           |1|
           |2|
           |3| 0
           |4| 0
        0 |5| 0
        0 |6| 0
        0 |7| 0
        0 |8|
        0 |9|
           ×10
```

(c) Comparison C

```
  Group 1|  |Group 2
  - - - - - - - - - - - - - - - - -
        0 |1|
          |2|
        0 |3|
          |4|
        0 |5| 0
          |6| 0
        0 |7| 0
          |8| 0
        0 |9| 0
           ×10
```

4.16 *Irish healthcare websites.* Table 3.1 in the prior chapter considered the reading levels of Irish healthcare websites. Here's a reissue of the stemplot for the data:

```
08 | 0
09 |
10 | 0
11 | 00
12 | 0
13 | 0000
14 | 000
15 | 0000000
16 | 000
17 | 00000000000000000000000000
×1
```

(a) Which measure of location and spread would you use to describe this distribution?

(b) Calculate the five-point summary for the distribution.

(c) Does this data set have any outside values? (Show all your work.)

4.17 *Health insurance by state.* Table 4.3 lists the percent of people without health insurance by state.

(a) Calculate the mean and median of these data. Compare these statistics. What does this tell you about the shape of the distribution?

(b) Determine the five-point summary for the data.

(c) Are there any outside values in this data set?

4.18 *Skinfold thickness.* Skinfold thickness over the triceps muscle in the arm is an anthropometric measure that varies with states of health. Table 4.4 lists skinfold measurements at the midpoint of the triceps in five men with chronic lung disease and six comparably aged controls. Compare the groups with side-by-side stemplots. Then calculate group means and standard deviations.

4.19 *What would you report?* A small data set ($n = 9$) has the following values {3.5, 8.1, 7.4, 4.0, 0.7, 4.9, 8.4, 7.0, 5.5}. Plot the data as a stemplot and then report an appropriate measures of central location and spread for the data.

4.20 $\bar{x}$ *and s by hand.* To assess the air quality in a surgical suite, the presence of colony-forming spores per cubic meter of air is measured on three successive days. The results are as follows: {12, 24, 30} (data are fictitious).

(a) Calculate the mean and standard deviation for these data by hand, using the step-by-step process demonstrated in Section 4.7 of this chapter.

(b) Enter the data into your calculator or computer program. Calculate the mean and standard deviation with this device. Do these results match your hand calculations?

TABLE 4.3   Percent of residents without health insurance by state, United States, 2004, $n = 51$, presented in rank order.

State	% w/o insurance	Depth	State	% w/o insurance	Depth
Minnesota	8.5	1	Kentucky	13.9	27
Hawaii	9.9	2	Maryland	14.0	28
Iowa	10.1	3	Illinois	14.2	29
Wisconsin	10.4	4	Washington	14.2	30
Rhode Island	10.5	5	New Jersey	14.4	31
Vermont	10.5	6	New York	15.0	32
Maine	10.6	7	West Virginia	15.9	33
New Hampshire	10.6	8	Wyoming	15.9	34
Kansas	10.8	9	Oregon	16.1	35
Massachusetts	10.8	10	Georgia	16.6	36
Connecticut	10.9	11	North Carolina	16.6	37
Nebraska	11.0	12	Arkansas	16.7	38
North Dakota	11.0	13	Colorado	16.8	39
Michigan	11.4	14	Arizona	17.0	40
Pennsylvania	11.5	15	Mississippi	17.2	41
Missouri	11.7	16	Idaho	17.3	42
Delaware	11.8	17	Montana	17.9	43
Ohio	11.8	18	Alaska	18.2	44
South Dakota	11.9	19	California	18.4	45
Tennessee	12.7	20	Florida	18.5	46
Utah	13.4	21	Louisiana	18.8	47
Alabama	13.5	22	Nevada	19.1	48
Dist. of Columbia	13.5	23	Oklahoma	19.2	49
Virginia	13.6	24	New Mexico	21.4	50
Indiana	13.7	25	Texas	25.1	51
South Carolina	13.8	26			

Data from DeNavas-Walt, C., Proctor, B. D., & Lee, C. H. (2005). *Income, Poverty, and Health Insurance Coverage in the United States: 2004* (No. P60-229). Washington, DC: U.S. Government Printing Office.

Data are stored in INC-POV-HLTHINS.SAV as the variable NOINS (no insurance).

TABLE 4.4   Data for Exercise 4.18; skinfold thickness over the triceps (mm).

Chronic lung disease:	9.1	10.9	11.4	15.3	18.4	
Controls:		10.4	19.6	20.6	23.8	24.7	32.8

Data are fictitious but realistic and are stored online in the file SKINFOLD.[*].

4.21 *Practicing docs (side-by-side boxplots).* In Exercise 3.15, you used side-by-side stemplots to compare the number of practicing medical doctors per 10,000 in the fifty United States and District of Columbia for the years 1975 and 2004. Let us now create side-by-side boxplots for the same purpose. Here are the data sorted in rank order.

Data for 1975

7.7	7.8	8.0	8.5	8.6	8.9	8.9	9.2	9.3	9.4
9.4	9.6	10.0	10.1	10.1	10.1	10.5	10.6	10.6	10.7
10.9	10.9	11.0	11.2	11.3	11.4	11.6	11.9	12.0	12.2
12.7	13.0	13.1	13.1	13.4	13.6	13.7	13.8	13.9	14.0
14.1	14.7	15.0	15.5	16.1	16.5	17.3	17.7	18.3	20.2
34.6									

Data for 2004

15.6	16.1	16.6	16.8	17.2	17.7	18.9	18.9	18.9	19.0
19.5	19.6	20.0	20.1	20.3	20.4	20.5	20.9	21.0	21.0
21.2	21.3	21.4	21.6	22.1	22.2	22.2	22.6	22.9	23.0
23.1	23.3	23.4	23.6	23.6	23.6	23.9	24.0	24.1	24.2
25.3	25.6	26.8	28.1	30.7	31.3	31.9	33.0	33.8	37.4
64.2									

Source: Adapted from Table 104 of National Center for Health Statistics. (2006). *Health United States 2006: Chartbook on Trends in the Health of Americans*, Retrieved from www.cdc.gov/nchs /data/hus/hus06.pdf on Nov 19, 2010. Data are also provided in SPSS and Excel formats on the companion website as HEALTHUNITEDSTATES2006TABLE104.*.

After creating the side-by-side boxplots, compare the shapes, locations, and spreads of the two distributions.

4.22 *Practicing docs (means and standard deviations).* Recall the data set used in Exercises 3.15 and 4.21. Download the data set (HEALTHUNITEDSTATES2006 TABLE104.*) from the companion website. Then, compute the mean and standard deviation for the 1975 and 2004 data using a statistical program of your choosing. How do these numerical summaries relate to the side-by-side boxplots completed in Exercise 4.21? (The side-by-side boxplots are shown in the answer key toward the back of the book.)

Computational note: SPSS will calculate these summary statistics with "Analyze > Descriptive Statistics > Explore > Descriptives" and "Analyze > Descriptive Statistics > Explore."

4.23 *Melanoma treatment.* A study by Morgan and coworkers used genetically modified white blood cells to treat patients with melanoma who had not responded to standard treatments.[k] In patients in whom the cells were cultured *ex vivo* for an extended period of time (cohort 1), the cell doubling times were {8.7, 11.9, 10.0} days. In a second group of patients in whom the cells were cultured for a shorter period of time (cohort 2), the cell doubling times were {1.4, 1.0, 1.3, 1.0, 1.3, 2.0, 0.6, 0.8, 0.7, 0.9, 1.9} days. In a third group of patients (cohort 3), actively dividing cells were generated by performing a second rapid expansion via active cell transfer. Cell doubling times for cohort 3 were {0.9, 3.3, 1.2, 1.1} days. Data are available in Excel and SPSS formats on the companion website as file MORGAN2006.*. Create side-by-side boxplots of these data. In addition, calculate the mean and standard deviations within each group. Comment on your findings.

---

[k]Morgan, R. A., Dudley, M. E., Wunderlich, J. R. Hughes, M. S., Yang, J. C., Sherry, R. M., et al. (2006). Cancer regression in patients after transfer of genetically engineered lymphocytes. *Science*, *314*(5796), 126–129.

5 | Probability Concepts

■ 5.1 What Is Probability?

Probability permits us to deal with random variability in constructive ways; a good deal of statistical reasoning depends on probability. This is a difficult topic, so let's start with several definitions:

> - A **random variable** is a numerical quantity that takes on different values depending on chance.
> - A **population** is the set of all possible outcomes for a random variable. (This refers to a hypothetical population of numbers, not a population of people.)
> - An **event** is an outcome or set of outcomes for a random variable.
> - **Probability** refers to the proportion of times an event is expected to occur in the long run. Probabilities are always numbers between 0 and 1 with 0 corresponding to "never" and 1 corresponding to "always."

The concept of probability is founded on our experience of the *relative frequency* of an event. When an event in the population has been defined in a way that allows for counting, we can determine its probability by counting the number of ways it can occur and dividing by the total number of events in the population. For example, if the event of interest is "traffic crash fatalities in the U.S. in the year 2004" (there were 42,636), and the population of U.S. residents that year was about 293,655,000, the probability that a resident selected at random from this population at the beginning of the year would experience a traffic fatality would be 42,636 ÷ 293,655,000 = 0.0001452 or about 1 in 6887.[a]

---

[a] Fatality Analysis Reporting System (FARS). Retrieved February 9, 2006, from www-fars.nhtsa.dot. gov/.

In practice, however, we do not often have the opportunity to enumerate all events in a population. Nonetheless, we can still understand how probabilities behave by addressing relative frequencies as part of a **repetitive process**.

ILLUSTRATIVE EXAMPLE

Probability as part of a repetitive process. Figure 5.1 shows the results of two trials in which we randomly selected 10,000 individuals from a population in which 20% of the individuals are smokers.[b] The proportion of the smokers in the sample is plotted as the sample size increases from $n = 0$ to $n = 10,000$.

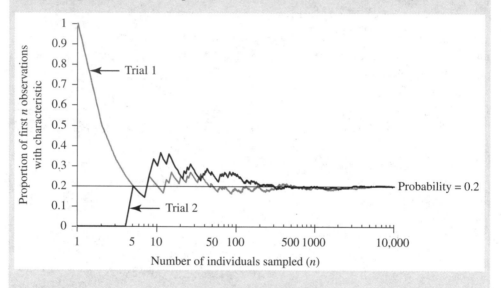

FIGURE 5.1   In the long run, the proportion approaches the true probability.

Trial 1 (wide blue line) begins with this sequence: smoker, nonsmoker, nonsmoker, and nonsmoker. This is why you see the line decreasing from 1.0 (one in one) to 0.5 (one in two) to 0.33 (one in three) to 0.25 (one in four).

Trial 2 (thin blue line) begins with four successive nonsmokers. The fifth selection was a smoker. This is why the line stays at 0.0 until $n = 5$, at which point the line rises to one in five (0.20).

Notice that the proportions derived in these trials are highly variable at first. However, as the number of replications increases, both trials converge on the true probability of selecting a smoker, which is 0.2.

---

[b] Values generated with SPSS (version 11) RV.BERNOULLI(0.2) random generator function.

Figure 5.1 demonstrates three important principles about probabilities. These are:

1. The proportion in all but a very large number of trials can be far from the actual probability.[c]
2. The proportion gradually closes in on the true probability as *n* increases.
3. Individual occurrences are uncertain, but occurrences are regular over a large number of repetitions.

These facts show that probability is unpredictable in the short run but is predictable in the long run. Probabilities *in the long run* are not haphazard.

Exercises

5.1 *Explaining probability.* A patient newly diagnosed with a serious ailment is told he has a 60% probability of surviving 5 or more years. Let's assume this statement is accurate. Explain the meaning of this statement to someone with no statistical background in terms he or she will understand.

5.2 *Roll a die.* A standard die has six faces: one with one spot on it, one with two spots, and so on. We can say that the chance the die lands on "one" is 1 in 6. How would you design an experiment to confirm this statement?

5.3 *February birthdays.* What is the probability of being born on . . .

(a) February 28?

(b) February 29?

(c) February 28 *or* 29?

5.4 *Childhood leukemia.* Compilation of results from a clinical trial reveals that 475 of 601 cases survive at least 5 years after diagnosis. Based on this information, estimate the probability of survival; then explain why this is only an estimate of the probability and is not the "true" probability of survival.

5.5 *N = 26.* Suppose a population has 26 members identified with the letters A through Z.

(a) You select one individual at random from this population. What is the probability of selecting individual A?

(b) Assume person A gets selected on an initial draw, you replace person A into the sampling frame, and then take a second random draw. What is the probability of drawing person A on the second draw?

---

[c] Even after $n = 100$ in trial 2 of Figure 5.1, the proportion is about 25% while the true probability is actually 20%.

(c) Assume person A gets selected on the initial draw and you sample again without replacement. What is the probability of drawing person A on the second draw?

5.6 **Random eights.** Chapter 2 introduced Table A, our table of random digits. This table was produced so that each digit 0 through 9 has an equal probability of appearing at any point in the table.

(a) Each row has 50 digits. How many 8s do you expect to encounter in the first *two* rows of the table?

(b) How many 8s appear in the first two rows of Table A?

(c) Why does the answer to (b) *not* match the answer to (a)?

5.7 **Personal expressions of probability.** The probability associated with a proposition can be used to quantify the confidence in a judgment. At one extreme, a probability statement of 0 represents the personal belief that the proposition can never happen. At the other extreme, a probability statement of 1 represents the personal belief that the proposition will always occur. Probabilities between these two extremes represent shades of gray. Match each of the following percentages with the narrative statement listed in (a)–(e) that most closely reflects its meaning.

95% 80% 50% 20% 5%

(a) This seldom happens. It has a very low chance of occurring.

(b) This event is infrequent; it is unlikely.

(c) This happens as often as not; chances are even.

(d) This is very frequent and occurs with high probability.

(e) This event almost always occurs; it has a very high probability of occurrence.

■ 5.2 Types of Random Variables

As previously noted, a **random variable** is a numerical quantity that takes on different values depending on chance. Two types of random variables are considered: discrete random variables and continuous random variables.

Discrete random variables exist as a countable set of possible outcomes. Examples of discrete random variables are:

- The variable number of leukemia cases in a geographic region in a given period
- The variable number of successes in *n* independent treatments
- The variable number of smokers in a simple random sample of size *n*

As an example, consider the number of leukemia cases in a community in a given year. There could be zero cases, one case, two cases, and so on. Fractional units are impossible. This random variable is discrete.

Continuous random variables address quantities that take on an unbroken continuum of possible values. Examples of continuous random variables are:

- The variable amount of time it takes to complete a task
- The average weight in a simple random sample of newborns in a particular community
- The height of an individual selected at random

What do we mean by an unbroken continuum of possible values? Consider the variable "amount of time it takes to complete a task." We can divide time intervals into infinitely smaller and smaller fractional units: 1 year, 0.5 years, 0.25 years, 0.125 years, 0.0625 years, and so on. Continuous random variables have this characteristic of being capable of infinite subdivision. We will discuss how to determine probabilities for continuous random variables in Section 5.4. For now, let us address the task of assigning probabilities for discrete random variables.

◼ 5.3 Discrete Random Variables

Properties of Probability Distributions

A **probability mass function (*pmf*)** is a mathematical relation that assigns probabilities to all possible outcomes for a discrete random variable. Probability mass functions can be displayed in a table, as a graph, or as a formula.

Let us introduce the following **terminology** and **notation** to discuss probabilities.

- *A* denotes "event *A*."
- Pr(*A*) denotes "the probability of event *A*."
- $\overline{A}$ denotes the *complement* of *A* (This means all events other than *A*; "not *A*.")
- *S* represents the "sampling universe" or "sample space" of all possible outcomes.

ILLUSTRATIVE EXAMPLE

Example of a probability mass function (the "four patients" illustrative example). Suppose we treat four patients with an intervention that is successful 75% of the time. The number of successes will vary randomly. Occasionally, no patients will respond, sometimes one will respond, and so on. Table 5.1 lists the

continues

probability mass function for this random variable. Figure 5.2 displays it in a graphical form. The formula for the probability mass function will be considered in Chapter 6.

TABLE 5.1 Probability model (probability mass function) for the four patients illustrative example.

No. of success	0	1	2	3	4
Probability	0.0039	0.0469	0.2109	0.4219	0.3164

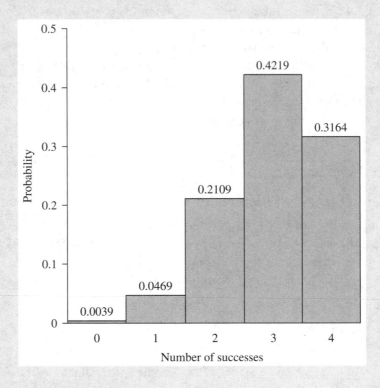

FIGURE 5.2 Probability histogram, "four patients" illustrative example.

Four properties of probability functions are:

- **Property 1 (Range of possible probabilities).** Individual probabilities are never less than 0 and never more than 1: $0 \leq \Pr(A) \leq 1$.
- **Property 2 (Total probability).** Probabilities in the sample space must sum to exactly 1: $\Pr(S) = 1$.

- **Property 3 (Complements).** The probability of a *complement* is equal to 1 minus the probability of the event: $\Pr(\overline{A}) = 1 - \Pr(A)$.
- **Property 4 (Disjoint events).** Events are **disjoint** if they cannot exist concurrently. If A and B are **disjoint**, then $\Pr(A \text{ or } B) = \Pr(A) + \Pr(B)$.

ILLUSTRATIVE EXAMPLE

Properties of probabilities. Table 5.1 lists the probability mass function for the four patients illustrative *pmf*. Notice that:

- **Property 1.** All probabilities in the *pmf* are between 0 and 1.
- **Property 2.** Probabilities in the sample space sum to exactly 1. $\Pr(X = 0)$ + $\Pr(X = 1)$ + $\Pr(X = 2)$ + $\Pr(X = 3)$ + $\Pr(X = 4)$ = 0.0039 + 0.0469 + 0.2109 + 0.4219 + 0.3164 = 1.000.
- **Property 3.** The probability of a complement is equal to 1 minus the probability of the event. As an example, let A represent "all four treatments are successful." Its complement $\overline{A}$ is "not all four are successful." Since $\Pr(A) = \Pr(X = 4) = 0.3164$, then $\Pr(\overline{A}) = 1 - \Pr(A) = 1 - 0.3164 = 0.6836$.
- **Property 4.** Zero successes and one success are disjoint events. Therefore, $\Pr(X = 0 \text{ or } X = 1) = \Pr(X = 0) + \Pr(X = 1) = 0.0039 + 0.0469 = 0.0508$.

Expectation and Variance

Probability functions form distributions that have shapes, locations, and spreads. The **shape** of the four "patients" illustrative example *pmf* is seen in Figure 5.2. This *pmf* is asymmetric, with a negative skew.

The location of the *pmf* is characterized by its mean (μ). The **mean μ** of a *pmf* (often referred to as its **expected value**) is a weighted average of values with weights based on probabilities:

$$\mu = \Sigma x \cdot \Pr(X = x)$$

where X denotes the random variable and x represents a given value.

The spread is characterized by its variance σ^2. The **variance σ^2** of a *pmf* is the weighted average of the squared distances around the mean with weights provided by probabilities:

$$\sigma^2 = \Sigma (x - \mu)^2 \cdot \Pr(X = x)$$

ILLUSTRATIVE EXAMPLE

Expectation and variance. What are the mean and variance of the four patients illustrative *pmf*, as presented in Table 5.1?

    Solution: Table 5.2 restates the *pmf* with an additional row listing values for $x \cdot \Pr(X = x)$. The expected value, calculated below the table, is $\mu = 3.00$. We expect three successes over the long run.

TABLE 5.2   Calculation of expected value for the "four patients" illustrative example.

x	0	1	2	3	4
$\Pr(X = x)$	0.0039	0.0469	0.2109	0.4219	0.3164
$x \cdot \Pr(X = x)$	0.0000[a]	0.0469[b]	0.4218[c]	1.2657	1.2656

Examples of calculations:

[a] $x \cdot \Pr(X = 0) = 0 \cdot 0.0039 = 0.0000$

[b] $x \cdot \Pr(X = 1) = 1 \cdot 0.0469 = 0.0469$

[c] $x \cdot \Pr(X = 2) = 2 \cdot 0.2109 = 0.4218$

Calculation of the expected value:

$\mu = \Sigma x \cdot \Pr(X = x) = 0.0000 + 0.0469 + 0.4218 + 1.2657 + 1.2656 = 3.00.$

    Table 5.3 restates the *pmf* with an additional row listing values for $(x - \mu)^2 \cdot \Pr(X = x)$. The variance of the model is calculated below the table revealing $\sigma^2 = 0.75$. This is a measure of the random variable's variability.

TABLE 5.3   Calculation of variance for the Four patients illustrative example.

x	0	1	2	3	4
$\Pr(X = x)$	0.0039	0.0469	0.2109	0.4219	0.3164
$(x - \mu)^2 \cdot \Pr(X = x)$	0.0351[a]	0.1876[b]	0.2109[c]	0.0000	0.3164

Recall that $\mu = 3$.

**Examples of calculations:

[a] $(x - \mu)^2 \cdot \Pr(X = x) = (0 - 3)^2 \cdot 0.0039 = 9 \cdot 0.0039 = 0.0351$

[b] $(x - \mu)^2 \cdot \Pr(X = x) = (1 - 3)^2 \cdot 0.0469 = 4 \cdot 0.0469 = 0.1876$

[c] $(x - \mu)^2 \cdot \Pr(X = x) = (2 - 3)^2 \cdot 0.2109 = 1 \cdot 0.2109 = 0.2109$

Calculation of variance:

$\sigma^2 = \Sigma(x - \mu)^2 \cdot \Pr(X = x) = 0.0351 + 0.1876 + 0.2109 + 0.0000 + 0.3164 = 0.75$

Area Under the "Curve"

Figure 5.3 shows the *pmf* for the four patients illustrative example in a graphical form with the histogram bar corresponding to $\Pr(X = 2)$ shaded. Notice that the area of

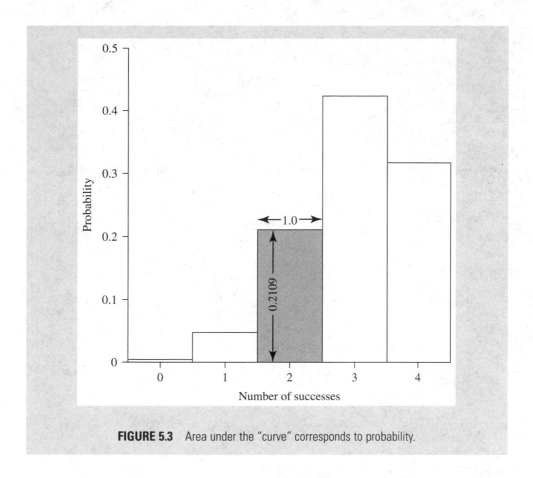

FIGURE 5.3   Area under the "curve" corresponds to probability.

this bar = height $\times$ width = 0.2109 $\times$ 1.0 = 0.2109, corresponding to Pr(X = 2). *This illustrates an important facet of working with probability functions: areas correspond to probabilities.* We refer to this as the **area under the curve concept**.

Figure 5.4 shows the area corresponding to the cumulative probability of two successes for the four patients illustrative *pmf*. This corresponds to Pr($X \le 2$) = Pr(X = 0) + Pr(X = 1) + Pr(X = 2) = 0.0039 + 0.0469 + 0.2109 = 0.2617.

Exercises

5.8 *Natality.* Table 5.4 is a *pmf* for the age of mothers of newborns in the United States.

(a) If we select a birth at random, what is the probability the mother is 19 years of age or younger?

(b) What is the probability she is 30 years of age or older?

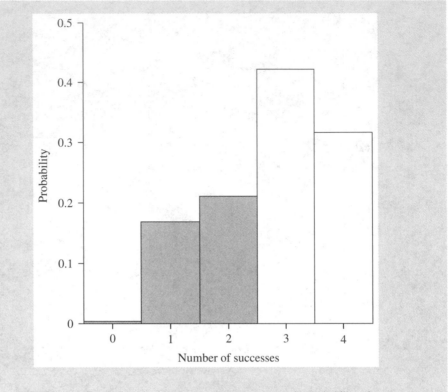

FIGURE 5.4   The shaded area corresponds to the *cumulative* probability of two.

TABLE 5.4   Probability mass function for Exercise 5.8, age of mother at birth of child.

Age	10–14	15–19	20–29	30–39	40–44	45–49
Age mid-point	12	17	24.5	34.5	42	47
$\Pr(X = x)$	0.003	0.100	0.500	0.300	0.097	0.000*

*Less than 0.0005

(c) What is the expected age of the mother at birth? (Use age mid-points for your calculations.)

(d) What is the variance of mothers' ages?

5.9   *Lottery.* A lottery offers a grand prize of $10 million. The probability of winning this grand prize is 1 in 55 million ($\approx 1.8 \times 10^{-8}$). There are no other prizes, so the probability of winning nothing $= 1 - (1.8 \times 10^{-8}) = 0.999999982$. Table 5.5 shows the probability mass function for the problem.

TABLE 5.5   Probability model for Exercise 5.9.

Winnings (X)	0	$\$10 \times 10^6$
$\Pr(X = x_i)$	0.999999982	1.8×10^{-8}

TABLE 5.6   Data for Exercise 5.10. Proportion of Population by Hispanic or Latino Origin and Race for the United States, 2010.

	Hispanic or Latino	**Non-Hispanic or Latino**
White	0.0866	0.6375
Black/Af Am	0.0040	0.1221
Am Ind/Alask Native	0.0022	0.0073
Asian	0.0007	0.0469
Native Haw/Pac Island	0.0002	0.0016
Other	0.0599	0.0019
Two or more races	0.0098	0.0193

Data from the U.S. Census Bureau and Humes et al. (2011). Overview of Race and Hispanic Origin: 2010, Table 2.

(a) What is the expected value of a lottery ticket?

(b) Fifty-five million lottery tickets will be sold. How much does the proprietor of the lottery need to charge per ticket to make a profit?

5.10 *U.S. Census.* Table 5.6 shows the proportions of individuals cross-classified according to two race/ethnic criteria: Hispanic/not Hispanic and Asian/Black/White/other. (The 2010 U.S. census allowed individuals to identify their ethnicity according to more than one criterion.) What is the probability an individual selected at random from this population self-identifies as Hispanic?

■ 5.4 Continuous Random Variables

Continuous Probability Distributions

Probabilities for discrete random variables are assigned in discrete "chunks." How do we assign probabilities for continuous random variables? Let's start by the "Random number spinner" illustrative example that starts on the next page. Because the number of potential outcomes for this random number spinner forms a continuum, we must address probabilities for ranges rather than considering probabilities for individual values. For example, the probability the spinner lands between 0 and 0.5 is one-half, since this constitutes half the *area* on the device.

Random number spinner. Consider spinning a pointer like the one depicted in Figure 5.5. The pointer can come to rest at any angle, on any value between 0 and 1. With a fair spin, the pointer is equally likely to stop at any point on the device, generating random numbers between 0 and 1 with uniform probability. There are an infinite number of such possibilities. For example, between 0.0 and 0.5, it can land on 0.25, 0.375, 0.4375, 0.46875, 0.484375, *ad infinitum.*

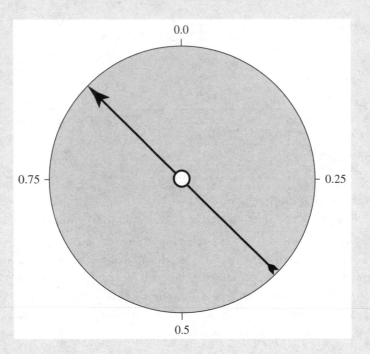

FIGURE 5.5   Spinning a pointer between 0 and 1.

Figure 5.6 depicts a probability model for the random spinner illustrative example. This is the **probability density functions (*pdf*)** for this random variable. As was true with *pmfs*, probabilities on *pdfs* are represented as areas under the "curve". The area corresponding to $\Pr(0.0 \le X \le 0.5) = 0.5$ in Figure 5.6 has been shaded.

Note that the probability density function depicted in Figure 5.6 meets all the required properties for probabilities stated earlier, in Section 5.3:

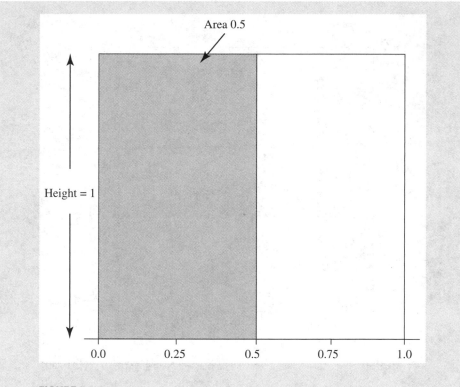

FIGURE 5.6   Probability curve (*pdf*) for Random Number Spinner showing Pr(0.0 ≤ X ≤ 0.5).

- **Property 1 (Range of possible probabilities).** Probabilities between any two points (represented by areas under the curve) are never less than 0 and never more than 1.

- **Property 2 (Total probability).** The total area under the curve sums to exactly 1.

- **Property 3 (Complements).** The probability of an event's complement is equal to 1 minus the probability of the event (see Exercise 5.12).

- **Property 4 (Disjoint events).** The probability of disjoint events can be added to determine the probability of their union. Figure 5.7 illustrates this principle by shading regions corresponding to $\Pr(X \leq 0.2)$ and $\Pr(X \geq 0.9)$. The combined area $\Pr(X \leq 0.2 \text{ or } X \geq 0.9) = \Pr(X \leq 0.2) + \Pr(X \geq 0.9) = 0.2 + 0.1 = 0.3$.

Notes on *pdfs*
1. Models for continuous random variables (i.e., *pdfs*) assign a **probability of 0 for observing any exact numerical outcome.**

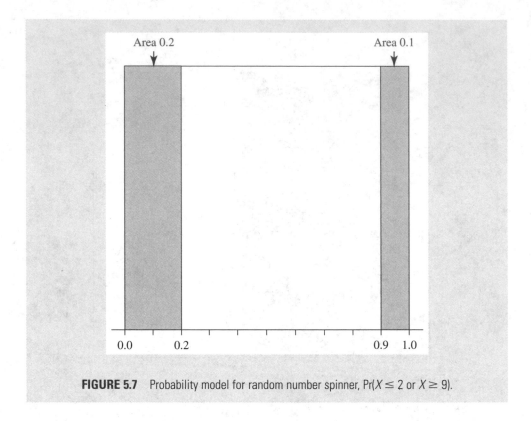

FIGURE 5.7   Probability model for random number spinner, Pr($X \le 2$ or $X \ge 9$).

2. **Intervals on *pdfs* have nonzero probabilities.** For example, Pr($0 \le X \le 0.2$) = 0.2 because this interval has a length of 0.2 and height of 1.

3. **Normal distributions** are a common type of *pdf* that describe many natural and mathematical phenomena. The next chapter introduces Normal distributions.

4. **Additional *pdfs*** that serve as statistical tools include Student's *t pdfs* (introduced in Chapter 11), *F pdfs* (introduced in Chapter 13), and χ^2 *pdfs* (introduced in Chapter 18).

Exercises

5.11 *Uniform (0,1) pdf.* Figure 5.7 depicts the probability density function for the random number spinner for values between 0.0 and 1.0. This is the uniform probability density function for the range 0 to 1, denoted X, $\sim$ uniform(0, 1). Use geometry to determine the following probabilities on this *pdf*.

(a) Pr($X \le 0.8$)

(b) Pr($X \le 0.2$)

(c) Pr($0.2 \le X \le 0.8$)

5.12 *Uniform (0,1) pdf, continued*. Again consider the uniform probability density function for the range 0.0 to 1.0. Find the following probabilities.

(a) $\Pr(X \geq 0.6)$

(b) $\Pr(X \leq 0.6)$

5.13 *The sum of two uniform (0, 1) random variables*. Figure 5.8 depicts the *pdf* for the sum of two uniform (0, 1) random variables.

(a) What is the probability of observing a value for this random variable that is less than 1?

(b) Use geometry to find the probability of a value that is less than 0.5.

(c) Find the probability of a value between 0.5 and 1.5.

5.14 *Bound for Glory*. Most people have an intuitive sense of how probabilities work. Here is a passage from Woody Guthrie's autobiography *Bound for Glory*[d] that demonstrates clear probabilistic reasoning:

> A kid named Bud run the gambling wheel. It was an old lopsided bicycle wheel that he had found in the dumps and tried to even up. He paid you ten to one if you called off the right spoke it would stop on. But there was sixty spokes.

(a) Draw a graph of the probability density function curve for Bud's gambling wheel. Assume the bicycle wheel has been evened up to create a uniform distribution for values between 1 and 60.

(b) What is the expected value on a dollar bet?

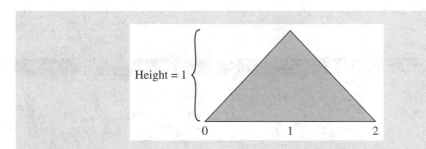

FIGURE 5.8   Probability model for the sum of two uniform(0,1) random variables, Exercise 5.13.

---

[d] Guthrie, W. (1943). *Bound for Glory*. London: Penguin, p. 1040.

■ 5.5 More Rules and Properties of Probability

This chapter began by introducing these four properties of probabilities:

1. Probabilities are always between 0 and 1.

2. The sum of all possible probabilities for a sample space is exactly 1.

3. The probability of a complement of an event is equal to 1 minus the probability of the event.

4. Probabilities of disjoint events can be added to determine their union.

Now six additional properties are introduced. These go by the following names:

5. Independent events

6. General rule of addition

7. Conditional probability

8. General rule of multiplication

9. Total probability rule

10. Bayes' theorem

Venn diagrams are used to visualize some of these principles. Figure 5.9 is a Venn diagram depicting disjoint events. The area of the entire sampling space (S) is 1. The sizes of A and B are proportional to their probabilities. In this diagram, A and B are disjoint, so $Pr(A \text{ or } B) = Pr(A) + Pr(B)$.

Figure 5.10 depicts events that are *not* disjoint. The area where events overlap is said to **intersect**. This is where both A and B occur. When A and B intersect, $Pr(A \text{ and } B) > 0$.

ILLUSTRATIVE EXAMPLE

Pet ownership illustration. Let A represent cat ownership and B represent dog ownership. Suppose 35% of households in a population own cats, 30% own dogs, and 15% own both a cat and a dog. In selecting a household at random:

- $Pr(A) = 0.35$

- $Pr(B) = 0.30$

- $Pr(A \text{ and } B) = 0.15$

Let us keep these probabilities in mind for future illustrations.

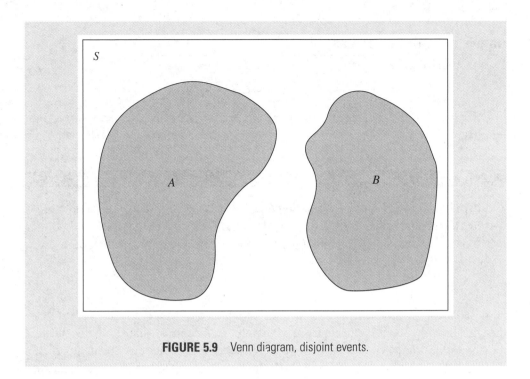

FIGURE 5.9 Venn diagram, disjoint events.

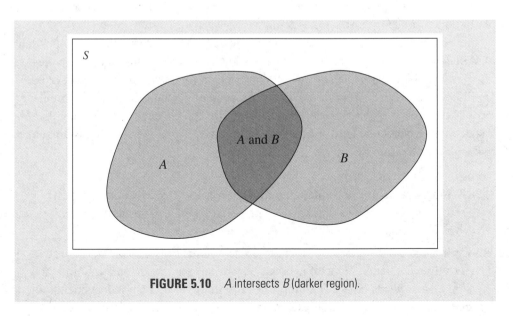

FIGURE 5.10 *A* intersects *B* (darker region).

Rule 5: Independent Events

Events A and B are **independent** *if and only if*:

$$\Pr(A \text{ and } B) = \Pr(A) \times \Pr(B)$$

For A and B to be independent, their joint probability must overlap just the right amount.

ILLUSTRATIVE EXAMPLE

Independence (Pet ownership example). *If* cat ownership and dog ownership were independent, then $\Pr(A \text{ and } B)$ would equal $\Pr(A) \times \Pr(B) = 0.35 \times 0.30 = 0.105$. However, in our example, $\Pr(A \text{ and } B) = 0.15$. Therefore, these events are *not* independent.

Statistical *independence* indicates that knowledge of one event provides no further information about the occurrence of the other. In contrast, statistical dependence indicates that knowledge of one event increases or decreases the likelihood of the associated event. In the "Pet ownership" illustrative example, the statistical dependence between cat ownership and dog ownership lets us know that cat owners are more likely than average to (also) own a dog.

Let us consider examples of characteristics that are independent and dependent of skin cancer. Knowing whether someone is left-handed or right-handed provides no information about their relative risk of skin cancer; "handedness" and "skin cancer" are independent. On the other hand, knowing whether someone has light-colored eyes lets us know that they have higher than average risk of skin cancer. Therefore, there is a statistical dependency between the light-colored eyes and skin cancer. Skin cancer risk and eye color are dependent.

Notes

1. **Disjoint events are not independent.** When A and B are disjoint, knowing that A has occurred lets you know (with certainty) that B has not occurred. Therefore, disjoint events are not independent.

2. **Coin flips.** Coin flips are independent. The outcome of one coin toss does not influence the outcome of another. Let A represent heads on a first toss and B represent heads on a second toss. Because these events are independent, you can multiply their probabilities to determine the probability of two heads in a row: $\Pr(A \text{ and } B) = 0.5 \times 0.5 = 0.25$.

3. **Sampling independence.** "Sampling independence" implies that of selecting a given individual from a population does not influence the probability of selecting any other.

4. **Contagious events are not independent.** The idea of contagion corresponds to statistical dependence. Being in the vicinity to one contagious event increases the likelihood of another.

Rule 6: General Rule of Addition

The general additional rule is:

$$\Pr(A \text{ or } B) = \Pr(A) + \Pr(B) - \Pr(A \text{ and } B)$$

Figure 5.10 provides insight into this rule. The area "A or B" is equal to their sum minus the amount they overlap.

ILLUSTRATIVE EXAMPLE

General rule of addition (Pet ownership). We have established that the probability of cat ownership $\Pr(A) = 0.35$, the probability of dog ownership $\Pr(B) = 0.30$, and the probability of joint cat and dog ownership $\Pr(A \text{ and } B) = 0.15$. By the general rule of addition, the probability of owning a dog or a cat $\Pr(A \text{ or } B) = \Pr(A) + \Pr(B) - \Pr(A \text{ and } B) = 0.35 + 0.30 - 0.15 = 0.50$.

The rule for adding probabilities for disjoint events (property 4) is a special case of the general rule of addition in which $\Pr(A \text{ and } B) = 0$.

Rule 7: Conditional Probability

Let $\Pr(B \mid A)$ represent the conditional probability of B given A. This denotes the probability of B given that A is evident. By definition,

$$\Pr(B \mid A) = \frac{\Pr(A \text{ and } B)}{\Pr(A)}$$

as long as $\Pr(A) > 0$.

ILLUSTRATIVE EXAMPLE

Conditional probability (Pet ownership). Suppose you know a household owns a cat. What is the probability that it also owns a dog? In other words, what is the probability of dog ownership given cat ownership?

Solution: $\Pr(B \mid A) = \dfrac{\Pr(A \text{ and } B)}{\Pr(A)} = \dfrac{0.15}{0.35} = 0.4286$. The probability of owing a dog given cat ownership is about 43%.

Notes

1. $\Pr(B \mid A)$ and $\Pr(A \mid B)$ are distinct.

2. When A and B are independent, $\Pr(B \mid A) = \Pr(B \mid \overline{A}) = \Pr(B)$. Proof: When A and B are independent, $\Pr(A \text{ and } B) = \Pr(A) \times \Pr(B)$. Therefore,

$$\Pr(B \mid A) = \frac{\Pr(A \text{ and } B)}{\Pr(A)} = \frac{\Pr(A) \times \Pr(B)}{\Pr(A)} = \Pr(B)$$

3. **Risk** is the probability of developing an adverse condition in a specified time period. Let D represent the event of developing a disease. The risk of the disease is $\Pr(D)$.

4. The **relative risk (RR)** is the probability of disease given exposure divided by the probability of disease given nonexposure:

$$RR = \frac{\Pr(D \mid E)}{\Pr(D \mid \overline{E})}$$

5. When D and E are independent, $\Pr(D \mid E) = \Pr(D \mid \overline{E}) = \Pr(D)$ and

$$RR = \frac{\Pr(D \mid E)}{\Pr(D \mid \overline{E})} = \frac{\Pr(D)}{\Pr(D)} = 1$$

ILLUSTRATIVE EXAMPLE

Relative risk (Older formulation oral contraceptives). Oral contraceptives are associated with an increased risk of thromboembolic diseases. Let D represent thromboembolic diseases and E represent exposure to oral contraceptives. Suppose $\Pr(D \mid E) = 0.00040$ and $\Pr(D \mid \overline{E}) = 0.00010$. Therefore,

$$RR = \frac{0.0004}{0.0001} = 4.0. \text{ Exposure quadrupled risk.}$$

Rule 8: General Rule for Multiplication

Start with the definition of conditional probability:

$$\Pr(B \mid A) = \frac{\Pr(A \text{ and } B)}{\Pr(A)}$$

then rearrange the formula as follows:

$$\Pr(A \text{ and } B) = \Pr(A) \times \Pr(B \mid A)$$

This is the general rule for multiplication.

ILLUSTRATIVE EXAMPLE

General rule for multiplication (Pet ownership example). The probability of cat ownership $\Pr(A) = 0.35$. The probability of dog ownership given cat ownership $\Pr(B|A) = 0.4286$. By the general rule for multiplication, the probability of owning a cat and a dog $\Pr(A \text{ and } B) = \Pr(A) \times \Pr(B|A) = 0.35 \times 0.4286 = 0.15$.

Rule 9: Total Probability Rule

From Figure 5.11, we can see that event B is made up of two components: $(B \text{ and } A)$ and $(B \text{ and } \overline{A})$.

Therefore,

$$\Pr(B) = [\Pr(B \text{ and } A)] + [\Pr(B \text{ and } \overline{A})]$$

By the general rule of multiplication, $\Pr(B \text{ and } A) = \Pr(B \mid A) \times \Pr(A)$ and $\Pr(B \text{ and } \overline{A}) = \Pr(B \mid \overline{A}) \times \Pr(\overline{A})$. It follows that:

$$\Pr(B) = [\Pr(B \mid A) \times \Pr(A)] + [\Pr(B \mid \overline{A}) \times \Pr(\overline{A})]$$

This is the total probability rule.

ILLUSTRATIVE EXAMPLE

Total probability rule (Pet ownership). Recall that event A is cat ownership and event B is dog ownership. We have established $\Pr(B \mid A) = 0.4286$, $\Pr(\overline{A}) = 1 - \Pr(A) = 1 - 0.35 = 0.65$, and $\Pr(B \text{ and } \overline{A}) = 0.15$. By conditional probability,

$\Pr(B|\overline{A}) = \dfrac{\Pr(B \text{ and } \overline{A})}{\Pr(\overline{A})} = \dfrac{0.15}{0.65} = 0.2308$. By the total probability rule, $\Pr(B)$

$= [\Pr(B \mid A) \times \Pr(A)] + [\Pr(B \mid \overline{A}) \times \Pr(\overline{A})] = [0.4286 \times 0.35] + [0.2308 \times 0.65] = 0.15 + 0.15 = 0.30$.

Rule 10: Bayes' Theorem

We combine rule 8 (general rule of multiplication) and rule 9 (total probability rule) to establish **Bayes' theorem** as follows:

$$\Pr(A \mid B) = \frac{\Pr(A \text{ and } B)}{\Pr(B)} = \frac{\Pr(B|A) \times \Pr(A)}{[\Pr(B|A) \times \Pr(A)] + [\Pr(B|\overline{A}) \times \Pr(\overline{A})]}$$

Bayes' rule has many applications, one of which pertains to population-based disease screening tests. Let D represent a disease (its complement, being disease free

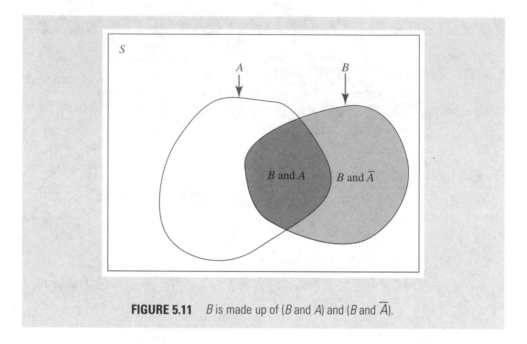

FIGURE 5.11   *B* is made up of (*B* and *A*) and (*B* and *Ā*).

is denoted $\overline{D}$). Let T represent a positive test result ($\overline{T}$ is a negative test result). Pr(*D*) represents the **prevalence** (probability of disease when selecting someone at random) in the population. The probability of a positive test in a diseased individual, Pr(*T* | *D*), is a characteristic of the test known as **sensitivity**. The probability of a positive test in a disease-free individual, Pr(*T* | $\overline{D}$), is a characteristic of the test known as the **false-positive rate**. The complement of the false-positive rate is the **specificity** of the test: $\text{Pr}(\overline{T} \mid \overline{D}) = 1 - \text{Pr}(T \mid \overline{D})$.

We can use Bayes' theorem to derive the probability of disease given a positive test (**predictive value of a positive test**) as follows:

$$\text{Pr}(D \mid T) = \frac{\text{Pr}(T \mid D) \times \text{Pr}(D)}{[\text{Pr}(T \mid D) \times \text{Pr}(D)] + [\text{Pr}(T \mid \overline{D}) \times \text{Pr}(\overline{D})]}$$

$$= \frac{\text{sensitivity} \times \text{prevalence}}{[\text{sensitivity} \times \text{prevalence}] + [\text{false positive rate} \times 1 - \text{prevalence}]}$$

ILLUSTRATIVE EXAMPLE

Predictive value positive. Suppose a screening test has a sensitivity of 0.98 and a false-positive rate of 0.01. The test is used in a population that has a disease prevalence of 0.001. We can use Bayes' theorem to determine the predictive value of a positive test under these circumstances as follows:

$\Pr(D \mid T) =$

$$\frac{\text{sensitivity} \times \text{prevalence}}{[\text{sensitivity} \times \text{prevalence}] + [\text{false positive rate} \times 1 - \text{prevalence}]}$$

$$= \frac{0.98 \times 0.001}{[0.98 \times 0.001] + [0.01 \times (1 - 0.001)]} = 0.0893$$

The probability of having the disease given a positive test result is 8.93%.

The prior example provides results that may seem counterintuitive. How can a test that is 98% sensitive with a false-positive rate of only 1% provide such unreliable positive results? The answer to this lies in the fact that the prevalence of disease in the population is very low. Consider using this test in 100,000 individuals from this population. Of these, 0.1% or 100 will actually have the disease. Because the test is 98% sensitive, we expect 98 of these individuals to test positive—these are the true positive—leaving two false negatives. At the same time, there are $100,000 - 100 = 99,900$ disease-free individuals in the population. The false-positive rate is 1%, and there are $1\% \times 99,900 = 999$ false positives. Overall, there are 98 true positives and 999 false positives, so of $98 + 999 = 1097$ positive testers, only 98 or 8.93% are true positives. Most of the positive test results are false positives.

Summary Points (Probability Concepts)

1. **Probability** is a mathematical tool that helps us understand and address chance. With probabilistic events, individual occurrences are uncertain, but occurrences over the long term are predictable.

2. A **random variable** is a numerical quantity that takes on different values depending on chance. There are two types of random variables. **Discrete random variables** form a countable set of possible values. **Continuous random variables** form an unbroken continuum of possible values.

3. The **notation** $\Pr(A)$ is read as "the probability of event A."

4. The **four most basic properties (rules) of probability are:**

   (a) Probabilities are always between 0 and 1 inclusive: $0 \le \Pr(A) \le 1$.

   (b) Probabilities for a sample space must sum to 1: $\Pr(S) = 1$. [A sample space (S) identifies all possible outcomes for a random variable.]

   (c) Law of complements: $\Pr(\overline{A}) = 1 - \Pr(A)$. (The complement of an event is "the event not occurring.")

   (d) You can add the probability of disjoint events $\Pr(A \text{ or } B) = \Pr(A) + \Pr(B)$. (Events are disjoint if they cannot exist concurrently.)

5. A **probability mass function (*pmf*)** assigns probabilities to all possible outcomes for a discrete random variable. *pmf*s are displayed in tabular or graphical form in this chapter.

6. A **probability density function (*pdf*)** assigns probabilities to all possible outcomes for a continuous random variable. *pdf*s are displayed in graphical form in this chapter.

7. The **area under the curve (AUC)** of *pmf* and *pdf* graphs is equal to the probability of the corresponding range.

8. **More advanced rules of probability:**

   (a) Definition of independence: Events A and B are independent *if and only if* $\Pr(A \text{ and } B) = \Pr(A) \times \Pr(B)$.

   (b) General addition rule: $\Pr(A \text{ or } B) = \Pr(A) + \Pr(B) - \Pr(A \text{ and } B)$.

   (c) Conditional probability of B given A: $\Pr(B \mid A) = \dfrac{\Pr(A \text{ and } B)}{\Pr(A)}$.

   (d) General rule of multiplication: $\Pr(A \text{ and } B) = \Pr(A) \times \Pr(B \mid A)$.

   (e) $\Pr(A \mid B) = \dfrac{\Pr(B \mid A) \times \Pr(A)}{\Pr(B \mid A) \times \Pr(A) + \Pr(B \mid \overline{A}) \times \Pr(\overline{A})}$

   (f) Bayes' theorem: $\Pr(B) = [\Pr(B \mid A) \times \Pr(A)] + [\Pr(B \mid \overline{A}) \times \Pr(\overline{A})]$.

9. Two **epidemiologic applications** of conditional probabilities (relative risk and principles of screening for disease) are presented toward the end of the chapter.

Vocabulary

Area under the curve	Prevalence
Complement $(\overline{A})$	Probability
Continuous random variables	Probability density function (*pdf*)
Discrete random variables	Probability mass function (*pmf*)
Disjoint	Random variable
Event	Relative risk
Expected value (μ)	Repetitive process
False-positive rate	Risk
Independent	Sample space (S)
Intersect	Sensitivity
Population	Specificity
$\Pr(\overline{A})$	Variance (σ^2)
Predictive value of a positive test	

Review Questions

5.1   Fill in the blank: The probability of an event is its relative _____ in an infinitely long series of repetitions.

5.2   What is a numerical quantity that takes on different values depending on chance?

5.3   The two types of random variables are continuous and _____.

5.4   What type of random variable takes on a countable set of possible outcomes?

5.5   What type of random variable forms an unbroken chain of possible outcomes?

5.6   Select the best response: This is a mathematical function that assigns probabilities for discrete random variables.

(a) sample space

(b) probability mass function (*pmf*)

(c) probability density function (*pdf*)

5.7   Select the best response: This is a mathematical function that assigns probabilities for a continuous random variable.

(a) sample space

(b) probability mass function (*pmf*)

(c) probability density function (*pdf*)

5.8   Select the best response: This is a description of all possible outcomes for a random variable.

(a) an event

(b) the sample space

(c) probability

5.9   What is the *complement* of an event?

5.10 List the first four elementary properties of probability presented in this text.

5.11 What symbol is used to denote the expected value of a *pmf* or *pdf*?

5.12 What symbol is used to denote the variance of a *pmf* or *pdf*?

5.13 What symbol is used to denote the standard deviation of a *pmf* or *pdf*?

5.14 What does AUC stand for?

5.15 Select the best response: The area under the curve of *pmf*s and *pdf*s corresponds to _____.

(a) events

(b) probabilities

(c) cumulative probabilities

5.16 Select the best response: The area under a *pmf* or *pdf* curve to the *left* of a point corresponds to its _____.

(a) event

(b) probability

(c) cumulative probability

5.17 Select the best response: The total area under a *pmf* or *pdf* curve is always equal to

(a) 0

(b) 1

(c) something between 0 and 1

5.18 Select the best response: The probability of landing on any exact value on a *pdf* curve is always equal to

(a) 0

(b) 1

(c) something between 0 and 1

Exercises

5.15 *Uniform distribution of highway accidents*. Accidents occur along a 5-mile stretch of highway at a uniform rate. The following "curve" depicts the probability density function for accidents along this stretch:

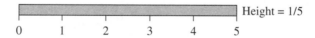

Notice that this "curve" demonstrates the first two basic properties of probability density functions:

- Property 1: The area under the curve between any two points is never less than 0 and is never more than 1. (Probabilities are always between 0 and 1, inclusive.)

- Property 2: The total area under the curve sums exactly to 1.

Respond to each of the following questions using the "area under the curve" principles introduced in the chapter.

(a) What is the probability that an accident occurred in the first mile along this stretch of highway?

(b) What is the probability that the accident did *not* occur in the first mile?

(c) What is the probability that an accident occurred between miles 2.5 and 4?

(d) What is the probability that an accident occurred either within the first mile or between miles 2.5 and 4?

5.16 *Nonuniform distribution of highway accidents.* There is a particularly treacherous stretch on a 3-mile stretch of a different highway between miles 1 and 2 where accidents occur at twice the rate of the other two miles.

(a) Sketch the probability density curve that describes the occurrence of accidents along this 3-mile stretch.

(b) What is the probability that an accident occurred between miles 1 and 2?

(c) What is the probability that an accident occurred between miles 2 and 3?

5.17 *Bound for Glory (variance).* Exercise 5.14 addressed the probability density function of a crude roulette wheel with values 1 to 60. A bet of a dollar pays $10 on a winning spin. Let X represent the payoff on a one-dollar wager. The probability mass function of potential winnings is therefore:

X	*Probability*
0	59/60
10	1/60

The expected value of a one dollar bet $\mu = \Sigma x_i \cdot \Pr(X = x_i) = [0 \times (59/60)] + [10 \times (1/60)] = 0 + 0.1667 = 0.1667$, or slightly less than 17 cents. What is the variance σ^2 of a payoff?

5.18 *The sum of two uniform (0, 1) random variables (μ and median).* Exercise 5.13 addressed the *pdf* for the sum of two uniform (0,1) random numbers (Figure 5.8).

(a) The expected value μ of a *pdf* can be visualized as its gravitational center—the point where the *pdf* would balance if put on a fulcrum. Use Figure 5.8 to determine the value of μ for this *pdf*.

(b) The median of a *pdf* is the point that cuts the area under the curve into two so that 0.50 of the area under the curve is to the right of this points and 0.50 is to its left. What is the median of the *pdf* displayed in Figure 5.8?

5.19 *The sum of two uniform (0, 1) random variables.* Let X represent the sum of these two uniformly distributed (0,1) random variables, as shown in Figure 5.8. Recall that the area of a triangle $= \frac{1}{2}hb$. Use Figure 5.8 to determine:

(a) $\Pr(X < 1)$.

(b) $\Pr(X > 1.5)$.

(c) $\Pr(1 < X < 1.5)$.

6 | Binomial Probability Distributions

6.1 Binomial Random Variables

A researcher takes a simple random sample of 20 elementary school students. How many students in the sample will have asthma? A new treatment for breast cancer is used in 150 cases. How many of these individuals will survive five or more years? In a random sample of 100 leukemia cases, how many will have a history of exposure to high-tension electric transmission lines? The random number of "successes" in each of these scenarios is described by **binomial** probability mass functions (*pmfs*).

Binomial *pmf*s are a family of probability models that apply to random variables in which:

- There are n independent observations.
- Each observation can be characterized as a "success" or "failure."
- The probability of "success" for each observation is a constant p.

Single random occurrences that can be characterized as either a "success" or "failure" are called **Bernoulli trials**. The total number of successes in a series of n independent Bernoulli trials when each trial has probability of success p is a binomial random variable. Thus, binomial random variables are characterized by two **parameters**:

1. n (the number of independent Bernoulli trials)
2. p (the probability of success for each trial)

ILLUSTRATIVE EXAMPLE

Four patients binomial example. Four patients are treated with an intervention that is successful 75% of the time. The random number of successes is a binomial random variable with parameters $n = 4$ and $p = 0.75$.

133

ILLUSTRATIVE EXAMPLE

Asthma survey binomial example. Four percent of the children in a large school population have asthma. The number of asthmatics in a simple random sample of $n = 20$ from this population is a binomial random variable with $n = 20$ and $p = 0.04$.

Exercises

6.1 **Tay-Sachs.** Tay-Sachs is a metabolic disorder that is inherited as an autosomal recessive trait. Both recessive alleles are necessary for expression of the disease. Therefore, when each parent is a carrier, there is a one in four chance of transmitting the genetic disorder to each offspring.

(a) Let X represent the number of offspring affected in three consecutive conceptions from Tay-Sachs carrier parents. Is X a binomial random variable? Explain your response.

(b) A carrier couple is unaware of their carrier state. Let X represent the number of children conceived before an affected offspring is encountered. Is X a binomial random variable? Explain.

6.2 **Breast cancer.** Say the lifetime probability of developing female breast cancer in a population is 1 in 10. Let X represent the number of women among 5102 women, selected randomly from this population, who ultimately develop breast cancer. Explain why X is a binomial random variable.

■ 6.2 Calculating Binomial Probabilities

We will use the notation $X \sim b(n, p)$ to refer to a binomial random variable with parameters n and p. The tilda ($\sim$) is read "distributed as" so "$X \sim b(n, p)$" is read "X is distributed as a binomial random variable with parameters n and p." For example, $X \sim b(4, 0.75)$ is read "X is a binomial random variable with $n = 4$ and $p = 0.75$" (like the four patients binomial example).

Binomial probabilities are calculated with the formula:

$$\Pr(X = x) = {}_nC_x p^x q^{n-x}$$

where ${}_nC_x$ is the binomial coefficient (described below), n is the number of independent Bernoulli trials, p is the probability of success for each trial, and $q = 1 - p$.

The **binomial coefficient** $_nC_x$ (loosely called the "choose function") determines the number of different ways to choose x items out of n ("n choose x"). Its formula is:

$$_nC_x = \frac{n!}{x!(n - x)!}$$

where the **factorial function** $x! = x \cdot (x - 1) \cdot (x - 2) \cdot \ldots \cdot 1$. For example, $4! = 4 \cdot 3 \cdot 2 \cdot 1 = 24$. By definition $0! = 1$.

ILLUSTRATIVE EXAMPLES

Binomial coefficients. Three examples of $_nC_x = \dfrac{n!}{x!(n - x)!}$ are presented.

- How many ways can you choose two items out of three? *Solution:*

  $_3C_2 = \dfrac{3!}{2!(3 - 2)!} = \dfrac{3 \cdot 2 \cdot 1}{2 \cdot 1(1)} = \dfrac{6}{2} = 3$. This means that you can choose two items out of the three different ways—label items A, B, and C.

  You can choose {A, B}, {A, C}, or {B, C}.

- How many ways can we choose two items out of four? *Solution:*

  $_4C_2 = \dfrac{4!}{2!(4 - 2)!} = \dfrac{4 \cdot 3 \cdot 2 \cdot 1}{(2 \cdot 1)(2 \cdot 1)} = 6$

- How many ways can we choose five items out of seven?

  $_7C_5 = \dfrac{7!}{5!(7 - 5)!} = \dfrac{7 \cdot 6 \cdot 5!}{5! \cdot 2!} = 21.$

We are now ready to calculate binomial probabilities.

ILLUSTRATIVE EXAMPLE

Binomial probabilities (Four patients). We have established a scenario in which four patients are treated with an intervention that is successful 75% of the time. The number of patients who respond to treatment is a binomial random variable with parameters $n = 4$ and $p = 0.75$. It follows that probability of a failure $q = 1 - p = 1 - 0.75 = 0.25$.

continues

What is the probability of observing no successes among the four treatments?

Solution: $\Pr(X = 0) = {}_nC_x \cdot p^x \cdot q^{n-x} = {}_4C_0 \cdot 0.75^0 \cdot 0.25^{4-0} = 1 \cdot 1 \cdot 0.0039$
$= 0.0039$

What is the probability of one success?

Solution: $\Pr(X = 1) = {}_nC_x \cdot p^x \cdot q^{n-x} = {}_4C_1 \cdot 0.75^1 \cdot 0.25^{4-1}$
$= 4 \cdot 0.75 \cdot 0.015625 = 0.0469$

The probability of two successes is

$\Pr(X = 2) = {}_4C_2 \cdot 0.75^2 \cdot 0.25^{4-2} = 6 \cdot 0.5625 \cdot 0.0625 = 0.2109$

The probability of three successes is

$\Pr(X = 3) = {}_4C_3 \cdot 0.75^3 \cdot 0.25^{4-3} = 4 \cdot 0.4219 \cdot 0.25 = 0.4219$

The probability of four successes is

$\Pr(X = 4) = {}_4C_4 \cdot 0.75^4 \cdot 0.25^{4-4} = 1 \cdot 0.3164 \cdot 1 = 0.3164$

Table 6.1 lists this *pmf* in tabular form.

TABLE 6.1   Probability mass function for $X \sim b(4, 0.75)$ in tabular form.

No. of successes (x)	0	1	2	3	4
$\Pr(X = x)$	0.0039	0.0469	0.2109	0.4219	0.3164

ILLUSTRATIVE EXAMPLE

Binomial probabilities (Asthma survey). We take a simple random sample of $n = 20$ from a population in which the prevalence of asthma is 0.04. The number of asthmatics in a given random sample is denoted $X \sim b(20, 0.04)$. Since $p = 0.04$, $q = 1 - 0.04 = 0.96$.

What is the probability of observing no asthmatics in a sample?

Solution: $\Pr(X = 0) = {}_nC_x \cdot p^x \cdot q^{n-x} = {}_{20}C_0 \cdot 0.04^0 \cdot 0.96^{20-0}$
$= 1 \cdot 1 \cdot 0.4420 = 0.4420$

What is the probability of observing one asthmatic?

Solution: $\Pr(X = 1) = {}_nC_x \cdot p^x \cdot q^{n-x} = {}_{20}C_1 \cdot 0.04^1 \cdot 0.96^{20-1}$
$= 20 \cdot 0.04 \cdot 0.4604 = 0.3683$

What is the probability of observing two asthmatics?

Solution: $\Pr(X = 2) = {}_{20}C_2 \cdot 0.04^2 \cdot 0.96^{20-2} = 190 \cdot 0.0016 \cdot 0.4796$
$= 0.1458$

To complete the *pmf*, we would need to calculate $\Pr(X = 3)$, $\Pr(X = 4)$, ... , $\Pr(X = 20)$. We will not demonstrate all these calculations, but do show the results as Figure 6.1.

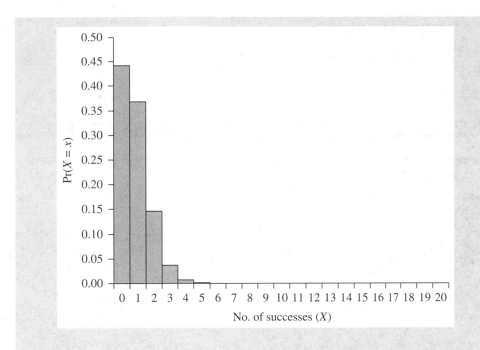

FIGURE 6.1   Probability mass function for binomial random variable with $p = 0.04$ and $n = 20$.

■ 6.3 Cumulative Probabilities

A **cumulative probability** is the probability of observing a certain value *or less*. For example, the cumulative probability of 2 for a discrete random variable $\Pr(X \leq 2)$ $= \Pr(X = 0) + \Pr(X = 1) + \Pr(X = 2)$.

ILLUSTRATIVE EXAMPLE

Cumulative probability (Asthma survey). The number of asthmatics in the asthma survey discussed in the prior illustration varies according to a binomial distribution with $n = 20$ and $p = 0.04$. What is the *cumulative* probability of observing two asthmatics in a sample? We established in the prior illustration box that $\Pr(X = 0) = 0.4420$, $\Pr(X = 1) = 0.3683$, and $\Pr(X = 2) = 0.1458$.

Solution: $\Pr(X \leq 2) = \Pr(X = 0) + \Pr(X = 1) + \Pr(X = 2)$
$= 0.4420 + 0.3683 + 0.1458 = 0.9561$

■ 6.4 Probability Calculators

Because binomial probabilities can be tedious to calculate, some textbooks will include tables of binomial probabilities for selected values of n and p. These tables are limited in that they are very bulky and cannot show all possible combinations of n and p.

ILLUSTRATIVE EXAMPLE

Cumulative probability (Four patients). The number of patients responding to treatment in the four patients illustration is a binomial random variable with $n = 4$ and $p = 0.75$. What is the cumulative probability of three successes? Table 6.1 lists the *pmf* for the random variable.

  Solution: $\Pr(X \leq 3) = \Pr(X = 0) + \Pr(X = 1) + \Pr(X = 2) + \Pr(X = 3)$
  $= 0.0039 + 0.0469 + 0.2109 + 0.4219 = 0.6836$. Figure 6.2 shows this cumulative probability as the shaded region in the left "tail" of the distribution.

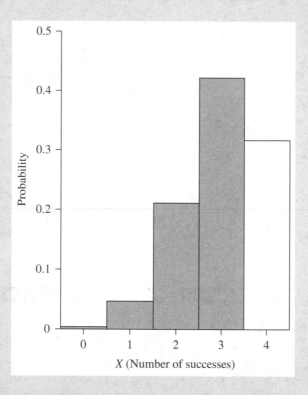

FIGURE 6.2   Cumulative probabilities correspond to areas under the curve to the left of the value. This *pmf* shows the cumulative probability of 3 for $X \sim b(4, 0.75)$.

> The **cumulative distribution function** (*cdf*) of a random variable is the cumulative probabilities for all values of the variable. Table 6.2 lists the *cdf* for the four patients example in tabular form.
>
> **TABLE 6.2** Cumulative probability function for $X \sim b(4, 0.75)$ in tabular form.
>
> | No. of successes (x) | 0 | 1 | 2 | 3 | 4 |
> |---|---|---|---|---|---|
> | $\Pr(X \leq x)$ | 0.0039 | 0.0508 | 0.2617 | 0.6836 | 1.0000 |

It is for this reason that this text does *not* include binomial tables; instead, it occasionally uses freeware probability calculators (such as StaTable®)[a] for this purpose. In addition, Microsoft Excel® will calculate binomial probabilities.

For example, Excel's[b] BINOMDIST function can be used to calculate the probability of three successes for $X \sim b(4, 0.75)$ with this argument: = BINOMDIST(3,4,0.75,0). The *cumulative* probability of 3 uses the argument = BINOMDIST(3,4,0.75,1).

Exercises

6.3 *Tay-Sachs inheritance.* Exercise 6.1 introduced facts about Tay-Sachs inheritance. If both parents are carriers, there is a one in four chance the offspring inherits both alleles necessary for expression of the disease. Suppose a carrier couple plans on having three children. Build the probability mass function (*pmf*) for the number of conceptions that will receive Tay-Sachs genes from both parents.

6.4 *Tay-Sachs couples.* Approximately 1 in 28 people of Ashkenazi Jewish descent are Tay-Sachs carriers. In randomly sampling one man and one woman from this population, what is the probability neither is a Tay-Sachs carrier? What is the probability both are carriers?

6.5 *Telephone survey.* A telephone survey uses a random digit dialing machine to call subjects. The random digit dialing machine is expected to reach a live person 15% of the time. In eight attempts, what is the probability of achieving exactly two successful calls?

6.6 *Telephone survey, two or fewer.* In the telephone survey technique introduced in Exercise 6.5, what is the probability of two or fewer successful calls in eight attempts?

---

[a] Cytel Inc., 675 Massachusetts Avenue, Cambridge, MA 02139-3309. Available: www.cytel.com
[b] See Microsoft Excel, www.microsoft.com.

■ 6.5 Expected Value and Variance of a Binomial Random Variable

There is a general formula for calculating the mean (expected value) and variance of a discrete random variable. Those formulas can also be applied to binomial random variables. Shortcut formulas also exist. The shortcut formula for the **mean** of a binomial random variable is:

$$\mu = np$$

The shortcut formula for the **variance** of a binomial random variable is:

$$\sigma^2 = npq$$

where $q = 1 - p$.

The **standard deviation** is the square root of the variance:

$$\sigma = \sqrt{\sigma^2} = \sqrt{npq}$$

ILLUSTRATIVE EXAMPLE

Expected value and variance of a binomial random variable (Asthma survey). The asthma survey considered in previous illustrative examples addresses a binomial random variable with $n = 20$ and $p = 0.04$. It follows that $q = 1 - 0.04 = 0.96$. The expected value of this random variable $\mu = np = 20 \cdot 0.04 = 0.8$. It has variance $\sigma^2 = npq = 20 \cdot 0.04 \cdot 0.96 = 0.768$ and standard deviation $\sigma = \sqrt{npq} = \sqrt{20 \cdot 0.04 \cdot 0.96} = 0.8764$.

ILLUSTRATIVE EXAMPLE

Expected value and variance of a binomial random variable (Four patients example). The four patients illustrative example has established a binomial random variable in which $n = 4$ and $p = 0.75$. Therefore, this distribution has $\mu = np = 4 \cdot 0.75 = 3$ and $\sigma^2 = npq = 4 \cdot 0.75 \cdot (1 - 0.75) = 0.75$. Its standard deviation $\sigma = \sqrt{0.75} = 0.8660$. Figure 6.3 depicts μ as the balancing point of the *pmf*.

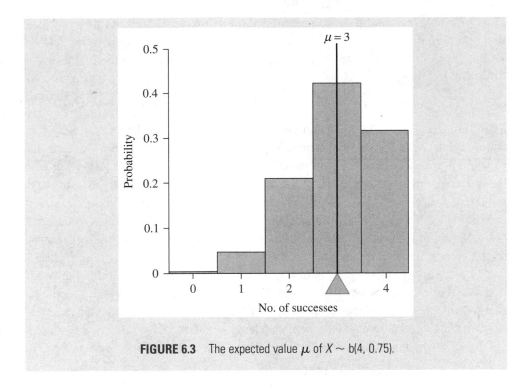

FIGURE 6.3 The expected value μ of $X \sim b(4, 0.75)$.

Exercises

6.7 *Tay-Sachs.* Exercises 6.1 and 6.3 considered the random number of Tay-Sachs cases in three pregnancies from a carrier couple ($X \sim b(3, 0.25)$). What is the expected value and variance of this random variable?

6.8 *Telephone survey.* Exercise 6.5 introduced a random digit dialing machine with $p = 0.15$ for each call attempt.

(a) What is the expected value and variance for the number of contacts in eight attempts?

(b) What is the expected value and variance for 50 attempts?

■ 6.6 Using the Binomial Distribution to Help Make Judgments About the Role of Chance

The binomial distribution can be used to assess whether a given number of successes in a sample would be surprising under specified conditions. As an example, consider that 1 in 10 women will develop breast cancer during their lifetime. In a simple random sample of $n = 3$, the number of cases $X \sim b(3, 0.10)$. Suppose we take a simple random sample of three women and all three ultimately develop breast cancer. Is this just "bad luck" or should we suspect something is awry? To address this problem, first ask, "What is the probability of seeing three cases under these conditions?" If the sampling process is truly random, the number of cases will follow a binomial distribution with $n = 3$ and $p = 0.10$, and the probability of observing three cases is:

$$\Pr(X = 3) = {_nC_x} \cdot p^x \cdot q^{n-x} = {_3C_3} \cdot 0.1^3 \cdot 0.9^{3-3} = 1 \cdot 0.001 \cdot 1 = 0.001$$

Therefore, only 1 in 1000 such samples will have three cases. This is unlikely (but could still happen). This is the "chance explanation."[c] At least two other explanations can be entertained. These are the "assumed value of p is wrong explanation" and the "biased selection explanation." The "assumed value of p is wrong explanation" suggests that p is greater than anticipated. The initial probability model assumed p was equal to 0.10. If we had made our selection from a subgroup with greater underlying risk—for example, from a family with a higher than typical risk of breast cancer—then $X \sim b(3, 0.1)$ would not hold.

The last explanation is the "biased selection explanation." The binomial distribution assumes that the observations were derived by a simple random sample. If the sample was not random, but instead had intentionally over-sampled breast cancer cases, the sample would not be random and the binomial model would no longer hold.

In summary, three alternative explanations are presented for the observed finding.

1. The chance explanation
2. The assumed value of p is wrong explanation
3. The biased selection explanation

Given the limited available information, all are good explanations for the observation. Let's apply this line of reasoning to another example.

ILLUSTRATIVE EXAMPLE

Using the binomial model to question the role of chance (Asthma survey). Start with the assumption that the prevalence of asthma in a school population is 0.04. Select at random 20 students from the school. Therefore, the random number of asthmatics in a given sample $X \sim b(20, 0.04)$.

Suppose we find two asthmatics in a sample. The expected number of asthmatics in the sample $\mu = np = (20)(0.04) = 0.8$, so we have seen more than expected. We are aware that some samples are going to randomly capture more cases than others. What is the probability of seeing two *or more* cases under these conditions?

---

[c] We can also call this the "bad luck explanation" because it was a just a matter of bad luck to have selected three cases.

To assess the role of chance, we determine the probability of seeing at least two cases. This is $\Pr(X \geq 2) = \Pr(X = 2) + \Pr(X = 3) + \cdots + \Pr(X = 20)$. Because this is a lengthy series of calculation, we make use of the fact that $\Pr(X \geq 2) = 1 - \Pr(X \leq 1)$ and calculate:

$$\Pr(X = 0) = {}_nC_x \cdot p^x \cdot q^{n-x} = {}_{20}C_0 \cdot 0.04^0 \cdot 0.96^{20-0} = 1 \cdot 1 \cdot 0.4420 = 0.4420$$

$$\Pr(X = 1) = {}_nC_x \cdot p^x \cdot q^{n-x} = {}_{20}C_1 \cdot 0.04^1 \cdot 0.96^{20-1} = 20 \cdot 0.04 \cdot 0.4604$$
$$= 0.3683$$

Therefore, $\Pr(X \leq 1) = \Pr(X = 0) + \Pr(X = 1) = 0.4420 + 0.3683 = 0.8103$ and $\Pr(X \geq 2) = 1 - \Pr(X \leq 1) = 1 - 0.8103 = 0.1897$ (nearly 19%). Figure 6.4 displays this as the shaded region in the right tail of the distribution.

The chance of seeing two or more cases in a sample is pretty high. Therefore, chance is a good explanation of the finding.

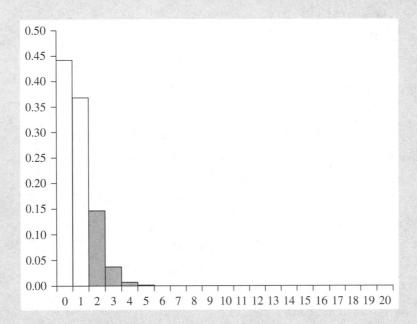

FIGURE 6.4 Probability of 2 or more on $X \sim b(20, 0.04)$.

Summary Points (Binomial Probability Distributions)

1. **Binomial distributions** are a family of probability mass functions (*pmf*s) that describe the random number of "successes" among n independent trials, where the probability of "success" in each trial is consistently p.

2. Binomial random variables have two **parameters**: n (number of observations) and p (probability of success for each observation).

3. The **notation** $X \sim b(n, p)$ is read as "X is distributed as a binomial random variable with parameters n and p."

4. The probability of observing x successes for a binomial random variable X is given by $\Pr(X = x) = {}_nC_x p^x q^{n-x}$, where ${}_nC_x = \dfrac{n!}{x!(n-x)!}$, $x! = x \cdot (x - 1) \cdot (x - 2) \cdot \ldots \cdot 1$, and $q = 1 - p$.

5. The **cumulative probability** of an event is the probability of observing given value x *or less*, that is, $\Pr(X \leq x)$.

Vocabulary

Bernoulli trial	Factorial function ($x!$)
Binomial	Parameters
Binomial coefficient (${}_nC_x$)	Probability mass function (*pmf*)
Cumulative distribution function (*cdf*)	$X \sim b(n, p)$

Review Questions

6.1 Binomial distributions have two parameters. Name them.

6.2 What does the symbol X represent in the statement $X \sim b(n, p)$?

6.3 What does the symbol $\sim$ represent in the statement $X \sim b(n, p)$?

6.4 What does the symbol n represent in the statement $X \sim b(n, p)$?

6.5 What does the symbol p represent in the statement $X \sim b(n, p)$?

6.6 What is a Bernoulli trial?

6.7 What does "${}_4C_2 = 6$" mean in plain language?

6.8 What does X represent in the statement $\Pr(X = x)$?

6.9 What does x represent in the statement $\Pr(X = x)$?

6.10 How do you read the statement "$\Pr(X = x)$"?

6.11 How do you read the statement "$X \sim b(n, p)$"?

6.12 By definition, $0! = $ ___

6.13 Determine the value of $7!/6!$ without using a calculator.

6.14 What does q represent in the context of binomial distributions?

6.15 Fill in the blank: $\Pr(X \leq x)$ represents the _____ probability of x.

6.16 What symbol is used to represent the mean (expected value) of a binomial distribution?

6.17 What symbol is used to represent the variance of a binomial distribution?

6.18 Fill in the blank: The expected number of successes μ for a binomial random variable $X \sim (n, p)$ is equal to $n \times$ ___.

Exercises

6.9 *Prevalence 76.8%.* The prevalence of a trait is 76.8%.

   (a) In a simple random sample of $n = 5$, how many individuals are expected to exhibit this characteristic?

   (b) How many would you expect to see with this characteristic in a simple random sample of $n = 10$?

   (c) What is the probability of seeing nine or more individuals with this characteristic in an SRS of $n = 10$?

6.10 *Smoking on campus.* Suppose 20% of the students on campus smoke. You select two students at random. In what percentage of samples will both students be smokers?

6.11 *Prevalence 10%.* The prevalence of a condition in a population is 10%. You take a simple random sample of 15 people from this population. Let X represent the number of individuals in the sample with the condition in question.

   (a) Describe the distribution X.

   (b) What is the probability of seeing no cases in a sample?

   (c) What is the probability of seeing one case?

   (d) What is the probability of seeing one or fewer cases?

   (e) What is the probability of *at least* two cases?

6.12 *Herpes simplex-2.* Suppose 7.5% of a population is infected with Herpes simplex-2 virus (HSV2). You select seven individuals at random from the population. What is the probability of finding at least one HSV2-positive individual in your sample? [Hint: First find $\Pr(X = 0)$. Then make use of the fact that $(X \geq 1) = 1 - \Pr(X = 0)$.]

6.13 *Linda's omelets.* Linda hears a story on National Public Radio stating that one in six eggs in the United States are contaminated with *Salmonella*. If *Salmonella* contamination occurs independently within and between egg cartons and Linda makes a three-egg omelet, what is the probability that her omelet will contain at least one *Salmonella*-contaminated egg?

6.14 *Electromagnetic fields.* Twenty-five percent of the children in a community are exposed to high levels of electromagnetic field (EMF) radiation. You select at random a control series of 20 children from this neighborhood. Construct the *pmf* that describes the number of children in the sample that are exposed to high EMF levels.

6.15 *Decayed teeth.* A child has 20 deciduous teeth. Two of her teeth are decayed. Given that this is all that you currently know about the child's dentition, how many different possible combinations of decayed teeth might she have?

6.16 *False positives.* A rapid screening test for HIV has a false positive rate of 0.5%. This means that the probability of a false positive test is 0.005. If the test is used in 500 HIV-free individuals, what is the expected number of false positives in the sample? In addition, what is the probability of encountering no false positives in the sample?

6.17 *Human papillomavirus.* In a particular population, 20% of the individuals are human papillomavirus carriers. Select four individuals at random from this population.

(a) Build the *pmf* for the number of HPV+ individuals in the sample.

(b) What is the probability of finding at least one carrier in the sample?

6.18 *Random 7s.* The digits (0 through 9) in Appendix Table A occur randomly throughout the table. Thus, a value of 7 has a 0.1 chance of occurring in any single slot anywhere in the table.

(a) Each line in Appendix Table A has 50 slots. On average, how many 7s do we expect to find in a given line?

(b) What is the probability of finding exactly five 7s in a given line?

7

Normal Probability Distributions

The prior chapter considered the most popular type of probability distribution that applies to discrete random variables—the binomial distributions. This chapter considers the most popular type of probability distribution that applies to continuous random variables—the Normal distributions. The popularity of the Normal distribution can be explained by four facts: (a) some natural phenomena follow Normal distributions; (b) some natural phenomena are approximately Normal; (c) some can be transformed to approximate Normality; and (d) the variability associated with random sampling tends toward Normality even in non-Normal populations (Section 8.2).

The mathematics of the Normal distribution was developed in the 18th century by Abraham de Moivre and Pierre-Simon Laplace. Johann Carl Friedrich Gauss popularized the use of the distribution in the early 19th century, which explains why it is sometimes called the Gaussian distribution.

■ 7.1 Normal Distributions

A Heuristic Example

While discrete random variables were described with chunky mass functions, continuous random variables are described with smooth **probability density functions (*pdf*)**. Figure 7.1 depicts a histogram with an overlying Normal probability density function curve. Normal distributions are recognized by their bell shape. Notice that a large percentage of the curve's area is located near its center and that its tails approach the horizontal axis as asymptotes.

Figure 7.2 displays the same distribution with the histogram bars shaded for subjects that are less than 9 years of age.[a] The area of the shaded bars makes up approximately one third of the histogram; about one third of the sample is less than 9. When working with Normal curves, we drop the histogram and look only at the **area under**

---
[a] This includes those that are up to 8.9999... years of age.

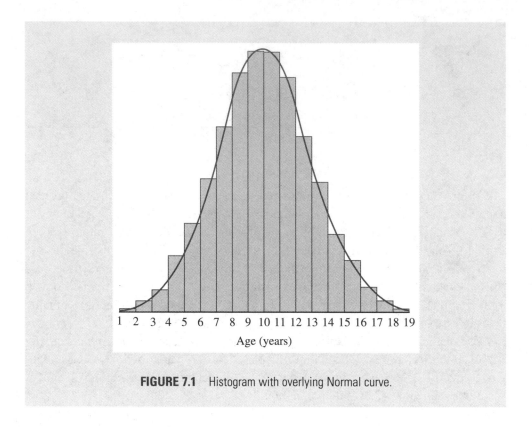

FIGURE 7.1   Histogram with overlying Normal curve.

the curve (Figure 7.3). As introduced in Section 5.4, the area under the *pdf*'s curve corresponds to the probability of the specified range.

Characteristics of Normal Distributions

Normal distributions are defined by this mathematical function:

$$f(x) = \frac{1}{\sqrt{2\pi}\,\sigma}\, e^{-\frac{1}{2}\left(\frac{x-\mu}{\sigma}\right)^2}$$

where μ represents the **mean** of the distribution and σ represents its **standard deviation**.[b] This function applies to a family of distributions with each family member identified by parameters μ and σ. Because there are many different Normal distributions, each with its own μ and σ, let $X \sim N(\mu, \sigma)$ represent a specific member of the Normal distribution family. As before, the symbol "$\sim$" is read as "distributed as." For example, $X \sim N(100, 15)$ is read as "X is a Normal random variable with mean 100 and standard deviation 15.

---

[b] μ and σ are the parameters of the distribution.

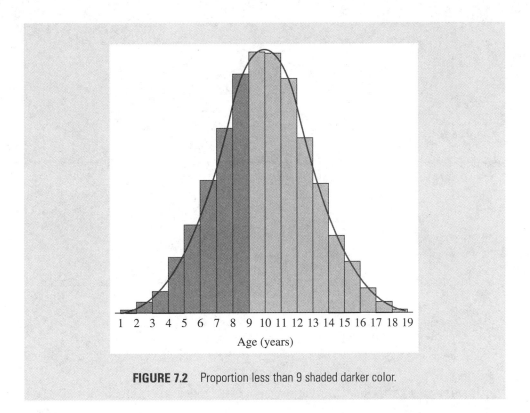

FIGURE 7.2   Proportion less than 9 shaded darker color.

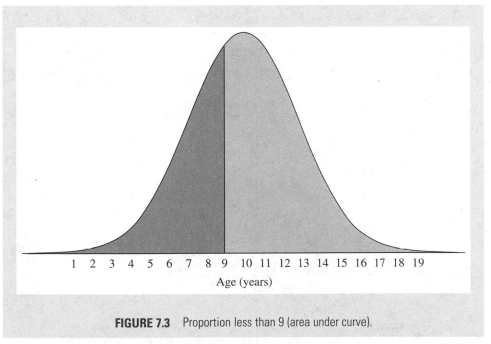

FIGURE 7.3   Proportion less than 9 (area under curve).

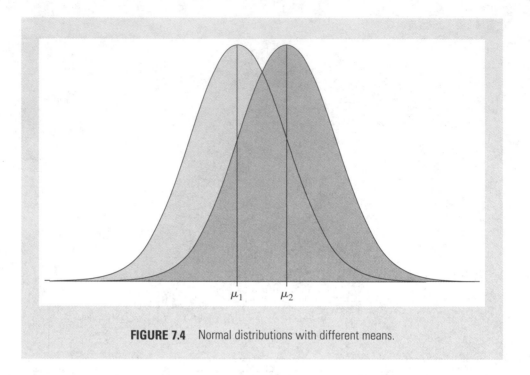

FIGURE 7.4   Normal distributions with different means.

Changing μ shifts the distribution on its horizontal axis. Figure 7.4 displays Normal curves with different means.

The standard deviation σ determines the spread of the distribution. Figure 7.5 depicts two Normal curves with different standard deviations.

You can get a rough idea of the size of the distribution's standard deviation σ by identifying its **points of inflection**. This is where the Normal curve begins to

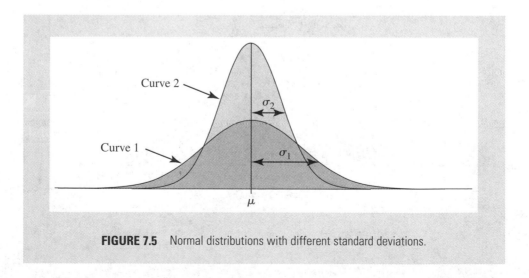

FIGURE 7.5   Normal distributions with different standard deviations.

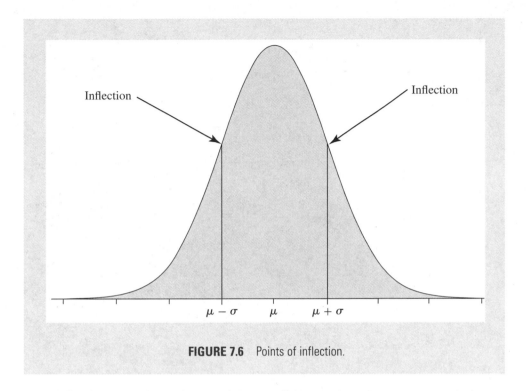

FIGURE 7.6 Points of inflection.

change slope. Figure 7.6 shows the location of these inflection points on a Normal curve. Trace the curve with your finger; a point of inflection is where you feel the curve begin to change slope. Once you've identified the inflection points, use these landmarks to identify points that are one standard deviation above and below the mean ($\mu \pm \sigma$) on the curve's horizontal axis.

The 68–95–99.7 Rule

The **68–95–99.7 rule** is used as a guide when working with Normal random variables. This rule says that:

- 68% of the area under the Normal curve lies in the region $\mu \pm \sigma$.
- 95% of the area under the Normal curve lies in the region $\mu \pm 2\sigma$.
- 99.7% of the area under the Normal curve lies in the region $\mu \pm 3\sigma$.

Figure 7.7 shows this visually.

Although μ and σ vary from Normal random variable to Normal random variable, you can always depend on the 68–95–99.7 rule to predict the percentage of individuals who fall in these ranges. Here is an example.

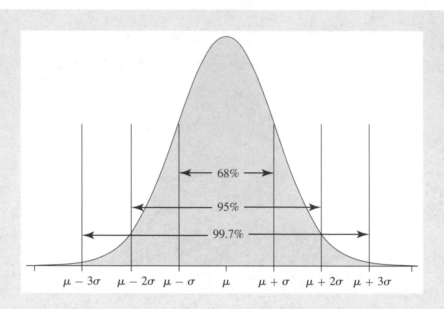

FIGURE 7.7   The 68–95–99.7 rule.

ILLUSTRATIVE EXAMPLE

68–95–99.7 rule (WAIS). The Wechsler Adult Intelligence Scale (WAIS) is a commonly used intelligence test that is calibrated to produce a Normal distribution of scores with $\mu = 100$ and $\sigma = 15$ for various age groups.[c] Based on the 68–95–99.7 rule, we can say that:

- 68% of the scores lie in the range $100 \pm 15 = 85$ to 115.
- 95% lie in the range $100 \pm (2)(15) = 70$ to 130.
- 99.7% lie in the range $100 \pm (3)(15) = 55$ to 145.

Figure 7.8 depicts these landmarks.

Focus on the middle 95% of scores in Figure 7.8. This is defined by the range 70 and 130. Five percent of the scores lie outside this range. Because the distribution is symmetrical, half of the remaining 5.0% (2.5%) of the scores are below 70 (in the left tail of the distribution) and the other 2.5% are above 130, in the right tail. Figure 7.9 depicts these tails.

---

[c]WAIS "IQ" scores follow a bell curve, but it is not necessarily true that "intelligence" has a Normal distribution; what we call intelligence is not likely to be captured by a single test score.

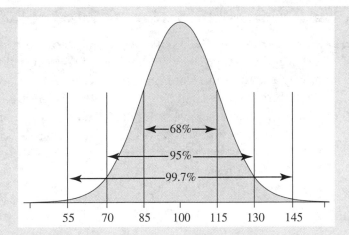

FIGURE 7.8   Distribution of Wechsler adult intelligence scale.

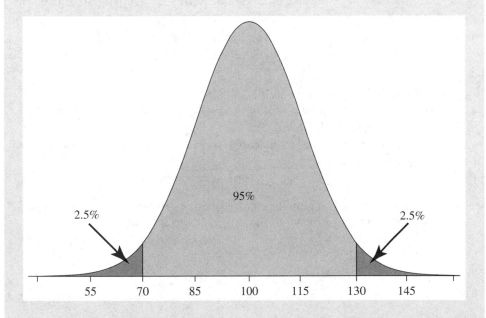

FIGURE 7.9   The symmetry of the Normal curve allows us to make statements about probabilities above and below certain cut points.

> **ILLUSTRATIVE EXAMPLE**
>
> *Tails of the Normal distribution (Gestational length).* Gestation is the period of pregnancy from conception to delivery.[d] Uncomplicated human gestations (without intervention) vary according to a Normal distribution with $\mu = 39$ weeks and $\sigma = 2$ weeks.[e] According to the "95 part" of the 68-95-99.7 rule, we can say that 95% of uncomplicated human gestations will fall in the range $\mu \pm 2\sigma = 39 \pm 2 \cdot 2 = 39 \pm 4 = 35$ to 43 weeks. This leaves 5% of the values outside of this range. Since the distribution is symmetrical, we can say that 2.5% of gestations will be fewer than 35 weeks (in the left tail of the distribution) and 2.5% will be more than 43 weeks (in the right tail).

Reexpression

Many random variables encountered in nature are not Normal.[f] We can, however, often make random variables more Normal by reexpressing them with a mathematical transformation. Many types of mathematical transforms are available for this purpose (e.g., logs, exponents, powers, roots, and quadratics). Logarithmic transformations are particularly useful for bringing in the right tails of distributions with positive skews. The use of logarithmic transformations is common, so let us take this opportunity for a brief review of the subject.

There are different kinds of logarithms, depending on their base. The two most popular logarithms are *common logarithms* (base 10) and natural logarithms (base e; e is approximately equal to 2.71828...).[g]

Logarithms are exponents of their base. For example, the common log (base 10) of 100 is 2 because $10^2 = 100$.

$$\log_{10}(100) = 2$$

By the same token, the natural logarithm (base e) of 100 is approximately 4.60517 because $e^{4.60517...} \approx 2.71828^{4.60517...} \approx 100$. This is expressed as:

$$\ln(100) = 4.60517...$$

No matter the base, $\log_{\text{base}}(\text{base}) = 1$ and $\log_{\text{base}}(1) = 0$, since $\text{base}^1 = \text{base}$ and $\text{base}^0 = 1$.

---

[d] It may also be defined as the period between the last menstrual period to delivery.

[e] Mittendorf, R., Williams, M. A., Berkey, C. S., & Cotter, P. F. (1990). The length of uncomplicated human gestation. *Obstetrics & Gynecology, 75*(6), 929–932; Durham, J. (2002). *Calculating due dates and the impact of mistaken estimates of gestational age.* Retrieved February 2006 from http://transitiontoparenthood.com/ttp/birthed/duedatespaper.htm.

[f] Elveback, L. R., Guillier, C. L., & Keating, F. R., Jr. (1970). Health, normality, and the ghost of Gauss. *JAMA, 211*(1), 69–75.

[g] We will work on the natural log scale (base e) unless otherwise specified.

Here is an example of how a logarithmic transformation is used in practice:

ILLUSTRATIVE EXAMPLE

Log transformation (Prostate-specific antigen). Prostate-specific antigen (PSA) and its isoforms are used to screen for prostate cancer in men. Reference ranges for PSA do not vary Normally, but their logarithms do so approximately. A reference lab finds that natural logarithmic transformed PSA values for 50- to 60-year-old disease-free men vary according to a Normal with a mean of -0.3 and standard deviation of 0.80.[h] According to the 68–95–99.7 rule, 95% of the ln(PSA) values in this population will be in the interval $-0.3 \pm (2)(0.80)$ $= -1.9$ to 1.3. Figure 7.10 depicts this distribution. Notice that 2.5% of men will have ln(PSA) values less than -1.90 and 2.5% will have values above 1.30. We exponentiate these limits to convert them back to their initial scale: $e^{-1.9} = 0.15$ and $e^{1.3} = 3.67$. The upper limit can now serve as a cutoff for PSA screening.[i]

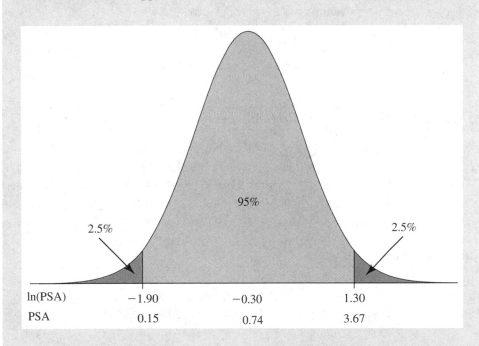

FIGURE 7.10 PSA levels in 50- to 60-year-old men.

---

[h] Sibley, P. E. C. (2001). *Reference range analysis—Lessons from PSA.* Retrieved December 10, 2005, from www.dpcweb.com/documents/news&views/tech_reports_pdfs/zb204-a.pdf5700.
[i] Low levels of PSA are of no health concern.

Section 7.4 includes additional examples of logarithmic reexpressions.

Exercises

7.1 **Heights of 10-year-olds.** Heights of 10-year-old male children follow a Normal distribution with $\mu = 138$ centimeters and $\sigma = 7$ centimeters.[j] Create a sketch of a Normal curve depicting this distribution. Mark the horizontal axis with values that are one standard deviation above and below the mean. (Locate points of inflection on the curve as accurately as possible.) Then, use the 68–95–99.7 rule to determine the middle 68% of values. What percent of values fall below this range? What percent fall above this range?

7.2 *Height of 10-year-olds.* This exercise continues the work we began in the prior exercise. What range of values will capture the middle 95% of heights? How tall are the tallest 2.5%?

7.3 *Visualizing the distribution of gestational length.* The gestation length illustrative example presented earlier in this chapter established that uncomplicated pregnancies vary according to a Normal distribution with $\mu = 39$ weeks and $\sigma = 2$ weeks. Sketch a Normal curve depicting this distribution. Label its horizontal axis with landmarks that are $\pm\sigma$ and $\pm2\sigma$ on either side of the mean. Remember to use the curve's inflection points to establish distances for landmarks.

■ 7.2 Determining Normal Probabilities

We often need to determine probabilities for Normal values that do not fall exactly $\pm1\sigma$, $\pm2\sigma$, or $\pm3\sigma$ from the mean. To accomplish this, we first **standardize** the values and then use a **Standard Normal table** to look up the associated probability. A Standard Normal table lists cumulative probabilities for a Normal random variable with $\mu = 0$ and $\sigma = 1$.

Standardizing Values

Standardizing a Normal value transforms it to a Normal scale with mean 0 and standard deviation 1. This special Normal distribution is called a Z-distribution, and the transformed value is called a *z***-score**; $Z \sim N(0,1)$. The formula is:

$$z = \frac{x - \mu}{\sigma}$$

---

[j] *United States Growth Charts.* Retrieved February 24, 2006, from www.cdc.gov/nchs/about/major /nhanes/growthcharts/zscore/zscore.htm. Values have been rounded and fit to a Normal distribution.

where x is the value you want to standardize, μ is the mean of the distribution, and σ is its standard deviation. The z-score tells you the distance the value falls from the mean in standard deviation units. Values that are larger than the mean will have positive z-scores. Values that are smaller than the mean will have negative z-scores. For example, a z-score of 1 tells you that the value is one standard deviation *above* the mean. A z-score of -2 tells you that the value is two standard deviations *below* the mean.

ILLUSTRATIVE EXAMPLE

Standardization (WAIS). Recall that Wechsler adult intelligence scores vary according to a Normal distribution with $\mu = 100$ and $\sigma = 15$. What is the z-score of a value of 95?

Solution: $z = \dfrac{95 - 100}{15} = -0.33$. This indicates the score is 0.33 standard deviations *below* the mean.

ILLUSTRATIVE EXAMPLE

Standardization (Gestational length). We have established that uncomplicated human pregnancies have a gestation period that is approximately Normal with $\mu = 39$ weeks and $\sigma = 2$ weeks. What is the z-score for a pregnancy that lasts 36 weeks?

Solution: A pregnancy that is 36 weeks in length corresponds to

$z = \dfrac{36 - 39}{2} = -1.5$. This gestation is 1.5 standard deviations *below* the mean.

The Standard Normal Table

After the Normal random variable has been standardized, we can find its cumulative probability with a **Standard Normal table (z table)**. Our Standard Normal (z) table appears as Appendix Table B. It is also available online. Figure 7.11 shows a portion of this table.

The setup of this table may seem strange. The one and tens places for Normal z-scores appear in the left column of the table. The hundredths place for the Normal z-scores appears in the top row of the table. Table entries represent *cumulative probabilities*, or areas under the curve to the left of the Normal z-score. As an example,

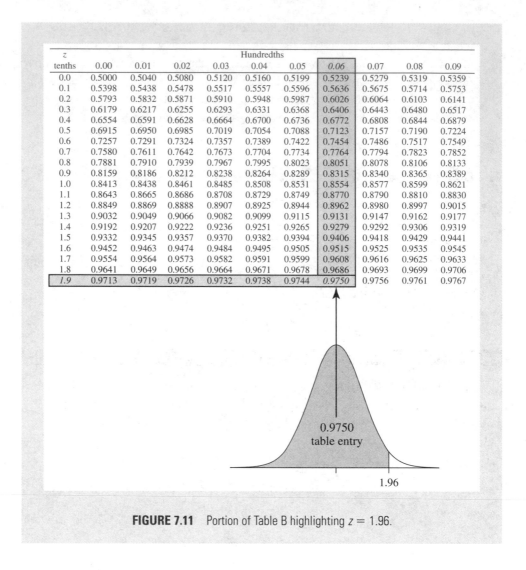

z tenths	0.00	0.01	0.02	0.03	Hundredths 0.04	0.05	0.06	0.07	0.08	0.09
0.0	0.5000	0.5040	0.5080	0.5120	0.5160	0.5199	0.5239	0.5279	0.5319	0.5359
0.1	0.5398	0.5438	0.5478	0.5517	0.5557	0.5596	0.5636	0.5675	0.5714	0.5753
0.2	0.5793	0.5832	0.5871	0.5910	0.5948	0.5987	0.6026	0.6064	0.6103	0.6141
0.3	0.6179	0.6217	0.6255	0.6293	0.6331	0.6368	0.6406	0.6443	0.6480	0.6517
0.4	0.6554	0.6591	0.6628	0.6664	0.6700	0.6736	0.6772	0.6808	0.6844	0.6879
0.5	0.6915	0.6950	0.6985	0.7019	0.7054	0.7088	0.7123	0.7157	0.7190	0.7224
0.6	0.7257	0.7291	0.7324	0.7357	0.7389	0.7422	0.7454	0.7486	0.7517	0.7549
0.7	0.7580	0.7611	0.7642	0.7673	0.7704	0.7734	0.7764	0.7794	0.7823	0.7852
0.8	0.7881	0.7910	0.7939	0.7967	0.7995	0.8023	0.8051	0.8078	0.8106	0.8133
0.9	0.8159	0.8186	0.8212	0.8238	0.8264	0.8289	0.8315	0.8340	0.8365	0.8389
1.0	0.8413	0.8438	0.8461	0.8485	0.8508	0.8531	0.8554	0.8577	0.8599	0.8621
1.1	0.8643	0.8665	0.8686	0.8708	0.8729	0.8749	0.8770	0.8790	0.8810	0.8830
1.2	0.8849	0.8869	0.8888	0.8907	0.8925	0.8944	0.8962	0.8980	0.8997	0.9015
1.3	0.9032	0.9049	0.9066	0.9082	0.9099	0.9115	0.9131	0.9147	0.9162	0.9177
1.4	0.9192	0.9207	0.9222	0.9236	0.9251	0.9265	0.9279	0.9292	0.9306	0.9319
1.5	0.9332	0.9345	0.9357	0.9370	0.9382	0.9394	0.9406	0.9418	0.9429	0.9441
1.6	0.9452	0.9463	0.9474	0.9484	0.9495	0.9505	0.9515	0.9525	0.9535	0.9545
1.7	0.9554	0.9564	0.9573	0.9582	0.9591	0.9599	0.9608	0.9616	0.9625	0.9633
1.8	0.9641	0.9649	0.9656	0.9664	0.9671	0.9678	0.9686	0.9693	0.9699	0.9706
1.9	0.9713	0.9719	0.9726	0.9732	0.9738	0.9744	0.9750	0.9756	0.9761	0.9767

0.9750
table entry

1.96

FIGURE 7.11   Portion of Table B highlighting $z = 1.96$.

Figure 7.11 highlights the entry for $z = 1.96$. This point has a cumulative probability of 0.9750. The figure below the table depicts this graphically.

Table B lists cumulative probabilities for Normal z-scores between -3.49 and 3.49. Values less than or equal to -3.50 have a cumulative probability less than 0.0002, and values greater than 3.50 have a cumulative probability of more than 0.9998.

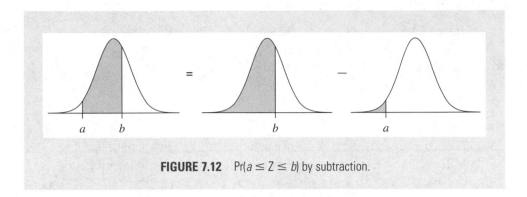

FIGURE 7.12 Pr($a \leq Z \leq b$) by subtraction.

Probabilities for Ranges of Normal Random Variables

You can determine probabilities for ranges of Normal random variables by standardizing the range and using the z table to determine the enclosed area under the curve. Here is a procedure you can use for this purpose:

1. **State** the problem.
2. **Standardize** values.
3. **Sketch** the curve and shade the probability area.
4. **Use Table B** to determine the probability.

You can determine probabilities between any two points on a Normal distribution as follows: Let a represent the lower boundary of the interval and b represent the upper boundary. The probability of seeing a value between a and b is:

$$Pr(a \leq Z \leq b) = Pr(Z \leq b) - Pr(Z \leq a)$$

Figure 7.12 is a schematic of this approach.

ILLUSTRATIVE EXAMPLE

Probabilities for ranges (Gestational length). A prior illustrative example established that uncomplicated human gestation varies according to a Normal distribution with $\mu = 39$ weeks and $\sigma = 2$ weeks. What is the probability that an uncomplicated pregnancy selected at random lasts 41 weeks or fewer? What is the probability it lasts more than 41 weeks?

continues

1. **State.** Let X represent gestational length: $X \sim N(39, 2)$. We want to determine $\Pr(X \leq 41)^k$ and $\Pr(X > 41)$.

2. **Standardize.** The z-score associated with a 41-week gestation is

$$z = \frac{41 - 39}{2} = 1.$$

3. **Sketch.** Figure 7.13 shows the sketch of the Normal distribution for this problem. The horizontal axis is scaled with values for both the original variable and Standard z-scores. The region corresponding to $\Pr(X \leq 41)$ is shaded dark blue. The region corresponding to $\Pr(X > 41)$ is shaded light blue.

4. **Use Table B.** Table B tells us that $\Pr(Z \leq 1) = 0.8413$. Therefore, 84.13% of pregnancies last 41 weeks or fewer. Because the area under the curve sums to 1: (Area under the curve to the right) = 1 − (Area under the curve to the left). Therefore, $\Pr(X > 41) = 1 - \Pr(X \leq 41) = 1 - 0.8413 = 0.1587$ and 15.87% of the pregnancies last more than 41 weeks.

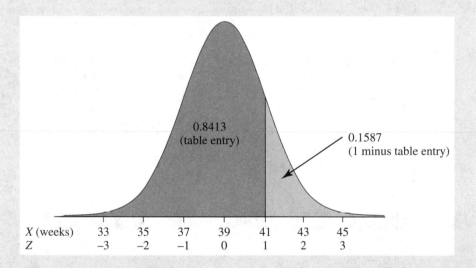

FIGURE 7.13   Distribution of gestational length. About 84% of pregnancies are less than or equal to 41 weeks. About 16% are at least 41 weeks in length.

---

^{k} It makes no difference whether we state this probability as $\Pr(X \leq 41)$ or $\Pr(X < 41)$ because $\Pr(X = 41) = 0$. See Section 5.4.

Exercises

7.4 *Standard Normal probabilities.* Use Table B to find the probabilities listed here. In each instance, sketch the Normal curve and shade the area under the curve associated with the probability.

(a) $\Pr(Z < -0.64)$

(b) $\Pr(Z > -0.64)$

(c) $\Pr(Z < 1.65)$

(d) $\Pr(-0.64 < Z < 1.65)$

7.5 *Heights of 10-year-old boys.* We've established that heights of 10-year-old boys vary according to a Normal distribution with $\mu = 138$ cm and $\sigma = 7$ cm.

(a) What proportion of this population is less than 150 cm tall?

(b) What proportion is less than 140 cm in height?

(c) What proportion is between 150 and 140 cm?

7.6 *Heights of 20-year-olds.* Heights of 20-year-old men vary approximately according to a Normal distribution with $\mu = 176.9$ cm and $\sigma = 7.1$ cm.

(a) What percentage of U.S. men are at least 6 feet tall? (Six feet $\approx$ 183 cm.)

(b) Heights of 20-year-old women vary approximately according to a Normal with $\mu = 163.3$ cm and $\sigma = 6.5$ cm. What percentage of U.S. women are at least 6 feet tall?

(c) Why does a 6-foot-tall woman appear unusually tall while a 6-foot-tall man is barely out of the ordinary?

ILLUSTRATIVE EXAMPLE

Normal probability between points (Gestational length). Births that occur before 35 weeks of gestation are considered premature. Those more than 40 weeks are considered "postdate." What proportion of pregnancies are neither premature nor postdate? Previous illustrative examples established that gestational lengths for uncomplicated pregnancies vary according to a Normal distribution with $\mu = 39$ weeks and $\sigma = 2$ weeks.

   Solution:

   1. **State.** We propose to find $\Pr(35 \leq X \leq 40)$.
   2. **Standardize.** The z-score for the lower boundary is $z = (35 - 39)/2 = -2.00$. The z-score for the upper boundary is $z = (40 - 39)/2 = 0.50$.
   3. **Sketch.** Figure 7.14 shows the Normal sketch with landmarks indicated.

continues

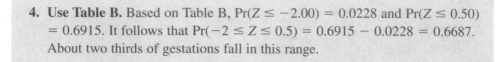

4. Use Table B. Based on Table B, $\Pr(Z \le -2.00) = 0.0228$ and $\Pr(Z \le 0.50)$ $= 0.6915$. It follows that $\Pr(-2 \le Z \le 0.5) = 0.6915 - 0.0228 = 0.6687$. About two thirds of gestations fall in this range.

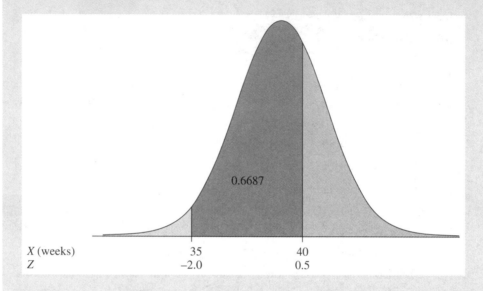

FIGURE 7.14   Gestational length between 35 and 40 weeks.

7.3 Finding Values that Correspond to Normal Probabilities

Table B can also be used to find values that correspond to Normal probabilities. Here is a four-step procedure that can be used for this purpose:

1. **State** the problem.
2. **Use Table B** to look up the z-score for the given probability.
3. **Sketch** the distribution with associated landmarks.
4. **Unstandardize** the z-score using the formula $x = \mu + z\sigma$. (We have merely rearranged $z = \dfrac{x - \mu}{\sigma}$ to solve for x.)

Terminology and Notation

Step 2 of this process requires us to find the z-score associated with a stated probability. It is helpful to establish notation and terminology when discussing this part of the procedure. Let z_p denote a z-score with cumulative probability p. This z-score value is greater than $p \times 100\%$ of the z-scores and is thus called a **z-percentile**. For example, $z_{0.90}$ is the 90th percentile of the Standard Normal distribution. To find this z-score, scan the entries in Table B for the cumulative probability closest to 0.9000. In this instance, the closest cumulative probability is 0.8997. This has an associated z-score of 1.28; therefore, $z_{0.90} = 1.28$.

Exercises

7.7   *45th percentile on a Standard Normal curve.* What is the 45th percentile on a Standard Normal distribution? (Use Table B to look up $z_{0.45}$.)

7.8   *64th percentile on a Normal z-curve.* What is the 64th percentile on a Standard Normal distribution? What notation denotes this value?

7.9   *Middle 50% of WAISs.* Recall that the Wechsler Adult Intelligence Scale scores are calibrated to vary according to a Normal distribution with $\mu = 100$ and $\sigma = 15$. What Wechsler scores cover the middle 50% of the population? In other words, identify the 25th percentile and 75th percentile of the population.

7.10   *Top 10 and 1% of the WAIS.* How high must a WAIS score be to be in the top 10% of scores? How high must it be to rank in the top 1%?

7.11   *Death row inmate.* An inmate on death row in the state of Illinois has a WAIS score of 51. What percentage of people have a score below this level?

ILLUSTRATIVE EXAMPLE

Finding values that correspond to Normal probabilities (Heights of women). What height does a woman have to be in order to be in the 90th percentile of heights? In other words, how tall does a women have to be to be taller than 90% of women?

Solution:

1. **State the problem.** We have established in a prior exercise that heights of women in the United States vary according to a Normal distribution with $\mu = 163.3$ cm and $\sigma = 6.5$ cm (approximately so). Let X represent heights of U.S. women: $X \sim N(163.3, 6.5)$. We want to find the value of x such that $\Pr(X \le x) = 0.9000$.

2. **Use Table B** to scan for a cumulative probability that is closest to 0.9000. As noted, $z_{0.90} = 1.28$.

3. **Sketch.** Figure 7.15 is a drawing for this problem. The value we are looking for is 1.28 standard deviations above mean.

4. **Unstandardize.** $x = \mu + z_{0.90}\sigma = 163.3 + (1.28)(6.5) = 171.62$.
   Therefore, a woman that is 171.62 cm tall (about 5'7½") is taller than 90% of women in the population.

continues

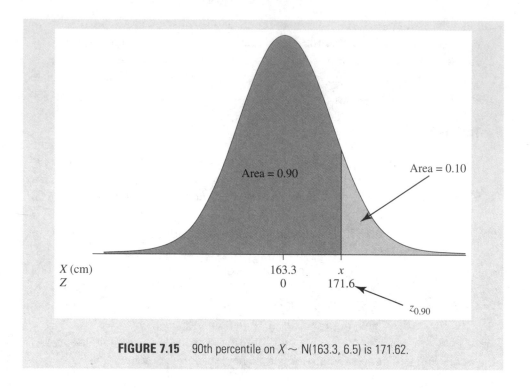

FIGURE 7.15   90th percentile on $X \sim N(163.3, 6.5)$ is 171.62.

■ 7.4 Assessing Departures from Normality

It is important to establish that the random variable being assessed is approximately Normal before applying methods in this chapter. There are several ways to make this assessment.

First *look* at the shape of the distribution with a **stemplot** or **histogram**. If the distribution is asymmetrical or otherwise clearly departs from the typical bell-shape of a Normal curve, avoid application of the methods in this chapter.

We may also examine the distribution with a **Normal probability (Q-Q) plot**. The idea of a Q-Q plot is to graph observed values against expected Normal z-scores. If the distribution is approximately Normal, points on the Q-Q plot will form a diagonal line. Deviations from the diagonal indicate departures from Normality. **Expected z-scores** are derived by finding the percentile rank of each value and converting these to Normal z-percentiles. For example, the median of a data set has a percentile rank of 50, which converts to $z_{0.5} = 0.00$. The 25th percentile of a data set (Q1) converts to $z_{0.25} = -0.67$. The 75th percentile (Q3) converts to $z_{0.75} = 0.67$ (and so on). More generally, we need a rule to interpolate percentiles for empirical distributions. One such rule is to rank data $1, 2, \ldots, r$ and then use z_p as its z-percentile, where $p = (r - 1/3)/(n + 1/3)$. For example, if a data set has 10 observations, the lowest ranking value has $p = (1 - 1/3)/(10 + 1/3)$ = 0.0645 with an expected z-score of $z_{0.0645} = -1.51$.

Here are examples of Q-Q plots:

- Figure 7.16 demonstrates an approximately Normal distribution. Points on the Q-Q plot adhere well to the diagonal line.

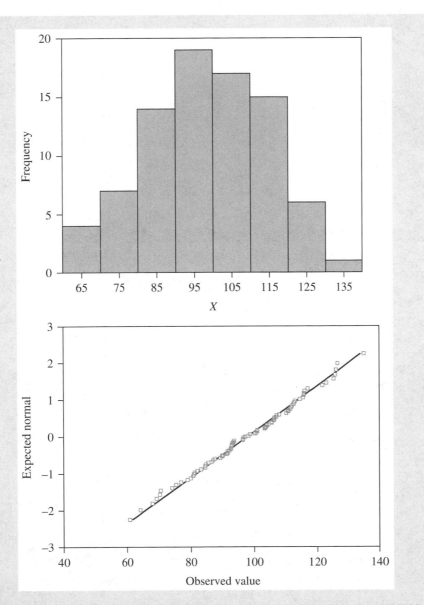

FIGURE 7.16 Histogram and Q-Q plot, approximately Normal data. Graph produced with SPSS for Windows, Rel. 11.0.1.2001. Chicago: SPSS Inc. Reprint Courtesy of International Business Machines Corporation.

• Figure 7.17 depicts a distribution with a pronounced negative skew. The Q-Q plot forms an upward curve.

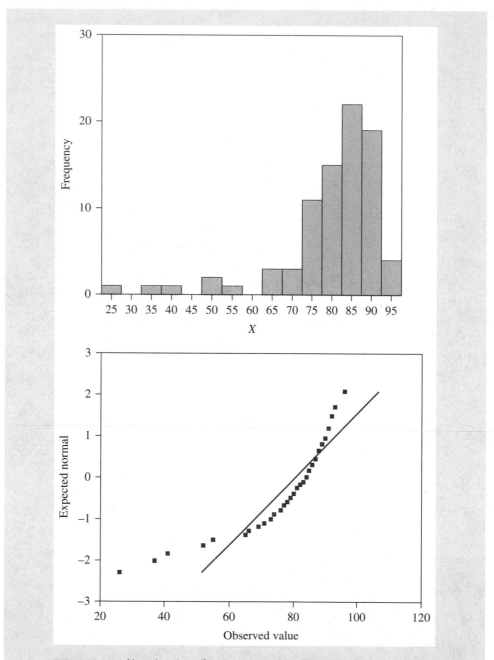

FIGURE 7.17   Negative skew. Graph produced with SPSS for Windows, Rel. 11.0.1.2001. Chicago: SPSS Inc. Reprint Courtesy of International Business Machines Corporation.

• Figure 7.18 depicts a distribution with a positive skew. Positive skews form downward-curving Q-Q plots.

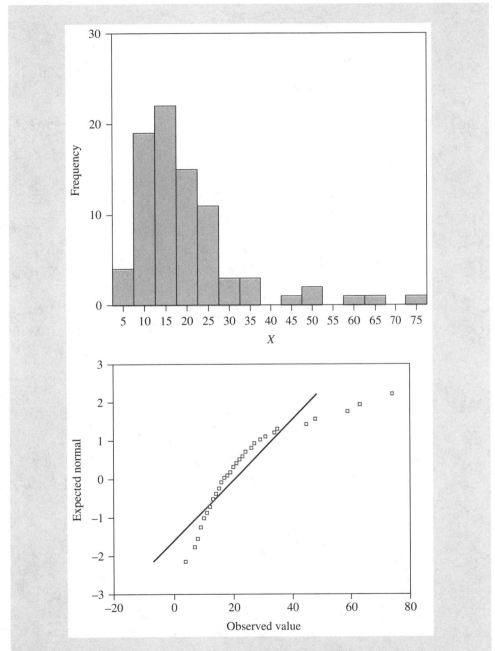

FIGURE 7.18   Positive skew, histogram, and Q-Q plot. Graph produced with SPSS for Windows, Rel. 11.0.1.2001. Chicago: SPSS Inc. Reprint Courtesy of International Business Machines Corporation.

- Figure 7.19 shows a leptokurtic distribution (i.e., a distribution with long skinny tails). This forms an S-shaped Q-Q plot.

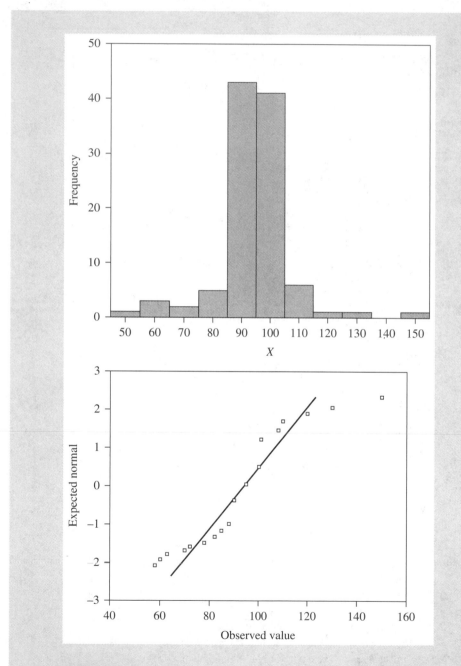

FIGURE 7.19   Leptokurtic = high peak with skinny tails. Graph produced with SPSS for Windows, Rel. 11.0.1.2001. Chicago: SPSS Inc. Reprint Courtesy of International Business Machines Corporation.

A platykurtic distribution (broad fat tails) is not illustrated, but would form a reverse S on the Q-Q plot.

Figure 7.20 has reexpressed the skewed data initially presented in Figure 7.18 on a natural logarithmic scale. This distribution is approximately Normal, now permitting the use of Normal probability methods with these data.

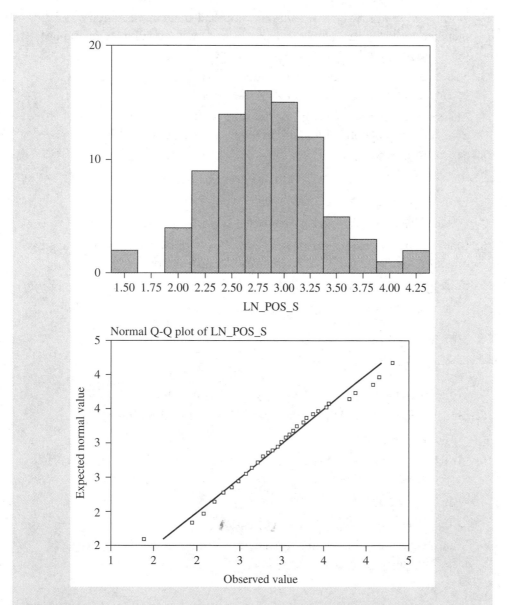

FIGURE 7.20   Same data as in Figure 7.18 but with data re-expressed on a log scale. Graph produced with SPSS for Windows, Rel. 11.0.1.2001. Chicago: SPSS Inc. Reprint Courtesy of International Business Machines Corporation.

Summary Points (Normal Probability Distributions)

1. **Normal probability distributions** are a family of probability density functions (*pdf*s) characterized by their symmetry, bell shape, points of inflection at $\mu - \sigma$ and $\mu + \sigma$, and horizontal asymptotes as they approach the X axis on either side.

2. Normal *pdf*s have two **parameters**: mean μ and standard deviation σ. Each member of the Normal family is distinguished by its μ and σ. μ determines the location of a Normal *pdf* and σ determines its spread.

3. The **notation** $X \sim N(\mu, \sigma)$ is read "random variable X is distributed as a Normal random variable with mean μ and standard deviation σ."

4. The **area under the curve (AUC)** between any two points on a Normal curve corresponds to the probability of observing a value between these two points.

5. The **68–95–99.7 rule**:

   (a) 68% of the area under a Normal curve lies within $\mu \pm \sigma$.

   (b) 95% of the area under a Normal curve lies within $\mu \pm 2\sigma$.

   (c) 99.7% of the area under the Normal curve lies within $\mu \pm 3\sigma$.

6. Many phenomena in nature are *not* Normally distributed but can be re-expressed on a different scale to approximate a Normal distribution.

7. To determine **Normal probabilities** for a given range of values for $X \sim N(\mu, \sigma)$:

   (a) State the problem.

   (b) Standardize: $z = \dfrac{x - \mu}{\sigma}$.

   (c) Sketch the curve and shade the appropriate areas.

   (d) Use Table B or a software application to determine the AUC.

8. To determine **values that correspond to Normal probabilities**:

   (a) State the problem.

   (b) Use Table B or a software application to look up the z_p value–associated desired probability.

   (c) Sketch the curve.

   (d) Unstandardize the value: $x = \mu + \sigma z_p$.

9. Major departures from Normality in data can often be detected by visually inspecting histograms and/or **Q-Q plots**.

Vocabulary

Area under the curve	Standard Normal table (z table)
Expected z-scores	z-percentile (z_p)
Normal probability (Q-Q) plots	z-score
Points of inflection	μ (mean)
Probability density function (*pdf*)	σ (standard deviation)
Standardize	68–95–99.7 rule

Review Questions

7.1 Fill in the blank: Normal distributions are centered on the value of _____.

7.2 What is an inflection point?

7.3 Fill in the blank: Normal curves have inflection points that are one _____ above and below μ.

7.4 Fill in the blanks: ____% of area under a Normal curve is within $\mu \pm \sigma$, ____% is within $\mu \pm 2\sigma$, and ____% is within $\mu \pm 3\sigma$.

7.5 The total area under a Normal curve sums to exactly _____.

7.6 Normal *pdf*s have two parameters. Name them.

7.7 What parameter controls the location of the Normal curve?

7.8 What parameter controls the spread of the Normal curve?

7.9 Fill in the blank: The area under the curve between any two points on a Normal density curve corresponds to the _____ of values within that range.

7.10 Fill in the blank: The area under the curve to the left of a point on a Normal density corresponds to the _____ probability of that value.

7.11 How many different Normal distributions are there?

7.12 How many different Standard Normal distributions are there?

7.13 What does "$X \sim N(\mu, \sigma)$" mean?

7.14 What is the mean of the Standard Normal distribution?

7.15 What is the standard deviation of the Standard Normal distribution?

7.16 Fill in the blank: The Standard Normal random variable is often referred to as a ____ variable. (Answer is a letter.)

7.17 Fill in these blanks: $Z \sim N(\text{____}, \text{____})$

7.18 Use the 68–95–99.7 rule to determine $\Pr(Z < -2)$. (Z table not required.)

7.19 Use the 68–95–99.7 rule to determine $\Pr(Z > -2)$. (Z table not required.)

7.20 In the notation z_p, what does the subscript p represent?

7.21 $z_{0.50} = ?$ (Z table not required; use your knowledge of Standard Normal curves.)

7.22 $z_{0.025} = ?$ (Z table not required.)

7.23 $z_{0.16} = ?$ (Z table not required.)

7.24 $z_{0.84} = ?$ (Z table not required.)

Exercises

7.12 *Standard Normal proportions.* Use Table B to find the proportion of a Standard Normal distribution that is:

    (a) below -1.42

    (b) above 1.42

    (c) below 1.25

    (d) between 1.42 and 1.25

7.13 *Alzheimer brains.* The weight of brains from Alzheimer cadavers varies according to a Normal distribution with mean 1077 g and standard deviation 106 g.[1] The weight of an Alzheimer-free brain averages 1250 g. What proportion of brains with Alzheimer disease will weigh more than 1250 g?

7.14 *Coliform levels.* Water samples from a particular site demonstrate a mean coliform level of 10 organisms per liter with standard deviation 2. Values vary according to a Normal distribution. What percentage of samples will contain more than 15 organisms?

7.15 *Z-percentiles.* Find the following z-percentiles:

    (a) $z_{0.10}$

    (b) $z_{0.35}$

    (c) $z_{0.74}$

    (d) $z_{0.85}$

    (e) $z_{0.999}$

7.16 *Gestation (99th percentile).* Recall that gestation in uncomplicated human pregnancies from conception to birth varies according to a Normal distribution (approximately so) with a mean of 39 weeks and standard deviation of 2 weeks (Figure 7.3). What is the 99th percentile on this distribution? That is, what gestational length is greater than or equal to 99% of the other normal gestations?

---

[1]Dusheiko, S. D. (1973). [The pathologic anatomy of Alzheimer's disease]. *Zhurnal nevropatologii i psikhiatrii imeni S.S. Korsakova, 73*(7), 1047–1052.

7.17 *Gestation less than 32 weeks.* Recall that uncomplicated human gestational length is approximately Normally distributed with $\mu = 39$ weeks and $\sigma = 2$ weeks. What percentage of gestations are less than 32 weeks long?

7.18 *Coliform levels (90th percentile).* Exercise 7.14 addressed coliform levels in water samples from a particular site. The coliform levels were assumed to vary according to a Normal distribution with a mean of 10 organisms per liter and a standard deviation of 2 organisms per liter. What is the 90th percentile on this distribution?

7.19 *A six-foot seven-inch tall man.* Have you ever wondered why a man who is $6' 7''$ tall ($79''$) seems so much taller than a man who is $5' 10''$ ($70''$) even though he is only 13% taller in relative terms? [$(79'' - 70'')/70'' = 0.13 = 13\%$.] Let us assume that male height is Normally distributed with $\mu = 70$ inches and $\sigma = 3$ inches. What proportion of men are $5' 10''$ or taller? What proportion of men are $6' 7''$ or taller?

7.20 *College entrance exams.* The SAT and ACT are standardized tests for college admission in the United States. Both tests include components that measure reading comprehension. Suppose that SAT critical reading scores are Normally distributed with a mean of 510 and standard deviation of 115. In contrast, ACT reading scores are Normally distributed with a mean of 20.5 and standard deviation of 5. Sam takes the SAT reading test and scores 660. Dave takes the ACT test and scores 28. Who had the superior score, Sam or Dave?

7.21 *|Z| ≥ 2.56.* What proportion of Standard Normal Z-values are greater than 2.56? What proportion are less than -2.56? What proportion are either below -2.56 or above 2.56?

7.22 *BMI.* Body mass index (BMI) is equal to "weight in kilograms" divided by "height in meters squared." A study by the National Center for Health Statistics suggested that women between the ages of 20 and 29 in the United States have a mean BMI of 26.8 with a standard deviation of 7.4. Let us assume that these BMIs are Normally distributed.

(a) A BMI of 30 or greater is classified as being overweight. What proportion of women in this age range are overweight according to this definition?

(b) A BMI less than 18.5 is considered to be underweight. What proportion of women are underweight?

7.23 *MCATs.* Suppose that scores on the biological sciences section of the Medical College Admissions Test (MCAT) are Normally distributed with a mean of 9.2 and standard deviation of 2.2. Successful applicants to become medical students had a mean score of 10.8 on this portion of the test. What percentage of applicants had a score of 10.8 or greater?

8 | Introduction to Statistical Inference

Statistical **inference** is the act of using data in a particular sample to make generalizations about the population from which it came. R. A. Fisher[a] put it this way:

> For everyone who does habitually attempt the difficult task of making sense of figures is, in fact, assaying a logical process of the kind we call induction, in that he is attempting to draw inferences from the particular to the general; or, as we more usually say in statistics, from the sample to the population.[b]

Before addressing specific inferential techniques, certain concepts about the variability of means and proportions must be established.

■ 8.1 Concepts

Sampling Variability

Suppose you want to learn about the prevalence of asthma in a population. You conduct a survey of 100 individuals selected at random and identify three asthmatics in the bunch. You conclude that the *sample* has a prevalence of 3 per 100 (3%) but you are still uncertain about the prevalence of asthma in the *population*. You are aware that the next 100 individuals selected at random from the same population may have 0, 1, 2, 3, or some other number of asthmatics. How are you going to deal with the element of chance introduced by sampling?

A similar problem occurs when performing an experiment. You randomly assign a treatment to half the subjects. If the trial is properly randomized, each group will represent a simple random sample of available study participants. The laws of chance will encourage groups to resemble each other but do not guarantee perfect comparability. Therefore, differences observed at the end of the trial can be due to the effects of the treatment *or* random differences in the groups. How can you assess the role of chance in this situation?

---

[a] Ronald Aylmer Fisher (1890–1962)—"The father of modern statistics"—Made discoveries in statistical theory, experimental design, analysis of variance, maximum likelihood estimation, exact testing, and applied biostatistics.
[b] Box, J. F. (1978). *R. A. Fisher, the Life of a Scientist.* New York: John Wiley, p. 448.

Parameters and Statistics

We start by distinguishing between parameters and statistics. The term **parameter** refers to a numerical characteristic of a *statistical population*. In contrast, the term **statistic** refers to a value calculated in a *sample*. We are particularly interested in a type of statistic known as an **estimate**. A statistical estimate is a direct reflection of an underlying population parameter.

The **statistical population** is the entire collection of values (the "universe" of values) about which we want to draw conclusions. Because populations are often very large, we often draw a subset or **sample** from the population from which to draw our conclusions. For example, if we wish to draw conclusions about U.S. residents, it would not be practical to study all 315 million-plus U.S. residents. Instead, a small proportion of the population would be selected to serve as a sample for the study.

In addition to studying large groups of individuals, we are also often interested in studying natural phenomena. When this is the case, we must imagine an infinitely large **hypothetical population** of potential values that could ensue following the study. Thus, in attempting to draw a conclusion from the study, the current data should be viewed as one potential result out of an infinite number of similar observations. We thereby consider a population of potential results, not just the single set of results from the current study.

ILLUSTRATIVE EXAMPLE

Parameters and statistics (Body weight). The prevalence of being overweight and obese has increased in the United States. The National Health Exam Survey in the 1960s estimated that the average body weight of men was about 166.3 pounds. About 30 years later, the National Health and Nutrition Exam Survey estimated an average weight of 189.8.[c] Both surveys were based on samples of between 3000 and 4000 men. The numbers 166.3 and 189.8 are *statistics* because they are based on a sample of the population. These statistics estimate population means (parameters) at the time they were taken, but the actual values of the parameters are not truly known.

The distinction between statistics and parameters is essential to the understanding of statistical inference. To reflect this distinction, we use different symbols to represent each. For example, the population mean (parameter) is denoted μ, while the sample mean (statistic) is denoted $\bar{x}$. As another example, the population proportion is denoted p, while the sample proportion is denoted $\hat{p}$.

---

[c] Ogden, C. L., Fryar, C. D., Carroll, M. D., & Flegal, K. M. (2004). Mean body weight, height, and body mass index, United States 1960–2002. *Advance Data* (347), 1–17. See Table 6.

Parameters and statistics (Youth risk behavior). The Youth Risk Behavior Surveillance System monitors six categories of health behaviors in U.S. youth and adolescents. Many types of behaviors that contribute to adverse health consequences are monitored. As examples, the proportion of high school students who had ridden with a driver who had been drinking alcohol during the preceding 30-day period was 30.2%, the proportion who had smoked cigarettes was 21.9%, and the proportion who were overweight was 13.5%.[d] These values are statistical estimates of population parameters.

Note that population parameters are constants, while sample statistics are random variables. The values of parameters do not change from sample to sample. In contrast, statistics change whenever the population is resampled. The following table summarizes distinctions you should keep in mind when learning about parameters and statistics:

	Parameters	Statistics
Source	Population	Sample
Able to calculate?	Not usually	Yes
Random variable?	No (mathematical constant)	Yes
Type of notation	Often uses Greek characters (e.g., μ)	Roman letters, sometimes with overhead "hats" (e.g., $\bar{x}$)

Exercises

8.1 *Breast cancer survival.* A study of 1225 incident breast cancer cases found that survival varied greatly by stage of disease at diagnosis. The median 7-year survival rate for stage I breast cancer was **92%**. For stage II breast cancer, the 7-year survival rate was **71%**. For stage II breast cancer, the survival rate was **39%**. For stage IV, the survival rate was **11%**. Say whether each of the boldface numbers in this exercise is a parameter or a statistic.

8.2 *Pancreatic cancer survival.* Whether a value is a parameter or a statistic often depends on how we state the research question. For example, suppose that we review all the pancreatic cancer cases in a specified region and find that in those cases where surgical resection was performed, the median survival time was 18 months.

---

[d] Grunbaum, J. A., Kann, L., Kinchen, S., Ross, J., Hawkins, J., Lowry, R., et al. (2004). Youth risk behavior surveillance—United States, 2003. *MMWR Surveillance Summary, 53*(2), 1–96.

(a) If we wish to use the number "18 months" to make a statement about the median survival time for cases in our region during the time interval of the study, would this number represent a parameter or a statistic?

(b) Now suppose we want to use this number to make a statement about the effect of surgical resection in other similar cases. Would "18 months" now represent a parameter or a statistic?

8.3 *Parameter or statistic?* Say whether each of the boldface numbers is a parameter or a statistic.

(a) *Insured?* A data set based on 168 hospital discharge summaries shows that **20%** of patients were uninsured. (The review takes place in a large referral hospital.)

(b) *Percent African American.* Data from the complete enumeration of a standard metropolitan area (SMA) indicate that **12%** of the inhabitants are African American. A telephone survey based on random-digit–dialing of individuals from this SMA found that **8%** of the respondents were African American.

(c) *Cost of antihypertensive medication.* Data from 10 online pharmacies reveal that the average cost of a 1-month supply of a particular medication is **$31.20** with standard deviation **$7.75**. Data from 10 community pharmacies shows an average cost of **$33.18** with standard deviation **$7.88**.

■ 8.2 Sampling Behavior of a Mean

How well does a given sample mean $\bar{x}$ reflect the underlying population mean μ? To address this question, we must consider what would happen if you took repeated simple random samples each of the same size n from the source population. The $\bar{x}$s from these repeated samples will vary from sample to sample and ultimately form a distribution of their own.

> The **sampling distribution of a mean** is the *hypothetical* distribution of means from all possible samples of size n taken from the same population.

In practice we would not take repeated samples from the same population to build the sampling distribution of a mean. This would be impractical. We could, however, *simulate* this experience.

Simulation Experiment

This simulation experiment starts with a population of $N = 10,000$ body weights. This population distribution, which was constructed to mimic the body weight distribution

TABLE 8.1   Five samples selected at random from the population SIMULATION.

i	Sample 1	Sample 2	Sample 3	Sample 4	Sample 5
1	162	203	153	171	227
2	199	178	217	156	186
3	155	186	151	151	180
4	145	144	146	190	186
5	180	135	160	183	191
6	170	199	136	162	214
7	145	160	138	145	236
8	182	153	179	151	220
9	187	172	163	175	175
10	171	174	149	182	127
$\bar{x}$	169.6	170.4	159.2	166.6	194.2
s	18.0	22.5	23.9	15.7	31.8

of 20- to 29-year-old men in the 1975 U.S. population,[e] has a log-normal distribution (positive skew) with $\mu = 173$ pounds and $\sigma = 30$ pounds. **Graph A** in Figure 8.1 shows the population distribution.

We take repeated independent SRSs, each of size $n = 10$, from this population. Table 8.1 lists our first five samples. The means from these samples vary considerably. (This should encourage us to view any single sample mean $\bar{x}$ as an example of any of a series of means that could have been drawn from this population.)

The experiment continues by taking an additional 9995 SRSs, each of $n = 10$, from the population "sampling frame." We calculate the $\bar{x}$ for each of these samples and create with them a histogram to show their distribution.[f] **Graph B** in Figure 8.1 shows this "sampling distribution of means."

What can we learn by comparing Graphs A and B from this simulation experiment? Three general findings emerge:

1. The sampling distribution of $\bar{x}$s is more symmetrical than the distribution of the population.

2. Both the population distribution and the sampling distribution of $\bar{x}$s are centered on μ.

3. The sampling distribution of $\bar{x}$s is less spread out than the population distribution.

---

[e] Burmaster, D. E., & Crouch, E. A. (1997). Log-normal distributions for body weight as a function of age for males and females in the United States, 1976–1980. *Risk Analysis, 17*(4), 499–505. The natural log of the weight (in kilograms) is assumed to vary according to a Normal distribution with mean 4.35 and standard deviation 0.17. Ten thousand data points were randomly generated with SPSS version 11 using RV.NORMAL(4.35,0.17) on March 2, 2006. Data are stored in the file SIMULATION.* as the variable LBS.

[f] This is not the full sampling distribution of the mean—there are $_{10,000}C_{10} \approx 2.74 \times 10^{33}$ possible samples of $n = 10$ from a population of 10,000—but it is a good representation of the distribution.

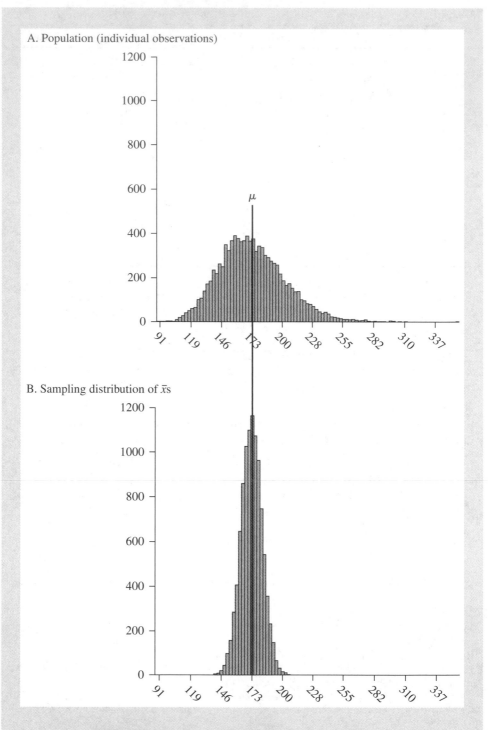

FIGURE 8.1   Results from the sampling simulation experiment.

Finding 1 is due to the **central limit theorem**, which states that the sampling distribution of $\bar{x}$s tends toward Normality even when the underlying population is not Normal. The influence of the central limit theorem becomes strong as n gets large. In fact, when the sample is large (say, at least 100), we are almost assured that the sampling distribution of $\bar{x}$s will be Normal.

Finding 2 states that the expected value of the sampling distribution of $\bar{x}$s is population mean μ. This is not to suggest that all sample means will equal the population mean; some will be greater than μ and some will be less than μ. However, in the long run, these discrepancies will balance out so that the average of the $\bar{x}$s will equal μ. This indicates that the sample mean is an **unbiased** estimator of the population mean.

Finding 3 shows that the sampling distribution of a mean clusters more tightly around the value of μ than the distribution of individual values. That is, the sample means are less variable than individual observations. In fact, the standard deviation of a sample mean is inversely proportional to the square root of its sample size. This is the **square root law**: When individual observations have standard deviation σ, the sample mean has standard deviation $\dfrac{\sigma}{\sqrt{n}}$. Because the standard deviation of $\bar{x}$ is such an important statistic, it has been given a name: it is called the **standard error of the mean**.

$$\sigma_{\bar{x}} \equiv \mathrm{SE}_{\bar{x}} = \frac{\sigma}{\sqrt{n}}$$

"Standard deviation of a mean" ($\sigma_{\bar{x}}$) and "standard error of a mean" ($\mathrm{SE}_{\bar{x}}$) are synonyms.[g]

The Sampling Distribution of a Mean

Putting together the findings from the simulation experiment permits us to say that the sampling distribution of a mean tends to be Normal, has expected value μ, and has standard deviation $\dfrac{\sigma}{\sqrt{n}}$. Using the notation established in the prior chapter:

$$\bar{x} \sim \mathrm{N}\left(\mu, \frac{\sigma}{\sqrt{n}}\right)$$

---

[g] An old-fashioned term for "standard error of the mean" is "probable error of the mean."

ILLUSTRATIVE EXAMPLE

Sampling distribution of a mean (WAIS). Wechsler Adult Intelligence Scale (WAIS) scores vary according to a Normal distribution with $\mu = 100$ and $\sigma = 15$. What can we say about the sampling distribution of a mean based on an SRS of 10 such scores?

Given what we know about sampling theory, the sampling distribution of this mean will vary according to a Normal distribution with mean $\mu = 100$ with standard deviation given by $SE_{\bar{x}} = \dfrac{\sigma}{\sqrt{n}} = \dfrac{15}{\sqrt{10}} \approx 4.74$. In symbols, $\bar{x} \sim N(100, 4.74)$.

Figure 8.2 compares the population distribution of WAIS scores with the sampling distribution of $\bar{x}$s based on $n = 10$. The distributions have the same mean, but the sampling distribution of the $\bar{x}$s is much skinnier.

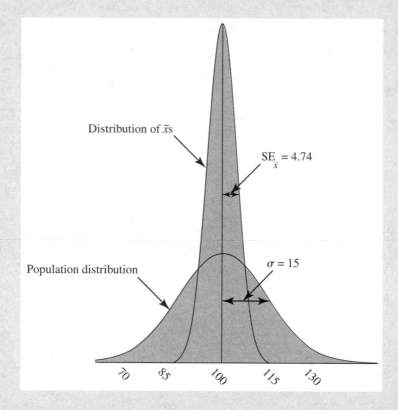

FIGURE 8.2   Sampling distribution of the mean based on $n = 10$ compared to distribution of population values, Wechsler Adult Intelligence Scale scores.

By applying what we know about Normal probability distributions, we can now make inferences about μ based on findings in a sample. For example, because of mean WAIS scores based on $n = 10$ vary according to a Normal distribution with $\mu = 100$ and $\sigma_{\bar{x}} = 4.74$, we can predict that 95% of the $\bar{x}$s will fall in the range $100 \pm (2)(4.74) = 90.6$ to 109.4 (based on the 68–95–99.7 rule).

We can also infer the probability of observing means in any range using the method of calculating Normal probabilities presented in Section 7.2. For an SRS of WAIS scores based on $n = 10$, for instance, we can ask "What is the probability of getting an $\bar{x}$ less than 90?" Here's the four-step solution:

1. **State** the problem. We start with the assumption that $\bar{x} \sim N(100, 4.74)$ and seek to determine $\Pr(\bar{x} \leq 90)$.

2. **Standardize** the mean. We standardize the mean of 90 as follows:

$$z = \frac{\bar{x} - \mu}{\sigma_{\bar{x}}} = \frac{90 - 100}{4.74} = -2.11.$$

3. **Sketch** the Normal curve and shade the area associated with $\Pr(Z \leq -2.11)$.

4. **Use Table B** to determine $\Pr(Z \leq -2.11) = 0.0174$. Therefore, the probability of observing a mean score of 90 or less under these conditions is 0.0174.

The Effect of Increasing the Sample Size

Because of the square root law, the standard error of the mean gets smaller and smaller as the sample size gets larger and larger. Therefore, sample means based on large n are more likely to fall close to the true value of μ than means based on small n (all other things being equal).

ILLUSTRATIVE EXAMPLE

Standard error of the mean and sample size (WAIS). What effect does increasing the sample size have on the precision of a mean? We have established that Wechsler Adult Intelligence Scale (WAIS) scores vary according to a Normal distribution with $\mu = 100$ and $\sigma = 15$. Let us consider the standard errors of means for SRSs of WAIS scores with $n = 1$, $n = 4$, and $n = 16$.

- The standard error of the mean when $n = 1$ is $\text{SE}_{\bar{x}} = \dfrac{\sigma}{\sqrt{n}} = \dfrac{15}{\sqrt{1}} = 15$.

- The standard error of the mean when $n = 4$ is $\text{SE}_{\bar{x}} = \dfrac{15}{\sqrt{4}} = 7.5$.

- The standard error of the mean when $n = 16$ is $\text{SE}_{\bar{x}} = \dfrac{15}{\sqrt{16}} = 3.75$.

Each time, quadrupling the sample size cuts the $\text{SE}_{\bar{x}}$ in half.

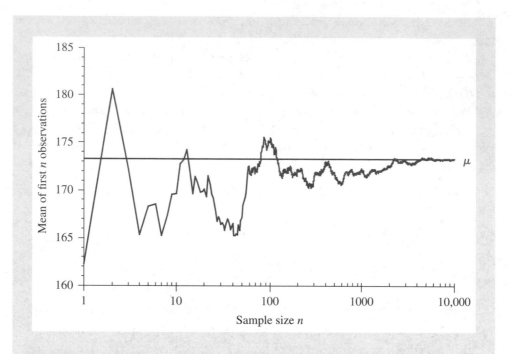

FIGURE 8.3   As the sample gets larger and larger, the sample mean tends to get closer and closer to the true value of μ (law of large numbers).

Figure 8.3 depicts the results of an experiment in which observations are drawn at random from a population in which $\mu = 173.30$. In this experiment, the first observation was 162. The next observation was 199, so the average of the first two observations $= (162 + 199)/2 = 180.5$. We continued to select individuals at random and determine the mean after each draw.

Initially, sample means bounce around with each additional draw. After a large number of draws, $\bar{x}$ begins to stabilize, eventually honing in on the true value of μ. This phenomenon is the **law of large numbers**, which states that as an SRS gets larger and larger, its sample mean ($\bar{x}$) is likely to get closer and closer to the true value of μ.

Exercises

8.4   *Tiny population.* A tiny population consists of the following values:

1   3   5   7   9

This population has $\mu = 5$ and $\sigma = 2.8$.[h] The distribution is flat, as seen in this stemplot:

---

[h] Population standard deviation $\sigma = \sqrt{\frac{\Sigma(x - \mu)^2}{n}}$. This differs from the sample standard deviation formula (which has a denominator of $n - 1$) because it does not lose a degree of freedom in estimating μ.

```
* | 1
T | 3
F | 5
S | 7
. | 9
×10
```

A total of $_5C_2 = 10$ unique samples with $n = 2$ that can be selected from this population. These samples are:

Sample 1: {1, 3}   Sample 2: {1, 5}   Sample 3: {1, 7}   Sample 4: {1, 9}
Sample 5: {3, 5}   Sample 6: {3, 7}   Sample 7: {3, 9}   Sample 8: {5, 7}
Sample 9: {5, 9}   Sample 10: {7, 9}

Calculate the mean of each of these samples and then construct a stemplot of the means. Compare the location, spread, and shape of the distribution of sample means to that of the original population.

8.5 *Survey of health problems.* A survey selects an SRS of $n = 500$ people from a town of 55,000. The sample shows a mean of 2.30 health problems per person (standard deviation $= 1.65$). Based on this information, say whether each of the following statements is *true* or *false*. Explain your reasoning in each instance.

(a) The standard deviation of the sample mean is 0.074.

(b) It is reasonable to assume that the number of health problems per person will vary according to a Normal distribution.

(c) It is reasonable to assume that the sampling distribution of the mean will vary according to a Normal distribution.

8.6 *Cholesterol in undergraduate men.* Suppose the distribution of serum cholesterol values in undergraduate men is approximately Normally distributed with mean $\mu = 190$ mg/dL and standard deviation $\sigma = 40$ mg/dL.

(a) What is the probability of selecting someone at random from this population who has a cholesterol value that is less than 180?

(b) You take a simple random sample of $n = 49$ individuals from this population and calculate the mean cholesterol of the sample. Describe the sampling distribution of $\bar{x}$.

(c) Regarding the mean derived from a sample of $n = 49$, what is the probability of getting a sample mean that is less than 180? [Determine $\Pr(\bar{x} \leq 180)$.]

8.7 *Repeated lab measurements.* A laboratory kit states that the standard deviation of its results can be expected to vary with $\sigma = 1$ unit. A lab technician takes four measurements using this kit.

(a) What is the value of the standard deviation of the $\bar{x}$ of the four measurements taken with this kit?

(b) Explain to a lay person the advantage of reporting the average of four measurements rather than using a single measurement.

(c) How many times must we repeat the measurement before we obtain a standard deviation of the mean measurement of 0.2 units? [Hint: Rearrange the $SE_{\bar{x}}$ formula to solve for n.]

8.8 *Sampling behavior of a mean.* Suppose you could take all possible samples of size $n = 25$ from a Normal population with mean $\mu = 50$ and $\sigma = 5$.

(a) Sketch or describe in words the sampling distribution of $\bar{x}$. Identify landmarks on the horizontal axis of this sampling distribution that are ± 1 and ± 2 standard deviations around its center.

(b) Would you be surprised to find a sample mean of 47 under these conditions? Explain your reasoning.

(c) Would you be surprised to find a sample mean of 51.5? Explain your reasoning.

■ 8.3 Sampling Behavior of a Count and Proportion

Section 8.2 began with the question "How well does a given sample mean $\bar{x}$ reflect underlying population mean μ?" Similarly, we can ask "How well does a given sample proportion $\hat{p}$ reflect underlying population proportion p?" Sample proportion $\hat{p}$ is analogous to sample mean $\bar{x}$, while population proportion p is analogous to population mean μ.

To answer this question, consider taking an SRS of size n from a population in which $p \times 100\%$ of the individuals are classified as "successes." The number of successes in a sample (call this x) will vary according to a binomial distribution with parameters n and p. Recall from Chapter 6 that:

- The random number of successes X in n independent Bernoulli trials follows a binomial distribution with parameters n and p.[i]
- The probability of seeing x successes out of n is $\Pr(X = x) = {}_nC_x p^x q^{n-x}$ where $q = 1 - p$.
- The expected number of successes $\mu = np$.
- The standard deviation $\sigma = \sqrt{npq}$.

---

[i] X denotes the *random variable*; x denotes the *observed* number of successes in a sample.

Sample proportion $\hat{p}$ is a mathematical reexpression of the random number of success (x) as $\dfrac{x}{n}$. We can show that the sample proportion will have a binomial distribution with $\mu = p$ and $\sigma = \sqrt{\dfrac{pq}{n}}$.

ILLUSTRATIVE EXAMPLE

Sampling distribution of a count and proportion (Smoking survey). We take an SRS of $n = 10$ from a population in which 20% of individuals smoke. The number of smokers in SRSs from this population will vary according to a binomial distribution with parameters $n = 10$ and $p = 0.2$. The proportion of success in a sample will vary between 0 and 1 with expectation p. Figure 8.4 depicts this distribution graphically. The horizontal axis is labeled with both the number (x) and proportion $(\hat{p})$ of successes that could be observed. Probabilities are based on the binomial distribution $X \sim b(10, 0.2)$. For example, the probability of seeing no (0) smokers $\Pr(X = 0) = {}_{10}C_0\,(0.2^0)\,(0.8^{10-0}) = 1(1)(0.1074) = 0.1074$. Therefore, 10.74% of the samples will show no smokers. By the same token, 26.84% will show one smoker, 30.20% will show two smokers, and so on. This sampling distribution will allow us to make inferences about observed counts and

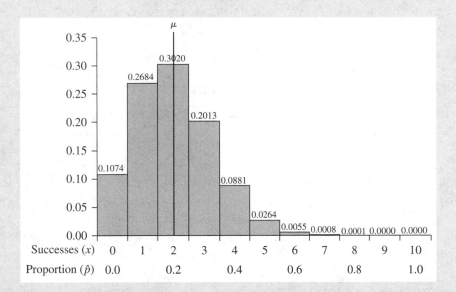

FIGURE 8.4 Sampling distribution of a count and proportion, illustrative example.

continues

proportions. For example, in this illustration, we can say that most of the samples will have zero, one, two, three, or four smokers, corresponding to proportions of 0.0, 0.1, 0.2, 0.3, and 0.4. It would be surprising to see five or more smokers in a sample under these conditions, because this will occur so rarely.

The Normal Approximation to the Binomial

Figure 8.5 displays sampling distributions of three different binomial random variables. In each instance, population proportion $p = 0.2$. Figure A shows the binomial distribution for $n = 5$ and $p = 0.2$. Figure B shows the binomial distribution for $n = 10$ and $p = 0.2$. Figure C shows the binomial distribution for $n = 100$ and $p = 0.2$.

The first two distributions have positive skews. This is typical of binomial random variables when n is small. However, the third distribution is symmetrical and bell shaped. In fact, when n is large, binomial random variables take on the characteristics of Normal random variables. This phenomenon is called the **Normal approximation to the binomial**. As a rough guide, we can rely on the Normal approximation to the binomial when $npq \geq 5$ (let us call this the "npq rule") so that the count of successes in such samples will be approximately Normal with $\mu = np$ and $\sigma = \sqrt{npq}$ and the sample proportion $\hat{p}$ will be approximately Normal with $\mu = p$ and $\sigma = \sqrt{\frac{pq}{n}}$.

> When n is large X, $\mathrm{N}\left(np, \sqrt{npq}\right)$ and $\hat{p} \sim \mathrm{N}\left(p, \sqrt{\frac{pq}{n}}\right)$.

Because of the Normal approximation to the binomial, we can use the 68–95–99.7 rule to make inferences about the counts and proportions in large samples. In the following illustrative example, for instance, we know that 95% of the time the sample counts will fall within two standard deviations of the mean—this range is $20 \pm (2)(4) = 20 \pm 8 = 12$ to 28. Equivalently, we can say that the sample proportion will fall in the range $0.2 \pm (2)(0.04) = 0.2 \pm 0.08 = 0.12$ to 0.28.

We can also use the fact that $\dfrac{x - \mu}{\sigma}$ will have a Standard Normal distribution. For example, in this illustration, we can determine the probability of seeing 30 or more smokers as $\Pr\left(Z \geq \dfrac{30 - 20}{4}\right) = \Pr(Z \geq 2.5) = 0.0062$.

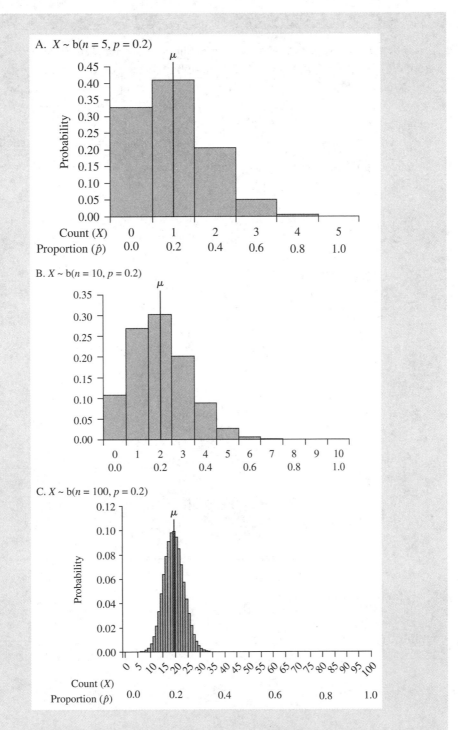

FIGURE 8.5   Three binomial distributions. Figure C is approximately Normal.

ILLUSTRATIVE EXAMPLE

Normal approximation to the binomial. Suppose we take an SRS of $n = 100$ from a population in which 20% of the individuals smoke. Because $npq = (100)(0.2)(0.8) = 16$, which is greater than 5, the Normal approximation to the binomial may be used. Therefore, the count of smokers in samples will vary approximately according to a Normal distribution with $\mu = np = (100)(0.2) = 20$ and $\sigma = \sqrt{npq} = \sqrt{100 \cdot 0.2 \cdot 0.8} = 4$; in symbols, $X \sim N(20, 4)$. In addition, sample proportion $\hat{p}$ will vary according to a Normal distribution with $\mu = p$ and $\sigma = \sqrt{\dfrac{pq}{n}} = \sqrt{\dfrac{(0.20)(0.80)}{100}} = 0.04$; in symbols, $\hat{p} \sim N(0.2, 0.04)$. Figure 8.6 depicts this sampling distribution graphically.

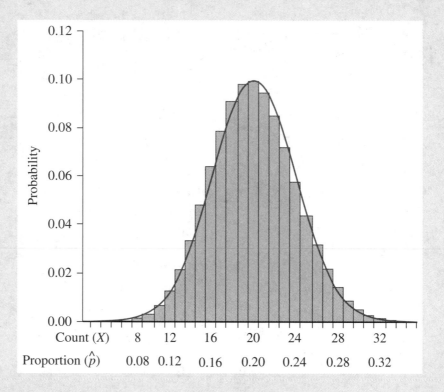

FIGURE 8.6   Normal approximation to the binomial, illustrative example.

Exercises

8.9 *Pediatric asthma survey, n = 50.* Suppose that asthma affects 1 in 20 children in a population. You take an SRS of 50 children from this population. Can the Normal approximation to the binomial be applied under these conditions? If not, what probability model can be used to describe the sampling variability of the number of asthmatics?

8.10 *Pediatric asthma survey, n = 50, continued.* In the survey introduced in Exercise 8.9, what is the probability of observing at least five cases in a sample?

8.11 *Pediatric asthma survey, n = 250.* Suppose we redo the survey addressed in the prior two exercises, now using $n = 250$. Can we use the Normal approximation to the binomial in this instance? What is the probability of seeing at least 25 cases in a sample?

Summary Points (Introduction to Statistical Inference)

1. A **parameter** is a number that characterizes a *pdf*, *pmf*, or population distribution. For example, population mean μ is a parameter.

2. An **estimator** is a statistic calculated from data that reflects the value of a population parameter. For example, sample mean $\bar{x}$ is the point estimator of population mean μ.

3. **Statistical inference** is the process of using sample statistics to generalize about population parameters with calculated degree of certainty.

4. When conducing statistical inference, be aware of these three different types of distributions:

   (a) **Population distributions** describe the frequency of values in a *pdf*, *pmf*, or population.

   (b) **Sample distributions** describe the frequency of values in a data set.

   (c) **Sampling distributions of statistics** describe the hypothetical frequency distribution of a statistic derived from all possible samples of size n taken from a population.

5. The standard deviation of the sampling distribution of $\bar{x}$ is often referred to as the **standard error of the mean** and is denoted with either $\sigma_{\bar{x}}$ or $SE_{\bar{x}}$. This statistic is an estimate of the sample mean's precision as an estimate of the population mean.

6. The formula for the standard error of the mean is $SE_{\bar{x}} = \dfrac{\sigma}{\sqrt{n}}$. Notice that the standard error of the mean is inversely proportional to the square root of the sample size. This is known as the **square root law**.

7. When taking an SRS from a Normal distribution, the sampling distribution of the mean is Normal with an expected value of μ and standard deviation equal to the standard deviation of the population divided by the square root of the sample size:

$$\bar{x} \sim N\left(\mu, \frac{\sigma}{\sqrt{n}}\right)$$

8. When sample size n is large, the sampling distribution of $\bar{x}$ tends toward Normality even when the population is not Normal. This phenomenon is known as the **central limit theorem**.

9. The **count of "successes"** (X) based in an SRS of size n will follow a binomial distribution with parameters n and p: $X \sim b(n, p)$.

10. In large samples, the count of successes in an SRS of size n can be modeled with this **Normal approximation to the binomial**: $X \sim N\left(np, \sqrt{npq}\right)$.

Vocabulary

Central limit theorem	Sampling distribution of a mean
Estimate	Square root law
Inference	Standard error of the mean ($\sigma_{\bar{x}}$ or $SE_{\bar{x}}$)
Law of large numbers	Standard error of the proportion
Normal approximation to the binomial	Statistic
Parameter	
Sampling distribution of a count (or proportion)	

Review Questions

8.1   What is *statistical inference*?

8.2   What is a *statistical population*?

8.3   What is a *sample*?

8.4   Suppose you take an SRS from a population and find that 28% of the people in the study have BMIs in excess of 30. You redo the study the following day in the same population and find that this time 35% of the people in the survey have this characteristic. What is the most plausible explanation for the inconsistent results?

8.5   Select the best response: Which of these is a parameter?

(a) $\bar{x}$

(b) s

(c) μ

8.6 What symbol is used to represent the population mean?

8.7 What symbol is used to represent the sample mean?

8.8 Is the population mean a statistic?

8.9 Is the sample mean a statistic?

8.10 Fill in the cells in this table with these choices: Population, Sample, Not usually, Yes, Constant, Random variable, Greek character, or Roman character sometimes modified with overhead "hats."

	Parameters	Statistics
Source		
Calculated?		
Mathematical		
Notation		

8.11 What does it mean when we say that $\bar{x}$ is an unbiased estimator of μ?

8.12 What is a *sampling distribution of a mean*?

8.13 What is the name of the statistical principle that indicates that sampling distributions of means tend toward Normality even when the population is not normal when the sample size is large?

8.14 What is the name of the statistical principle that states "the value of $\bar{x}$ will approach the value of μ as n gets larger and larger"?

8.15 What is another name for "the standard deviation of the sample mean"?

8.16 The standard error of the mean is inversely proportional to the _____ _____ of the sample size. (Two words.)

8.17 Select the best response: The "standard error of the mean" is the standard deviation of the

(a) sample distribution.

(b) population distribution.

(c) sampling distribution of the mean.

8.18 True or false? The precision of a sample mean as a reflection of μ is partially determined by the study's sample size.

8.19 Select the best response: The standard error of the mean is

(a) proportional to the sample size.

(b) inversely proportional to the sample size.

(c) inversely proportional to the square root of the sample size.

8.20 Complete this sentence: The point estimator of population proportion p is

_____.

8.21 The number of "successes" among n independent Bernoulli trials is governed by this *pmf*.

8.22 Select the best response: The binomial distribution can be approximated with a Normal distribution when the sample is

(a) small.

(b) moderate.

(c) large.

8.23 When is a sample is large enough to use a Normal approximation to a binomial?

(a) when $n \geq 5$

(b) when $npq \geq 5$

(c) when $npq < 5$

Exercises

8.12 *Fill in the blanks.* A particular random sample of n observations can be used to calculate a sample mean. We can determine the characteristics of the distribution of means derived by other samples of the size n taken from the same populations without taking additional samples. This distribution is called the sampling distribution of the (*a*) _____. It is centered on (*b*) _____ and has standard deviation equal to (*c*) _____. The theorem that postulates that the distribution will tend to be Normal is called the (*d*) _____.

8.13 *Fill in the blanks.* A particular random sample of n observations can be used to calculate a sample proportion. The count of successes in the sample will vary according to a (*a*) _____ probability distribution with parameters n and (*b*) _____. When the sample is large, the number of success will vary according to a Normal distribution with $\mu = $ (*c*) _____ and $\sigma = $ (*d*) _____. At the same time, the sampling distribution of proportion $\hat{p}$ in large samples will be Normally distributed with mean p with standard deviation equal to (*e*) _____.

8.14 *Undercoverage? Undercoverage* is a problem that occurs in surveys when some groups in the population are underrepresented in the sampling frame used to select the sample. We can check for undercoverage by comparing the sample with known facts about the population.

(a) Suppose we take an SRS of $n = 500$ people from a population that is 25% Hispanic. How many Hispanic are expected in a given sample?

(b) What is the standard deviation for the number of Hispanics in a sample?

(c) Can the Normal approximation to the binomial be used to help make probabilistic statements about samples from this population?

(d) Determine the probability that a sample contains 100 or fewer Hispanics under the stated conditions.

(e) Would you suspect undercoverage in a sample with 100 Hispanics? Explain your reasoning.

8.15 *Patient preference.* Ten people are given a choice of two treatments. Let p represent the proportion of patients in the patient population who prefer treatment A. Among the 10 patients asked, 7 preferred method A. Assuming there is no preference in the patient population (i.e., $p = 0.5$), calculate $\Pr(X \geq 7)$.

9 | Basics of Hypothesis Testing

This chapter introduces one of the two primary methods of statistical inference: **hypothesis testing** (also called **significance testing**); the other major method of statistical inference called **estimation**, is introduced in the next chapter. In both hypothesis testing and estimation, the population is sampled, data are used to calculate statistics, and statistics are used to help infer parameters (Figure 9.1).

Hypothesis testing uses a deductive procedure to judge claims about parameters. We break down the procedure into the following components:

A. State the statistical hypotheses in null and alternative forms.

B. Calculate the appropriate test statistic.

C. Convert the test statistic to a *P*-value.

D. Assess the statistical significance level of results.

E. Derive a conclusion in the context of the data and research question.

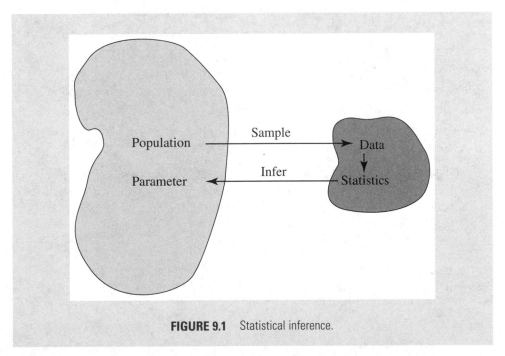

FIGURE 9.1 Statistical inference.

■ 9.1 The Null and Alternative Hypotheses

The first step in hypothesis testing is to state the hypotheses in null and alternative forms. The **null hypothesis** (H_0) is a statement of "no difference." The **alternative hypothesis** (H_a) contradicts the null hypothesis. The researcher then seeks evidence *against* the null hypothesis as a way of bolstering the alternative hypothesis.

ILLUSTRATIVE EXAMPLE

Null and alternative hypotheses (Body weight of 20–29-year-old U.S. men). In the late 1970s, the United States male population between 20 and 29 years of age had body weights with an approximate lognormal distribution with mean μ = 170 pounds and standard deviation σ = 40.[a] To test whether body weight has increased since that time, we take an SRS of men from the current population, calculate their mean weight, and compare it to the historical mean. Under the null hypothesis, there is no difference in the mean body weights between then and now. Let μ represent the current population mean. Therefore, the null hypothesis is H_0: μ = 170. Under the alternative hypothesis, the mean weight has increased. In other words, H_a: μ > 170 pounds.

This alternative hypothesis is one-sided, looking only for values larger than the hypothesized population mean. However, we may also consider a two-sided alternative that looks for both higher- and lower-than-expected population means. For this example, the two-sided alternative hypothesis would be H_a: $\mu \neq$ 170.

Notes

1. The correct form of the null hypothesis is H_0: μ = "some number." It would be incorrect to state H_0: $\bar{x}$ = "some number."

2. The null and alternative hypotheses are set up in such a way so that one or the other must be true and the other must be false.

3. The alternative hypothesis is closely aligned with the research hypothesis. The objective of the test is to seek evidence *against* the null hypothesis as a way of bolstering the alternative hypothesis, and thus the research hypothesis.

---

[a] Ogden, C. L., Fryar, C. D., Carroll, M. D., & Flegal, K. M. (2004). Mean body weight, height, and body mass index, United States 1960–2002. *Advance Data from Vital and Health Statistics* (347), 1–17; Table 7.

Exercises

9.1 *Misconceived hypotheses*. What is wrong with each of the following hypothesis statements?

(a) H_0: $\mu = 100$ vs. H_a: $\mu \neq 110$

(b) H_0: $\bar{x} = 100$ vs. H_a: $\bar{x} < 100$

(c) H_0: $\hat{p} = 0.50$ vs. H_a: $\hat{p} \neq 0.50$

9.2 *Hypothesis statements*. Set up null and alternative hypotheses for each of these claims. Use two-sided alternative hypotheses in each instance.

(a) A counselor claims that a new method of conflict resolution is less prone to interruption than an old method. Prior experience suggests that clients interrupted each other an average of 10 times per session.

(b) A drug company claims that a single dose of a new drug relieves symptoms for a longer period of time than the current formulation. The current formulation relieves symptoms for eight hours, on average.

(c) Americans gain an average of one pound per year as they age between the ages of 25 and 45. A public health campaign aims to decrease the amount of weight gained during this interval.

■ 9.2 Test Statistic

In the illustrative example concerning body weight, let us temporarily accept the notion that the null hypothesis is true and assume that $\mu = 170$ pounds. We are aware that repeated SRSs of a set size (say, $n = 64$) taken from such a population will yield different sample means. Figure 9.2 depicts this situation.

In this schematic, the first SRS derives a mean of 173 (pounds), the second derives a sample mean of 185, and so on. Even though the population mean is a constant 170 pounds, the $\bar{x}$s differ. The question becomes "At what point will $\bar{x}$ be sufficiently different from 170 to suspect that the claim H_0: $\mu = 170$ is false?" Inspecting the **sampling distribution** of $\bar{x}$ will provide some clues about how to answer this question.

Under the null hypothesis, the population distribution is lognormal with $\mu = 170$ and $\sigma = 40$. Based on what we know, the sampling distribution of $\bar{x}$ based on $n = 64$ will be approximately Normal (central limit theorem) with $\mu = 170$ and $\sigma_{\bar{x}} = \dfrac{\sigma}{\sqrt{n}} = \dfrac{40}{\sqrt{64}} = 5$ (square root law). Figure 9.3 is a schematic of this sampling distribution. The location of $\bar{x}$s of 173 and 185 are shown on its horizontal axis. The sample mean of 173 is fairly close to 170 and does *not* provide good evidence against H_0. In contrast, the $\bar{x}$ of 185 is far from 170, out in the far right-tail. Thus, the sample

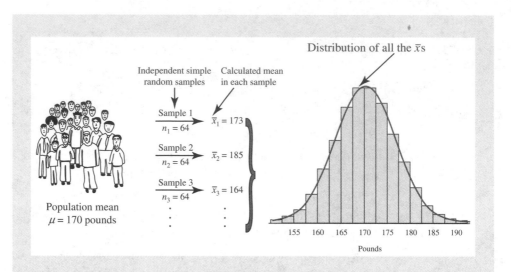

FIGURE 9.2   Sampling distribution of a mean. This model is based on taking all possible samples of $n = 64$ from a population with $\mu = 170$ and $\sigma = 40$.

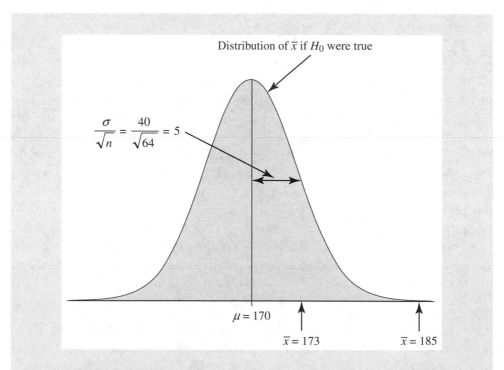

FIGURE 9.3   Sampling distribution of the mean for the illustrative example.

mean of 185 is unlikely to have come from this sampling distribution and therefore *does* provide good evidence against H_0.

The statistical distance of sample mean $\bar{x}$ from the hypothesized value of μ thus provides the weight of evidence against the null hypothesis. This statistical distance is reported in the form of the test statistic:

$$z_{\text{stat}} = \frac{\bar{x} - \mu_0}{\text{SE}_{\bar{x}}}$$

where μ_0 represents the value of the population mean specified by the null hypothesis and $\text{SE}_{\bar{x}} = \dfrac{\sigma}{\sqrt{n}}$. Large values of the z_{stat} represent large statistic distances and, thus, good evidence against H_0. Small values of the z_{stat} represent small statistical distances and, thus, weak evidence against H_0.

ILLUSTRATIVE EXAMPLE

z-statistic (Body weight of 20–29-year-old men). We have established a test of H_0: $\mu = 170$ versus H_a: $\mu > 170$ in a population with $\sigma = 40$ pounds. A simple random sample of $n = 64$ is associated with

$$\text{SE}_{\bar{x}} = \frac{40}{\sqrt{64}} = 5.$$

Sample 1. What is the z-statistic for a sample in which $\bar{x} = 173$?

Solution: $z_{\text{stat}} = \dfrac{\bar{x} - \mu_0}{\text{SE}_{\bar{x}}} = \dfrac{173 - 170}{5} = 0.6$

Sample 2. What is the z-statistic in a sample with $\bar{x} = 185$?

Solution: $z_{\text{stat}} = \dfrac{\bar{x} - \mu_0}{\text{SE}_{\bar{x}}} = \dfrac{185 - 170}{5} = 3.0$

Because the distribution of the z_{stat} is Normal, we can apply the 68–95–99.7 rule in establishing probabilities. For example, we know that 95% of the z-statistics will fall between -2 and 2. z-statistics that are less than -2 or greater than 2 would be unlikely to come from this distribution, casting doubt on H_0.

Exercises

9.3 *Patient satisfaction.* Scores derived from a patient satisfaction survey are Normally distributed with $\mu = 50$ and $\sigma = 7.5$, with high scores indicating high satisfaction. An SRS of $n = 36$ is taken from this population.

(a) What is the SE of $\bar{x}$ for these data?

(b) We seek evidence *against* the hypothesis that a particular group of patients comes from a population in which $\mu = 50$. Sketch the curve that describes the sampling distribution of $\bar{x}$ under the null hypothesis. Mark the horizontal axis with values that are ± 1, ± 2, and ± 3 standard errors above and below the mean.

(c) Suppose a sample of $n = 36$ finds an $\bar{x}$ of 48.8. Mark this finding on the horizontal axis of your sketch. Then calculate the z_{stat} for the result. Does this observation provide strong evidence against H_0?

9.4 *Patient satisfaction (cont.).* Exercise 9.3 considered an SRS of $n = 36$ in which $\bar{x}$ was 48.8. A different sample derives $\bar{x} = 46.5$. Mark this result on the horizontal axis of your sketch of the sampling distribution of $\bar{x}$. Calculate the z_{stat} for this finding, and explain why this result provides good evidence against H_0.

9.3 *P*-Value

We convert the z-statistic to a probability statement called the **P-value**. The *P*-value answers the question: "If the null hypothesis were true, what is the probability of the observed test statistic or one that is more extreme?" A small *P*-value indicates that observed data are unlikely to have come from the distribution suggested by H_0.

> **Small *P*-value provides good evidence against H_0.**

For z-statistics, *P*-values correspond to areas under the curve in one tail (one-sided H_a) or two tails (two-sided H_a) of the Standard Normal distribution. A Standard Normal table (Table B) or software utility is used to find this probability. Figure 9.4 depicts the *P*-value for the illustrative example when $\bar{x} = 173$ and $z_{stat} = 0.6$. The size of the shaded area corresponding to $P = 0.2743$ was found with Table B. This *P*-value indicates that a result this extreme will occur 27.4% of the time when the null hypothesis is true. Because this is not particularly surprising, the evidence against H_0 is weak.

Figure 9.5 depicts the *P*-value for a sample in which $\bar{x} = 185$ and $z_{stat} = 3.0$. Because the area in this tail is 0.0013, the evidence against H_0 is strong.

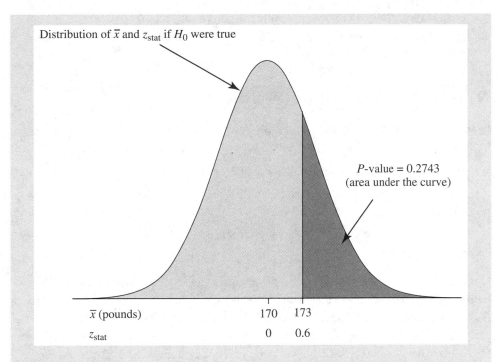

FIGURE 9.4 *P*-value for $z_{stat} = 0.6$ (corresponds to a sample mean of 173 in the illustrative example).

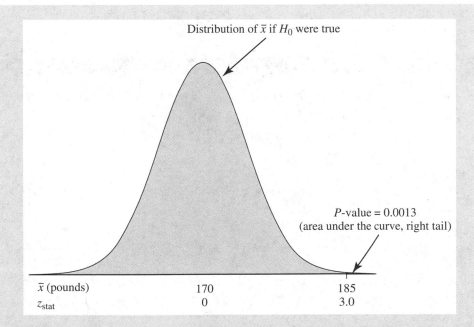

FIGURE 9.5 *P*-value for $z_{stat} = 3.0$ (corresponds to a sample mean of 185 in the illustrative example).

Be aware that the P-value rests on the assumption that H_0 is true. It is the probability of the data given the null hypothesis. It is *not* the probability that the null hypothesis is true given the data.

■ 9.4 Significance Level and Conclusion

Researchers will often refer to the **significance level** of a test. Although it is unwise to draw too firm a line for "significance," the following benchmarks are often applied:

- When $P > 0.10$, the observed difference is said to be *not statistically significant.*
- When $0.05 < P \leq 0.10$, the observed difference is said to be *marginally statistically significant.*
- When $0.01 < P \leq 0.05$, the observed difference is said to be *statistically significant.*
- When $P \leq 0.01$, the observed difference is said to be *statistically highly significant.*

Another way to look at this is to define **alpha** (α) as the chance you are willing to take in mistakenly rejecting a true null hypothesis. If we agree to reject H_0 only when $P \leq \alpha$, we limit the false rejection rate of the null hypothesis to no more than α. For example, by setting α to 0.05, we are willing to make a false rejection of H_0 no more frequently than 1 in 20 tests. If this is not cautious enough, we may draw the line at 1 in 50 (0.02) or 1 in 100 (0.01), or whatever. The jargon is to say that results are statistically significant when $P \leq \alpha$. Conversely, when $P > \alpha$, the results are *not* statistically significant. Beware, though, that failing to find statistical significance is *not* the same as accepting H_0 as true. It is merely saying the evidence is not strong enough to say it is unequivocally false.

ILLUSTRATIVE EXAMPLE

Statistical significance and conclusion (Body weight of 20–29-year-old men).

Sample 1. Our first sample produced a sample mean of 173 pounds. In conducting a one-sided test directed against H_0: $\mu = 170$ (the average weight of the population in the earlier period), we determined $P = 0.27$. This P-value is greater than conventional levels for α and therefore does *not* provide statistically significant evidence against H_0. Thus, the observed sample mean $\bar{x}$ of 173 in this sample is not significantly different from the population mean weight μ from the earlier period ($P = 0.27$).

Sample 2. Our second sample produced a sample mean of 185 pounds. In conducting a one-sided test directed against H_0: $\mu = 170$, we derived $P = 0.0013$. This P-value is less than an α level of 0.01. In fact, it is less than an α level of 0.002. Therefore, this second sample provides a highly significant evidence against the H_0. Thus, the observed sample mean $\bar{x}$ of 185 is significantly greater than the population mean weight μ from the earlier period ($P = 0.0013$).

Notes

1. Use at least two significant digits when reporting *P*-values (e.g., $P = 0.041$, $P = 0.0062$). This will provide enough information for readers to judge the level of statistical significance of the test for themselves.

2. *Avoid* reporting the *P*-value as an inequality (e.g., $P < 0.05$), when possible.

3. There are no sharp distinctions between *P*-value increments. For example, a *P*-value of 0.06 provides about the same degree of evidence against H_0 as a *P*-value of 0.05.

4. The *P*-value does not indicate the magnitude or importance of an observed difference. It merely tells you only about the role of random sampling chance as an explanation for the observed difference.

◼ 9.5 One-Sample *z*-Test

Prior sections in this chapter introduced the elements of hypothesis testing using a one-sample *z*-test by example. This section summarizes the procedure more concisely. The one-sample *z*-test compares a mean from a single sample to an expected value. Several conditions must exist for the test to be used. These include:

*   The variable must be quantitative.

*   Measurements must be valid.

*   Data must be derived by an SRS or reasonable approximation thereof.

*   The standard deviation of the population (σ) must be known prior to data collection.

*   The sampling distribution of $\bar{x}$ is approximately Normal.

Here are the testing steps involved in the procedure:

A. Hypotheses. The null hypothesis is H_0: $\mu = \mu_0$, where μ_0 represents the population mean specified by the null hypothesis.[b] The alternative hypothesis is either H_a: $\mu > \mu_0$ (one sided to the right), H_a: $\mu < \mu_0$ (one sided to the left), or H_a: $\mu \neq \mu_0$ (two sided).

B. Test statistic. $z_{\text{stat}} = \dfrac{\bar{x} - \mu_0}{\text{SE}_{\bar{x}}}$ where $\text{SE}_{\bar{x}} = \dfrac{\sigma}{\sqrt{n}}$.

C. *P*-value. The z_{stat} is converted to a *P*-value with the help of Table B or a software utility. For one-sided alternatives to the right, $P = \Pr(z \geq z_{\text{stat}})$. For one-sided

---

[b] The mean in the reference population under the null hypothesis can be called the *null value*.

alternatives to the left, $P = \Pr(z \leq z_{\text{stat}})$. For two-sided alternative hypotheses, $P = 2 \times \Pr(z \geq |z_{\text{stat}}|)$. Small P-values suggest that either a rare event has occurred or the null hypothesis is false.

D. **Significance level.** When the P-value is less than the specified α level, the observed difference is said to be statistically significant at the α-level and H_0 is rejected. Conventional language can be used to describe the P-value when first learning the procedure (e.g., when $P \leq 0.05$, the observed difference can be said to be statistically significant, and so on). Keep in mind that there are no sharp distinctions between P-value increments.

E. **Conclusion.** The test results are related back to the research question to derive a conclusion.

ILLUSTRATIVE EXAMPLE

z-test ("Lake Wobegon").[c] A citizen claims that the intelligence of children in a particular community is above average. To test this claim, he selects a simple random sample consisting of $n = 9$ children from this population. Scores of the intelligence test typically vary according to a Normal distribution with mean $\mu = 100$ and standard deviation $\sigma = 15$. The sample produces these scores $\{116, 128, 125, 119, 89, 99, 105, 116, 118\}$. Thus, $\bar{x} = \dfrac{116 + 128 + \ldots + 118}{9} = 112.8$. Does this sample provide statistically significant evidence in support of the citizen's claim?

A. **Hypotheses.** Since 100 represents the expected population average if no difference were to exist, H_0: $\mu = 100$. The one-sided alternative hypothesis is H_a: $\mu > 100$. The two-sided alternative is H_a: $\mu \neq 100$.

B. **Test statistic.** The $\text{SE}_{\bar{x}} = \dfrac{\sigma}{\sqrt{n}} = \dfrac{15}{\sqrt{9}} = 5$ and the

$$z_{\text{stat}} = \frac{\bar{x} - \mu}{\text{SE}_{\bar{x}}} = \frac{112.8 - 100}{5} = 2.56.$$ Figure 9.6 is a schematic of the sampling distribution of $\bar{x}$ and the z_{stat} under H_0.

C. **P-value.** Using Table B, we find that the area to the *left* of the z_{stat} of 2.56 is 0.9948. Therefore, the area to the right is $1 - 0.9948 = 0.0052$. This is the one-sided P-value. The two-sided P-value $= 2 \times 0.0052 = 0.0104$.

---

[c] With apologies to Garrison Keillor's *A Prairie Home Companion* (American Public Media).

D. Significance. The two-sided P-value provides significant evidence against the null hypothesis (at $\alpha = 0.05$ but not at $\alpha = 0.01$).

E. Conclusion. The observed sample mean ($\bar{x}$) of 112.8 is significantly greater than the hypothesized population mean μ of 100. Thus, the children of Lake Wobegon are significantly smarter than average.

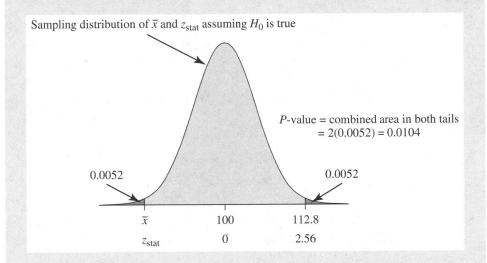

FIGURE 9.6 Two-sided P-value, "Lake Wobegon" illustrative example.

The Normality Assumption

The z-test assumes that the sampling distribution of $\bar{x}$ is Normal. This will be true when the population is Normal. It will also be true when the population is not Normal and the sample is large enough to impart Normality through the central limit theorem. The question becomes "How large does the sample need to be in order for the central limit theorem to impart Normality to the sampling distribution of the mean? This will depend on the extent of the skew in the population. When the population is slightly skewed, a moderately sized sample will impart Normality to the sampling distribution of $\bar{x}$. When the population is severely skewed, a large sample is needed to impart Normality. Here are guidelines for making this type of judgment:

- When the population is Normal, we can use a z procedure on samples of any size.
- When the population is mound shaped and symmetrical, we can use a z procedure on samples of 10 or more.
- When the population is prominently skewed, z procedures should be reserved for large samples of 40 or more.

The paradox is that the sample provides only a rough reflection of the population's shape, especially when n is small, yet this is precisely when information about the population's shape is most needed. When doubt exists, it is probably wise to take a larger sample. When this is not possible, the consultation of an experienced statistician may be required.

Exercises

9.5   *P from z.* What is the one-sided P-value for $z_{stat} = -2.45$? What is the two-sided P-value? (*Suggestion*: Sketch the Normal curve and shade the P-value region.)

9.6   *P from z.* What is the one-sided P-value for a z_{stat} of 1.72? What is the two-sided P-value?

9.7   *Patient satisfaction (sample mean of 48.8).* In Exercise 9.3, you tested H_0: $\mu = 50$ based on a sample of $n = 36$ showing $\bar{x} = 48.8$. The population had standard deviation $\sigma = 7.5$.

   (a)   What is the one-sided alternative hypothesis for this test?

   (b)   Calculate the z-statistic for the test.

   (c)   Convert the z_{stat} to a P-value, and interpret the results.

9.8   *Patient satisfaction survey (sample mean of 46.5).* Carry out a hypothesis test for the conditions laid out in Exercise 9.7 for an observed $\bar{x}$ of 46.5. Is this result statistically significant at $\alpha = 0.05$?

9.9   *LDL and fiber.* A cross-over trial compared serum cholesterol in 13 subjects while on moderate-fiber diets and while on high-fiber diets. The study concluded that the high-fiber diet reduced very-low-density lipoprotein cholesterol by 12.5% ($P = 0.01$).[d] A dietician criticizes the study by saying that very-low-density lipoproteins vary greatly from day to day and that the results could merely reflect a random variation. Explain how the P-value of 0.01 addresses this objection.

9.10   *Lithium.* Lithium carbonate is a drug used to treat bipolar mental disorders. The average dose in well-maintained patients is 1.3 mEq/L with standard deviation 0.3 mEq/L. A random sample of 25 patients on lithium demonstrates a mean level ($\bar{x}$) of 1.4 mEq/L. Test to see whether this mean is significantly higher than that of a well-maintained patient population. Use a two-sided alternative hypothesis for your test. Show all hypothesis testing steps. Comment on your findings.

---

[d]Chandalia, M., Garg, A., Lutjohann, D., von Bergmann, K., Grundy, S. M., & Brinkley, L. J. (2000). Beneficial effects of high dietary fiber intake in patients with type 2 diabetes mellitus. *New England Journal of Medicine, 342*(19), 1392–1398.

■ 9.6 Power and Sample Size

Types of Errors

There are two types of errors we can make when testing a null hypothesis. These are:

Type I error = an erroneous rejection of a true null hypothesis

Type II error = an erroneous retention of a false null hypothesis

		Truth	
		H_0 True	H_0 False
	Retain H_0	Correct retention of H_0	Type II error
Decision			
	Reject H_0	Type I error	Correct rejection of H_0

So far in this chapter, we have focused on avoiding type I errors. A balanced approach also guards against type II errors. Therefore, we provide labels for probabilities associated with each type of error. The Greek letter α (alpha) will represent the probability of a type I error. The Greek letter β (beta) will represent the probability of a type II.

$\alpha \equiv \Pr(\text{type I error})$

$\beta \equiv \Pr(\text{type II error})$

Confidence is the complement of α, and *power* is the complement of β.

Confidence $\equiv 1 - \alpha$

Power $\equiv 1 - \beta$

After retaining a null hypothesis, we are compelled to ask whether the study was powerful enough to avoid a type II error.

Calculating Power

The power of a z-test can be calculated with this formula:

$$1 - \beta = \Phi\left(-z_{1-\frac{\alpha}{2}} + \frac{|\mu_0 - \mu_a|\sqrt{n}}{\sigma}\right)$$

where $\Phi(z)$ represents the cumulative probability of z on a Standard Normal distribution (e.g., $\Phi(0) = 0.5000$; Figure 9.7), μ_0 is the population mean under the null hypothesis, and μ_a is the population mean under an alternative hypothesis. This formula applies to a two-sided test. For one-sided tests, replace $-z_{1-\alpha/2}$ with $-z_{1-\alpha}$.

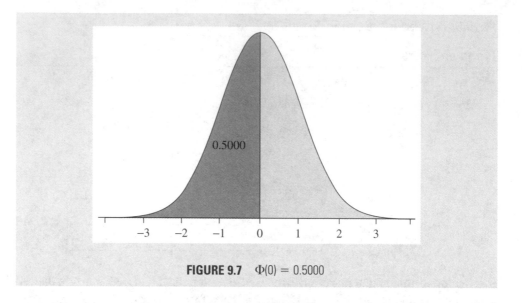

FIGURE 9.7 $\Phi(0) = 0.5000$

ILLUSTRATIVE EXAMPLE

Power of a z-test. A study of $n = 16$ retains H_0: $\mu = 170$ at an α of 0.05 (two sided). Based on prior study, the standard deviation of this variable is 40. What was the power of the test to identify a population mean of 190?

Solution:

$$1 - \beta = \Phi\left(-z_{1-\frac{\alpha}{2}} + \frac{|\mu_0 - \mu_a|\sqrt{n}}{\sigma}\right)$$

$$= \Phi\left(-1.96 + \frac{|170 - 190|\sqrt{16}}{40}\right)$$

$$= \Phi(0.04) = 0.5160 \text{ (from Table B)}$$

The test had about a 50/50 chance of rejecting the H_0 under the stated conditions. This is low. Hypothesis tests should strive for at least 80% power.

Figure 9.8 illustrates the reasoning behind this power calculation. In this figure, both sampling distribution curves are Normal with standard error $\sigma_{\bar{x}} = \dfrac{\sigma}{\sqrt{n}} = \dfrac{40}{\sqrt{16}} = 10$. The sampling distribution for H_0 is centered on 170, while the sampling distribution for H_a is centered on 190. Based on a two-tailed α-level of 0.05 (0.025 in each tail), the sampling distribution under the null hypothesis has a rejection region above 189.6 (dark blue–shaded areas of top curve). The sampling distribution of means under the alternative hypothesis is shaded for $\bar{x}$s greater than 189.6 (dark blue–shaded area of bottom curve), corresponding to a power region of 0.5160.

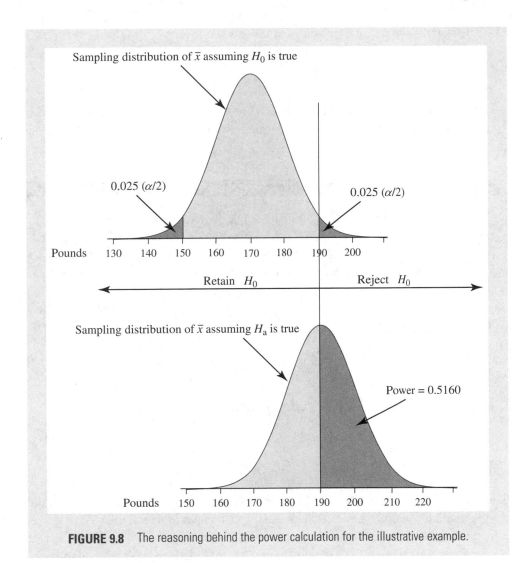

FIGURE 9.8   The reasoning behind the power calculation for the illustrative example.

Sample Size Requirements

You can increase the power of the test by increasing the sample size of the study. Sample size requirements will depend on:

- The desired power of the test $(1 - \beta)$
- The desired significance level of the test (α)
- The standard deviation of the variable (σ)
- The difference between the means under H_0 and H_a. This is the **difference worth detecting**: $\Delta = \mu_0 - \mu_a$.

The sample size required to achieve these conditions is:

$$n = \frac{\sigma^2(z_{1-\beta} + z_{1-\frac{\alpha}{2}})^2}{\Delta^2}$$

This formula applies to two-sided tests. For one-sided tests, replace $z_{1-\frac{\alpha}{2}}$ with $z_{1-\alpha}$. Results are "rounded up" to the next integer so that power does not slip below the stated level.

ILLUSTRATIVE EXAMPLE

Sample size requirement. How large a sample is needed for a z-test with 90% power and $\alpha = 0.05$ (two sided)? The variable is known to have $\sigma = 40$. The null hypothesis assumes $\mu = 170$ and the alternative assumes $\mu = 190$ (i.e., the difference worth detecting $\Delta = \mu_0 - \mu_a = 170 - 190 = -20$).

Solution: $n = \dfrac{\sigma^2(z_{1-\beta} + z_{1-\frac{\alpha}{2}})^2}{\Delta^2} = \dfrac{40^2(1.28 + 1.96)^2}{(-20)^2} = 41.99.$ Round this up to 42 to ensure an adequate sample size.

Figure 9.9 depicts the sampling distributions of the mean under H_0 and under H_a for these stated conditions. The sample size of 42 has shrunk the standard error

$\sigma_x = \dfrac{\sigma}{\sqrt{n}} = \dfrac{40}{\sqrt{42}} = 6.17$, creating less overlap between the competing curves.

Exercises

9.11 ***Gestational length, African American women, hypothesis test.*** Studies in the general population suggest that the gestational length of uncomplicated pregnancies varies according to a Normal distribution with $\mu = 39$ weeks with $\sigma = 2$ weeks.[e] A sample of 22 middle-class African American women demonstrates an average gestation length ($\bar{x}$) of 38.5 weeks. Test whether the mean gestation period in the African American women is significantly different from the expected value of 39 weeks. Use a two-sided alternative hypothesis. Show all hypothesis testing steps.

9.12 ***Gestational length, African American women, power.*** Suppose the average gestational length in African American women is actually 38.5 weeks. What was the power of the test described in Exercise 9.11 to detect this difference at $\alpha = 0.05$ (two-sided)?

---

[e] Mittendorf, R., Williams, M. A., Berkey, C. S., & Cotter, P. F. (1990). The length of uncomplicated human gestation. *Obstetrics & Gynecology, 75*(6), 929–932.

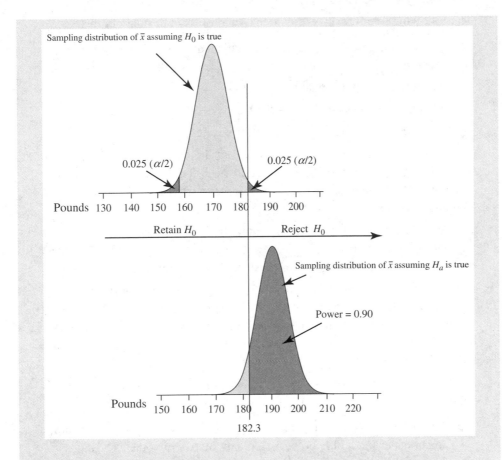

FIGURE 9.9   Sample size requirement illustration. Increasing the sample size shrinks the standard error (skinnies the sampling distributions) and increases power.

9.13 *Gestational length, African American women, sample size.* Let us return to the conditions specified in the previous exercise. How large a sample would be needed to detect the proposed difference in gestational lengths with 90% power at $\alpha = 0.05$ (two sided)?

Summary Points (Introduction to Hypothesis Testing)

1. The **two forms of statistical inference** are hypothesis testing and estimation. This chapter introduces hypothesis testing. The next chapter introduces estimation.

2. **Hypothesis tests** (also called **significance tests**) are used to assess evidence against a claim. Effective use of statistical hypothesis requires an understanding of the reasoning behind the method.

3. The following **hypothesis testing steps** are promoted throughout this text:

   (a) Convert the research question into statistical null and alternative hypotheses.

   (b) Calculate the appropriate test statistics.

   (c) Convert the test statistics to a *P*-value.

   (d) Consider the significance level of the results.

   (e) Formulate a conclusion in the context of the data and research question.

4. The main "output" of hypothesis testing is the **P-value**. The *P*-value is the probability of the data or data that are more extreme assuming the null hypothesis is correct.

5. *P*-values should be viewed as **a continuous measure of evidence**: smaller-and-smaller *P*-values provide stronger-and-stronger evidence against the null hypothesis and in favor of the alternative hypothesis.

6. Although it is *unwise* to use firm cutoff points for determining statistical significance, these **benchmarks** are presented for beginners: $0.05 \leq P \leq 0.10$ provides a marginally significant evidence against H_0, $0.01 \leq P \leq 0.05$ provides a significant evidence against H_0, and $P \leq 0.01$ provides a highly significant evidence against H_0.

7. This chapter considers **one-sample z-tests of means**, which have the following steps:

   (a) **Hypotheses.** $H_0: \mu = \mu_0$ versus $H_a: \mu \neq \mu_0$, where μ_0 is the population mean under the null hypothesis. The alternative hypothesis can also be stated in a one-sided way ($H_a: \mu < \mu_0$ or $H_a: \mu > \mu_0$), depending on how the research question is phrased.

   (b) **Test statistic.** $z_{stat} = \dfrac{\bar{x} - \mu_0}{\sigma/\sqrt{n}}$.

   (c) *P*-value $= 2 \times \Pr(z > |z_{stat}|)$ for two-sided tests use Appendix Table B, Appendix Table F, or a computer application to look up this probability.

   (d) **Significance level.** Compare the *P*-value to various α levels.

   (e) **Conclusion.** Formulate a conclusion in the context of the data and research question.

8. One-sample *z*-tests require the following **conditions**:

   (a) SRS (or reasonable approximation thereof)

   (b) Normal population (or a sample large enough to invoke the central limit theorem)

   (c) Population standard deviation σ known before collecting data

   (d) Data are accurate

9. Rejecting a null hypothesis that is true results in a **type I error**. Failure to reject a false null hypothesis is a **type II error**. The probability of a type I error is called α ("alpha"). The probability of a type II error is called β ("beta"). The probability of avoiding a type II error is called **power** $(1 - \beta)$.

10. The **power of a one-sample z-test** depends on sample size n, the acceptable type I error rate α, standard deviation σ, and the difference worth detecting $|\mu_0 - \mu_a|$ according to the formula $\quad 1 - \beta = \Phi\left(-z_{1-\frac{\alpha}{2}} + \dfrac{|\mu_0 - \mu_a|\sqrt{n}}{\sigma}\right).$

11. The **sample size requirements of a one-sample z-test** depends on the desired level of power $(1 - \beta)$, the α level of the test, the standard deviation in the population σ, and the difference worth detecting $\Delta = |\mu_0 - \mu_a|$ according to the formula $n = \dfrac{\sigma^2(z_{1-\beta} + z_{1-\frac{\alpha}{2}})^2}{\Delta^2}.$

Vocabulary

Alpha (α)	P-value
Alternative hypothesis (H_a)	Sampling distribution of $\bar{x}$
Beta (β)	Significance level
Confidence $(1-\alpha)$	Two-sided alternative hypotheses
Hypotheses testing	Type I error
Null hypothesis (H_0)	Type II error
One-sided alternative hypotheses	z-statistic (z_{stat})
Power $(1-\beta)$	

Review Questions

9.1   Select the better response: This term is used interchangeably with *statistical hypothesis test*.

    (a) Significance test

    (b) Statistical inference

9.2   Select the better response: This is "the hypothesis of no difference."

    (a) H_0

    (b) H_a

9.3   Select the better response: This is the hypothesis that declares a (nonrandom) difference between the observed results and the hypothesized value.

    (a) H_0

    (b) H_a

9.4   Select the better response: This statistical hypothesis contradicts the research hypothesis.

(a) H_0

(b) H_a

9.5   Select the better response: This statistical hypothesis parallels the research hypothesis.

(a) H_0

(b) H_a

9.6   Select the better response: This hypothesis is presumed to be true until proven otherwise.

(a) H_0

(b) H_a

9.7   What is wrong with this sentence? A statistical hypothesis addresses whether a statement about a statistic is true.

9.8   What is wrong with this statement? A one-sided alternative hypothesis considers possible outcomes that are either higher or lower than expected.

9.9   Select the better response: The value of the population mean under the null hypothesis (μ_0) comes from the _____.

(a) data

(b) research question

9.10 Fill in the blank: A mean from a Normal population has a Normal sampling distribution with mean μ and a standard deviation equal to the population standard deviation σ divided by the square root of _____ .

9.11 Select the better response: The P-value refers to the probability of the data or data more extreme assuming the null hypothesis is _____.

(a) false

(b) true

9.12 Select the better response: The evidence against the null hypothesis mounts as the P-value gets _____.

(a) larger and larger

(b) smaller and smaller

9.13 This is the risk a researcher is willing to take in mistakenly rejecting a true null hypothesis.

(a) α

(b) β

(c) $1 - \beta$

9.14 Define β.

9.15 Define *power*.

9.16 Fill in the cells in this hypothesis testing decision table with "correct rejection of H_0," "correct retention of H_0," "type I error," and "type II error."

	Retain H_0 Decision	Reject H_0 Decision
H_0 is actually true		
H_0 is actually false		

Exercises

9.14 *NHES*. A National Center for Health Statistics survey suggested that, in the late 1970s, the mean serum cholesterol level in men was $\mu = 210$ mg/dL with standard deviation $\sigma = 90$ mg/dL. You sample $n = 36$ men from the population and are willing to assume that the sampling distribution of $\bar{x}$ is approximately Normal. What is the probability of finding an $\bar{x}$ of 240 or greater in your sample? (*Suggestions:* Calculate the standard error of the mean. Then sketch the sampling distribution of $\bar{x}$ showing the location of $\bar{x} = 240$ on the horizontal axis. Shade the region corresponding to the area under the curve beyond 240.)

9.15 *Female administrators*. The average annual salary of 20 female hospital administrators is \$80,900. These administrators believe they make less than their male counterparts because of sex bias. Published results suggest that hospital administrators have an average salary of $\mu = \$85,100$ ($\sigma = \$10,000$).

(a) What conditions (assumptions) are needed to perform a z-test to address this question?

(b) Sketch the sampling distribution of $\bar{x}$ based on an SRS of $n = 20$. Assume $\mu = \$85,100$ and $\sigma = \$10,000$. Mark the horizontal axis with landmarks that are ± 1 and ± 2 standard errors around μ. Locate $\bar{x} = \$80,900$ on the horizontal axis of the curve.

(c) Calculate the z_{stat} and P-value for an observed $\bar{x}$ of \$80,900.

(d) Come up with four explanations for the lower-than-expected observed sample mean.

9.16 *Fathers had heart attacks*. Suppose the mean fasting cholesterol of teenage boys in the United States is $\mu = 175$ mg/dL with $\sigma = 50$ mg/dL. An SRS of 39 boys whose fathers had a heart attack reveals a mean cholesterol $\bar{x} = 195$ mg/dL. Use a two-sided test to determine if the sample mean is significantly higher than expected. Show all hypothesis testing steps.

9.17 *University men.* An SRS of 18 male students at a university has an average height of 70 inches. The average height of men in the general population is 69 inches. Assume that male height is approximately Normally distributed with $\sigma = 2.8$ inches. Conduct a two-sided hypothesis test to determine whether the male students are significantly taller than expected. Show all hypothesis testing steps.

9.18 *Diet and bowel cancer.* Histologically confirmed cases of colorectal adenoma were randomly assigned to a treatment group (low-fat, high-fiber diet) or to a control group (regular diet). Participants were screened for new occurrences of colon polyps over the next 4 years. The mean number of new polyps by the end of the follow-up period was *not* significantly different.[f] Explain the meaning of this finding in terms a layman would understand.

9.19 *The criminal justice analogy.* A jury having heard the evidence in a criminal case is debating whether to render a verdict of *guilty* or *not guilty*. Because it is worse to convict an innocent person of a crime they did not commit than let a guilty person go free, each case starts with a presumption of innocence. This is analogous to the presumption that H_0 is true in a statistical hypothesis test. Other analogies exist. To make these connections, fill in the cells of a table with the labels: (a) Acquittal of an innocent person; (b) conviction of an innocent person; (c) acquittal of a guilty person; and (d) declaring a criminal guilty. Which type of decision is analogous to a type I error? Which is analogous to a type II error?

9.20 *Lab reagent, hypothesis test.* A reference solution used as a lab reagent is purported to have a concentration of 5 mg/dL. Six samples are taken from this solution and the following concentrations are recorded: {5.32, 4.88, 5.10, 4.73, 5.15, 4.75} mg/dL. These six measurements are assumed to be an SRS of all possible measurements from the solution. They are also assumed to have a standard deviation σ of 0.2, a Normal distribution, and a mean concentration equal to the true concentration of the solution. Carry out a significance test to determine whether these six measurements provide reliable evidence that the true concentration of the solution is actually not 5 mg/dL.

9.21 *Lab reagent, power analysis.* Exercise 9.20 failed to show a significant difference in the mean concentrations in a lab reagent and the specified concentration of 5 mg/dL. Now let us assume that the true concentration of the reagent is 4.75. What was the power of the test performed in Exercise 9.20 to reject H_0: $\mu = 5$ at $\alpha = 0.05$ two-sided?

---

[f] Schatzkin, A., Lanza, E., Corle, D., Lance, P., Iber, F., Caan, B., et al. (2000). Lack of effect of a low-fat, high-fiber diet on the recurrence of colorectal adenomas. Polyp Prevention Trial Study Group. *New England Journal of Medicine, 342*(16), 1149–1155.

10 | Basics of Confidence Intervals

■ 10.1 Introduction to Estimation

The prior chapter introduced the form of statistical inference known as hypothesis testing. This chapter introduces the other common form of statistical inference: estimation. There are two forms of estimation: point estimation and interval estimation. **Point estimation** provides a single estimate of the parameter; **interval estimation** provides a range of values (confidence interval) that seeks to capture the parameter.

Figure 10.1 is a schematic of a **confidence interval**. The interval extends a **margin of error** above and below the point estimate. We may think of the margin of error as the "wiggle room" surrounding the point estimate. Therefore, a general formula for a confidence interval is:

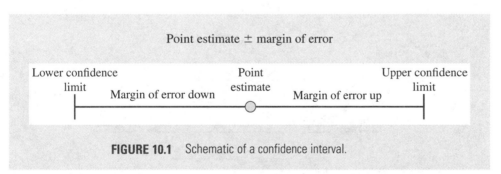

Point estimate $\pm$ margin of error

FIGURE 10.1 Schematic of a confidence interval.

ILLUSTRATIVE EXAMPLE

Point and interval estimation (Body weight, 20–29-year-old U.S. men, 2000). The National Health and Nutrition Examination Survey (NHANES)[a] assesses the health and nutritional status of adults and children in the United States using data from interviews and direct physical examinations. From 1999 to 2002,

continues

---

[a] National Center for Health Statistics, Centers for Disease Control.

continued

712 men between the ages of 20 and 29 were examined. The mean body weight in this sample ($\bar{x}$) was 183.0 pounds.[b] This is the **point estimate** for the mean body weight in the population of men in this age group. The margin of error associated with this estimate for 95% confidence was ±3 pounds. Therefore, the 95% confidence interval for μ is 183 pounds ± 3 pounds = 180 to 186 pounds.

The **confidence level** of a confidence interval refers to the success rate of the method in capturing the parameter it seeks. For example, a 95% confidence interval for μ is constructed in such a way that 95% of like intervals will capture μ and 5% will fail to capture μ.

Figure 10.2 depicts five 95% confidence intervals for μ constructed from five independent SRSs of the same size from the same source population. It so happens

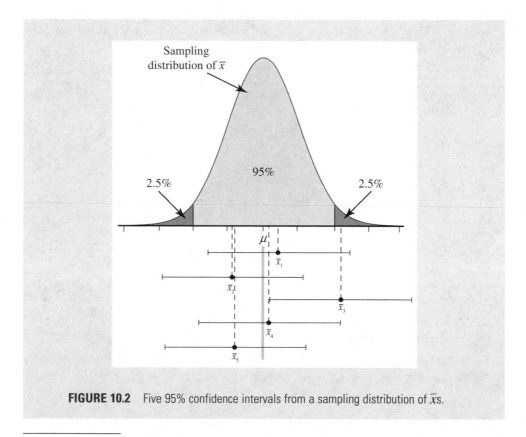

FIGURE 10.2   Five 95% confidence intervals from a sampling distribution of $\bar{x}$s.

---

[b] Ogden, C. L., Fryar, C. D., Carroll, M. D., & Flegal, K. M. (2004). Mean body weight, height, and body mass index, United States 1960–2002. *Advance Data from Vital and Health Statistics, 347,* 1–17. See Table 6. Values have been rounded for illustrative purposes.

that four of the five intervals captured μ in this schematic. (The third one missed.) In practice, we would not know which of the confidence intervals were successful and which were not. We would only know that, in the long run, 95% of the intervals will capture the parameter.

◼ 10.2 Confidence Intervals for μ When σ Is Known

The Reasoning Behind Confidence Intervals

To understand how a confidence interval is constructed, we return to the notion of the **sampling distribution of $\bar{x}$** (Section 8.2). Figure 10.3 depicts repeated SRSs, each of $n = 712$, taken from the same population. Each independent sample calculates a mean ($\bar{x}$) that contributes to a distribution of $\bar{x}$s. The value of population mean μ is not known, but the standard deviation of the population is known and is equal to 40.[c] Based on the reasons discussed in Section 8.2, the sampling distribution of $\bar{x}$ will

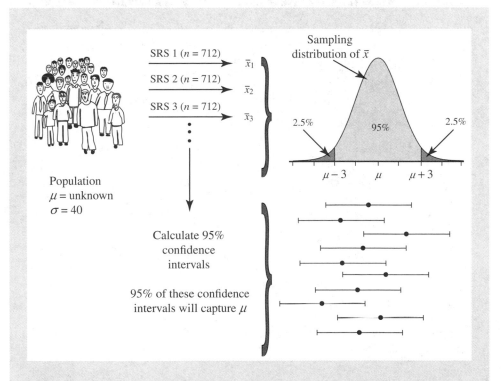

FIGURE 10.3   Schematic of repeated independent samples of size $n = 712$ from the same source population with confidence intervals calculated with each sample mean.

[c] Chapter 11 will eliminate the need to specify the value of the population standard deviation ahead of time.

be approximately Normal with mean μ and $SE_{\bar{x}} = \dfrac{\sigma}{\sqrt{n}} = \dfrac{40}{\sqrt{712}} \approx 1.5$. (Recall that the $SE_{\bar{x}}$ is the standard deviation of the sampling distribution of $\bar{x}$.) Because of the 68–95–99.7 rule, we know that 95% of the $\bar{x}$s will fall within two standard errors of μ. In this instance, two standard errors = 2×1.5 pounds = 3 pounds. It follows that the interval $\bar{x} \pm 3.0$ will capture the value of population mean μ 95% of the time. For the current example, a sample mean of 183 generates a 95% confidence for μ of (183 − 3) to (183 + 3) = 180 to 186 pounds.

In general, the confidence interval for μ is given by:

$$\bar{x} \pm m$$

where m is the margin of error. For 95% confidence, $m \approx 2 \times$ the standard error of the mean.

ILLUSTRATIVE EXAMPLE

95% confidence intervals for μ, σ known (Body weight, 20–29-year-old males). We wish to estimate the mean weight of a population in which body weight has a Normal distribution with $\sigma = 40$ pounds. Three independent SRSs each of $n = 712$ are selected from the population. These samples derive $\bar{x}_1 = 183$, $\bar{x}_2 = 180$, and $\bar{x}_3 = 184$. Calculate 95% confidence intervals for μ based on each of these samples.

In each sample, $SE_{\bar{x}} = \dfrac{40 \text{ pounds}}{\sqrt{712}} = 1.5$ pounds and $m \approx 2 \cdot SE = 2 \cdot 1.5$ = 3 pounds.

- For sample 1, the confidence interval is $\bar{x}_1 \pm m = 183 \pm 3 = 180$ to 186 (pounds).
- For sample 2, the confidence interval is $\bar{x}_2 \pm m = 180 \pm 3 = 177$ to 183 (pounds).
- For sample 3, the confidence interval is $\bar{x}_3 \pm m = 184 \pm 3 = 181$ to 187 (pounds).

Each of these confidence intervals has a 95% chance of capturing the population's true mean weight μ.

Other Levels of Confidence

To calculate a $(1 - \alpha)100\%$ confidence for μ, use this formula:

$$\bar{x} \pm z_{1-\alpha/2} \cdot SE_{\bar{x}}$$

where $z_{1-\alpha/2}$ represents a Standard Normal value with cumulative probability $1 - \alpha/2$ and $\text{SE}_{\bar{x}} = \dfrac{\sigma}{\sqrt{n}}$. In this formula, α represents the probability a confidence interval will fail to capture μ. Common levels of α and confidence are:

α-level	$1 - \alpha$ (confidence level)	$z_{1-(\alpha/2)}$
0.10	90%	$z_{1-(.10/2)} = 1.645$
0.05	95%	1.960
0.01	99%	2.576

Figure 10.4 depicts these levels of confidence on standard Normal curves.

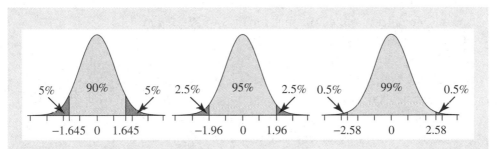

FIGURE 10.4 z values for various levels of confidence; curves represent sampling distributions of means.

ILLUSTRATIVE EXAMPLE

Confidence intervals, various levels of confidence (Body weight, 20–29-year-old U.S. males). We select an SRS of $n = 712$ from a population of 20–29-year-old men and calculate $\bar{x} = 183$ pounds. The population standard deviation (σ) is 40. We propose to calculate 90%, 95%, and 99% confidence intervals for μ based on this information.

In each instance, $\text{SE}_{\bar{x}} = \dfrac{\sigma}{\sqrt{n}} = \dfrac{40}{\sqrt{712}} = 1.50$.

- The 90% confidence interval for μ is $\bar{x} \pm (z_{1-.10/2})(\text{SE}_{\bar{x}}) = 183.0 \pm (1.645)(1.5) = 183.0 \pm 2.5 = 180.5$ to 185.5 (pounds).
- The 95% confidence interval for μ is $\bar{x} \pm (z_{1-.05/2})(\text{SE}_{\bar{x}}) = 183.0 \pm (1.960)(1.5) = 183.0 \pm 2.9 = 180.1$ to 185.9 (pounds).
- The 99% confidence interval for μ is $\bar{x} \pm (z_{1-.01/2})(\text{SE}_{\bar{x}}) = 183.0 \pm (2.576)(1.5) = 183.0 \pm 3.9 = 179.1$ to 186.9 (pounds).

continues

continued

Thus, we can be 90% confident that the true mean weight in the population is between 180.5 and 185.5 pounds, we can be 95% confident that it is between 180.1 and 185.9 pounds, and we can be 99% confident that it is between 179.1 and 186.9 pounds. Notice that we "pay" for each additional step up in the confidence level with a broader confidence interval.

Notes

1. **Conditions for valid inference.** This procedure assumes data were collected by an SRS (sampling independence), population standard deviation σ is known before data are collected, and the sampling distribution of $\bar{x}$ is approximately Normal.[d]

2. **Confidence level and confidence interval width.** Figure 10.5 plots the 90%, 95%, and 99% confidence intervals for the illustrative example. Higher levels of confidence are associated with wider intervals.

3. **What *confidence* means.** The confidence level of an interval tells you how often the method will succeed in capturing μ in the long run. With repeating samples, $(1 - \alpha)100\%$ of the intervals will capture μ and $(\alpha)100\%$ will not.

4. **Confidence intervals address random error only.** Confidence intervals do not control for nonrandom sources of error, such as those that might be due to biased sampling, poor-quality information, and confounding. The method addresses random sampling error only.

5. **Table C.** The last line of Appendix Table C lists $z_{1-\alpha/2}$ critical values for various levels of confidence, thus providing a handy way to look up z values for confidence intervals with four significant-digit accuracy.

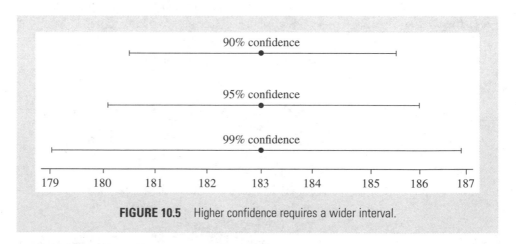

FIGURE 10.5   Higher confidence requires a wider interval.

---

[d] See Section 9.5 for comments about the Normality assumption.

6. **Interpretation.** Keep in mind that confidence intervals seek to capture population mean μ, *not* sample mean $\bar{x}$.

Exercises

10.1 *Misinterpreting a confidence interval*. A pharmacist reads that a 95% confidence interval for the average price of a particular prescription drug is $30.50 to $35.50. Asked to explain the meaning of this, the pharmacist says "95% of all pharmacies sell the drug for between $30.50 and $35.50." Is the pharmacist correct? Explain your response.

10.2 *Newborn weight*. A study reports that the mean birth weight of 81 full-term infants is 6.1 pounds. The study also reports "SE = 0.22 pounds."

   (a) What is the margin of error of this estimate for 95% confidence?

   (b) What is the 95% confidence interval for μ?

   (c) What does it mean when we say that we have 95% confidence in the interval?

   (d) Recall that $SE_{\bar{x}} = \dfrac{\sigma}{\sqrt{n}}$. Rearrange this equation to solve for σ. What was the standard deviation of birth weights in this population?

10.3 *Newborn weight*. A study takes an SRS from a population of full-term infants. The standard deviation of birth weights in this population is 2 pounds. Calculate 95% confidence intervals for μ for samples in which:

   (a) $n = 81$ and $\bar{x} = 6.1$ pounds

   (b) $n = 36$ and $\bar{x} = 7.0$ pounds

   (c) $n = 9$ and $\bar{x} = 5.8$ pounds

10.4 *90% confidence intervals*. Calculate 90% confidence intervals for μ based on the information reported in Exercise 10.3a–c.

10.5 *SIDS*. A sample of 49 sudden infant death syndrome (SIDS) cases had a mean birth weight of 2998 g. Based on other births in the county, we will assume $\sigma = 800$ g. Calculate the 95% confidence interval for the mean birth weight of SIDS cases in the county. Interpret your results.

10.6 *99% confidence interval*. Use the information in Exercise 10.5 to calculate a 99% confidence interval for μ.

■ 10.3 Sample Size Requirements

The length of a confidence interval is equal to its **upper confidence limit (UCL)** minus its **lower confidence limit (LCL)**.

$$\text{Confidence interval length} = \text{UCL} - \text{LCL}$$

Ɪ   **Confidence interval length** reflects the precision of the estimate. Narrow intervals reflect precision; wide intervals reflect imprecision.

The margin of error is also a reflection of the estimate's precision. Margin of error m is equal to

$$m = z_{1-\frac{\alpha}{2}} \frac{\sigma}{\sqrt{n}}$$

where $z_{1-\alpha/2}$ is a Standard Normal deviate with cumulative probability $1 - \alpha/2$, σ is the population standard deviation, and n is the sample size. This formula is rearranged to determine the sample size needed to achieve margin of error m:

$$n = \left(z_{1-\frac{\alpha}{2}} \frac{\sigma}{m} \right)^2$$

Results are rounded up to ensure that we achieve the stated precision.

ILLUSTRATIVE EXAMPLE

Sample size requirement (Body weight, 20–29-year-old males). We have established that the body weights of 20–29-year-old U.S. males are $\sigma = 40$ pounds. How many observations do we need in order to estimate mean body weight μ with 95% confidence and a margin of error of plus or minus 10 pounds?

Solution: Recall that for 95% confidence, $m = z_{1-\frac{\alpha}{2}} = z_{1-\frac{0.05}{2}} = z_{0.975} = 1.96$ (from Appendix Table B). Therefore, the required sample size is $n = \left(z_{1-\frac{\alpha}{2}} \frac{\sigma}{m} \right)^2 = \left(1.96 \cdot \frac{40}{10} \right)^2 = 61.5$. Round this up to 62.

What size sample is needed to estimate μ with margin of error plus or minus 5 pounds?

Solution: $n = \left(1.96 \cdot \frac{40}{5} \right)^2 = 245.9$. Round this to 246.

What size sample is needed to cut the margin down to 1 pound?

Solution: $n = \left(1.96 \cdot \frac{40}{1} \right)^2 = 6146.6$. Round this up to 6147.

The formula $m = z_{1-\frac{\alpha}{2}} \frac{\sigma}{\sqrt{n}}$ lets us know that three factors contribute to the margin of error. These are:

1. $z_{1-\frac{\alpha}{2}}$ which determines the **confidence level** of the interval. Decreasing confidence will shrink $z_{1-\frac{\alpha}{2}}$ and the margin of error.

2. σ, which is the **standard deviation** of the variable. A variable with high variability will obscure population mean μ.

3. n, which is the **sample size**. Increasing the sample size decreases the margin of error according to the square root of n. For example, quadrupling the sample size will cut the margin of error in half.

Exercises

10.7 *Hemoglobin*. Hemoglobin levels in 11-year-old boys vary according to a Normal distribution with $\sigma = 1.2$ g/dL.

(a) How large a sample is needed to estimate mean μ with 95% confidence so the margin of error is no greater than 0.5 g/dL?

(b) How large a sample is needed to estimate μ with margin of error 0.5 g/dL with 99% confidence?

10.8 *Sugar consumption*. Based on prior studies, a dental researcher is willing to assume that the standard deviation of the weekly sugar consumption in children in a particular community is 100 g.

(a) How large a sample is needed to estimate mean sugar consumption in the community with a margin of error 10 g at 95% confidence?

(b) How many kids should be studied if the researcher is willing to accept a margin of error of 25 g at 95% confidence?

■ 10.4 Relationship Between Hypothesis Testing and Confidence Intervals

You can use a $(1 - \alpha)100\%$ confidence interval for μ to predict whether a two-sided test of $H_0: \mu = \mu_0$ will be significant at the α-level of significance. When the value of the parameter identified in the null hypothesis (μ_0) falls outside the interval, the results will be statistically significant (reject H_0). When the value of the parameter identified in the null hypothesis is captured by the interval, the results will *not* be statistically significant (retain H_0).

ILLUSTRATIVE EXAMPLE

Significance test from confidence interval. A prior illustrative example calculated confidence intervals for the mean body weight of the U.S. 20–29-year-old male population. We use this information to test $H_0: \mu = 180$ pounds at differing α levels.

continues

continued

- The 90% confidence interval for μ was 180.5 to 185.5 (pounds). In testing H_0: $\mu = 180$ pounds at $\alpha = 0.10$, results are statistically significant (reject H_0).
- The 95% confidence interval for μ was 180.1 to 185.9 pounds. In testing H_0: $\mu = 180$ pounds at $\alpha = 0.05$, these results are statistically significant (reject H_0).
- The 99% confidence interval for μ was 179.1 to 186.9 pounds. In testing H_0: $\mu = 180$ pounds at $\alpha = 0.01$, these results are not statistically significant (retain H_0).

Figure 10.6 illustrates the 95% and 99% confidence intervals for this problem in relation to the sampling distribution of $\bar{x}$ assuming H_0: $\mu = 180$. Notice that the 95% confidence interval fails to capture μ_0, while the 99% confidence interval captures it.

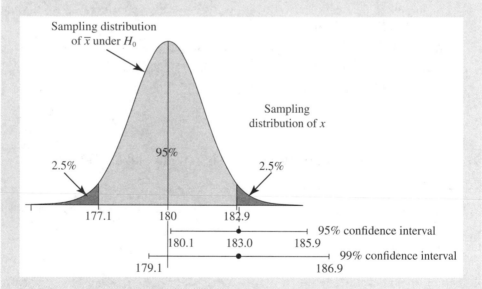

FIGURE 10.6   Relation between confidence intervals and hypothesis test. $(1 - \alpha)100\%$ confidence intervals that exclude μ_0 are significant at the α level of significance.

Exercise

10.9 *P-value and confidence interval*. A two-sided test of H_0: $\mu = 0$ yields a *P*-value of 0.03. Will the 95% confidence interval for μ include 0 in its midst? Will the 99% confidence interval for μ include 0? Explain your reasoning in each instance.

Summary Points (Introduction to Confidence Intervals)

1. This chapter introduces the second form of statistical inference: **estimation**.

2. Sample mean $\bar{x}$ is the **point estimator** of population parameter μ.

3. **Interval estimates** are called confidence intervals. Confidence intervals can be calculated at almost any level of confidence. The most common levels of confidence are 90%, 95%, and 99%.

4. The $(1 - \alpha)100\%$ **confidence interval** has a $(1 - \alpha)100\%$ chance of capturing the true value of the parameter and an α chance of not capturing the true value.

5. **A $(1 - \alpha)100\%$ confidence interval for $\mu = \bar{x} \pm z_{1-\frac{\alpha}{2}}\frac{\sigma}{\sqrt{n}}$**, where $\bar{x}$ is the sample mean and $z_{1-\frac{\alpha}{2}}$ is the $1 - \alpha/2$ percentile on a Standard Normal distribution.

   (a) This confidence interval consists of the point estimate and surrounding margin of error (m): $\bar{x} \pm m$.

   (b) Margin of error m is a measure of the precision of the estimate.

   (c) The confidence interval is written as (LCL to UCL) or (LCL, UCL), where LCL represents the lower confidence limit and UCL represents the upper confidence limit.

6. z confidence interval procedure for μ requires the following **conditions**:

   (a) Data are an SRS from the population (or reasonable approximation thereof).

   (b) The population is Normal or the sample is large enough to invoke the central limit theorem.

   (c) Population standard deviation σ is known before collecting data.

   (d) The data are accurate.

7. The **sample size requirement to** achieve margin of error m when estimating μ is $n = \left(z_{1-\frac{\alpha}{2}}\frac{\sigma}{m}\right)^2$.

   (a) Use an educated guess as standard deviation σ. If an educated guess for σ is not available, then first do a pilot study.

   (b) m is the margin of error you can "live with." Smaller-and-smaller margins of error require larger-and-larger sample sizes.

   (c) Increasing the required level of confidence mandates larger-and-larger sample sizes.

Vocabulary

Confidence interval	Interval estimation
Confidence interval length	Level of confidence

Lower confidence limit   Point estimation
Margin of error   Upper confidence limit (UCL)
Point estimate

Review Questions

10.1 Fill in the blank: The two forms of estimation are point estimation and
_____ estimation.

10.2 While the goal of hypothesis testing is to test a claim, the goal of estimation is
to estimate a _____.

10.3 Select the best response: $\bar{x}$ is a(n) _____ estimator of μ because as a
sample size gets larger and larger, we expect $\bar{x}$ to get closer and closer to the true
value of μ.

(a) biased

(b) reliable

(c) unbiased

10.4 Select the best response: This is the "wiggle room" placed around the point estimate to derive a confidence interval.

(a) standard deviation

(b) standard error

(c) margin of error

10.5 Select the best response: This is the larger of the two numbers listed when reporting a confidence interval.

(a) point estimate

(b) lower confidence limit

(c) upper confidence limit

10.6 Select the best response: The confidence interval for the mean seeks to capture
the _____.

(a) sample mean

(b) population mean

(c) population standard deviation

10.7 Select the best response: The standard error of the mean is inversely proportional
to the _____.

(a) standard deviation

(b) sample size

(c) square root of the sample size

10.8 What percentage of 90% confidence intervals for μ will fail to capture μ?

10.9 What percentage of $(1 - \alpha)100\%$ confidence intervals for μ will fail to capture μ?

10.10 Select the best response: Which of the following reflects the precision of $\bar{x}$ as an estimate of μ?

(a) the margin of error m

(b) the confidence interval length

(c) both (a) and (b)

10.11 Select the best response: The margin of error m for the $(1 - \alpha)100\%$ confidence interval for μ is equal to $z_{1-\alpha/2}$ times the _____.

(a) sample mean

(b) standard deviation

(c) standard error of the mean

10.12 Select the best response: To cut the margin of error in half, we must _____ the sample size.

(a) double

(b) triple

(c) quadruple

10.13 Select the best response: For the same set of data, which confidence interval will be the longest?

(a) the 90% confidence interval for μ

(b) the 95% confidence interval for μ

(c) the 99% confidence interval for μ

10.14 Select the best response: Which of the following α levels is associated with 95% confidence?

(a) 1%

(b) 5%

(c) 10%

10.15 Select the best response: Which of the following z critical values is used to achieve 95% confidence?

(a) $z_{0.90}$

(b) $z_{0.95}$

(c) $z_{0.975}$

Exercises

10.10 *Laboratory scale*. The manufacturer of a laboratory scale claims their scale is accurate to within 0.0015 g. You read the documentation for the scale and learn that this means that the standard deviation of an individual measurement (σ) is equal to 0.0015 g. Assume measurements vary according to a Normal distribution with μ equal to the actual weight of the object. You weigh the same specimen twice and get readings of 24.31 and 24.34 g. Based on this information, calculate a 95% confidence interval for the true weight of the object.

10.11 *Antigen titer*. A vaccine manufacturer analyzes a batch of product to check its titer. Immunologic analyses are imperfect, and repeated measurements on the same batch are expected to yield slightly different titers. Assume titer measurements vary according to a Normal distribution with mean μ and $\sigma = 0.070$. (The standard deviation is a characteristic of the assay.) Three measurements demonstrate titers of 7.40, 7.36, and 7.45. Calculate a 95% confidence interval for true concentration of the sample.

10.12 *Newborn weight*. The 95% confidence interval for the mean weight of infants born to mothers who smoke is 5.7 to 6.5 pounds. The mean weight for all newborns in this region is 7.2 pounds. Is the birth weight of the infants in this sample significantly different from that of the general population at $\alpha = 0.05$? Explain your response.

10.13 *Reverse engineering the confidence interval*. The 95% confidence interval in Exercise 10.12 (5.7 to 6.5 pounds) was calculated with the usual formula $\bar{x} \pm z_{1 - \frac{\alpha}{2}} \frac{\sigma}{\sqrt{n}}$ Confidence intervals constructed in this way are symmetrical around the mean; $\bar{x}$ is the mid-point of the confidence interval.

(a) What is the value of the sample mean used to calculate the confidence interval?

(b) What is the margin of error of the confidence interval?

(c) What is the standard error of the estimate?

(d) Calculate a 99% confidence interval for μ.

(e) The sample mean is significantly different from 7.2 pounds at $\alpha = 0.05$. Is the difference significant at $\alpha = 0.01$?

10.14 *True or false?* Answer "true" or "false" in each instance. Explain each response.

(a) 95% of 95% confidence intervals for a mean will fail to capture $\bar{x}$.

(b) 95% of 95% confidence intervals for a mean will fail to capture μ.

(c) 95% of 95% confidence intervals for a mean will capture μ.

10.15 ***True or false?*** *A confidence interval for* μ *is 13* $\pm$ *5.*

    (a) The value 5 in this expression is the estimate's standard error.

    (b) The value 13 in this expression is the estimate's margin of error.

    (c) The value 5 in this expression is the estimate's margin of error.

10.16 ***True or false?***

    (a) Populations with high variability will produce estimates with large margins of error (all other things being equal).

    (b) Studies with small sample sizes will tend to have large margins of error.

10.17 ***Lab reagent, 90% confidence interval for true concentration***. Exercise 9.20 presented six measurements of a reagent in which the true concentration of the reagent was uncertain. The sample mean $\bar{x}$ based on these six measurements was 4.9883 mg/dL. The distribution of an infinite number of concentration measurements taken from the solution is assumed to be Normal with μ equal to the true concentration of the solution and $\sigma = 0.2$ mg/dL. Calculate a 90% confidence interval for the true concentration of the solution based on the current set of observations. What do you conclude?

10.18 ***Lab reagent, 99% confidence interval for true concentration***. Calculate a 99% confidence interval for the true concentration of the solution considered in the prior exercise. Interpret the results. How does this confidence interval compare the 90% confidence interval?

Part II

Quantitative Response Variable

11 | Inference About a Mean

■ 11.1 Estimated Standard Error of the Mean

The prior three chapters relied on z-procedures to help infer population means. z-procedures, in turn, rely on knowing the value of the population standard deviation (σ) before data are collected. This chapter introduces methods that do away with this often unrealistic condition.

With z-procedures, population standard deviation σ is used to calculate the standard error of the mean using the formula $\frac{\sigma}{\sqrt{n}}$. In this chapter, we replace population standard deviation σ with sample standard deviation s to derive the **standard error of mean**:

$$\mathrm{SE}_{\bar{x}} = \frac{s}{\sqrt{n}}$$

Both $\frac{\sigma}{\sqrt{n}}$ and $\frac{s}{\sqrt{n}}$ are referred to as *standard errors*, even though they have slightly different formulas.[a]

Exercises

11.1 *Blood pressure.* A study of 35 individuals found mean systolic blood pressure $\bar{x} = 124.6$ mmHg with sample standard deviation $s = 10.3$ mmHg.

(a) Calculate the standard error of the mean based on this information.

(b) How many individuals would you need to study to decrease the standard error of the mean to 1 mmHg? [*Hint*: Rearrange the standard error formula to solve for n.]

---

[a] Perhaps it would be more accurate to call $\frac{s}{\sqrt{n}}$ the *estimated* standard error, but this distinction often gets lost in practice.

11.2 *Published report*. An article published in the *American Journal of Public Health* that reported the relation between tall stature and cardiovascular disease mortality included a table with the column heading "Mean Height, cm. (SE)." One entry in the table based on $n = 1243$ individuals was "173.2 (0.2)."[b] From this information, determine the standard deviation of height in this group. [*Hint: Rearrange the standard error formula to solve for s*.]

■ 11.2 Student's *t*-Distributions

Using $\dfrac{s}{\sqrt{n}}$ to estimate the standard error of the mean tacks on an additional element of uncertainty to inferential procedures. To accommodate this additional uncertainty, we use a *t*-distribution instead of a Standard Normal *z*-distribution when making inferences. *t*-distributions were introduced by William Sealy Gosset (1876–1937) in 1908 writing under the pseudonym "Student," and are hence referred to as **Student's *t*-distributions**.[c]

t-distributions resemble the Standard Normal distribution. They are bell shaped and centered on 0. However, *t*-distributions have more area in their tails than the Standard Normal *z*-distribution. These broader tails accommodate the additional uncertainty that comes from estimating σ with *s*.

The *t*-distributions are a family of distributions with family members sharing common characteristics. Each member of *t* is identified by its **degree of freedom (df)**. Figure 11.1 displays *t*-distributions with 1, 9, and ∞ degrees of freedom. As the number of degrees of freedom increases, *t*-distributions become increasingly like a Standard Normal distribution. A *t*-distribution with ∞ degrees of freedom *is* a Standard Normal *z*-distribution. *t*-distributions with (say) 60 or more degrees of freedom are nearly indistinguishable from the Standard Normal (*z*) distribution. Therefore, with large samples, it really doesn't matter much if you use a *t*-procedure or *z*-procedure.

Table C in the appendix of this book lists landmarks (*critical values*) on *t*-distributions. Each row in this table refers to a *t*-distribution with a particular df. Each column corresponds to a cumulative probability on the distribution. Entries in the table are values for these **_t_-percentiles**.

---

[b] Langenberg, C., Shipley, M. J., Batty, G. D., & Marmot, M. G. (2005). Adult socioeconomic position and the association between height and coronary heart disease mortality: Findings from 33 years of follow-up in the Whitehall Study. *American Journal of Public Health, 95*(4), 628–632.
[c] Student. (1908). The probable error of a mean. *Biometrika, VI*, 1–25.

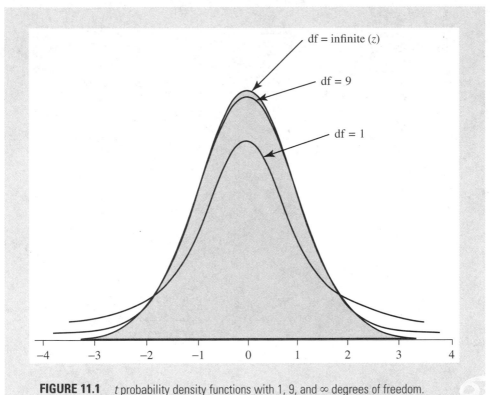

FIGURE 11.1   *t* probability density functions with 1, 9, and ∞ degrees of freedom.

Figure 11.2 illustrates how Table C is used. We focus on the row for 9 degrees of freedom and column for a cumulative probability of 0.975. The table entry at the intersection of these points is 2.262, indicating that the 97.5th percentile on a *t*-distribution with 9 degrees of freedom ($t_{9,0.975}$) is equal to 2.262. A visual depiction of this is shown in Figure 11.2.

Notation: Let $t_{\mathrm{df},p}$ denote a *t* critical value with df degrees of freedom and cumulative probability *p*. Figure 11.2 highlights $t_{9,0.975} = 2.262$. Notice that this *t*-value has a right-tail probability of 0.025.

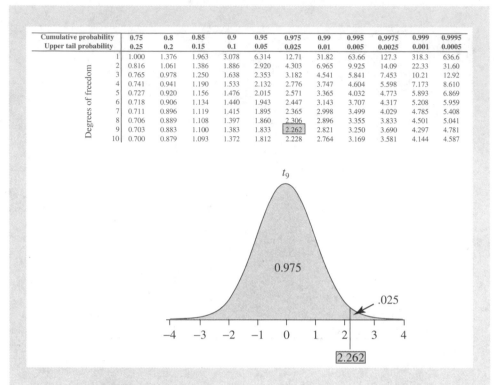

Cumulative probability	0.75	0.8	0.85	0.9	0.95	0.975	0.99	0.995	0.9975	0.999	0.9995
Upper tail probability	0.25	0.2	0.15	0.1	0.05	0.025	0.01	0.005	0.0025	0.001	0.0005
1	1.000	1.376	1.963	3.078	6.314	12.71	31.82	63.66	127.3	318.3	636.6
2	0.816	1.061	1.386	1.886	2.920	4.303	6.965	9.925	14.09	22.33	31.60
3	0.765	0.978	1.250	1.638	2.353	3.182	4.541	5.841	7.453	10.21	12.92
4	0.741	0.941	1.190	1.533	2.132	2.776	3.747	4.604	5.598	7.173	8.610
5	0.727	0.920	1.156	1.476	2.015	2.571	3.365	4.032	4.773	5.893	6.869
6	0.718	0.906	1.134	1.440	1.943	2.447	3.143	3.707	4.317	5.208	5.959
7	0.711	0.896	1.119	1.415	1.895	2.365	2.998	3.499	4.029	4.785	5.408
8	0.706	0.889	1.108	1.397	1.860	2.306	2.896	3.355	3.833	4.501	5.041
9	0.703	0.883	1.100	1.383	1.833	2.262	2.821	3.250	3.690	4.297	4.781
10	0.700	0.879	1.093	1.372	1.812	2.228	2.764	3.169	3.581	4.144	4.587

Degrees of freedom

t_9

0.975

.025

2.262

FIGURE 11.2   Table C and $t_{9,0.975} = 2.262$, illustrative example.

ILLUSTRATIVE EXAMPLE

Understanding Table C. Suppose we want to identify the values of a *t*-random variable with 9 degrees of freedom that captures the middle 80% of values on t_9. Start by sketching the curve, which is bell shaped with a mean of 0 and inflection points *approximately* one standard deviation above and below the mean.[d] Because the curve is symmetrical, we look for the critical values in Table C that cut off the bottom 10% and top 10% of the curve. These are the 10th percentile and 90th percentile of the t_9 distribution. Table C lets us know that $t_{9,0.90} = 1.383$.

---

[d]The standard deviation of a *t*-distribution is actually a little more than one, but this won't be detectable in a sketch. The standard deviation of a *t*-distribution is $\sqrt{\dfrac{df}{df - 2}}$. For example, a *t*-distribution with 10 df has $\sigma = \sqrt{\dfrac{10}{10 - 2}} = 1.118$.

The symmetrical point on the left-hand aspect of the curve corresponding to $t_{9,0.10} = -1.383$.[e] Figure 11.3 depicts these relationships graphically.

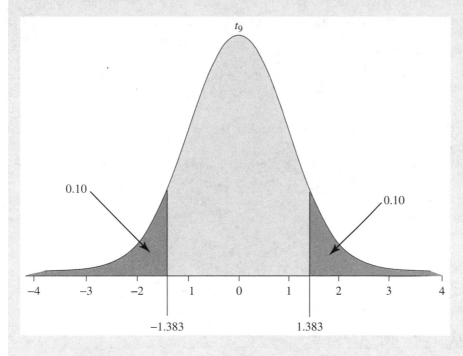

FIGURE 11.3   The 10th and 90th percentiles on t_9.

Exercises

11.3 *Sketch a curve*. Use Table C to determine the values on a t_{22} distribution that captures the middle 95% of the area under the curve. Sketch the curve showing the *t*-values on the horizontal axis and associated tail areas.

11.4 *t-percentiles*. Use Table C to determine the value of a *t*-random variable with 19 degrees of freedom and a cumulative probability of 0.95. In addition, determine the value of $t_{19,0.05}$.

---

[e] Because *t* tables do not have negative *t*-values, you must use your knowledge about the symmetry of the curve to determine lower percentile points.

11.5 *Probabilities not in Table C*. There are times you may need to determine a probability for a *t*-random variable that does *not* appear in Table C. For example, you may need to determine the probability a *t*-random variable with 8 degrees of freedom is greater than 2.65. Even though this value is not in Table C, you can still get a good idea of its probability by bracketing it between two landmarks in the table. In this case, 2.65 is bracketed between $t_{8,0.975}$ (2.306) and $t_{8,0.99}$ (2.896). Therefore, it has a cumulative probability that is a little bigger than 0.975 and a little smaller than 0.99. What is the probability of $t_8 > 2.65$? In other words, what is the probability of observing a *t* random variable with 8 df that is greater than 2.65?

11.6 *Upper tail*. Determine the probability that a *t*-random variable with 8 df is greater than 2.98.

11.7 *Software utility programs*. The Internet has many free applets that can determine exact probabilities for *t*-random variables.[f] You can also use *WinPepi*[g] WhatIs.exe or Microsoft Excel's TDIST function for this purpose. Use one of these software applications to find:

(a) $\Pr(T_8 \geq 2.65)$

(b) $\Pr(T_8 \geq 2.98)$

(c) $\Pr(T_{19} \leq 2.98)$

■ 11.3 One-Sample *t*-Test

When σ is known, $\dfrac{\bar{x} - \mu}{\sigma / \sqrt{n}}$ has a Standard Normal (z) distribution. When σ is *not* known, $\dfrac{\bar{x} - \mu}{s / \sqrt{n}}$ has a *t*-distribution with $(n - 1)$ degrees of freedom. This is the basis of the **one-sample *t*-test.** Here are the steps of the procedure:

A. **Hypothesis statements.** The null hypothesis is H_0: $\mu = \mu_0$, where μ_0 represents the mean under the null hypothesis. The alternative hypothesis is either H_a: $\mu > \mu_0$ (one sided to the right), H_a: $\mu < \mu_0$ (one sided to the left), or H_a: $\mu \neq \mu_0$ (two sided).

B. **Test statistic.** $t_{\text{stat}} = \dfrac{\bar{x} - \mu_0}{\text{SE}_{\bar{x}}}$ where $\text{SE}_{\bar{x}} = \dfrac{s}{\sqrt{n}}$. This *t*-statistic has df $= n - 1$.[h]

---

[f] See, for example, www.stat.tamu.edu/applets/tdemo.html.
[g] Abramson, J. H. (2004). *WINPEPI* (PEPI-for-Windows): Computer programs for epidemiologists. *Epidemiologic Perspectives & Innovations, 1*(1), 6.
[h] You lose the 1 degree of freedom in using *s* as an estimate for σ.

C. *P*-value. Use Table C or a software utility to convert the t_{stat} to a *P*-value. For one-sided alternatives, $P = \Pr(t \geq |t_{stat}|)$. For two-sided alternatives, $P = 2 \times \Pr(t \geq |t_{stat}|)$. Recall that small *P*-values provide good evidence against H_0 (Section 9.3).

D. Significance level. The test is said to be significant at the α-level when $P \leq \alpha$ (Section 9.4).

E. Conclusion. The test results are interpreted in the context of the data and research question.

As was the case with one-sample *z*-tests, the one-sample *t*-test assumes that the data were generated by an SRS and that the sampling distribution of $\bar{x}$ is Normal.

ILLUSTRATIVE EXAMPLE

One-sample t-test (SIDS birth weights). We want to know whether birth weights of full-term infants who ultimately died of SIDS is significantly different from that of other full-term births. A sample of $n = 10$ SIDS cases demonstrated the following birth weights:

```
2998   3740   2031   2804   2454   2780   2203   3803   3948   2144
```

Based on this information, we calculate $\bar{x} = 2890.5$ g and $s = 720.0$ g. By comparison, the mean weight of other full-term births in the region during this period was 3300 g. We test whether the birth weight in this sample is significantly different from a population mean of 3300 g. A two-sided test is demonstrated.

A. Hypotheses. H_0: $\mu = 3300$ g versus H_a: $\mu \neq 3300$ g.

B. Test statistic. The estimated standard error of the mean

$$SE_{\bar{x}} = \frac{s}{\sqrt{n}} = \frac{720}{\sqrt{10}} = 227.7.$$

$$t_{stat} = \frac{\bar{x} - \mu_0}{SE_{\bar{x}}} = \frac{2890.5 - 3300}{227.7} = -1.80$$

with df $= n - 1 = 10 - 1 = 9$.

C. *P*-value. The two-sided *P*-value is twice the area under the curve to the right of $|-1.80|$ on a *t*-distribution with 9 df. Because there is no entry for 1.80 in Table C, we bracket the t_{stat} between 1.383 (right tail = 0.10) and 1.833 (right tail = 0.05). Therefore, the one-sided *P*-value is between 0.10 and 0.05 and the two-sided *P*-value is between 0.20 and 0.10. Using a software utility, we determine $P = 0.105$. Figure 11.4 depicts the test statistic and associated *P*-value regions for the problem.

continues

continued

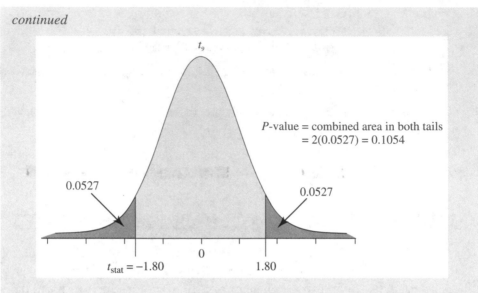

FIGURE 11.4   Two-tailed *P*-value, SIDS illustrative example.

D. **Significance level**. The observed difference is not significant at $\alpha = 0.10$. Results fall just short of "marginal significance."

E. **Conclusion**. The mean birth weight in this sample of 10 SIDS infants ($\bar{x} = 2890.5$ g) was not significantly different from that of the general population of infants ($\mu = 3300$ g) at conventional levels of α ($P = 0.105$).

Exercises

11.8   ***P-value from* t_{stat}**. A test of H_0: $\mu = 0$ based on $n = 16$ calculates $t_{stat} = 2.44$.

   (a) Determine the degrees of freedom for the test statistic.

   (b) Provide the *t*-values from the Table C that bracket the t_{stat}.

   (c) What is the approximate one-sided *P*-value for the problem?

   (d) What is the two-sided *P*-value?

11.9   ***Critical values for a t-statistic***. The term **critical value** is often used to refer to the value of a test statistic that determines statistical significance at some fixed α level for a test. For example, ± 1.96 are the critical values for a two-tailed *z*-test at $\alpha = 0.05$. In performing a *t*-test based on 21 observations, what are the critical values for a one-tailed test when $\alpha = 0.05$? That is, what values of the t_{stat} will give a one-sided *P*-value that is less than or equal to 0.05? What are the critical values for a two-tailed test at $\alpha = 0.05$?

11.10 **BMI**. Body mass index $(BMI) = \dfrac{weight\ in\ lbs}{height\ inches^2} \times 703$. An adult BMI of 24 is considered desirable.[i] A study of 12 adults that used a one-sample t-test to address H_0: $\mu = 24$ reported a t_{stat} of 2.16.

(a) What is the one-sided P-value for this test?

(b) What is the two-sided P-value?

11.11 **Menstrual cycle length**. Menstrual cycle lengths (days) in an SRS of nine women are as follows: {31, 28, 26, 24, 29, 33, 25, 26, 28}. Use this data to test whether mean menstrual cycle length differs significantly from a lunar month. (A lunar month is 29.5 days.) Assume that population values vary according to a Normal distribution. Use a two-sided alternative. Show all hypothesis-testing steps.

■ 11.4 Confidence Interval for μ

The t confidence interval formula is similar to the z confidence interval formula (Section 10.2) except that it uses s in place of σ and $t_{n-1,1-\alpha/2}$ in place of $z_{1-\alpha/2}$. A $(1 - \alpha)100\%$ confidence interval for μ is provided by:

$$\bar{x} \pm t_{n-1,1-\frac{\alpha}{2}} \cdot \text{SE}_{\bar{x}}$$

where $\bar{x}$ is the sample mean, $t_{n-1,1-\alpha/2}$ is the value of a t random variable with $n - 1$ df and a cumulative probability of $1-(\alpha/2)$,[j] and $\text{SE}_{\bar{x}} = \dfrac{s}{\sqrt{n}}$.

ILLUSTRATIVE EXAMPLE

Confidence intervals for μ (SIDS). Let us return to the birth weight illustrative data concerning 10 SIDS cases. We have established that $\bar{x} = 2890.5$ g and $s = 720.0$. Confidence intervals for μ at 90%, 95%, and 99% levels of confidence will be calculated.

- The standard error of the mean $\text{SE}_{\bar{x}} = \dfrac{720}{\sqrt{10}} = 227.684$.

- df = $10 - 1 = 9$.

continues

---

[i]CDC. (2006). About BMI for Adults. Retrieved on July 15, 2006, from www.cdc.gov/NCCdphp/dnpa/bmi/adult_BMI/about_adult_ BMI.htm.
[j]Table C list confidence levels for t random variables in its bottom row.

continued

- For 90% confidence, use $t_{9,0.95} = 1.833$ (Table C), so the 90% confidence interval for μ is $\bar{x} \pm (t_{n-1,1-\frac{\alpha}{2}})(SE_{\bar{x}}) = 2890.5 \pm (1.833)(227.684) = 2890.5 \pm 417.3 = (2473.2$ to $3307.8)$ g.
- For 95% confidence, use $t_{9,0.975} = 2.262$, so the 95% confidence interval for μ is $2890.5 \pm (2.262)(227.684) = 2890.5 \pm 515.0 = (2375.5$ to $3405.5)$ g.
- For 99% confidence, use $t_{9,0.995} = 3.250$, so the 99% confidence interval for μ is $2890.5 \pm (3.250)(227.7) = 2890.5 \pm 740.0 = (2150.5$ to $3630.5)$ g.

Keep in mind that the confidence interval is used to help infer *population* mean μ, *not* sample mean $\bar{x}$.

Exercises

11.12 *t-values for confidence*. What is the value of $t_{n-1,1-\alpha/2}$ when calculating a 95% confidence interval for μ based on $n = 28$? What is $t_{n-1,1-\alpha/2}$ for 90% confidence?

11.13 *Menstrual cycle length*. Exercise 11.11 calculated the mean length of menstrual cycles in an SRS of $n = 9$ women. The data revealed $\bar{x} = 27.78$ days with standard deviation $s = 2.906$ days.

(a) Calculate a 95% confidence interval for the mean menstrual cycle length.

(b) Based on the confidence interval you just calculated, is the mean menstrual cycle length significantly different from 28.5 days at $\alpha = 0.05$ (two sided)? Is it significantly different from $\mu = 30$ days at the same α-level? Explain your reasoning. (Section 10.4 considered the relationship between confidence intervals and significance tests. The same rules apply here.)

■ 11.5 Paired Samples

Data

With **paired samples**, each data point in one sample is matched to a unique point in a second sample. Here are examples of studies that employ paired samples:

- Pretest/posttest studies in which a factor is measured before and after an intervention in the same set of individuals.

- Cross-over trials, in which subjects start on one treatment and then switch to a different treatment.
- Pair-matches, in which subjects in one sample are matched to subjects in a separate sample based on specific criteria.

Here is an illustrative example of a cross-over trial.

ILLUSTRATIVE EXAMPLE

Data (Oat bran and LDL cholesterol). A cross-over trial sought to learn whether oat bran cereal lowered low-density lipoprotein (LDL) cholesterol in hypercholesterolemic men. Twelve subjects completed the study. Half were randomly assigned a diet that included oat bran cereal. The other half were assigned a diet that included corn flakes. Dietary interventions were applied for 2 weeks, after which LDL levels (mmol/L) were recorded. Subjects were then *crossed-over* to the alternative diet for 2 weeks. LDL was once again measured. Table 11.1 lists data from the study.

TABLE 11.1 "Oat bran" illustrative data. LDL cholesterol (mmol/L) on the corn flake diet (CORNFLK), oat bran diet (OATBRAN), and their difference (DELTA).

Observation	CORNFLK	OATBRAN	DELTA
1	4.61	3.84	0.77
2	6.42	5.57	0.85
3	5.40	5.85	−0.45
4	4.54	4.80	−0.26
5	3.98	3.68	0.30
6	3.82	2.96	0.86
7	5.01	4.41	0.60
8	4.34	3.72	0.62
9	3.80	3.49	0.31
10	4.56	3.84	0.72
11	5.35	5.26	0.09
12	3.89	3.73	0.16

Data from Anderson, J. W., Spencer, D. B., Hamilton, C. C., Smith, S. F., Tietyen, J., Bryant, C. A., et al. (1990). Oat-bran cereal lowers serum total and LDL cholesterol in hypercholesterolemic men. *American Journal of Clinical Nutrition, 52*(3), 495–499. Data are stored in the file oatbran.sav.

continues

continued

Paired samples are analyzed by creating a new variable to hold within pair differences. Call this new variable DELTA. Table 11.1 lists DELTA values created by subtracting each oat bran value from each corn flake value for each individual. Positive DELTA values reflect lower LDL on the oat bran diet.[k]

Exploration and Description

It is often a good idea to start the analysis by plotting the data. A stemplot of the DELTA values for the oat bran illustrative data looks like this:

```
−0|24
 0|0133
 0|667788
×1
```

This shows that DELTA values range from -0.4 to 0.8 mmol/L. Ten of the 12 subjects (83%) lowered their LDL on oat bran. There are no apparent outliers.

We will attach the subscript $_d$ to descriptive statistics to denote application to the DELTA variable. The illustrative data show a mean decline of $\bar{x}_d = 0.3808$ mmol/L with standard deviation $s_d = 0.4335$ ($n_d = 12$) while on oat bran.

Hypothesis Test

When samples are paired, t-procedures introduced earlier in the chapter are applied toward difference variable DELTA. Here's the procedure for testing a mean difference:

A. **Hypotheses.** The population mean difference is denoted μ_d. In testing "no mean difference," the null hypothesis is H_0: $\mu_d = 0$. The alternative hypothesis is one of the following: H_a: $\mu_d > 0$ (one sided to the right), H_a: $\mu_d < 0$ (one sided to the left), or H_a: $\mu_d \neq 0$ (two sided). In practice, most tests are two sided.

B. **Test statistic.** The test statistic is $t_{stat} = \dfrac{\bar{x}_d - \mu_0}{\text{SE}_{\bar{x}_d}}$ where $\mu_0 = 0$ and $\text{SE}_{\bar{x}_d} = \dfrac{s_d}{\sqrt{n_d}}$. This test statistic has df $= n_d - 1$, where n_d represents the number of paired observations.

C. **P-value.** The t_{stat} is converted to a P-value with Table C or a software utility. Small P-values provide good evidence against H_0 (Section 9.3).

---

[k]When creating the DELTA variable, it makes no difference which sample is subtracted from which. You must, however, be consistent and keep track of the direction of differences.

D. Significance level. The difference is said to be statistically significant at the α-level of significance when $P \le \alpha$. By convention, P-values less than 0.10 are said to be marginally significant, those less than 0.05 are said to be significant, and those less than 0.01 are said to be highly significant (Section 9.4).

E. Conclusion. The test results are interpreted in the context of the data and research question.

ILLUSTRATIVE EXAMPLE

Paired t-test (Oat bran). We test the oat bran illustrative data for significance. We have already established that $n_d = 12$, $\bar{x}_d = 0.3808$, and $s_d = 0.4335$.

A. Hypotheses. H_0: $\mu = 0$ versus H_a: $\mu_d \ne 0$

B. Test statistic. $\mathrm{SE}_{\bar{x}_d} = \dfrac{s_d}{\sqrt{n_d}} = \dfrac{0.4335}{\sqrt{12}} = 0.1251$ and $t_{\mathrm{stat}} =$

$\dfrac{\bar{x}_d - \mu_0}{\mathrm{SE}_{\bar{x}_d}} = \dfrac{0.3808 - 0}{0.1251} = 3.04$ with df $= n_d - 1 = 12 - 1 = 11$.

C. P-value. Figure 11.5 illustrates the sampling distribution of the test statistic under the null hypothesis. Under this hypothesis, both the t_{stat} and $\bar{x}_d$ have an

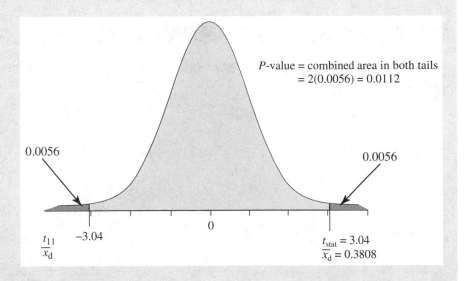

FIGURE 11.5 Two-tailed P-value, oat bran illustrative example. Sampling distribution under the null hypothesis.

continues

continued

expected value of 0. The observed mean difference of 0.3808 is 3.04 standard errors above 0. The one-tailed P-value is between 0.005 and 0.01 (Table C). The two-tailed P-value is twice this: $0.01 < P < 0.02$. Statistical software derives $P = 0.011$ (two tailed). This provides good evidence against the null hypothesis.

D. Significance. The results are statistically significant at $\alpha = 0.05$ (reject H_0) but not quite at $\alpha = 0.01$ (retain H_0).

E. Conclusion. LDL levels on the oat bran diet were significantly lower than that on the cornflake diet ($P = 0.011$).

Confidence Interval for μ_d

A $(1 - \alpha)100\%$ confidence interval for μ_d is provided by:

$$\bar{x}_d \pm t_{n_d - 1, 1 - \frac{\alpha}{2}} \cdot SE_{\bar{x}_d}$$

where $SE_{\bar{x}_d} = \dfrac{s_d}{\sqrt{n_d}}$.

ILLUSTRATIVE EXAMPLE

Confidence intervals for μ_d (Oat bran). 90%, 95%, and 99% confidence intervals for the oat bran illustrative data set are calculated. We have already established $n_d = 12$, $\bar{x}_d = 0.3808$, $s_d = 0.4335$, $SE_{\bar{x}_d} = \dfrac{0.4335}{\sqrt{12}} = 0.1251$, and df $= 12 - 1 = 11$.

- For 90% confidence, use $t_{11, 0.95} = 1.796$ (Table C); the 90% confidence interval for population mean difference μ_d is $\bar{x}_d \pm t_{n_d - 1, 1 - \frac{\alpha}{2}} \cdot SE_{\bar{x}_d} = 0.3808 \pm (1.796)(0.1251) = 0.3808 \pm 0.2247 = (0.1561 \text{ to } 0.6055)$ mmol/L.

- For 95% confidence, use $t_{11, 0.975} = 2.201$ (Table C); the 95% confidence interval for population mean difference μ_d is $0.3808 \pm (2.201)(0.1251) = 0.3808 \pm 0.2753 = (0.1055 \text{ to } 0.6561)$ mmol/L.

- For 99% confidence, use $t_{11, 0.995} = 3.106$; the 99% confidence interval for μ_d is $0.3808 \pm (3.106)(0.1251) = 0.3808 \pm 0.3886 = (-0.0078 \text{ to } 0.7694)$ mmol/L.

Exercises

11.14 *Placebo effect in Parkinson's disease patients*. The placebo effect occurs when a patient experiences a perceived benefit after receiving an inert substance. To help understand the mechanism behind this phenomenon in Parkinson's disease patients, investigators measured striatal RAC binding at a key point in the brains in six subjects. RAC binding was reduced by an average of 0.326 units on a placebo in the six subjects ($s_d = 0.181$).[1] Test this difference for statistical significance.

11.15 *Water fluoridation*. A study looked at the number of cavity-free children per 100 in 16 North American cities BEFORE and AFTER public water fluoridation projects. Table 11.2 lists the data.

   (a) Calculate DELTA values for each city. Then construct a stemplot of these differences. Interpret your plot.

   (b) What percentage of cities showed an improvement in their cavity-free rate?

   (c) Estimate the mean change with 95% confidence.

TABLE 11.2   Cavity-free children per 100 in 16 North American cities before and after public water fluoridation projects.

AFTER	BEFORE
49.2	18.2
30.0	21.9
16.0	5.2
47.8	20.4
3.4	2.8
16.8	21.0
10.7	11.3
5.7	6.1
23.0	25.0
17.0	13.0
79.0	76.0
66.0	59.0
46.8	25.6
84.9	50.4
65.2	41.2
52.0	21.0

Source unknown. Data stored online in FLUORIDE.SAV.

[1] de la Fuente-Fernández, R., Ruth, T. J., Sossi, V., Schulzer, M., Calne, D. B., & Stoessl, A. J. (2001). Expectation and dopamine release: Mechanism of the placebo effect in Parkinson's disease. *Science, 293*(5532), 1164–1166.

■ 11.6 Conditions for Inference

The *t*-procedures in this chapter rely on the following underlying conditions:

- Data are derived by an SRS of individual or paired observations.
- Measurements of the response are valid.
- The sampling distribution of the mean or mean difference is Normal.

Numerous studies have shown that *t*-procedures are *robust*[m] against the Normality condition, especially when a two-sided alternative hypothesis is used and the sample is large. This can be traced in part to the central limit theorem (Section 8.2). Rough guidelines for how large a sample needs to be to compensate for non-Normality in the population (Section 9.5) are:

- When the population is Normal, you can use *t*-procedures on samples of any size.
- When the population is mound shaped and symmetrical, you can use *t*-procedures on samples as small as 5 to 10.
- When the population is skewed, *t*-procedures should be reserved for large samples (roughly 30 to 100 observations, depending on the severity of the skew).[n]

ILLUSTRATIVE EXAMPLE

Can a t-procedure be used? Figure 11.6 displays stemplots for three data sets. Which of these data sets can support *t*-procedures?
- Stemplot A has a positive skew and outlier. It has only six observations. In this situation, *t*-procedures should be avoided.
- Stemplot B has $n = 25$ observations, is mound shaped, and has a modest negative skew. There are no outliers. It is okay to use the *t*-procedures on these data.
- Stemplot C is highly skewed with a high outlier. There are only 13 observations; it would be imprudent to use *t*-procedures on these data.

---

[m] This refers to the fact that the results remain substantially true even when the condition is not perfectly met.
[n] Skewed distributions may be mathematically transformed to Normalize the distribution (Sections 7.1 and 7.4).

```
A. Ages of Grad students, n = 6

2│1478
3│0
4│
5│3
×10
```

```
B.  Coliform levels in water samples, n = 25

1│4
1│789
2│2234
2│66789
3│00012344
3│5678
×1
```

```
C. Months between acute infection and first seizure, n = 13

0│0004
1│22
2│44
3│16
4│2
5│5
6│
7│
8│
9│6
×10  (months)
```

FIGURE 11.6   "Can a *t*-procedure be used?" illustrative examples.

■ 11.7 Sample Size and Power

The sample size requirements of a study can be approached from a confidence interval or hypothesis testing perspective. Let us start by considering the sample requirements for confidence intervals.

Sample Size for a Confidence Interval

To limit the margin of error of a $(1 - \alpha)100\%$ confidence interval for μ (or μ_d) to m, the sample size should be no less than:

$$n = \left(z_{1 - \frac{\alpha}{2}} \frac{\sigma}{m} \right)^2$$

where σ is the population standard deviation,[o] $z_{1-(\alpha/2)}$ is the Standard Normal deviate for $(1 - \alpha)100\%$ confidence, and m is the desired margin of error. Results from this formula should be rounded up to the next integer to achieve the stated level of precision.

This formula is accurate when $n \geq 30$ because $t_{30+, 1-(\alpha/2)} \approx z_{1-(\alpha/2)}$. When $n < 30$, apply adjustment factor $f = (df + 3)/(df + 1)$ to compensate for the difference between z and t.[p]

ILLUSTRATIVE EXAMPLE

Sample size, confidence interval. The oat bran illustrative example calculated a 95% confidence interval for μ that had a margin of error of 0.2753 mmol/L. How large a study is needed to achieve a margin of error 0.2 with 95% confidence? We will use the sample standard deviation from the study as our estimate of σ.

Solution: $n = \left(z_{1-\frac{\alpha}{2}} \dfrac{\sigma}{m} \right)^2 = \left(1.96 \dfrac{0.4335}{0.2} \right)^2 = 18.50$. Round this up to the next integer, so $n = 19$. Because this number is less than 30, apply adjustment factor $f = (df + 3)/(df + 1) = (18 + 3)/(18 + 1) = 1.105$ to the final result. Therefore, use $n = 1.105 \times 19 = 21$.

Sample Size for a Hypothesis Test

The sample size required to test $H_0: \mu = \mu_0$ against $H_a: \mu = \mu_a$ depends on the desired power of the test $(1 - \beta)$, desired level of significance (α), size of the mean difference worth detecting $(\Delta = \mu_0 - \mu_a)$, and standard deviation of the response variable (σ). To achieve these conditions use:

$$n = \frac{\sigma^2 (z_{1-\beta} + z_{1-\frac{\alpha}{2}})^2}{\Delta^2}$$

For one-sided tests, use $z_{1-\alpha}$ in place of $z_{1-(\alpha/2)}$ in the formula. Results from this equation should be rounded up to the next integer to achieve the stated level of power. For paired t-tests, use σ_{DELTA} in place of σ. Apply adjustment factor $f = (df + 3)/(df + 1)$ when $n \leq 30$ to compensate for the difference between z and t.

---

[o] You may need to estimate σ with a value of s from published sources or a pilot investigation.
[p] Lachin, J. M. (1981). Introduction to sample size determination and power analysis for clinical trials. *Controlled Clinical Trials, 2*(2), 93–113.

ILLUSTRATIVE EXAMPLE

Sample size requirement, one-sample t-test (SIDS). How large a sample is needed to test the SIDS data presented earlier in this chapter with 90% power at $\alpha = 0.05$ two sided? We want to detect a mean difference in birth weight (Δ) of 300 g (about 2/3 pound). Let us use sample standard deviation s (720.0) as a reasonable estimate of σ.

$$\text{Solution: } n = \frac{\sigma^2(z_{0.9} + z_{0.975})^2}{300^2} = \frac{720^2(1.28 + 1.96)^2}{300^2} = 60.47.$$ Round up to 61 to ensure adequate power. Because the sample size exceeds 30, there is no need to apply adjustment factor f.

ILLUSTRATIVE EXAMPLE

Sample size requirement, paired t-test (Oat bran). How large a sample is needed to test the oat bran data presented earlier in the chapter with 80% power at $\alpha = 0.05$ two sided? We want to detect a mean change of 0.2 mmol/L and will assume a standard deviation of 0.4 mmol/L.

$$\text{Solution: } n = \frac{\sigma^2(z_{0.8} + z_{0.975})^2}{0.2^2} = \frac{0.4^2(0.84 + 1.96)^2}{0.2^2} = 31.36.$$ Round up to 32 to ensure adequate power. Because the sample size exceeds 30, there is no need to apply adjustment factor f.

Power

The method to determine the power of the hypothesis initially presented in Section 9.6 applies with minor modification. The power of the test is approximately:

$$1 - \beta = \Phi\left(-z_{1-\frac{\alpha}{2}} + \frac{|\Delta|\sqrt{n}}{\sigma}\right)$$

where $\Phi(z)$ is the cumulative probability of a Standard Normal random variable (Table B), α is the desired significance level, Δ is the difference worth detecting, and σ is the standard deviation of the response variable.

ILLUSTRATIVE EXAMPLE

Power of t-test (SIDS). What is the probability a study with $n = 10$ will detect a mean difference of 300 g in birth weight in the SIDS population compared to the general population? Let us assume $\sigma = 720$ and use a two-sided α-level of 0.05.

continues

continued

Solution: $1 - \beta = \Phi\left(-z_{1-\frac{0.5}{2}} + \frac{|\Delta|\sqrt{n}}{\sigma}\right) =$

$\Phi\left(-1.96 + \frac{|300|\sqrt{10}}{720}\right) = \Phi(-0.64) = 0.2611$. The power of this test is about 26%.

Summary Points (Inference about a Mean)

1. Data are a **quantitative response variable** derived by a single SRS or match-pair sample.

2. Begin the analysis by **exploring and describing** the data with graphical techniques (e.g., stemplot and boxplot) and summary statistics (e.g., mean, standard deviation, and sample size).

3. In most practical situations, population standard deviation σ is not known. This invalidates z-procedures and requires the use of Student t-**procedures** instead.

   (a) t-probability density functions (*pdf*s) look like a Standard Normal "z" curve except for the fact that they have slightly (almost imperceptibly) broader tails.

   (b) t-distributions are a family of *pdf*s distinguished by their degrees of freedom (df). As the df increases, the distribution becomes more and more Normal.

   (c) A t-distribution with infinite degrees of freedom *is* a Standard Normal z-distribution.

4. **One-sample and paired-sample t-tests**

   (a) $H_0: \mu = \mu_0$, where μ_0 represents the population mean or paired mean difference under the null hypothesis. The alternative hypothesis may be stated in a two-sided ($H_a: \mu \neq \mu_0$) or one-sided way ($H_a: \mu < \mu_0$ or $H_a: \mu > \mu_0$).

   (b) $t_{stat} = \dfrac{\bar{x} - \mu_0}{s/\sqrt{n}}$ with df $= n - 1$.

   (c) Use Table C or a computer applet to convert the t_{stat} to a P-value. The one-sided P-value is the area under the curve to the right of the $|t_{stat}|$. The two-sided P-value is twice this amount.

   (d) Consider the level of statistical significance.

   (e) Formulate a conclusion in the context of the data and research question.

5. A $(1 - \alpha)100\%$ **confidence interval for μ** is given by $\bar{x} \pm t_{n-1,1-\frac{\alpha}{2}}\frac{s}{\sqrt{n}}$.

   (a) Keep in mind that the confidence interval seeks to capture μ, not $\bar{x}$.

   (b) The interval has $(1 - \alpha)100\%$ chance of capturing μ and an α chance of not capturing μ.

   (c) The margin of error of the confidence interval is given by the "$\pm$ value" in the formula.

   (d) Narrow confidence intervals indicate that the sample mean as an estimate of the population mean is precise.

6. Inferential methods for **matched-pair data** are the same as that for single-sample data except that inferences are directed against the differences variable DELTA.

7. t-procedures require the following conditions:

   (a) SRSs or a reasonable approximation thereof.

   (b) Source population is Normal *or* the sample is large. t-procedures are known to be robust when the sample is large because of the central limit theorem.

   (c) The measurements arc valid.

8. The **power and sample size methods** for z-procedures introduced in the prior chapter can be used in this chapter after applying an adjustment factor to compensate for the differences between z and t.

Vocabulary

Critical value	Student's t-distributions
Degrees of freedom (df)	t-distribution (t-probability density
DELTA	function)
One-sample t-test	t-percentiles
Paired samples	t-statistic
Standard error of the mean	

Review Questions

11.1 When do you use a t-procedure instead of a z-procedure to help infer a mean?

11.2 Describe the shape, location, and spread of t-distributions.

11.3 How many different t-distributions are there?

11.4 The mean of a t-distribution is equal to ___.

11.5 How do t-distributions differ from Standard Normal z-distributions?

11.6 Select the best response: The total area under a t-curve is equal to

(a) -1

(b) 0

(c) 1

11.7 Select the best response: In the notation $t_{df,p}$, the subscript $_{df}$ represents the

(a) degrees of freedom for the t-distribution.

(b) probability of t.

(c) cumulative probability of t (AUC to the left of the t-value).

11.8 Select the best response: In the notation $t_{df,p}$, the subscript $_p$ represents the

(a) degrees of freedom for the t-distribution.

(b) probability of t.

(c) cumulative probability of t (AUC to the left of the t-value).

11.9 Determine the value of $t_{8,0.50}$ without the aid of a t-table.

11.10 $t_{9,0.90} = 1.383$; therefore, $t_{9,0.10} = ?$ (t-table not required)

11.11 A t-distribution with 60 or more degrees of freedom is very nearly a _____distribution.

11.12 The standard error of the mean is equal to the sample standard deviation divided by the square root of ___.

11.13 Select the best response: In the statement H_0: $\mu = \mu_0$, μ_0 represents the value of the population mean when the null hypothesis is ___.

(a) true

(b) false

(c) either true or false

11.14 Select the best response: With matched paired sample, the null hypothesis is most often H_0: $\mu = $ ___.

(a) -1

(b) 0

(c) 1

11.15 A one-sample t-procedure with 35 observations has this many degrees of freedom.

11.16 The P-value for a two-sided t-test is equal to

(a) the area under the curve to the right of the t-statistic.

(b) twice the area under the curve to the right of the t-statistic.

(c) twice the area under the curve to the right of the absolute value of the t-statistic.

11.17 Select the best response: *P*-values assume that

(a) the null hypothesis is true.

(b) the null hypothesis is false.

(c) the null hypothesis is neither true nor false.

11.18 Select the best response: A *t*-test derives a *P*-value of 0.06. The *P*-value represents the probability that

(a) the null hypothesis is true.

(b) the null hypothesis is false.

(c) we would see the data or data that are more extreme assuming the null hypothesis is true.

11.19 Select the best response: A 95% confidence for μ is used to infer the value of the

(a) sample mean.

(b) population mean.

(c) population standard deviation.

11.20 Select the best response: A 95% confidence interval for a mean is -0.91 to 1.36. From this we can infer with 95% confidence that the

(a) population mean is 0.

(b) population mean is greater than 0.

(c) population mean is greater than -0.91.

11.21 Select the best response: A 95% confidence interval for a mean is 0.86 to 1.66. From this we can infer with 95% confidence that the

(a) population mean is 1.

(b) population mean is greater than 1.

(c) neither of the above

11.22 Select the best response: A 95% confidence interval for a mean is 0.91 to 1.36. From this we can infer with 95% confidence that the

(a) population mean is 0.

(b) population mean is greater than 0.

(c) population mean is less than 0.

11.23 Select the best response: A 95% confidence interval for a mean is 0.91 to 1.36. From this we can infer with 95% confidence that the population mean is

(a) not more than 0.91.

(b) not less than 0.91.

(c) not less than 1.36.

11.24 Select the best response. Paired samples can be achieved via

(a) pretest/posttest samples.

(b) matching closely on extraneous factors when sampling.

(c) both "a" and "b."

11.25 Select the best response: Paired *t*-procedures focus on the data in the

(a) first sample in the pair.

(b) second sample in the pair.

(c) differences between the first and second samples in the pair.

11.26 Select the best response: These conditions are needed for valid *t*-procedures:

(a) SRS of individual or paired difference (or reasonable approximation thereof)

(b) Normality of the sampling distribution of the mean

(c) both "a" and "b"

11.27 Select the best response: The sampling distribution of a mean will be approximately Normal even when the population is not exactly Normal as long as the sample is

(a) representative.

(b) large.

(c) small.

11.28 List the determinants of the sample size requirements for estimating μ with a margin of error m.

11.29 List the determinants of the sample size requirements when testing a mean at a given α level.

11.30 List the determinants of the power of a *t*-test.

Exercises

11.16 *t-percentiles.* Use Table C to determine the following values of *t*-values:

(a) $t_{24,0.975}$

(b) $t_{674,0.99}$ (*Suggestion:* Because there is no row for df = 674, use the row with df = 100 to derive a conservative estimate for the *t* critical value).

(c) $t_{24,0.05}$

11.17 *Large t-statistic*. A *t*-test calculates $t_{\text{stat}} = 6.60$. Assuming the study had more than just a few observations, you do not need a *t* table or software utility to draw a conclusion about the test. What is this conclusion, and why is a look-up table unnecessary?

11.18 *Sketch and shade*. In testing H_0: $\mu = 0$, you find $\bar{x} = 0.762$ and $s = 1.497$ based on $n = 50$. Calculate the t_{stat} for the test. Sketch a *t*-curve as accurately as possible. Then place this t_{stat} on your curve. Without using Table C, do you think these results would be surprising if H_0 were true?

11.19 *Vector control in an African village*. A study of vector control in an African village found that the mean sprayable surface area was 249 square feet with standard deviation 39.82 square feet in a simple random sample of $n = 100$ homes.

(a) Calculate a 95% confidence interval for μ.

(b) Would it be correct to say that 95% of all the homes in the village have sprayable surfaces between the lower confidence limit and upper confidence limit? Explain.

11.20 *Calcium in sound teeth*. The calcium content values in a sample $n = 5$ sound teeth (% calcium) are {33.4, 36.2, 34.8, 35.2, 35.5}. Provide a 99% confidence interval for μ. (Assume the data represent an SRS of healthy adult teeth.)

11.21 *Boy height*. An SRS of $n = 26$ boys between the ages of 13 and 14 has a mean height of 63.8 inches with a standard deviation 3.1 inches. Calculate a 95% confidence interval for the mean height of the population.

11.22 *Body weight, high school girls*. Body weights expressed as a percentage of ideal in an SRS of $n = 9$ girls selected at random are as follows {114, 100, 104, 94, 114, 105, 103, 105, 96}.

(a) Plot the data as a stemplot. (Use an axis multiplier of 10 and split stem values.) Are there any outliers or major departures from Normality in these data?

(b) Calculate a 95% confidence for population mean μ. Show all work.

(c) What is the margin of error of your confidence interval?

(d) How large a sample would be needed to reduce this margin of error to three?

11.23 *Faux pas*. Eight junior high school students were taken to a shopping mall. The number of socially inappropriate behaviors (*faux pas*) by each student was counted. The students were then enrolled in a program designed to promote social skills. After completing the programs, the subjects were again taken to the shopping mall and the number of social *faux pas* was again counted. Table 11.3 lists data from this experiment.

TABLE 11.3   Data for Exercise 11.23. Number of *faux pas* before and after an intervention.

OBS.	VISIT1	VISIT2
1	5	4
2	13	11
3	17	12
4	3	3
5	20	14
6	18	14
7	8	10
8	15	9

Data are fictitious. Data file = FAUXPAS.SAV.

(a) Calculate the change in the number of *faux pas* within individuals (i.e., calculate DELTA for each observation).

(b) Calculate the means and standard deviations for visit 1, visit 2, and their differences (DELTAS).

(c) Create a stemplot of the differences (DELTAS). Interpret your plot.

(d) Would you use *t*-procedures on these data?

(e) Test the mean decline for statistical significance. Use a two-sided test. Show all hypothesis-testing steps.

11.24   *Power*. A researcher fails to find a significant difference in mean blood pressure in 36 matched pairs. The standard deviation of the differences was 5 mmHg. What was the power of the test to find a mean difference of 2.5 mmHg at $\alpha = 0.05$ (two sided)?

11.25   *Beware $\alpha = 0.05$*. Two trials looked at red wine consumption in lowering cholesterol levels in hypercholesterolemic men. In each trial, 25 men consumed 8 ounces of red wine for 14 days.

(a) In trial A, the 25 subjects lowered their cholesterol by an average of 5% (standard deviation = 11.9%). In testing H_0: $\mu = 0$, $t_{stat} = 2.10$ with 24 df. Is this study statistically significant at $\alpha = 0.05$ (two sided)?

(b) In trial B, 25 different subjects lowered their cholesterol by 5% with standard deviation 12.2% ($t_{stat} = 2.05$ with 24 df). Is this result statistically significant at $\alpha = 0.05$?

(c) Is it reasonable to come to different conclusions for trial A and trial B?

11.26   *Benign prostatic hyperplasia, quality of life*. Benign prostatic hyperplasia is a noncancerous enlargement of the prostate gland that adversely affects the quality of life of millions of men. A study of a minimally invasive procedure for the treatment for this condition looked at pretreatment quality of life (QOL_BASE) and quality of life after 3 months on treatment (QOL_3MO). Table 11.4 lists data for 10 subjects chosen at random from this study.

TABLE 11.4   Data for Exercises 11.26 and 11.27. Variables are as follows:

QOL_BASE = quality of life at baseline (coded 0 = Delighted, 1 = Pleased, 2 = Mostly Satisfied, 3 = Mixed, 4 = Mostly Dissatisfied, 5 = Unhappy, 6 = Terrible)
QOL_3MO = Quality of life after 3 months of treatment (same codes)
MAXFLO_B = maximum urine flow at baseline (urine flow measurement scale misplaced)
MAXFLO3M = maximum urine flow after 3 months of treatment

i	ID	QOL_BASE	QOL_3MO	MAXFLO_B	MAXFLO3M
1	1	2	1	7	5
2	11	4	1	8	18
3	21	3	1	8	13
4	31	4	3	9	16
5	41	5	2	11	8
6	51	6	2	4	9
7	61	4	2	9	12
8	71	4	5	10	6
9	81	3	3	8	14
10	82	3	1	10	13

Source: Simple random sample of a data set provided by student Joanne Morales. Data are stored online in BPH-SAMP.SAV.

(a) Calculate differences in quality of life scores (DELTA) for each subject.

(b) Explore the differences with a stemplot. Discuss your exploration.

(c) Calculate the mean and standard deviation of the difference. Then test the mean difference for statistical significance. Use a two-sided alternative hypothesis.[q]

11.27 *Benign prostatic hyperplasia, maximum flow*. Table 11.4 also contains data for maximum urine flow at baseline (MAXFLO_B) and maximum urine flow after 3 months of treatment (MAXFLO3M). Test the mean difference in this outcome for statistical significance.

11.28 *NASA experiment*. A NASA study compared two methods of determining white blood cell counts in laboratory animals. Table 11.5 lists results for 42 paired observations. Calculate DELTA values for each observation and plot these differences as a stemplot. Based on this plot, do you think the methods are interchangeable?

---

[q]Because it may have been imprudent to use a *t*-procedure in this instance, I redid the analysis with a nonparametric (Wilcoxon signed rank) test that does not require Normality and $P = 0.014$, deriving a similar conclusion as the *t*-procedure.

TABLE 11.5   Data for Exercise 11.27. White blood cells counts (×1000 dL) by Celdyne method and Unopett method, $n = 42$.

CELDYNE	UNOPETT	CELDYNE	UNOPETT
8.2	8.6	13.0	20.4
9.7	11.0	12.7	10.4
5.6	8.1	14.1	13.0
14.0	15.7	12.9	14.5
5.7	6.3	7.4	6.8
10.8	9.1	9.1	10.5
10.5	11.3	9.5	7.4
7.9	9.3	14.4	15.9
12.7	11.0	8.8	9.3
3.6	2.6	13.1	18.0
10.4	9.6	10.3	9.5
13.6	10.3	9.4	11.3
11.3	10.6	9.8	9.5
10.3	7.8	11.5	9.3
8.3	8.4	11.8	10.4
6.3	6.5	12.5	9.6
23.9	27.6	6.1	6.0
16.0	7.5	10.4	9.0
10.4	9.9	10.9	12.3
9.5	8.4	7.8	7.8
13.8	13.8	11.4	8.8

Source: Data from student Adam Seddiqi. Data stored online in the file SEDDIQ.SAV.

11.29   ***Therapeutic touch.***[r] Proponents of an alternative medical treatment known as therapeutic touch claim that each person has a human energy field (HEF) that can be perceived and manipulated by touch. Therapists trained to recognize HEF-related perceptions are said to be particularly adept at manipulating HEFs. In an experiment that started out as a fourth-grade science fair project, therapeutic touch practitioners were tested under blind conditions to see whether they could correctly identify whether the HEF of an unseen hand hovered over their left or right hand (Figure 11.7). Fifteen therapeutic touch therapists underwent an initial set of 10 trials each. If HEF perception through therapeutic touch was possible, the therapists should have each been able to detect the experimenter's hand in 10 (100%) of 10 trials. Chance alone would produce a mean score of 5 (of 10). However, the $n = 15$ touch therapists

---

[r] Rosa, L., Rosa, E., Sarner, L., & Barrett, S. (1998). A close look at therapeutic touch. *JAMA*, *279*(13), 1005–1010.

FIGURE 11.7 Experimenter (right) hovers hand over one of the therapeutic touch practitioner's hands (left). The towel in the picture blinds the observation, preventing the therapeutic touch practitioner from seeing the location of the experimenter's hand. Drawing by Pat Linse. Published with the permission of the artist and *Skeptic* magazine.

correctly identified the location of the hand an average of 4.67 times (standard deviation 1.74). Calculate a 95% confidence interval for the mean number of correct identification of the HEF. Is the confidence interval compatible with random guessing?

11.30 ***Therapeutic touch, n = 28***. This exercise is an extension of Exercise 11.29. We add 8 observations to the initial 20, bringing the total sample size to 28. Each observation consists of 10 attempts to identify a human energy field, as previously discussed (see Figure 11.7). The number of correct identifications out of 10 was {1, 2, 3, 3, 3, 3, 3, 3, 3, 3, 4, 4, 4, 4, 4, 5, 5, 5, 5, 5, 5, 5, 6, 6, 7, 7, 7, 8}.

(a) Plot the data as a stemplot. Are there any clear departures from Normality? Can you use *t*-procedures on these data? Explain your reasoning.

(b) Provide a 95% confidence interval for the mean number of correct identifications.

12 Comparing Independent Means

12.1 Paired and Independent Samples

Data may be collected by a single sample, paired samples, or independent samples.

- **Single samples** reflect the experience of a single group. No concurrent control group is present, but results may be compared to previously established norms or expected values. The prior chapter (especially Section 11.3 and Section 11.4) addressed the analysis of single samples.

- **Paired samples** use data from two samples in which each data point in the first sample is uniquely matched to a data point in the second sample. Paired sample analysis was discussed in Section 11.5.

- **Independent samples** use SRSs from separate populations. We will address independent samples in this chapter.

Paired- and independent-sample study designs both collect data in two samples. With **paired samples**, each data point in the first sample is matched to a unique data point in the second sample. With **independent samples**, data points in the two samples are unrelated. To illustrate the difference between these two approaches, consider studying the effects of an LDL-cholesterol-lowering agent. We could take *paired* measurements in subjects, before and after use of the agent. Data would look something like this:

TABLE 12.1 Paired measurements.

Subject	Sample 1 Before treatment Cholesterol (mmol/L)	Sample 2 After treatment Cholesterol (mmol/L)
1	4.61	3.84
2	6.42	5.12
3	3.89	3.73
↓	↓	↓

In contrast, we could study *independent* groups, with data looking something like this:

TABLE 12.2   Independent groups.

Subject	Sample 1 = treatment 2 = control	Cholesterol (mmol/L)
1	1	4.61
2	2	6.42
3	1	3.89
↓	↓	↓

Theses approaches require different methods of analysis.

ILLUSTRATIVE EXAMPLE

Paired and independent samples (Effect of calcium supplementation on blood pressure). A randomized, double-blind, placebo-controlled trial examined the effects of calcium supplementation on blood pressure in normotensive men 19 to 52 years of age. After establishing baseline blood pressures, subjects were assigned to either a treatment or a control group. The treatment group was supplemented with 1500 mg of calcium per day for a 12-week period. The control group received an identical-appearing placebo. Table 12.3 lists systolic blood pressure measurements taken in the seated position for the African American men who took part in the study.

Each observation consists of paired measurements. In addition, two independent groups are present. The treatment group consists of $n_1 = 10$ observations. Here are the declines in blood pressure (mmHg) for this group:

7   −4   18   17   −3   −5   1   10   11   −2

The average decline in this group is 5.00 (standard deviation 8.743).

TABLE 12.3   Effect of calcium supplementation on blood pressure.

SUBJECT ID No.	GROUP 1 = treatment 2 = placebo	BEFORE mmHg	AFTER mmHg	DELTA mmHg
1	1	107	100	7
2	1	110	114	−4

continues

continued

SUBJECT ID No.	GROUP 1 = treatment 2 = placebo	BEFORE mmHg	AFTER mmHg	DELTA mmHg
3	1	123	105	18
4	1	129	112	17
5	1	112	115	−3
6	1	111	116	−5
7	1	107	106	1
8	1	112	102	10
9	1	136	125	11
10	1	102	104	−2
11	2	123	124	−1
12	2	109	97	12
13	2	112	113	−1
14	2	102	105	−3
15	2	98	95	3
16	2	114	119	−5
17	2	119	114	5
18	2	114	112	2
19	2	110	121	−11
20	2	117	118	−1
21	2	130	133	−3

Data from Dr. Roseann M. Lyle. Data are stored online in the file LYLE1987.SAV.

The placebo group consists of $n_2 = 11$. Here are the declines in blood pressure (mmHg) for this group:

−1 12 −1 −3 3 −5 5 2 −11 −1 −3

The average of these 12 values is − 0.27[a] (standard deviation 5.901).

These two samples are independent, as is the comparison of their means (5.00 mmHg vs. −0.27 mmHg).

---

[a] This "negative decline" represents a small net increase.

Exercises

12.1 *Sampling designs*. Identify whether the studies described here are based on (1) single samples, (2) paired samples, or (3) independent samples.

    (a) An investigator compares vaccination histories in 30 autistic schoolchildren to an SRS of nonautistic children from the same school district.

    (b) Cardiovascular disease risk factors are compared in husbands and wives.

    (c) A nutritional exam is applied to a random sample of individuals. Results are compared to expected means and proportions.

12.2 *Needle-stick injuries*. Healthcare workers are at risk of being exposed to blood-borne pathogens through needle-stick and other sharp object injuries. The pathogens of primary concern are the human immunodeficiency virus, hepatitis B virus, and hepatitis C virus. When a needle-stick injury occurs, workers report the incident to their supervisor. This information is forwarded to county health departments and ultimately to the Centers for Disease Control and Prevention (CDC). A CDC researcher uses these data to compare needle-stick injuries in community hospitals and tertiary-care hospitals. Is this a paired or independent comparison? Explain your answer.

12.3 *Facetious data*. Imagine a study of six monkeys that compares two treatments. Group 1 receives Monkey Tonic while group 2 receives Applied Monkey Training (AMT). Table 12.4 lists the performance scores BEFORE and AFTER these interventions.

TABLE 12.4 Data for Exercise 12.3.

Variables are:

ID	Identification number
GROUP	1 = Tonic; 2 = Training
BEFORE	Performance score before intervention
AFTER	Performance score after intervention

ID	GROUP	BEFORE	AFTER
1	1	100	104
2	1	88	93
3	1	106	109
4	2	116	117
5	2	102	104
6	2	106	106

Data are facetious.

(a) Calculate mean performance scores in each group BEFORE the intervention. Then calculate the mean difference. Is this mean difference based on paired or independent samples?

(b) Within group 1 (individuals 1 through 3), calculate the *difference* in scores from after to before the intervention (AFTER minus BEFORE). Then calculate the mean difference. Is this mean difference based on paired or independent samples?

(c) Within group 2 (individuals 4 through 6), calculate changes in scores (AFTER minus BEFORE). What is the mean difference in these scores?

(d) Compare mean improvements in the two groups. Is this a paired or independent comparison?

■ 12.2 Exploratory and Descriptive Statistics

Exploratory and descriptive techniques are applied before addressing inferential techniques. Let us start by presenting a new illustrative data set that will be used throughout the remainder of this chapter.

ILLUSTRATIVE EXAMPLE

Data (Cholesterol in two groups). This illustrative data set considers plasma cholesterol levels (mmol/L) in two groups. Group 1 consists of 12 individuals with the following values: {6.0, 6.4, 7.0, 5.8, 6.0, 5.8, 5.9, 6.7, 6.1, 6.5, 6.3, and 5.8}. Group 2 consists of these 7 values: {6.4, 5.4, 5.6, 5.0, 5.0, 4.5, and 6.0}. Table 12.5 also lists the data.

TABLE 12.5   "Cholesterol illustrative data." This is the primary illustrative data set in this chapter. Values are plasma cholesterol levels (mmol/L) in two independent groups.

Group 1	6.0 6.4 7.0 5.8 6.0 5.8 5.9 6.7 6.1 6.5 6.3 5.8
Group 2	6.4 5.4 5.6 5.0 5.0 4.5 6.0

Data from Group 1 data from Rassias, G., Kestin, M., & Nestel, P. J. (1991). Linoleic acid lowers LDL cholesterol without a proportionate displacement of saturated fatty acid. *European Journal of Clinical Nutrition, 45*(6), 315–320. Group 2 data were created with a Normal random number generator. Data are stored online in the file CHOLESTEROL.SAV.

continues

continued

Figure 12.1 is a screenshot of the data set in an SPSS data table. Notice that there are separate columns in this data table: one for the response variable (chol) and one for the explanatory variable (group).

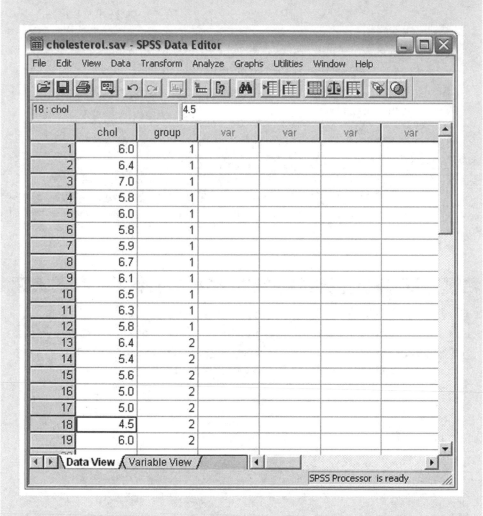

FIGURE 12.1   Screenshot of the "cholesterol in two groups" illustrative data. Notice the separate columns for the response variable and explanatory (group) variable. Graph reproduced with SPSS for Windows, Rel. 11.0.1.2001. Chicago: SPSS Inc. Reprint Courtesy of International Business Machines Corporation.

Exploratory Plots

Several techniques may be used to explore the data graphically. With small- to moderate-sized data sets, back-to-back stemplots (Section 3.1) are useful. Here is such a plot for the illustrative data:

```
Group 1 | | Group 2
--------|-|--------
        |4|
        |4|5
        |5|004
   9888|5|6
  43100|6|04
     75|6|
      0|7|
         ×1 (mmol/L)
```

Notice that the distribution for group 1 is farther down the axis than that of group 2, indicating that it has larger values on average. There are no apparent outliers in either group, and no major departures from Normality.

Side-by-side boxplots (Section 4.6) can also be used to good effect. Figure 12.2 displays side-by-side boxplots for the same data. The difference in locations is clearly visible.

Summary Statistics

Table 12.6 lists the notation we will use to denote group means and standard deviations. Table 12.7 lists results for the illustrative data, confirming our impression of higher average values with less variability in group 1.

TABLE 12.6 Notation used in comparing means and standard deviation from independent groups.

Group	Sample sizes	Sample mean	Sample deviation
1	n_1	$\bar{x}_1$	s_1
2	n_2	$\bar{x}_2$	s_2

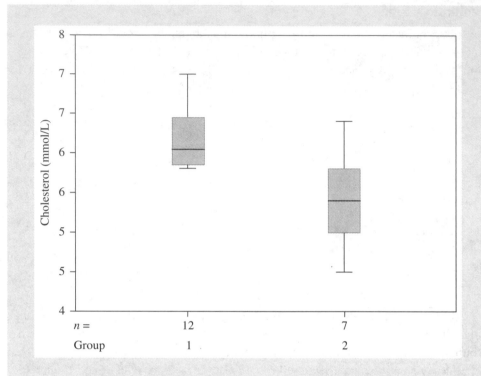

FIGURE 12.2   Side-by-side boxplots of the cholesterol illustrative data.

TABLE 12.7   Results for the illustrative data.

Group	Sample sizes	Mean (mmol/L)	SD (mmol/L)
1	12	6.192	0.3919
2	7	5.414	0.6492

Exercises

12.4 *Histidine excretion*. A study measures total histidine excretion (milligrams) in 24-hour urine samples in men and women on protein-restricted diets. The histidine values (mg) for men are {172, 204, 229, 236, and 256}. The values for women are {115, 135, 138, 174, 197, and 224}. Table 12.8 also lists the data.

(a) Create back-to-back stemplots of these data. Use an axis multiplier of ×100 and split stem values to create your plot. Discuss the results.

TABLE 12.8 "Histidine excretion data" for Exercises 12.4, 12.6, and 12.8. Data are milligrams of histidine in 24-hour urine samples.

Men	172	204	229	236	256	
Women	115	135	138	174	197	224

Source unknown. Data stored online in the file HISTIDINE.SAV.

    (b) Calculate means and standard deviations for each group. Relate these statistics to the stemplot created in part (a).

12.5 *Air samples*. A study of environmental air quality measured suspended particulate matter in air samples at two sites. Data ($\mu g/m^3$) for site 1 are {68, 22, 36, 32, 42, 24, 28, and 38}. Data for site 2 are {36, 38, 39, 40, 36, 34, 33, and 32}. Table 12.9 also lists the data.

TABLE 12.9 "Air samples data" for Exercises 12.5 and 12.7. Data are suspended particulate matter in air samples ($\mu g/m^3$).

Site 1	68	22	36	32	42	24	28	38
Site 2	36	38	39	40	36	34	33	32

Data are fictitious and are stored online in the file AIRSAMPLES.SAV.

    (a) Create back-to-back stemplots of data from these sites. Discuss the results.

    (b) Calculate group means and standard deviations.

    (c) Summarize your graphical and numerical findings.

■ 12.3 Inference About the Mean Difference

Estimation

Table 12.10 introduces the notation used to discuss population parameters in this chapter.

TABLE 12.10 Population parameters, notation.

Group	Population Mean	Population Std. deviation
1	μ_1	σ_1
2	μ_2	σ_2

The **point estimator** of population mean difference $\mu_1 - \mu_2$ is sample mean difference $\bar{x}_1 - \bar{x}_2$. In the illustrative example comparing cholesterol levels in two groups (Section 12.2), $\bar{x}_1 - \bar{x}_2 = 6.192 - 5.414 = 0.778$ (mmol/L). The precision of this estimate is quantified via its standard error $SE_{\bar{x}_1 - \bar{x}_2} = \sqrt{SE_{\bar{x}_1}^2 + SE_{\bar{x}_2}^2}$, where $SE_{\bar{x}_1}$ is the standard error of the mean for group 1 and $SE_{\bar{x}_2}$ is the standard error of the mean for group 2.[b] Equivalently, the **standard error of the mean difference** is:

$$SE_{\bar{x}_1 - \bar{x}_2} = \sqrt{\frac{s_1^2}{n_1} + \frac{s_2^2}{n_2}}$$

This standard error is used to build a **$(1 - \alpha)100\%$ confidence interval for $\mu_1 - \mu_2$** via the usual "point estimate $\pm\ t \cdot SE$" approach:

$$(\bar{x}_1 - \bar{x}_2) \pm t_{df, 1 - \frac{\alpha}{2}} \times SE_{\bar{x}_1 - \bar{x}_2}$$

The degrees of freedom for the t critical value in this equation is calculated with a method attributed to Welch[c]:

$$df_{welch} = \frac{(SE_{\bar{x}_1}^2 + SE_{\bar{x}_2}^2)^2}{\dfrac{SE_{\bar{x}_1}^4}{n_1 - 1} + \dfrac{SE_{\bar{x}_2}^4}{n_2 - 1}}$$

where $SE_{\bar{x}_1} = \dfrac{s_1}{\sqrt{n_1}}$ and $SE_{\bar{x}_2} = \dfrac{s_2}{\sqrt{n_2}}$.

Because calculation of df_{welch} by hand is tedious, we may use the smaller of $df_1 = n_1 - 1$ or $df_2 = n_2 - 1$ as a conservative approximation for the degrees of freedom[d]:

$$df_{conserv} = \text{the smaller of } df_1 \text{ or } df_2$$

ILLUSTRATIVE EXAMPLE

Confidence interval for $\mu_1 - \mu_2$ (Cholesterol in two groups). We calculate the 95% confidence interval for $\mu_1 - \mu_2$ for the cholesterol illustrative data (see Table 12.5).

---

[b] Notice that the standard errors of group means are added in a "Pythagorean way" to get the standard error of the mean difference: $c^2 = a^2 + b^2$, so $c = \sqrt{a^2 + b^2}$.

[c] Welch, B. L. (1936). The specification of rules for rejecting too variable a product, with particular reference to an electric lamp problem. *Supplement to the Journal of the Royal Statistical Society, 3*(1), 29–48. Also see Davenport, J. M., & Webster, J. T. (1975). The Behrens-Fisher problem, an old solution revisited. *Metrika, 22,* 47–54.

[d] The degrees of freedom are never less than the smaller of df_1 and df_2 (Welch, 1938, p. 356). Using $df_{conserv}$ will create a $(1 - \alpha)100\%$ confidence interval that will capture the parameter more than $(1 - \alpha)100\%$ of the time.

- $\bar{x}_1 - \bar{x}_2 = 6.192 - 5.414 = 0.778$ (mmol/L)
- $SE_{\bar{x}_1 - \bar{x}_2} = \sqrt{\frac{s_1^2}{n_1} + \frac{s_2^2}{n_2}} = \sqrt{\frac{0.3919^2}{12} + \frac{0.6492^2}{7}} = 0.2702$ (mmol/L).
- Using the conservative method to calculate the df, $df_1 = 12 - 1 = 11$ and $df_2 = 7 - 1 = 6$. Therefore, $df_{conserv} = 6$. For 95% confidence, use $t_{6, 0.975} = 2.447$. The 95% confidence interval for $\mu_1 - \mu_2$ is $(\bar{x}_1 - \bar{x}_2) \pm t_{df, 1 - \frac{\alpha}{2}}(SE_{\bar{x}_1 - \bar{x}_2})$ $= (0.778) \pm (2.447)(0.2702) = 0.778 \pm 0.661 = 0.117$ to 1.439 (mmol/L).
- To derive df by the Welch method, note that $SE_{\bar{x}_1} = \frac{s_1}{\sqrt{n_1}} = \frac{0.3919}{\sqrt{12}} = 0.11313$, $SE_{\bar{x}_2} = \frac{s_2}{\sqrt{n_2}} = \frac{0.6492}{\sqrt{7}} = 0.24537$, and

$$df_{welch} = \frac{(SE_{\bar{x}_1}^2 + SE_{\bar{x}_2}^2)^2}{\frac{SE_{\bar{x}_1}^4}{n_1 - 1} + \frac{SE_{\bar{x}_2}^4}{n_2 - 1}} = \frac{(0.11313^2 + 0.24537^2)^2}{\frac{0.11313^4}{12 - 1} + \frac{0.24537^4}{7 - 1}} = 8.61$$

The value of $t_{8.61, 0.975}$ via a software utility is 2.306. The 95% confidence interval is $(\bar{x}_1 - \bar{x}_2) \pm (t_{8.61, 0.975})(SE_{\bar{x}_1 - \bar{x}_2}) = (0.778) \pm (2.306)(0.2702) = 0.778 \pm 0.623 = 0.155$ to 1.401 (mmol/L).

After calculating the confidence interval, return to the research question to interpret the results in this setting. For this illustration, we can conclude with 95% confidence that the difference in the mean cholesterol levels in the two populations $\mu_1 - \mu_2$ is between 0.15 and 1.40 mmol/L.

Test of H_0: $\mu_1 = \mu_2$ (Independent t Test)

The procedure to test the means for inequivalence is:

A. **Hypotheses.** The null hypothesis is H_0: $\mu_1 - \mu_2 = 0$, or equivalently H_0: $\mu_1 = \mu_2$. The alterative hypothesis is either H_a: $\mu_1 \neq \mu_2$ (two sided), H_a: $\mu_1 > \mu_2$ (one sided to right), or H_a: $\mu_1 < \mu_2$ (one sided to the left).

B. **Test statistic.** The test statistics is $t_{stat} = \frac{\bar{x}_1 - \bar{x}_2}{SE_{\bar{x}_1 - \bar{x}_2}}$, where $SE_{\bar{x}_1 - \bar{x}_2} = \sqrt{\frac{s_1^2}{n_1} + \frac{s_2^2}{n_2}}$. Equivalently, $t_{stat} = \frac{\bar{x}_1 - \bar{x}_2}{\sqrt{\frac{s_1^2}{n_1} + \frac{s_2^2}{n_2}}}$.

When working by hand, use the conservative estimate for the degrees of freedom $df_{conserv}$ = the smaller of $(n_1 - 1)$ or $(n_2 - 1)$. When access to software is available, the more sophisticated df_{welch} presented earlier is recommended.

C. **P-value.** The one-sided $P = \Pr(t \geq |t_{stat}|)$. The two-sided $P = 2 \times \Pr(t \geq |t_{stat}|)$. These probabilities can be determined with Table C or with a software utility such as *StaTable*.[e] The *P*-value is interpreted in the usual fashion (see Section 9.3).

D. **Significance (level).** The difference is said to be significant at the a-level of significance when $P \leq \alpha$. See Section 9.4 for comments about the interpretation of statistical significance.

E. **Conclusion.** A conclusion is drawn in the context of the data and original research question.

ILLUSTRATIVE EXAMPLE

Independent t test (Cholesterol in two groups). Let us submit the cholesterol illustrative data to an independent *t* test. We want to determine whether there is a significant difference in the mean cholesterol levels in the two populations. Recall that $n_1 = 12$, $\bar{x}_1 = 6.192$ mmol/L and $s_1 = 0.3919$. In group 2, $n_2 = 7$, $\bar{x}_2 = 5.414$, and $s_2 = 0.6492$. We've already established that the degrees of freedom are either $df_{conserv} = 6$ or $df_{Welch} = 8.61$.

A. **Hypotheses.** H_0: $\mu_1 = \mu_2$ against H_a: $\mu_1 \neq \mu_2$ (two-sided).

B. **Test statistic.** The test statistic is $t_{stat} = \dfrac{\bar{x}_1 - \bar{x}_2}{\sqrt{\dfrac{s_1^2}{n_1} + \dfrac{s_2^2}{n_2}}}$

$$= \dfrac{6.192 - 5.414}{\sqrt{\dfrac{0.3919^2}{12} + \dfrac{0.6492^2}{7}}} = 2.88$$

C. **P-value.** Using $df_{conserv} = 6$, the one-sided *P*-value is between 0.025 and 0.01 (via Appendix Table C) and the two-sided *P*-value is between 0.05 and 0.02. Using $df_{Welch} = 8.61$, we use a computer program to calculate $P = 0.019$ (two-sided).

D. **Significance level.** The results are significant at the $\alpha = 0.05$ level but not at the $\alpha = 0.01$ level.

E. **Conclusion.** The data provide reliable ("significant") evidence that the mean cholesterol levels in the two populations differ ($P = 0.019$).

Figure 12.3 shows the output for the problem from SPSS. The line labeled "equal variance not assumed" replicates our results.

---

[e] Cytel Corporation. (2006). *StaTable*. www.cytel.com/Products/StaTable/.

Independent Samples Test

	Levene's Test for Equality of Variances		*t*-test for Equality of Means					95% Confidence Interval of the Difference	
	F	Sig.	*t*	df	Sig. (2-tailed)	Mean Difference	Std. Error Difference	Lower	Upper
CHOL Equal variances assumed	1.999	0.175	3.282	17	0.004	0.778	0.2369	0.2776	1.2772
Equal variances not assumed			2.877	8.610	0.019	0.778	0.2702	0.1619	1.3928

FIGURE 12.3 SPSS Independent *t* procedures output, cholesterol illustrative data. Output includes results for Levene's test of equal variance (H_0: $\sigma_1{}^2 = \sigma_2{}^2$), which is explained in Section 13.5. Graph produced with SPSS for Windows, Rel. 11.0.1.2001. Chicago: SPSS Inc. Reprint Courtesy of International Business Machines Corporation.

■ 12.4 Equal Variance *t* Procedure (Optional)

Inferential comparisons presented in the prior section (Section 12.3) made no assumption about whether the variances in underlying populations were equal. Now we present a procedure that does have this requirement. This procedure is called the **equal variance *t* procedure** or **pooled variance *t* procedure**.[f] Many biostatistics textbooks still use this method as the standard version of the two-sample *t* procedure. However, we recommend against its routine use. Verifying the equality of population variances is difficult, and the equal variance *t* procedure will produce unreliable results when unequal variances exist. Since there is little downside in using the Welch **unequal variance procedure** even when population variances are equal, this should be our first choice when comparing means. The equal variance *t* procedure is presented here because (1) it is rooted in statistical history and (2) most statistical software calculates both "equal variance" and "unequal variance" *t* procedures.

The equal variance *t* procedure pools the variances from the independent samples to calculate a **pooled estimate of variance** as follows:

$$s^2_{pooled} = \frac{df_1 \cdot s^2_1 + df_2 \cdot s^2_2}{df_1 + df_2}$$

where s^2_1 *and* s^2_2 are group variances, $df_1 = n_1 - 1$, and $df_2 = n_2 - 1$. An equivalent formula is $s^2_{pooled} = \dfrac{(n_1 - 1)s^2_1 + (n_2 - 1)s^2_2}{n_1 + n_2 - 2}$. The denominator of s^2_{pooled} is the **pooled degrees of freedom**, sometimes denoted df (without a subscript).

---

[f] It is called the pooled variance *t* procedure because data from the two samples are pooled to derive an estimate of common population variance σ^2.

Note that $df = df_1 + df_2 = (n_1 - 1) + (n_2 - 1) = (n_1 + n_2 - 2)$.

The pooled estimate of variance is used to calculate the **pooled estimate of standard error**:

$$SE_{pooled} = \sqrt{s^2_{pooled}\left(\frac{1}{n_1} + \frac{1}{n_2}\right)}$$

Procedures for confidence intervals and hypothesis tests are now applied using the pooled standard error with $df = n_1 + n_2 - 2$. The $(1 - \alpha)100\%$ confidence interval for $\mu_1 - \mu_2$ is $(\bar{x}_1 - \bar{x}_2) \pm t_{df,1-\frac{\alpha}{2}} \times SE_{pooled}$. The $t_{stat} = \dfrac{\bar{x}_1 - \bar{x}_2}{SE_{pooled}}$.

ILLUSTRATIVE EXAMPLE

Pooled variance t procedures. The cholesterol illustrative data is used to demonstrate pooled t procedures. The following summary statistics have been established (mmol/L) $\bar{x}_1 = 6.192$ and $s_1 = 0.3919$ based on $n_1 = 12$. For group 2 ($n_2 = 7$), $\bar{x}_2 = 5.414$ and $s_2 = 0.6492$. We calculate:

- $df_1 = n_1 - 1 = 12 - 1 = 11$, $df_2 = n_2 - 1 = 7 - 1 = 6$, and $df = df_1 + df_2$ $= 11 + 6 = 17$

- $s^2_{pooled} = \dfrac{(11)(0.3919^2) + (6)(0.6492^2)}{11 + 6} = 0.24813$

- $SE_{pooled} = \sqrt{0.24813\left(\frac{1}{12} + \frac{1}{7}\right)} = 0.2369$

- For 95% confidence, use $t_{17, 0.975} = 2.110$; the 95% confidence interval for $\mu_1 - \mu_2 = (\bar{x}_1 - \bar{x}_2) \pm t_{df,1-\frac{\alpha}{2}}(SE_{pooled}) = (6.192 - 5.414) \pm (2.110)(0.2369)$ $= 0.778 \pm 0.500 = (0.278 \text{ to } 1.278)$ mmol/L.

- To test $H_0: \mu_1 = \mu_2$, $t_{stat} = \dfrac{\bar{x}_1 - \bar{x}_2}{SE_{pooled}} = \dfrac{6.192 - 5.414}{0.2369} = 3.284$ with 17 df. The two-sided P-value $= 0.004$.

Figure 12.3 includes **computer** output from SPSS for both equal variance and unequal variance t procedures.

■ 12.5 Conditions for Inference

Two-sample t procedures (confidence intervals and hypothesis tests) are rooted in conditions of sampling independence and Normal distribution theory. The independence condition for two-sample procedures requires independence within samples and between groups; data should represent simple random samples from independent populations. The Normality assumption refers to the sampling distribution of the mean difference. As was the case with other t procedures, independent t procedures can be relied on in non-Normal populations when samples are moderate to large (Sections 9.5 and 11.6). Independent t procedures are particularly robust when $n_1 = n_2$ and two-sided alternatives are used. Equal variance (pooled) t procedures also require equal variances in the populations.[g]

In addition to the distributional assumptions considered here, t procedures rely on the accuracy of the data. No amount of mathematical manipulation can compensate for poor-quality information (Section 1.4). Finally, confounding by lurking variables must also be considered (Sections 2.2, 17.5, and Section 19.1).

Exercises

12.6 *Histidine excretion*. Exercise 12.4 introduced data in which 24-hour histidine excretion was compared in men and women on protein-restricted diets. Data are listed in Table 12.8. Table 12.11 lists means and standard deviations from this data set.

TABLE 12.11   Means and standard deviations for Exercise 12.6.

Histidine (mg), 24-hour urine samples

Group	n	Mean	Std. Dev.
1 (men)	5	219.40	32.370
2 (women)	6	163.83	41.730

(a) Calculate the standard error of the mean difference without assuming equal variances.

(b) A computer program calculates $df_{Welch} \approx 9$ for these data. Based on this df, calculate a 95% confidence interval for $\mu_1 - \mu_2$.

(c) Calculate the two-sided P-value for testing H_0: $\mu_1 = \mu_2$.

12.7 *Air samples*. The air quality data in Exercise 12.5 (Table 12.9) are associated with the summary statistics reported in Table 12.12.

---

[g] Section 5 in Chapter 13 considers the equal variance assumption in greater detail.

TABLE 12.12   Means and standard deviations for Exercise 12.7.

Particulate matter in air samples ($\mu g/m^3$)

Site	n	Mean	Std. Dev.
1	8	36.25	14.56
2	8	36.00	2.88

(a) Calculate the standard error of the mean difference without assuming equal variance.

(b) Calculate a 95% confidence interval for $\mu_1 - \mu_2$, again without assuming equal variance. Use the conservative degrees of freedom to help calculate the confidence interval.

(c) Determine the two-sided P-value for testing H_0: $\mu_1 = \mu_2$.

12.8 *Histidine, equal variance assumed*. Redo the histidine excretion analyses (Exercise 12.6) with equal variance t procedures.

■ **12.6 Sample Size and Power**

Procedures similar to the power and sample size methods for single- and paired-sample problems Sections 9.6, 10.3, and 11.7 introduced power and sample size methods for single- and paired-sample problems. Similar procedures apply when comparing independent means. Four questions are addressed:

1. What sample size is required to ensure that a confidence interval has a margin of error no greater than m?

2. What sample size is required to ensure the hypothesis test has power $1 - \beta$?

3. What is the minimal detectable difference Δ for a given sample size?

4. What is the power of a test to detect a mean difference of Δ?

Sample Size Requirement for a Confidence Interval

To limit the margin of error in a $(1 - \alpha)100\%$ confidence interval for $\mu_1 - \mu_2$, the number of individuals in *each group* should be no less than

$$n = \frac{2\sigma^2 z_{1-\frac{\alpha}{2}}^2}{m^2}$$

where m is the desired margin of error, σ is the common standard deviation of the response variable, and $z_{1-(\alpha/2)}$ is the value of a Standard Normal random variable with cumulative probability $1 - (\alpha/2)$. Round results from this formula up to the next integer to ensure that the margin of error does not exceed m.

Notes

1. Before applying this formula, considerable effort should be put into getting a good estimate of common variance σ^2. This is one of the more difficult aspects of sample size planning. It is often reasonable to use s^2_{pooled} from a previous study or from a pilot investigation as the estimate of σ^2.

2. This calculation can be adjusted to accommodate for the difference between t and z procedures by multiplying the result by this adjustment factor[h]:

$$f = \frac{df + 3}{df + 1}$$

In samples with more than 30 degrees of freedom, the value of f approaches 1, so making this adjustment is superfluous.

3. Maximum efficiency in a study is gained by having an equal number of observations in each group. There are times, however, when the size of one of the groups is constrained. When this is the case, let n_1 represent the size of the constrained group, and then determine the number of people needed in group 2 according to this formula:

$$n_2 = \frac{nn_1}{2n_1 - n}$$

ILLUSTRATIVE EXAMPLE

Sample size requirement for a confidence interval for $\mu_1 - \mu_2$. Suppose you want to estimate $\mu_1 - \mu_2$ with 95% confidence for a problem similar to the cholesterol illustrative example. You want your margin of error to be no greater than 0.25 mmol/L. We have established that $s^2_{pooled} = 0.24813$, and will use this as a reasonable estimate of σ^2.

Solution: For 95% confidence, use $z_{1 - (0.05/2)} = 1.96$. The sample size required per group is $n = \dfrac{2\sigma^2 z^2}{m^2} = \dfrac{2 \cdot 0.24813 \cdot 1.96^2}{0.25^2} = 30.5$. Round this up to the next integer and resolve to use 31 individuals per group.

Now suppose that you can recruit only 20 individuals into group 1. How many individuals will you now need in group 2?

Solution: $n_2 = \dfrac{nn_1}{2n_1 - n} = \dfrac{31 \cdot 20}{2 \cdot 20 - 31} = 68.9$. Round this up to 69.

---

[h] Lachin, J. M. (1981). Introduction to sample size determination and power analysis for clinical trials. *Controlled Clinical Trials, 2*(2), 93–113.

Sample Size Requirement to Test H_0: $\mu_1 = \mu_2$ with a Given Power

To test H_0: $\mu_1 = \mu_2$ with $1 - \beta$ power and α significance, study this number *per group*

$$n = \frac{2\sigma^2(z_{1-\beta} + z_{1-\frac{\alpha}{2}})^2}{\Delta^2}$$

where $\Delta = \mu_1 - \mu_2$ ("the difference worth detecting") and σ^2 is the variance of the response variable.

ILLUSTRATIVE EXAMPLE

Sample size requirement for hypothesis test. Suppose you want to test H_0: $\mu_1 = \mu_2$ for the cholesterol illustrative example at $\alpha = 0.05$ (two-sided) with 90% power. You are looking for a mean difference of 1 mmol/L and will use the pooled estimate of variance (about 0.25) as a reasonable estimate of σ^2. How many individuals must be studied to achieve these conditions?

  Solution: For 90% power, $z_{1-\beta} = z_{0.90} = 1.282$. For a two-sided test at $\alpha = 0.05$, $z_{1-(\alpha/2)} = z_{0.975} = 1.96$ (Appendix Table B). Given these conditions,

$$n = \frac{2\sigma^2(z_{1-\beta} + z_{1-\frac{\alpha}{2}})^2}{\Delta^2} = \frac{2 \cdot 0.25 \cdot (1.282 + 1.96)^2}{1^2} = 5.26 \quad \text{Because } n \text{ is}$$

small, apply correction factor $f = (df + 3)/(df + 1)$ to this result. With 5.26 per group, df_1 and $df_2 = 4.26$. Therefore, $df = df_1 + df_2 = 4.26 + 4.26 = 8.52$ and adjustment factor $f = (8.52 + 3)/(8.52 + 1) = 1.21$. Applying the correction derives $n = 1.21 \times 5.26 = 6.36$. Round this up to seven and resolve to use a minimum of seven individuals per group.

As was the case with estimation, maximum efficiency is gained by studying the same number of individuals per group. If for some reason this is not possible, apply the formula $n_2 = nn_1/(2n_1 - n)$ to determine the size of group 2. For example, if group 1 in the prior illustration was restricted to five individuals, then group 2 should include

$$n_2 = \frac{nn_1}{2n_1 - n} = \frac{(7)(5)}{(2)(5) - 7} = 11.7 \approx 12 \text{ individuals.}$$

Minimal Detectable Difference

The sample size formula for hypothesis testing can be rearranged to determine the **minimal detectable difference (Δ)** for a given sample size to derive:

$$\Delta = \sqrt{\frac{2 \cdot \sigma^2}{n}}(z_{1-\beta} + z_{1-\frac{\alpha}{2}})$$

ILLUSTRATIVE EXAMPLE

Minimal detectable difference. What is the minimal detectable difference when using $n = 30$ per group in the cholesterol illustrative example? Assume $\alpha = 0.05$ (two-sided) and power $= 0.90$. We have established 0.25 as a reasonable estimate for σ^2.

Solution: $\Delta = \sqrt{\frac{2 \cdot \sigma^2}{n}}(z_{1-\beta} + z_{1-\frac{\alpha}{2}}) = \sqrt{\frac{2 \cdot 0.25}{30}}(1.282 + 1.96) = 0.42.$ This study will be able to detect a mean difference of 0.42 mmol/L under the stated conditions.

Power of Test

The power of a test of H_0: $\mu_1 = \mu_2$ can be calculated with this formula:

$$1 - \beta = \Phi\left(-z_{1-\frac{\alpha}{2}} + \frac{\Delta}{\sqrt{\frac{2\sigma^2}{n}}}\right)$$

where $\Phi(z)$ is the cumulative probability of z on a Standard Normal distribution.

ILLUSTRATIVE EXAMPLE

Power. Again reconsider the conditions of the illustrative example that compared cholesterol in the two groups. What is the power of a test of H_0: $\mu_1 = \mu_2$ when $\alpha = 0.05$ (two-sided), $n = 30$ per group, $\mu_1 - \mu_2 = 0.25$, and $\sigma^2 = 0.25$?

Solution: $1 - \beta = \Phi\left(-z_{1-\frac{\alpha}{2}} + \frac{\Delta}{\sqrt{\frac{2\sigma^2}{n}}}\right).$

$$\Phi\left(-1.96 + \frac{0.25}{\sqrt{\frac{2(0.25)}{30}}}\right) = \Phi(-0.02) = 0.4920 \text{ (about 50/50).}$$

Exercises

12.9   *Sample size calculation.* We wish to detect a mean difference of 0.25 for a variable that has a standard deviation of 0.67. How large a sample is needed to detect the mean differences with 90% power at $\alpha = 0.05$ (two-sided)?

12.10 *Sample size requirement, cholesterol comparison.* A variable has a standard deviation of 40 mg/dL. We want to test a mean difference in two groups with $\alpha = 0.05$ (two-sided) and power $= 80\%$.

(a) How many observations are needed to detect a difference of 10 mg/dL?

(b) Suppose we can recruit only 150 subjects into group 1. How many individuals do we need in group 2?

Summary Points (Comparing Independent Means)

1. This chapter considers the analysis of a **quantitative response from two independent groups**. Data are computerized in the form of a binary explanatory variable and quantitative response variable.

2. We begin the analysis by **exploring and comparing** the distributions of values from the independent groups with graphical techniques (e.g., side-by-side stemplots or boxplots) and group summary statistics (means, standard deviations, and sample sizes).

3. The **parameter** of interest is the population mean difference $\mu_1 - \mu_2$. The sample mean difference $\bar{x}_1 - \bar{x}_2$ is the **point estimator** of this parameter.

4. Two different types of **independent t procedures** are presented in this chapter. One method does not assume that the populations have equal variances. The other method does. We rely on the former method.

5. A $(1 - \alpha)100\%$ **confidence interval for**
   $\mu_1 - \mu_2 = (\bar{x}_1 - \bar{x}_2) \pm t_{\mathrm{df},1-\frac{\alpha}{2}} \cdot SE_{\bar{x}_1 - \bar{x}_2}$, where $SE_{\bar{x}_1 - \bar{x}_2} = \sqrt{\frac{s_1^2}{n_1} + \frac{s_2^2}{n_2}}$.

   (a) When working by hand, use $\mathrm{df_{conserv}}$ = the smaller of $(n_1 - 1)$ or $(n_2 - 1)$. When statistical software is available, use the more sophisticated $\mathrm{df_{Welch}}$ estimate.

   (b) Keep in mind that the confidence interval seeks to capture $\mu_1 - \mu_2$, not $\bar{x}_1 - \bar{x}_2$.

6. **Independent t test**.

   (a) H_0: $\mu_1 - \mu_2 = 0$ ("no difference in the population means").

   (b) $t_{\mathrm{stat}} = \dfrac{\bar{x}_1 - \bar{x}_2}{\sqrt{\dfrac{s_1^2}{n_1} + \dfrac{s_2^2}{n_2}}}$ with $\mathrm{df_{conserv}}$ or $\mathrm{df_{Welch}}$, as described in point 5.

   (c) Convert the t statistic to a *P*-value using Table C or a computer applet.

   (d) Considered the level of statistical significance.

   (e) Formulate a conclusion in the context of the research question and data.

7. Our independent t procedure requires the following **conditions**:

    (a) SRSs from independent populations (or reasonable approximation thereof)

    (b) Populations Normal or samples large enough to invoke the central limit theorem

    (c) Data are accurate

8. **Sample size and power.**

    (a) **When estimating** $\mu_1 - \mu_2$, the number of individuals in each group should be $n = \dfrac{2\sigma^2 z_{1-\frac{\alpha}{2}}^2}{m^2}$, where σ^2 represents the variance of the response variable (assumes equal variance), $z_{1-\frac{\alpha}{2}}$ represents the z percentile needed for $1 - \alpha$ confidence, and m represents desired margin of error m.

    (b) In testing H_0: $\mu_1 - \mu_2 = 0$ at a given α level with $1 - \beta$ power, study this number per group $n = \dfrac{2\sigma^2\left(z_{1-\beta} + z_{1-\frac{\alpha}{2}}\right)^2}{\Delta^2}$, where σ^2 represents the variance of the response variable (assumes equal variance) and Δ represents the difference worth detecting ($\Delta = \mu_1 - \mu_2$).

    (c) See text for adaptation of these formulas for unequal group sizes and for the power formula.

Vocabulary

$(1 - \alpha)100\%$ confidence interval for $\mu_1 - \mu_2$	Pooled degrees of freedom (df)
Equal variance t procedure (pooled variance t procedure)	Pooled estimate of standard error (SE_{pooled})
Independent samples	Pooled estimate of variance
Minimum detectable difference	Single samples
Paired samples	Standard error of the mean difference
Point estimator	Unequal variance t procedure

Review Questions

12.1 Select the best response: We take an SRS from a population and a separate SRS from a population. Which t procedure should we use?

    (a) a one-sample t procedure

    (b) a paired t procedure

    (c) an independent t procedure

12.2 Select the best response: We have pretest and posttest samples. Which t procedure should we use?

(a) a one-sample t procedure

(b) a paired t procedure

(c) an independent t procedure

12.3 Identify two graphical methods that can be used to compare quantitative data from two independent groups.

12.4 This symbol represents the sample mean of group i.

12.5 This symbol represents the sample standard deviation of group i.

12.6 This symbol represents the sample variance of group i.

12.7 This symbol represents the population mean of group i.

12.8 This symbol represents the population standard deviation of group i.

12.9 This is the point estimator of $\mu_1 - \mu_2$.

12.10 The statistic quantifies the precision of the sample mean difference as an estimate of the population mean difference.

12.11 Select the best response: The degrees of freedom for independent t procedures that do not assume equal variance may be based on a conservative method or a method attributed to:

(a) Student

(b) Fisher

(c) Welch

12.12 Fill in the blank: The conservative estimate for the degrees of freedom for the unequal variance t procedures is the _____ of $(n_1 - 1)$ or $(n_2 - 1)$.

12.13 Select the best response: Independent t procedures (unequal variance option) require independent samples and

(a) equal variances in the populations.

(b) Normal populations.

(c) Normal populations or a large enough sample to compensate for non-Normality through the central limit theorem.

12.14 Homoscedasticity means the population

(a) has equal variances.

(b) has equal means.

(c) is Normal.

12.15 List the three factors that govern the sample size requirements for estimating population mean differences.

12.16 Select the best response: Maximum power of a statistical test is achieved when the sample size for group 1 is _____ the sample size for group 2.

(a) greater than

(b) less than

(c) equal to

12.17 List the four factors that determine the sample size requirements of an independent *t* test.

12.18 List the four factors that govern the power of an independent *t* test.

Exercises

12.11 *Testing a test kit*. Would you use a one-sample, paired-sample, or independent-sample *t* procedure in the following situations?

(a) A lab technician obtains a specimen of known concentration from a reference lab. He/she tests the specimen 10 times using an assay kit and compares the calculated mean to that of the known standard.

(b) A different technician compares the concentration of 10 specimens using 2 different assay kits. Ten measurements are taken with each kit. Results are then compared.

12.12 *Lactation and bone loss.* A study compares changes in bone density in 22 women who are neither pregnant nor breastfeeding to that of 47 comparably aged women who are breastfeeding. Table 12.13 lists percent changes in bone mineral content in the spines of these women.

(a) Compare these groups with back-to-back stemplots on a common stem. How do the groups differ? Does either group show a major departure from Normality?

(b) Calculate the 95% confidence interval for the mean difference in bone loss in the two groups. Discuss your finding.

TABLE 12.13   Percent change in bone mineral content over three months in the spines of women.

Controls

2.4	0.0	0.9	−0.2	1.0	1.7	2.9	−0.6	1.1	−0.1
−0.4	0.3	1.2	−1.6	−0.1	−1.5	0.7	−0.4	2.2	−0.4
−2.2	−0.1								

Breastfeeders

−4.7	−2.5	−4.9	−2.7	−0.8	−5.3	−8.3	−2.1	−6.8	−4.3
2.2	−7.8	−3.1	−1.0	−6.5	−1.8	−5.2	−5.7	−7.0	−2.2
−6.5	−1.0	−3.0	−3.6	−5.2	−2.0	−2.1	−5.6	−4.4	−3.3
−4.0	−4.9	−4.7	−3.8	−5.9	−2.5	−0.3	−6.2	−6.8	1.7
0.3	−2.3	0.4	−5.3	0.2	−2.2	−5.1			

Data from Y axis of Figure 3 in Laskey, M. A., Prentice, A., Hanratty, L. A., Jarjou, L. M., Dibba, B., Beavan, S. R., et al. (1998). Bone changes after 3 mo of lactation: Influence of calcium intake, breast-milk output, and vitamin D-receptor genotype. *American Journal of Clinical Nutrition, 67*(4), 685–692.

Data are stored online in the file LACTATION.SAV.

12.13   ***Risk-taking behavior in boys and girls***. A questionnaire measures an index of risk-taking behavior in respondents. Scores are standardized so that 100 represents the population average. The questionnaire is applied to a sample of teenage boy and girls. Data are:[i]

Girls: 72	73	86	95	95	95	96	97	99	125
Boys: 89	92	93	98	105	106	110	126	127	130

Explore group differences with side-by-side boxplots.

12.14   ***Scrapie treatment, delay of symptoms***. Scrapie is a prion disease similar in pathology to bovine spongiform encephalopathy (mad cow disease) and new variant Creutzfeldt-Jakob disease. In a study of 20 scrapie-infected hamsters, 10 were treated with a substance known as IDX and 10 were left untreated. The mean time before appearance of symptoms in the control group was 81.9 days ($SE_1 = 2.2$ days). The mean time to symptoms in the treated group was 102.8 days ($SE_2 = 3.8$ days).[j] Test this difference for significance. Note that group standard errors (not standard deviations) are given in this problem. To calculate the standard error of the mean difference, apply this formula:

$$SE_{\bar{x}_1 - \bar{x}_2} = \sqrt{SE_{\bar{x}_1}^2 + SE_{\bar{x}_2}^2}$$

---

[i] Data are fictitious but realistic. File is stored online in the file RISK_INDEX.SAV.
[j] Tagliavini, F., McArthur, R. A., Canciani, B., Giaccone, G., Porro, M., Bugiani, M., et al. (1997). Effectiveness of anthracycline against experimental prion disease in Syrian hamsters. *Science, 276*(5315), 1119–1122.

12.15   *Scrapie treatment, delay of death*. Refer to the previous exercise. The mean survival time in the IDX-treated group was 116 days ($SE_1 = 5.6$). The mean survival time in the control group was 88.5 ($SE_2 = 1.9$). Test H_0: $\mu_1 = \mu_2$.

12.16   *Cytomegalovirus and coronary stenosis*. Coronary stenosis (narrowing of the artery supply to the heart muscle) is a direct cause of heart disease. A theory suggests that chronic cytomegalovirus (CMV) infection narrows coronary vessels and leads to coronary heart disease. To test this theory, 75 patients undergoing angioplasty were followed for 6 months following their procedure. The 49 patients who were seropositive for CMV experienced an average luminal diameter reduction of 1.24 mm ($s_1 = 0.83$ mm). In contrast, the 26 patients who were seronegative for CMV experienced an average luminal reduction of 0.68 ($s_2 = 0.69$).[k] Test whether this mean difference in luminal reduction is significant. Show all hypothesis-testing steps. Do the data support the hypothesis that chronic CMV virus infection plays a role in coronary luminal reduction?

12.17   *Bone density in newborns*. In a study of maternal cigarette smoking and bone density in newborns, 77 infants of mothers who smoked had a mean bone mineral content of 0.098 g/cm^3 ($s_1 = 0.026$ g/cm^3). The 161 infants whose mothers did not smoke had a mean bone mineral content of 0.095 g/cm^3 ($s_2 = 0.025$ g/cm^3).[l]

(a) Calculate the 95% confidence interval for $\mu_1 - \mu_2$.

(b) Chapter 10 included a section on the relationship between hypothesis testing and confidence intervals. We know that if the $(1 - \alpha)100\%$ confidence interval excluded the value of the parameter specified under the null hypothesis, the null hypothesis would be rejected at the α-level of significance. If the value specified in the null hypothesis was included in the midst of the $(1 - \alpha)100\%$ confidence interval, the null hypothesis would be retained at the α threshold. Based on the confidence interval you just calculated, would you reject or retain H_0: $\mu_1 - \mu_2 = 0$ at $\alpha = 0.05$?

12.18   *Hemodialysis and anxiety*. Severe anxiety often accompanies patients who must undergo chronic hemodialysis. A study was undertaken to determine the effects of a set of progressive relaxation exercises on anxiety in hemodialysis

[k] Zhou, Y. F., Leon, M. B., Waclawiw, M. A., Popma, J. J., Yu, Z. X., Finkel, T., et al. (1996). Association between prior cytomegalovirus infection and the risk of restenosis after coronary atherectomy. *New England Journal of Medicine, 335*(9), 624–630.

[l] Venkataraman, P. S., & Duke, J. C. (1991). Bone mineral content of healthy, full-term neonates. Effect of race, gender, and maternal cigarette smoking. *American Journal of Diseases of Children, 145*(11), 1310–1312.

patients. The treatment group consisted of 38 subjects who were shown a set of progressive relaxation videotapes. The control group was made up of 23 patients shown a set of neutral videotapes. A psychiatric questionnaire that measured anxiety revealed the posttest results reported in Table 12.14.[m] Test the mean difference for significance.

TABLE 12.14   Posttest results.

Group	n	Mean	Std. dev.
Experimental	38	33.42	10.18
Control	23	39.71	9.16

Data from Alarcon, R. D., Jenkins, C. S., Heestand, D. E., Scott, L. K., & Cantor, L. (1982). The effectiveness of progressive relaxation in chronic hemodialysis patients, Table 1. *Journal of Chronic Diseases, 35*(10), 797–802.

12.19   *Efficacy of echinacea, severity of symptoms*. A randomized, double-blind, placebo-controlled study evaluated the effects of the herbal remedy *Echinacea purpurea* in treating upper respiratory tract infections in 2- to 11-year-olds. Each time a child had an upper respiratory tract infection, treatment with either echinacea or a placebo was given for the duration of the illness. One of the outcomes studied was "severity of symptoms." A severity scale based on four symptoms was monitored and recorded by the parents of subjects for each instance of upper respiratory infection. The peak severity of symptoms in the 337 cases treated with echinacea had a mean score of 6.0 (standard deviation 2.3). The peak severity of symptoms in the placebo group ($n_2 = 370$) had a mean score of 6.1 (standard deviation 2.4).[n] Test the mean difference for significance. Discuss your findings.

12.20   *Efficacy of echinacea, duration of symptoms*. The study introduced in the prior exercise also compared duration of peak symptoms in the treatment and control group. The treatment group ($n_1 = 337$) had a mean duration of 1.60 days (standard deviation 0.98 days). The control group ($n_2 = 370$) had a mean duration of peak symptoms of 1.64 days (standard deviation 1.14 days). Test the difference in means for significance.

12.21   *Calcium supplementation and blood pressure, exploration*. Table 12.3 contains data from a randomized, double-blind, placebo-controlled trial that examined the effects of calcium supplementation on blood pressure in

---

[m] Alarcon, R. D., Jenkins, C. S., Heestand, D. E., Scott, L. K., & Cantor, L. (1982). The effectiveness of progressive relaxation in chronic hemodialysis patients. *Journal of Chronic Diseases, 35*(10), 797–802. Data from Table 1.

[n] Taylor, J. A., Weber, W., Standish, L., Quinn, H., Goesling, J., McGann, M., et al. (2003). Efficacy and safety of echinacea in treating upper respiratory tract infections in children: A randomized controlled trial. *JAMA, 290*(21), 2824–2830.

normotensive men. After establishing a baseline blood pressure measurement, subjects were assigned to either a treatment group or placebo control group. The treatment group was supplemented with 1500 mg of calcium per day for a 12-week period. The control group received an identical-looking placebo. Systolic blood pressure measurements were taken at the beginning and end of a 12-week period. Explore the changes in blood pressure in the two groups with side-by-side boxplots. Discuss your findings.

12.22 *Calcium supplementation and blood pressure, hypothesis test*.

   (a) Calculate summary statistics for the groups described in Exercise 12.21.

   (b) Test the mean difference for significance.

12.23 *Delay in discharge*. Delays in discharges from intermediate healthcare facilities create unnecessary risks and expenses for patients and their families. Discharge delays (in days) from two healthcare facilities are shown in the following table. Assume that these data represent SRSs from their respective populations and that the populations demonstrate no major departures from Normality. Compare delay times in the facilities and test whether there was significant difference in the mean delay times.

Delays in discharge (days)

Facility A	12	9	13	16	7	19	20	16	15	14	12	16
Facility B	7	10	13	15	9	15	12	16	8	9	6	12

Fictitious data. Data file on companion website: DELAY.*.

12.24 *Time spent standing or walking*. A study was undertaken to compare daily activity levels in lean (group 1) and obese (group 2) study subjects. Sensors monitored the type and amount of routine daily activities such as walking, sitting, and lying down in each subject. Data are shown in Table 12.15. Students are also encouraged to download LEVINE2005.* from the companion website. We wish to compare time spent standing or walking (column 4) in lean and obese study subjects. Provide summary statistics and test whether the means differ significantly.

TABLE 12.15 Daily activity levels in lean and obese study subjects.

Identification number	Group 1 = lean; 2 = obese	Sitting (min/day)	Standing or walking (min/day)	Lying down (min/day)
1	1	370.3	511.1	555.5
2	1	374.5	607.9	450.7
3	1	582.1	319.2	537.4
4	1	357.1	584.6	489.3

continues

continued

Identification number	Group 1 = lean; 2 = obese	Sitting (min/day)	Standing or walking (min/day)	Lying down (min/day)
5	1	349.0	578.9	514.1
6	1	385.3	543.4	506.5
7	1	268.2	677.2	467.7
8	1	322.2	555.7	567.0
9	1	537.0	374.8	531.4
10	1	528.8	504.7	397.0
11	2	646.3	260.2	521.0
12	2	456.6	464.8	514.9
13	2	578.7	367.1	563.3
14	2	463.3	413.7	532.2
15	2	567.6	347.4	504.9
16	2	567.6	416.5	448.9
17	2	621.3	358.7	460.6
18	2	646.2	267.3	510.0
19	2	572.8	410.6	448.7
20	2	591.4	426.4	412.9

Data from Table 18.1 in Moore, D. S. (2010). *The Basic Practice of Statistics* (5th ed.). New York: W. H. Freeman. Originally from Levine, J. A., Lanningham-Foster, L. M., McCrady, S. K., Krizan, A. C., Olson, L. R., Kane, P. H., et al. (2005). Inter-individual variation in posture allocation: possible role in human obesity. *Science*, *307*(5709), 584–586. Data files: LEVINE2005.*.

12.25 ***Time spent standing or walking.*** This exercise continues our analysis of data from the study by Levine and coworkers (2005) introduced in Exercise 12.24. We now wish to test whether lean and obese differ in the amount they sit each day (column 3). Data appear in Table 12.15 and can be downloaded from the companion website as file LEVINE2005.*.

13

Comparing Several Means (One-Way Analysis of Variance)

In the prior chapter, we compared means from two groups. This chapter compares means from two or more groups. Exploratory, descriptive, and inferential methods are addressed.

ILLUSTRATIVE EXAMPLE

Data (Pets as moderators of a stress response). It has been suggested that pets provide a supportive social function in buffering adverse responses to physiological stressors. A study to address this topic monitored physiological responses to psychological challenges in pet owners. Human subjects, all of whom were self-described dog lovers, were randomly assigned to one of three groups:

Group 1 was monitored in the presence of their pet dog.

Group 2 was monitored in the presence of a human friend.

Group 3 was monitored with neither their pet dog nor friend present.

After being exposed to a psychological stressor (mental arithmetic), heart rates in subjects were monitored. Table 13.1 lists the data.

Figure 13.1 is a screenshot of the first 10 rows of the data table. The response variable in this data table is named HRT_RATE. The name of the explanatory variable is GROUP.

TABLE 13.1 Data for the "pets and stress" illustrative example. Data are mean heart rates of subjects (beats per minute) during the experiment.

Group 1 (Pet present)	Group 2 (Friend present)	Group 3 (Neither pet nor friend present)
69.17	99.69	84.74
68.86	91.35	87.23
70.17	83.40	84.88
64.17	100.88	80.37
58.69 —	102.15	91.75
79.66	89.82	87.45

continues

continued

Group 1 (Pet present)	Group 2 (Friend present)	Group 3 (Neither pet nor friend present)
69.23	80.28	87.78
75.98	98.20	73.28
86.45	101.06	84.52
97.54	76.91	77.80
85.00	97.05	70.88
69.54	88.02	90.02
70.08	81.60	99.05
72.26	86.98	75.48
65.45	92.49	62.65

Data from Allen, K. M., Blascovich, J., Tomaka, J., & Kelsey, R. M. (1991). Presence of human friends and pet dogs as moderators of autonomic responses to stress in women. *Journal of Personality and Social Psychology, 61*(4), 582–589.

Data are stored online in the file PETS.*.

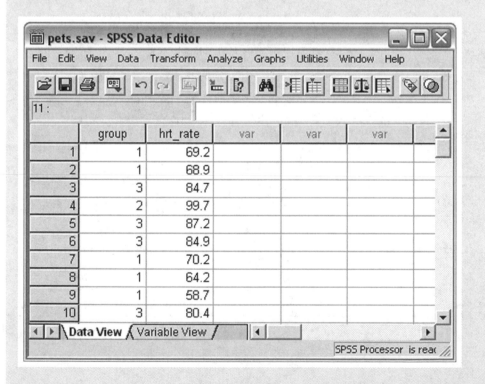

FIGURE 13.1   Screenshot of first 10 records in the pets and stress illustrative data. Also see Table 13.1. Graph produced with SPSS for Windows, Rel. 11.0.1.2001. Chicago: SPSS Inc. Reprint Courtesy of International Business Machines Corporation.

■ 13.1 Descriptive Statistics

Data analysis begins with graphical explorations and numerical summaries. Three graphical explorations are presented. Figure 13.2 displays side-by-side stemplots of the illustrative data set, Figure 13.3 displays side-by-side boxplots, and Figure 13.4 is a simple "dotplot" with the locations of group means indicated.

Group means and standard deviations are calculated in the usual manner. Here are these summary statistics for the illustrative data:

TABLE 13.2 Summary statistics pets and stress illustrative example.

Group	n_i	$\bar{x}_i$	s_i
1 (pets present)	15	73.4831	9.96983
2 (friends present)	15	91.3251	8.34115
3 (neither pets nor friends)	15	82.5241	9.24158
Total (pooled sample)	45	82.4441	11.62774

These graphs and summary statistics reveal group differences: group 1 (pets present) has the lowest mean heart rate and group 2 (friends present) has the highest. There are no clear departures from Normality except perhaps in group 1, which has an upper outside value. Variability (spread) of the response within groups does not seem to differ much.

```
5 |                5 |                5 |
5 | 8              5 |                5 |
6 | 4             6 |                6 | 2
6 | 58999         6 |                6 |
7 | 002           7 |                7 | 03
7 | 59            7 | 6             7 | 57
8 |               8 | 013           8 | 0444
8 | 56            8 | 689           8 | 777
9 |               9 | 12            9 | 01
9 | 7             9 | 789           9 | 9
10 |              10 | 012          10 |
  (×10)            (×10)             (×10)
```

| Pet present | Friend present | Neither friend nor pet present |

FIGURE 13.2 Mean heart rates of study subjects under stress (beats per minute), pets and stress illustrative example.

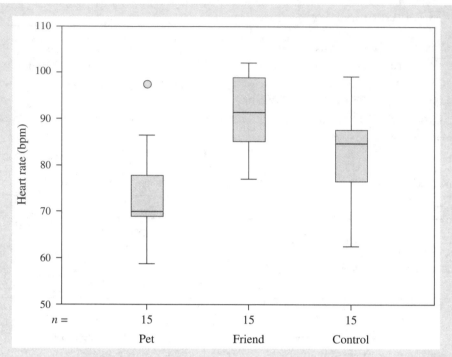

FIGURE 13.3   Boxplots, pets and stress illustrative example.

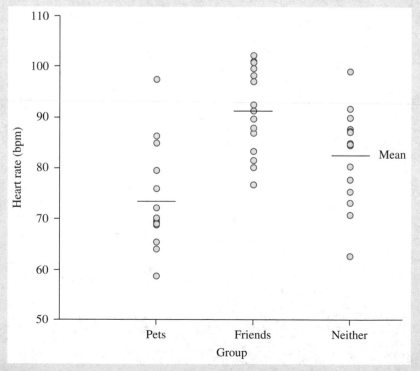

FIGURE 13.4   Dot plots with locations of means indicated, pets and stress illustrative example.

Exercises

13.1 *Birth weight, exploration*. Table 13.3 lists birth weights of infants based on the smoking status of mothers.

(a) Outline this study's design using a sketch of the type presented in Section 2.2 (see Figure 2.1) to depict this study's design. (*Comment:* In contrast to the pets and stress illustrative example, these data are nonexperimental.)

TABLE 13.3   Birth weights of infants in four groups of mothers. Data for Exercises 13.1, 13.3, 13.5, and 13.7.

Group 1 (nonsmokers)	8.56	8.47	6.39	9.26	7.98	6.84		
Group 2 (ex-smokers)	7.39	8.64	8.54	5.37	9.21			
Group 3 (<1/2 pack/day)	5.97	6.77	7.26	5.74	8.74	6.30		
Group 4 (≥1/2 pack/day)	7.03	5.24	6.14	6.74	6.62	7.37	4.94	6.34

Data were generated with the RV.NORMAL function in SPSS as follows: $X_1 \sim N(7.5, 1)$, $X_2 \sim N(7.5, 1)$, $X_3 \sim N(7.0, 1)$, $X_4 \sim N(6.5, 1)$.

Data are stored online in the file SMOK_BW.*.

(b) Figure 13.5 shows side-by-side boxplots of the response variable in the four groups. Interpret this plot.

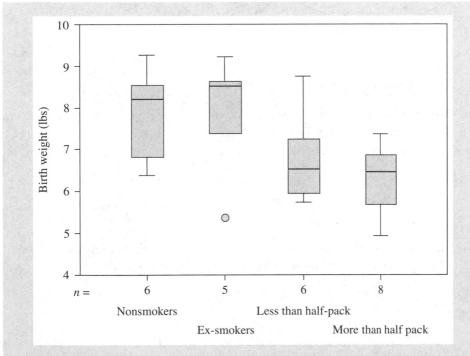

FIGURE 13.5   Side-by-side boxplots of birth weights in four groups of mothers, figure for Exercise 13.1.

13.2 *Laboratory experiment, exploration.* The ubiquity of junk food may contribute to the high prevalence of obesity in western societies. Table 13.4 lists data from a study in which 15 lab mice were randomly assigned to one of three groups: group 1 received a standard diet, group 2 received a diet of junk food, and group 3 received an organic diet. The response variable is weight gains (grams) over a 1-month period.

TABLE 13.4   Weight gain over a 1-month period (grams) in lab mice. Data for Exercises 13.2, 13.4, 13.6, and 13.8.

Group 1 (standard)	9.09	9.96	9.72	9.64	8.14
Group 2 (junk food)	10.21	10.48	13.01	12.74	12.58
Group 3 (organic)	9.03	9.55	12.35	9.33	9.51

Source: Data were generated with a random number generator so that $X_1 \sim N(10,1)$, $X_2 \sim (12,1)$, and $X_3 \sim N(10,1)$.

Data are stored online in the file LAB_EXPERIMENT.*.

(a) Outline the design of this study in schematic form. Is this study experimental or nonexperimental?

(b) Explore the data with side-by-side boxplots. Discuss the results of your graphical exploration.

(c) Calculate group means and standard deviations. Relate these summary statistics to your graphical exploration.

13.2 The Problem of Multiple Comparisons

The next step in data analysis involves questioning whether observed differences could be explained by random error. An impulse in this direction might be to test two groups at a time using separate *t*-statistics to address these three null hypotheses:

H_0: $\mu_1 = \mu_2$ (pet present vs. friend present)

H_0: $\mu_1 = \mu_3$ (pet present vs. neither present)

H_0: $\mu_2 = \mu_3$ (friend present vs. neither present)

This approach, however, bumps up against a problem known as the "problem of multiplicity" or "problem of multiple comparisons." This problem can be summarized with this colorful analogy[a]:

---

[a] Tukey, J. W. (1991). The philosophy of multiple comparisons. *Statistical Science, 6*(1), 103.

A man or woman who sits and deals out a deck of cards repeatedly will eventually get a very unusual set of hands. A report of unusualness would be taken quite differently if we knew it was the only deal ever made, or one of a thousand deals, or one of a million.

The problem boils down to identifying too many random differences when many comparisons are made. Consider testing three null hypotheses, all of which are (secretly) true. We will use an α-level of 0.05 in each instance. By the multiplicative rule for independent events (Section 5.5, Property 5), the probability of retaining (i.e., not rejecting) all three true null hypotheses is:

$$\text{Pr(retain all three true } H_0\text{s)} = (1 - 0.05)^3 = (0.95)^3 = 0.857$$

By the law of complements (Section 5.3, Property 3), the probability of rejecting at least one of the true null hypotheses is:

$$\text{Pr(reject at least one true } H_0) = 1 - 0.857 = 0.143$$

Therefore, the probability of making at least one false rejection is over 14%. This elevated type I error rate is called the **family-wise error rate**. $= FPR$

The family-wise error rate will get higher and higher as the number of comparisons is increased. Let c represent the number of comparisons in testing k groups. There are $c = {}_kC_2$ possible pairwise comparisons in testing k groups. For example, in comparing four groups, there are ${}_4C_2 = 6$ possible pair-wise comparisons.[b]

Here are family-wise error rates when comparing k groups in pairwise fashion when all the null hypotheses are true:

TABLE 13.5 Family-wise error rates for multiple hypothesis tests.

No. of groups (k)	2	3	4	5	6	7	8	9	10
No. of pair-wise comparisons (c)	1	3	6	10	15	21	28	36	45
Pr(at least one P-value < 0.05)[a]	0.050	0.143	0.265	0.401	0.537	0.659	0.762	0.842	0.901

[a]Pr(at least one P-value less than 0.05) $= 1 - \text{Pr(no } P\text{-value less than 0.05)} = 1 - 0.95^c$.

Small P-values are supposed to indicate that either a rare event has occurred or the null hypothesis is not true. In conducting multiple tests, neither of these conditions may be true. This violates the spirit of the hypothesis test and forms the basis of the problem of multiple comparisons.

The problem of multiple comparisons extends to confidence intervals. The probability that all 95% confidence intervals (CIs) in a family of confidence intervals will capture the parameter is:

---

[b]{1 v. 2}, {1 v. 3}, {1 v. 4}, {2 v. 3}, {2 v. 4}, {3 v. 4}.

TABLE 13.6   Family-wise confidence rates for multiple confidence intervals.

No. of groups (k)	2	3	4	5	6	7	8	9	10
No. of 95% CIs (c)	1	3	6	10	15	21	28	36	45
Pr(all capture parameter)[a]	0.950	0.857	0.735	0.599	0.463	0.341	0.238	0.158	0.099

[a]Pr(all the 95% confidence intervals capture their parameter) = 0.95^c.

Many methods have been developed to address this problem. These methods usually entail two steps. Step 1 is a test for *overall* significance. We use analysis of variance (ANOVA) for this purpose. Step 2 consists of procedures that compare groups two at time in order to identify specific areas of difference. We start by considering analysis of variance. *→ follow up comparisons —*

13.3 Analysis of Variance (ANOVA)

The method we are about to cover is called **one-way analysis of variance (ANOVA)**. It is *one-way* because one explanatory factor is considered.[c] It is *analysis of variance* because variation in the response is analyzed as a way of gaining insight into group differences. *albeit with multiple levels (values)*

Null and Alternative Hypotheses

The null hypothesis for one-way ANOVA in addressing k groups is

$$H_0: \mu_1 = \mu_2 = \cdots = \mu_k$$

where the subscript number is the group identifier. For example, the null hypothesis for addressing three groups is $H_0: \mu_1 = \mu_2 = \mu_3$. This is a statement of equality of population means.

The alternative hypothesis is "the null hypothesis is false" or

H_a: at least two of the population means differ $= not\ (all\ of\ them\ are\ equal)$

H_a can be true in multiple ways. In testing three groups, for example, all three population means may differ (i.e., $\mu_1 \neq \mu_2 \neq \mu_3$), or there may be one "odd-man out" (e.g., $\mu_1 \neq \mu_2 = \mu_3$ or $\mu_1 = \mu_2 \neq \mu_3$). The alternative is neither one sided nor two sided; it may be described as multisided.

Be aware that analysis of *variance* tests the equality of group means, not variances. Whether an observed mean difference is surprising depends on the variability of the response. Figure 13.6 demonstrates how this works. In this figure, the mean

---

[c]ANOVA techniques can be applied to analyze two explanatory factors (two-way ANOVA), repeated measurements (repeated measures ANOVA), multiple outcomes (multiple ANOVA), and other complex designs. Understanding one-way ANOVA is the gateway to these more advanced techniques.

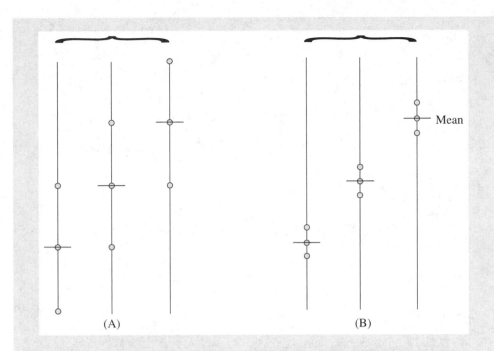

FIGURE 13.6 Data set (A) = high variance within groups obscures mean differences. Data set (B) = low variance within groups reveals mean differences.

differences in (A) and (B) are identical. However, the high degree of variability within groups in (A) obscures these mean differences, while the small amount of variability within (B) makes these differences stand out. When a response varies greatly, differences in means are likely to arise just by chance. When a response is consistent, group differences become clear.

ANOVA works by dividing the variance in a data set into these two components:

Variance *between* groups; also called the **mean square between**

Variance *within* groups; also called the **mean square within**

We start by considering the mean square between.

The Mean Square Between

The **mean square between (MSB)** quantifies the variance of group means around the grand mean. This is an estimate of the **variance between groups**.[d] Figure 13.7 depicts this variability graphically.

---

[d] We may use the symbol σ_B^2 to denote the variance between groups in the population.

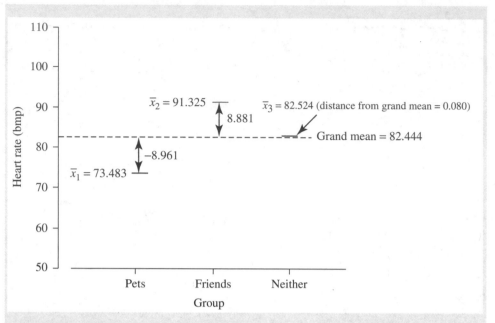

FIGURE 13.7   Variability of group means around the grand mean (basis of the MSB), pets and stress illustrative data.

The formula for the mean square between is

$$MSB = \frac{SS_B}{df_B}$$

where SS_B is the sum of squares between groups and df_B is the degrees of freedom between groups. The **sum of squares between** group $SS_B = \sum_{i=1}^{k} n_i(\bar{x}_i - \bar{x})^2$ and **degrees of freedom between** groups $df_B = k - 1$. In these formulas, k represents the number of groups, n_i represents the sample size of group i, $\bar{x}_i$ represents the sample mean of group i, and $\bar{x}$ represents the **grand mean** (the mean for all data combined). A concise formula for the mean square between is

$$MSB = \frac{\sum_{i=1}^{k} n_i(\bar{x}_i - \bar{x})^2}{k - 1}$$

Pooled (or grand) mean

For the illustrative data:

- $SS_B = \sum_{i=1}^{k} n_i(\bar{x}_i - \bar{x})^2 = [(15)(73.483 - 82.444)^2 + (15)(91.325 - 82.444)^2 +$

  $(15)(82.524 - 82.444)^2] = 2387.671$

- $df_B = 3 - 1 = 2$
- $MSB = \dfrac{2387.671}{2} = 1193.836.$

The Mean Square Within

The **mean square within (MSW)** quantifies the variability of data points in a group around its mean. This is the estimate of the **variance within groups**.[e] Figure 13.8 depicts this graphically.

The formula for the mean square within is:

$$MSW = \frac{SS_W}{df_W}$$

where SS_W is the sum of squares within groups and df_W is the degrees of freedom within groups. The **sum of squares within** is $SS_W = \sum_{i=1}^{k}(n_i - 1)s_i^2$ and **degrees of freedom within** is $df_W = N - k$. A concise formula for the mean square within is

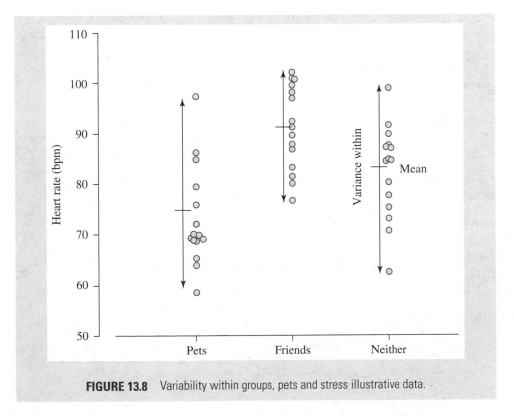

FIGURE 13.8   Variability within groups, pets and stress illustrative data.

---

[e] We may use the symbol σ^2_W to denote the variance within population groups.

$$MSW = \frac{\sum_{i=1}^{k} (n_i - 1)s_i^2}{N - k}.$$ In these formulas, s_i^2 represents the variance in group i and N is the total sample size ($N = n_1 + n_2 + \cdots + n_k$).

For the illustrative data

- $SS_W = [(15 - 1)(9.9698^2) + (15 - 1)(8.3412^2) + (15 - 1)(9.2416^2)] = 3561.316$
- $df_W = 45 - 3 = 42$
- $MSW = \dfrac{3561.316}{42} = 84.793.$

Equivalent formulas for the MSW are

$$MSW = \frac{(n_1 - 1)s_1^2 + (n_2 - 1)s_2^2 + \cdots + (n_k - 1)s_k^2}{(n_1 - 1) + (n_2 - 1) + \cdots + (n_k - 1)}$$

and $MSW = \dfrac{df_1 \cdot s_1^2 + df_2 \cdot s_2^2 + \cdots + df_k \cdot s_k^2}{df_1 + df_2 + \cdots + df_k}$. This last formula reveals that the MSW is a weighted average of group variances with weights provided by group degrees of freedom. When $k = 2$, $MSW = s^2_{pooled}$ (Section 12.3).

ANOVA Table and *F*-Statistic

ANOVA statistics are organized to form a table as follows:

TABLE 13.7   ANOVA table.

	Sum of Squares	df	Mean Square
Between groups	$SS_B = \sum_{i=1}^{k} n_i(\bar{x}_i - \bar{x})^2$	$df_B = k - 1$	$MSB = \dfrac{SS_B}{df_B}$
Within groups	$SS_W = \sum_{i=1}^{k} (n_i - 1)s_i^2$	$df_W = N - k$	$MSW = \dfrac{SS_W}{df_W}$
Total	$SS_T = SS_B + SS_W$	$df = df_B + df_W$	

For the pets and stress illustrative data:

TABLE 13.8   ANOVA table for pets and stress illustrative data.

	Sum of Squares	df	Mean Square
Between groups	2387.671	2	1193.836
Within groups	3561.316	42	84.793
Total	5948.987	44	

The ratio of the MSB and MSW is the *F*-statistic:

$$F_{stat} = \frac{MSB}{MSW}$$

This statistic has df_B numerator and df_W denominator degrees of freedom. For the pets and stress illustrative data, $F_{stat} = \dfrac{1193.836}{84.793} = 14.079$ with $df_B = 2$ and $df_W = 42$.

The F_{stat} can be thought of as a signal-to-noise ratio in which the "signal" comes from group differences and the "noise" comes from variation within groups. Large *F*-statistics suggest that the observed mean differences are *not* merely due to random noise.

P-Value and Conclusion

The F_{stat} is converted to a *P*-value with an *F* table of software utility. Table D in the Appendix is our *F* table. **F-distributions** have both numerator and denominator degrees of freedom, corresponding to df_B and df_W, respectively. We will use the notation F_{df_1, df_2} to denote an *F* random variable with df_1 numerator and df_2 denominator degrees of freedom. Unlike *z*- and *t*-distributions, *F*-distributions are asymmetrical. A picture of a typical *F*-distribution accompanies Table D.

The *P*-value for an analysis of variance corresponds to the area to the right of the F_{stat} under an F_{df_1, df_2} curve. To convert an F_{stat} to a *P*-value using Table D, find the numerator degrees of freedom near the top of the table; then find the denominator degrees of freedom in the left margin. *P*-values (right-tail regions) are listed in the second column. For the pets and stress illustrative example, the F_{stat} has $df_1 = 2$ and $df_2 = 42$. Because there is no entry for 42 denominator degrees of freedom, use the next smallest entry, which is 30 df.[f] The landmarks for $F_{2,30}$ are as follows:

TABLE 13.9   Critical values for $F_{2,30}$ from Table D.

	P	$df_1 = 2$
$df_2 = 30$	0.100	2.49
	0.050	3.32
	0.025	4.18
	0.010	5.39
	0.001	8.77

Therefore, an $F_{2,30}$ random variable with a value of 2.49 has a right tail of 0.100, an $F_{2,30}$ random variable with a value of 3.32 has a right tail of 0.050, and so on. The

---

[f]This will provide a conservatively calibrated *P*-value.

pets and stress data F_{stat} of 14.079 lies beyond the 0.001 point, indicating $P < 0.001$. Figure 13.9 depicts this graphically.

You can get a more precise P-value from an observed F_{stat} using a software utility such as *StaTable*[g] or *WinPepi*.[h] Figure 13.10 is a screenshot from www .statdistributions.com (a website created by Nathaniel Johnston, accessed on June 19, 2013) showing that an F_{stat} of 14.08 with 2 and 42 degrees of freedom has a right-tail (P-value) of about 0.

After deriving the P-value, the test results are summarized in the context of the original research question. Recall that the data compares heart rates among three experimental groups after study subjects were exposed to a psychological stressor. Study subjects with their pet present had a mean heart rate of 73.5 bpm (standard deviation 10.0), subjects with a friend present had a mean of 91.3 bpm (standard deviation 8.3), and subjects with neither a pet nor friends present had a mean of 82.5 bpm (standard deviation 9.2). These mean heart rates differed significantly according to the one-way ANOVA procedure ($P < 0.001$).

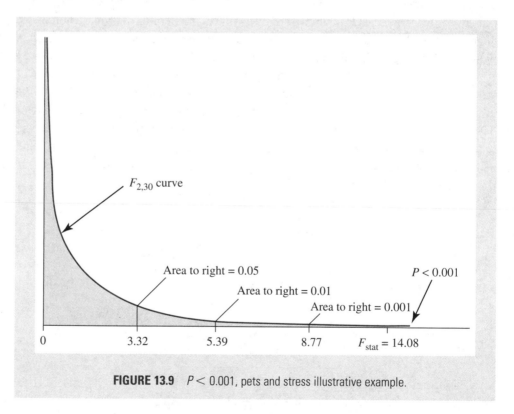

FIGURE 13.9 $P < 0.001$, pets and stress illustrative example.

[g]www.cytel.com/software/statable-apps/.
[h]Abramson, J. H. (2004). *WINPEPI* (PEPI-for-Windows): Computer programs for epidemiologists. *Epidemiologic Perspectives & Innovations, 1*(1), 6.

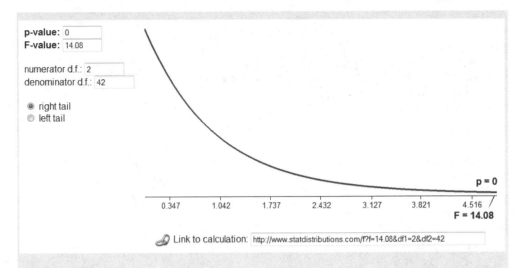

p-value: 0
F-value: 14.08

numerator d.f.: 2
denominator d.f.: 42

◉ right tail
◯ left tail

p = 0

0.347   1.042   1.737   2.432   3.127   3.821   4.516
F = 14.08

🔗 Link to calculation: http://www.statdistributions.com/f?f=14.08&df1=2&df2=42

FIGURE 13.10   Screenshot showing that an F_{stat} of 14.08 with 2 and 42 degrees of freedom has almost no area in its right tail and a P-value that is nearly 0. Screenshot from www.StatDistributions .com, Nathaniel Johnston.

Notes

1. **Software.** Because of the effort involved in calculating sums of squares and other ANOVA statistics, these are often carried out with a statistical package or utility.[i] Figure 13.11 displays output from SPSS's one-way ANOVA procedure for the pets and stress illustrative data.

2. **Relationships between one-way ANOVA and the equal variance independent t-test.** One-way ANOVA for two groups is equivalent to an equal variance independent t-test (Section 12.4).

    (a) Both address $H_0: \mu_1 = \mu_2$ against $H_a: \mu_1 \neq \mu_2$

    (b) ANOVA $df_W = df$ for the t-test (both equal $N - 2$)

    (c) $MSW = s^2_{pooled}$

    (d) $F_{stat} = (t_{stat})^2$

    (e) The F- and z-distributions are related as follows: $F_{1, df_2, 1-\alpha} = t^2_{df, 1-(\alpha/2)}$

3. **Plotting the means.** Figure 13.12 is a plot of group means with 95% confidence intervals for the pets and stress illustrative data set.[j] Plots like this are helpful in delineating group differences.[k]

---

[i] The website www.Statpages.org (maintained by John C. Pezzullo) lists many freeware and shareware programs that perform ANOVA calculations.

[j] Confidence intervals in this figure were calculated with the formula $\bar{x}_i \pm t_{n_i - 1, 0.975} \cdot \dfrac{s_i}{\sqrt{n_i}}$ (Section 11.4).

[k] More formal comparisons are also often necessary, as will be considered in Section 13.4.

ANOVA

HRT_RATE

	Sum of Squares	df	Mean Square	F	Sig.
Between groups	2387.671	2	1193.836	14.079	0.000
Within groups	3561.316	42	84.793		
Total	5948.987	44			

FIGURE 13.11 ANOVA table produced with SPSS for Windows, pets and stress illustrative data. Graph produced with SPSS for Windows, Rel. 11.0.1.2001. Chicago: SPSS Inc. Reprint Courtesy of International Business Machines Corporation.

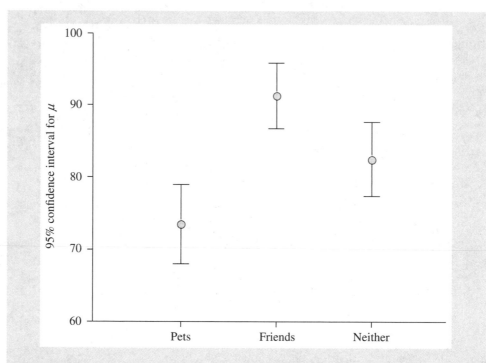

FIGURE 13.12 95% confidence intervals for population means, pets and stress illustrative data.

Exercises

13.3 *Smoking and birth weight, ANOVA.* Exercise 13.1 presented data on birth weights of infants based on the smoking status of mothers. Table 13.3 lists the data. Descriptive statistics are shown in Table 13.10. Conduct an ANOVA on these data. Show all hypothesis-testing steps.

TABLE 13.10   Descriptive statistics for Exercise 13.3.

Group	n	Mean	Std. Deviation
1 (nonsmoker)	6	7.9153	1.09947
2 (former smokers)	5	7.8296	1.52325
3 (< half pack)	6	6.7962	1.10107
4 (≥ half pack)	8	6.3024	0.84323
Total	25	7.1135	1.26907

13.4 *Laboratory experiment, ANOVA.* Recall the laboratory experiment described in Exercise 13.2. The response variable is weight gains (in grams). There are three groups of lab animals, each with five test animals. Data are shown in Table 13.4. Complete a one-way analysis of variance for this problem. Show all hypothesis-testing steps, and interpret the results.

■ 13.4 Post Hoc Comparisons

Rejection of the ANOVA null hypothesis indicates that at least one of the population means differs, but does not indicate which one(s). For example, in testing three groups, all the means may differ significantly, or the means from groups 1 and 2 may be similar, while the mean from group 3 is significantly larger (and so on). Although exploratory methods are useful in delineating group differences, sometimes formal tests are also needed. Procedures that address this need are called **post hoc comparisons**. Figure 13.13 is a screenshot from SPSS listing available post hoc comparison procedures. Eighteen different procedures are listed.[1] We will consider the first two on the list: the **least squares difference (LSD) method** and **Bonferroni method**.

Least Squares Difference Method

The least squares difference (LSD) procedure should be used only after a significant ANOVA test, and only for planned comparisons. Here are steps used in applying the LSD procedure.

A. **Hypotheses.** Set up multiple null hypotheses comparing means two at a time. Denote one group as group i and the other as group j; then test all possible combinations of i and j: H_0: $\mu_i = \mu_j$ versus H_a: $\mu_i \neq \mu_j$.

B. **Test statistic.** For each comparison calculate

$$t_{\text{stat}} = \frac{\bar{x}_i - \bar{x}_j}{\text{SE}_{\bar{x}_i - \bar{x}_j}}$$

---

[1] This large number of procedures reflects the lack of a one-size-fits-all method.

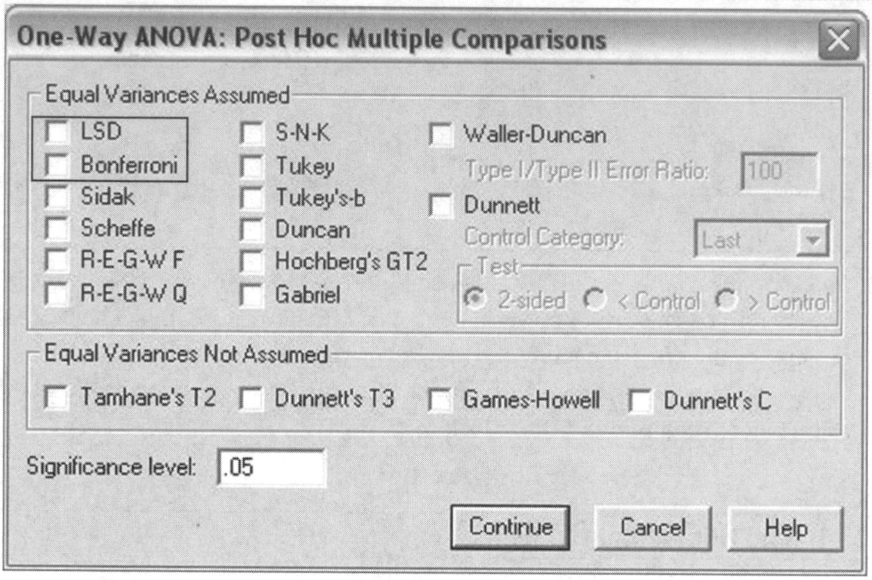

FIGURE 13.13   Multiple comparison procedures available in SPSS for Windows. The LSD (least square distance) and Bonferroni methods are covered in Section 13.4. Graph produced with SPSS for Windows, Rel. 11.0.1.2001. Chicago: SPSS Inc. Reprint Courtesy of International Business Machines Corporation.

where $\mathrm{SE}_{\bar{x}_i - \bar{x}_j} = \sqrt{\mathrm{MSW}\left(\frac{1}{n_i}\right) + \left(\frac{1}{n_j}\right)}$. In this formula, n_i represents the number of observations in group i and n_j represents the number of observations in group j. The test statistic has $N - k$ degrees of freedom, where N represents the total number of observations in all k groups combined (and not just the two groups).[m]

C. **P-value.** Convert the t-statistic to a P-value with Appendix Table C or statistical software.

D. **Significance level.** The P-value is compared to various α thresholds to assess the strength of evidence against the null hypothesis.

E. **Conclusion.** The test results are related back to the data in the context of the research question.

The procedure can also be adapted to calculate confidence intervals for mean differences. The $(1 - \alpha)100\%$ **confidence intervals for $\mu_i - \mu_j$** is given by:

$$(\bar{x}_i - \bar{x}_j) \pm t_{N-k,1-(\alpha/2)} \cdot \mathrm{SE}_{\bar{x}_i - \bar{x}_j}$$

where $\mathrm{SE}_{\bar{x}_i - \bar{x}_j} = \sqrt{\mathrm{MSW}\left(\frac{1}{n_i}\right) + \left(\frac{1}{n_j}\right)}$.

---

[m]All k groups contributed to the within group variance estimate, MSW.

ILLUSTRATIVE EXAMPLE

Post hoc comparisons, LSD method (Pets and stress). The pets and stress illustrative example found a significant difference among the three groups ($P <$ 0.001). Now we conduct post hoc comparisons with the LSD method. Recall that group 1 = pet present, group 2 = friend present, and group 3 = neither pet nor friend present. The three post hoc comparisons are

- H_0: $\mu_1 = \mu_2$ against H_a: $\mu_1 \neq \mu_2$
- H_0: $\mu_1 = \mu_3$ against H_a: $\mu_1 \neq \mu_3$
- H_0: $\mu_2 = \mu_3$ against H_a: $\mu_2 \neq \mu_3$

Only the first comparison is addressed by hand. The final two comparisons are addressed with a statistical package.

A. Hypotheses. H_0: $\mu_1 = \mu_2$ versus H_a: $\mu_1 \neq \mu_2$

B. Test statistics. $\text{SE}_{\bar{x}_1 - \bar{x}_2} = \sqrt{\text{MSW}\left(\frac{1}{n_1} + \frac{1}{n_2}\right)} =$

$\sqrt{84.793\left(\frac{1}{15} + \frac{1}{15}\right)} = 3.362$; $t_{\text{stat}} \dfrac{\bar{x}_1 - \bar{x}_2}{\text{SE}_{\bar{x}_1 - \bar{x}_2}} = \dfrac{73.483 - 91.325}{3.362} = -5.31$ with

df $= N - k = 45 - 3 = 42$

C. P-value. Using Table C, the two-sided P-value < 0.001. A statistical package determines that $P = 0.00000392$. This provides highly significant evidence against H_0.

D. Significance level. The observed difference is significant at the $\alpha = 0.001$ level.

E. Conclusion. The heart rates of study subjects with a pet present was significantly lower than that of study subjects with a friend present ($P = 0.0000039$).

The **95% confidence interval for $\mu_1 - \mu_2 = \bar{x}_i - \bar{x}_j \pm t_{N-k, 1-(\alpha/2)} \cdot \text{SE}_{\bar{x}_i - \bar{x}_j} =$** $(73.483 - 91.325) \pm (t_{42, 0.975})(3.362) = -17.842 \pm (2.021)(3.362) = -17.842$ $\pm 6.795 = (-24.6 \text{ to } -11.0)$ bpm.

Thus, the study subjects with their pet present had a heart rate that was on average 17.8 bpm less than the study subjects with their friend present (95% confidence interval for $\mu_1 - \mu_2$: 11.0 to 24.6 bpm).

Figure 13.14 shows SPSS output for the three possible post hoc comparisons. In testing, H_0: $\mu_1 = \mu_3$, $P = 0.010$. The 95% confidence interval for $\mu_1 - \mu_3$ is $(-15.8 \text{ to } -2.3)$. In testing, H_0: $\mu_2 = \mu_3$, $P = 0.012$. The 95% confidence interval for $\mu_2 - \mu_3$ is (2.0 to 15.6).

continues

continued

Multiple Comparisons

Dependent Variable: HRT_RATE

LSD

(I) Group	(J) Group	Mean Difference (I–J)	Std. Error	Sig.	95% Confidence Interval Lower Bound	Upper Bound
1	2	–17.842*	3.3624	0.000	–24.628	–11.056
	3	–9.041*	3.3624	0.010	–15.827	–2.255
2	1	17.842*	3.3624	0.000	11.056	24.628
	3	8.801*	3.3624	0.012	2.015	15.587
3	1	9.041*	3.3624	0.010	2.255	15.827
	2	–8.801*	3.3624	0.012	–15.587	–2.015

*The mean difference is significant at the 0.05 level.

FIGURE 13.14   LSD post hoc comparison, pets and stress illustrative example. Graph produced with SPSS for Windows, Rel. 11.0.1.2001. Chicago: SPSS Inc. Reprint Courtesy of International Business Machines Corporation.

Bonferroni's Method

Bonferroni's method ensures that the family-wise error rate is less than or equal to α after all possible pair-wise comparisons have been made. Recall that c represents the number of possible post hoc comparisons and that there are $c = {}_k C_2 = \dfrac{k!}{(k-2)!2!}$ such comparisons when k groups are assessed. To apply a Bonferroni's correction, simply multiply the P-value produced by the LSD method by c:[n]

$$P_{\text{Bonf}} = P_{\text{LSD}} \times c$$

For example, in completing an analysis of four groups, there are ${}_4 C_2 = \dfrac{4!}{(4-2)!2!} = \dfrac{4 \cdot 3 \cdot 2!}{2!/2 \cdot 1} = 6$ possible pair-wise comparisons. Therefore, you would multiply each P-value from the LSD procedure by 6 to get P_{Bonf}.

Bonferroni's adjustment can be applied to a construct **confidence intervals** for mean differences as follows:

$$(\bar{x}_i - \bar{x}_j) \pm t_{N-k,1-\frac{\alpha}{2c}} \cdot \text{SE}_{\bar{x}_i - \bar{x}_j}$$

---

[n] To be consistent with statistical software, we have applied Bonferroni's adjustment to the P-value. The more classical approach lowers the α-level for declaring "significance." To do this, just divide α by c. For example, to maintain a family-wise significance level of $\alpha = 0.05$ for six post hoc comparisons, the α-level would be lowered to $0.05 \div 6 = 0.0083$; then compare the P_{LSD} to the Bonferroni α-level.

where $(\bar{x}_i - \bar{x}_j)$ is the observed mean difference, $t_{N-k,1-\frac{\alpha}{2c}}$ is a t-value with $N - k$ degrees of freedom and cumulative probability $1 - \dfrac{\alpha}{2c}$, and $SE_{\bar{x}_i - \bar{x}_j} = \sqrt{MSW\left(\dfrac{1}{n_i} + \dfrac{1}{n_j}\right)}$.

ILLUSTRATIVE EXAMPLE

Post hoc comparisons, Bonferroni's methods (Pets and stress). To test H_0: $\mu_{\text{pet present}} = \mu_{\text{friend present}}$ for the illustrative example with application of a Bonferroni correction, recall that the LSD method produced $P = 0.00000392$ for this post hoc comparison. Because there are three post hoc comparisons in this analysis, $P_{\text{Bonf}} = P_{\text{LSD}} \times c = 0.00000392 \times 3 = 0.000012$.

To calculate the 95% confidence interval for $\mu_{\text{pet present}} - \mu_{\text{friend present}}$ with a Bonferroni correction, recall that $\bar{x}_1 = 73.483$, $\bar{x}_2 = 91.325$, and $SE_{\bar{x}_i - \bar{x}_j} = 3.362$.

For the 95% confidence interval, use $t_{N-k,1-\frac{\alpha}{2c}} = t_{45-3,1-\frac{0.05}{2\times3}} = t_{42,0.9917} = 2.51$.[o]

Therefore, the 95% confidence interval $= (\bar{x}_i - \bar{x}_j) \pm t_{N-k,1-\frac{\alpha}{2c}} \cdot SE_{\bar{x}_i - \bar{x}_j} =$ $(73.483 - 91.325) \pm (2.51)(3.362) = -17.842 \pm 8.439 = (-26.3 \text{ to } -9.4)$.

Figure 13.15 is a screenshot of SPSS output of the Bonferroni post hoc comparisons for these data. Notice that in testing, H_0: $\mu_1 = \mu_3$, $P_{\text{Bonf}} = 0.031$. In testing, H_0: $\mu_2 = \mu_3$, $P = 0.037$.

Multiple Comparisons

Dependent Variable: HRT_RATE
LSD

(I) Group	(J) Group	Mean Difference (I–J)	Std. Error	Sig.	95% Confidence Interval Lower Bound	95% Confidence Interval Upper Bound
1	2	−17.842*	3.3624	0.000	−26.227	−9.457
	3	−9.041*	3.3624	0.031	−17.426	−0.656
2	1	17.842*	3.3624	0.000	9.457	26.227
	3	8.801*	3.3624	0.037	0.416	17.186
3	1	9.041*	3.3624	0.030	0.656	17.426
	2	−8.801*	3.3624	0.037	−17.186	−0.416

*The mean difference is significant at the 0.05 level.

FIGURE 13.15 Bonferroni post hoc comparison, pets and stress illustrative example. Graph produced with SPSS for Windows, Rel. 11.0.1.2001. Chicago: SPSS Inc. Reprint Courtesy of International Business Machines Corporation.

---

[o] $t_{42,0.9917}$ determined with StaTable (Cytel Corp., www.cytel.com/Products/StaTable/).

The Bonferroni method is more conservative than the LSD method, favoring avoidance of type I errors at the cost of making more type II errors.

Exercises

13.5 *Smoking and birth weight, post hoc comparisons.* The ANOVA in Exercises 13.3 found that birth weight of four groups of infants differed significantly according to the smoking status of mothers ($P = 0.042$). Table 13.3 lists the data for the analysis. Data are also available online in the file SMOK_BW.SAV.

(a) Conduct post hoc comparisons using the LSD method. Summarize these results in concise narrative form.

(b) Incorporate Bonferroni's correction into each of the *P*-value calculated in part (a).

(c) Calculate 95% confidence intervals for all possible mean differences incorporating Bonferroni adjustments.

13.6 *Laboratory experiment, post hoc comparisons.* Exercises 13.2 and 13.4 considered weight gain in three groups of lab mice. Table 13.4 lists data for the problem. Results from the ANOVA were significant ($P = 0.015$). Complete post hoc comparisons of the groups using the Bonferroni method.

■ 13.5 The Equal Variance Assumption

ANOVA Assumptions

Conditions required for ANOVA mirror those of equal variance *t*-procedures (Section 12.5). These conditions include:

- sampling <u>I</u>ndependence within and among groups
- <u>N</u>ormality of sampling distribution of means
- <u>E</u>qual variances in the source populations

Notice that the letters underlined in the bulleted list form the mnemonic "INE," which stands for LINE minus the "L." This applies because ANOVA is a special type of (mathematical) linear model.

> Because we know the conditions of Independence and Normality (Sections 11.6 and 12.3), let us focus on evaluating the equal variance condition.

Assessing Group Variances

The term *scedastic* refers to the variance of a random variable. Groups may be either **homoscedastic** (equal in variance) or **heteroscedastic** (unequal in variance). Pooling group variances to form the MSW statistic of ANOVA (and the s_p^2 statistic of the equal variance *t*-test) relies on underlying homoscedasticity in the populations. We should therefore assess this condition before applying these procedures.

Three methods of assessing group variances are recommended. These are:

1. **Graphical exploration** with side-by-side stemplots or boxplots. Boxplots can be particularly enlightening. Widely discrepant hinge spreads (IQRs) warn the user of heteroscedasticity.

2. **Summary statistics,** especially standard deviations, can be used to assess group variances. One rule-of-thumb suggests if one group's standard deviation is more than double another's, heteroscedasticity should be suspected.

3. **Hypothesis tests of variance.** Many such tests exist,[p] but only **Levene's test** is widely applicable in non-Normal populations.[q]

Levene's Test of Variances

A. **Hypotheses.** Levene's procedure tests H_0: $\sigma_1^2 = \sigma_2^2 = \cdots = \sigma_k^2$ (homoscedasticity in the populations) against H_a: at least one of the populations has a different variance (heteroscedasticity in the populations). In testing two groups, for instance, H_0: $\sigma_1^2 = \sigma_2^2$ is tested against H_a: $\sigma_1^2 \neq \sigma_2^2$.

B. **Test statistic:** Levene's test is based on transforming each value to its absolute difference from its group mean. Data now consist of $a_{ij} = |x_{ij} - \bar{x}_i|$, where x_{ij} is the *j*th observation in group *i*.[r] A one-way ANOVA is then performed on the mathematically transformed data.

C. ***P*-value.** The *P*-value is derived from the one-way ANOVA *F*-statistic and its degrees of freedom.

D. **Significance level.** The *P*-value is compared to various α thresholds to assess the strength of evidence against the null hypothesis.

E. **Conclusion.** The test results are related back to the data in the context of the research question.

---

[p] For example *F* ratio test, Bartlett's test, Levene's test.
[q] Brown, M., & Forsythe, A. (1974). Robust tests for the equality of variances. *Journal of the American Statistical Association, 69*(346), 364–367.
[r] The procedure may be modified to use one of the following variable transformations: $a_{ij} = |x_{ij} - \bar{x}_i|$, where $\bar{x}_i$ is the median of group *i* ... or ... $a_{ij} = |x_{ij} - \bar{x}'_i|$, where $\bar{x}'$ is the 10% trimmed mean of group *i*.

Because calculations are tedious, a statistical package is used to conduct the test.[s]

ILLUSTRATIVE EXAMPLE

Assessing group variance (Pets and stress data). A multifaceted approach is used to assess group variances in the pets and stress illustrative data.

1. **Visual exploration.** Figure 13.3 shows side-by-side boxplots of the three groups. Notice that hinge spreads and whisker spreads do not differ greatly. The outside value in group 1 is of some concern, but is not sufficient by itself to declare population heteroscedasticity.

2. **Descriptive statistics.** Group 1 has the largest sample standard deviation ($s_1 = 9.97$) and group 2 has the smallest ($s_2 = 8.34$). Because there is *not* a twofold difference in standard deviations, evidence of population heteroscedasticity is absent.

3. **Levene's test.** Figure 13.16 is a screenshot of SPSS output showing Levene's test results for the pets and stress data. We test H_0: $\sigma_1^2 = \sigma_2^2 = \sigma_3^2$ against H_a: at least one of the σ_i^2s differ. The output shows that $F_{stat,\ Levene} = 0.06$ with 2 and 42 degrees of freedom, $P = 0.94$. Therefore, there is no significant difference among group variances. The null hypothesis of homoscedasticity is retained.

Test of Homogeneity of Variances			
HRT_RATE			
Levene Statistic	df_1	df_2	Sig.
0.059	2	42	0.943

FIGURE 13.16   Levene's test for pets and stress illustrative example. Graph produced with SPSS for Windows, Rel. 11.0.1.2001. Chicago: SPSS Inc. Reprint Courtesy of International Business Machines Corporation.

In summary, all three assessments suggest that data are consistent with homoscedasticity.

An illustration of heteroscedasticity is now presented.

---

[s] Most major statistical packages, including SAS (SAS Institute Inc., Cary, North Carolina), SPSS (SPSS Inc., Chicago, Illinois), and Stata (StataCorp LP, College Station, Texas), include a Levene's test procedure.

ILLUSTRATIVE EXAMPLE

Assessing group variance (Alcohol and income). Data are from a survey of socioeconomic status and alcohol consumption. The response variable is alcohol consumption graded on a 13-point scale as follows:

00 = nondrinker	01 = 1 drink per week
02 = 1–2 drinks per week	03 = 2 drinks per week
04 = 2–3 drinks per week	05 = 3 drinks per week
06 = 3–4 drinks per week	07 = 4 drinks per week
08 = 4–5 drinks per week	09 = 5 drinks per week
10 = 5–6 drinks per week	11 = 6 drinks per week
12 = 7–11 drinks per week	13 = 12+ drinks per week

The explanatory variable is income as five ordinal levels with 1 denoting the lowest income level and 5 denoting the highest. Because this is a large data set $n = 713$, analyses are done with a statistical package. Data are stored online in the file ALCOHOL.SAV as variables ALCS and INC.[t]

1. **Visual exploration.** Figure 13.17 shows side-by-side boxplots of alcohol consumption by income group. Hinge spreads in group 2 and group 3 are much larger than the hinge spread in group 5. Heteroscedasticity is evident. (We also note skews in all groups, and outside values in groups 1, 4, and 5.)

2. **Descriptive statistics.** Descriptive statistics for the groups (computed with a statistical package) are:

TABLE 13.11   Descriptive statistics for alcohol and income data.

INC	Mean	n	Std. Deviation
1	2.83	46	3.129
2	3.87	88	4.163
3	4.46	140	4.165
4	3.51	250	3.515
5	3.56	189	2.940
Total	3.71	713	3.593

Group standard deviations range from 2.940 (group 5) to 4.163 (group 2). It is difficult to tell whether these differences are random or whether they suggest population heteroscedasticity.

continues

[t] Source: Harvey Monder, Parklawn Computer Center, 1987.

continued

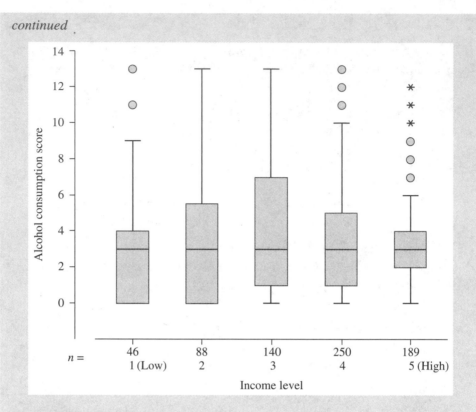

FIGURE 13.17   Side-by-side boxplots, alcohol consumption score by income level, illustrative example using the alcohol.* Graph produced with SPSS for Windows, Rel. 11.0.1.2001. Chicago: SPSS Inc. Reprint Courtesy of International Business Machines Corporation.

3. **Levene's test:** Figure 13.18 is SPSS output that tests H_0: $\sigma_1^2 = \sigma_2^2 = \sigma_3^2 = \sigma_4^2 = \sigma_5^2$ for these data. The $F_{\text{stat, Levene}} = 10.87$ with 4 and 708 degrees of freedom ($P = 0.000000015$). This provides good evidence against H_0. The populations appear to have unequal variances.

Test of Homogeneity of Variances

ALCS

Levene Statistic	df_1	df_2	Sig.
10.874	4	42	0.000000015

FIGURE 13.18   Results of Levene's test for equal variance, alcohol and income illustrative example. Graph produced with SPSS for Windows, Rel. 11.0.1.2001. Chicago: SPSS Inc. Reprint Courtesy of International Business Machines Corporation.

As a whole, evidence suggests group variances differ significantly. ANOVA should now be avoided.

Options for Comparing Groups in the Face of Non-Normality and/or Unequal Variance

How can we compare group means in the face of unequal variances? How can we compare small samples from non-Normal populations? Several analytic options exist. Here are specific recommendations:

1. **Stay "descriptive."** We may choose to avoid formal hypothesis tests and confidence intervals and instead rely on descriptive and exploratory analyses. You may be surprised to hear that *inferential statistics are not always necessary*. This is particularly true when samples are large and observed differences are obvious.

2. **Address outliers.** If unequal variances and non-Normality are caused by outliers, first determine the *cause* of the outliers. If the outliers are due to errors in the data, these errors must be corrected. If data errors are present and the errors cannot be corrected, consider removing these observations as faulty data. If the outliers appear to come from a different population, do not ignore the outliers, but consider analyzing them separately. If the outliers belong with the population of interest and are not the result of data errors, do *not* remove them from the data set and do not use Normal distribution-based methods.

3. **Mathematical transformations.** Reexpress the variable with a mathematical transformation (e.g., logarithm, root, and power) to make it more symmetrical and more nearly Normal (see Section 7.4, especially Figure 7.20).

4. **Use robust methods.** Consider using a robust inferential technique that is still valid under conditions of non-Normality or unequal variance. Earlier we opted for an independent *t*-test that did not require equal variance instead of the more traditional "equal variance" *t*-test. Examples of other robust techniques include *bootstrap methods*, *permutation tests*, *distribution-free methods*, and *nonparametric tests*. The next section of this chapter introduces a class of nonparametric tests called *rank* tests.

■ 13.6 Introduction to Rank-Based Tests

Background

Nonparametric tests encompass a broad array of statistical techniques used to analyze data.[u] This section introduces a class of nonparametric procedures called **rank tests**. Rank tests make fewer assumptions about distributional shape.

Table 13.12 lists the hypothesis-testing techniques we have covered or soon will. These tests are called Normal tests because they are based on Normal sampling theory. Nonparametric rank-based "equivalents" are listed in column 3 of the table. These

---

[u] There are many types of nonparametric test and there is no single definition of what constitutes a nonparametric procedure.

nonparametric tests work by replacing values in the data set with their associated ranks. This mathematical transformation allows us to dispense with assumptions about Normality and focus instead on relative positions within the ordered array. Table 13.13 lists null hypotheses for selected rank-based nonparametric tests.

TABLE 13.12   Normal tests and nonparametric rank test analogues.

Sample type	Normal test	Nonparametric test
One sample	One-sample t	Wilcoxon signed rank test
Paired samples	Paired t-test	Paired Wilcoxon signed rank test
Two independent samples	Independent t-test	Mann–Whitney (also called Wilcoxon rank sum test)
k independent samples	One-way ANOVA	Kruskal–Wallis test
Bivariate observations (Chapter 14)	Pearson's correlation coefficient	Spearman's rank correlation coefficient

TABLE 13.13.   Rank-based tests and their associated null hypotheses.

Sample Type	Rank-Based Nonparametric Test	Null Hypothesis
One sample	Wilcoxon signed rank	H_0: population median = specified value[a]
Matched pairs	Wilcoxon signed ranks on matched-paired differences	H_0: population median = 0[a]
Two independent samples	Mann–Whitney	H_0: population distributions are the same[b]
k independent samples	Kruskal–Wallis	H_0: population distributions are the same[b]
Bivariate observations	Spearman rank correlation	H_0: no correlation in the population

[a]Assumes population distribution is symmetrical.

[b]The null hypothesis can be made more specific by imposing distributional conditions such as symmetry and equal variance.

Kruskal–Wallis Test

The **Kruskal–Wallis test** is a nonparametric analogue of one-way ANOVA. However, unlike ANOVA, it can be used when equal variance and Normal sampling distribution conditions are absent. Here are the elements of the test:

A. **Hypotheses.** H_0: the population distributions are the same versus H_a: at least one of the populations differs.

Note: These hypotheses can be made more specific (e.g., H_0: population medians are the same). However, a more specific hypothesis statement would require additional distributional conditions such as "symmetry" or "equal variance." If the populations do not meet these additional requirements, the Kruskal–Wallis procedure will derive a *P*-value that does not reflect the true significance level of the test Fagerland, M. W., & Sandvik, L. (2009). The Wilcoxon–Mann–Whitney test under scrutiny. *Statistics in Medicine, 28*(10), 1487–1497.

B. **Test statistic.** The Kruskal–Wallis test statistic is based on assigning ranks to each observation and testing to see if mean ranks differ significantly. Calculation by hand is tedious, especially when tied ranks are encountered.[v] Fortunately, all major statistical packages include the Kruskal–Wallis procedure.[w] We will rely on SPSS for calculations.

Figure 13.19 shows the nonparametric menu of SPSS for Windows (Rel. 11.0.1. 2001. Chicago: SPSS Inc). The menu choice for the Kruskal–Wallis

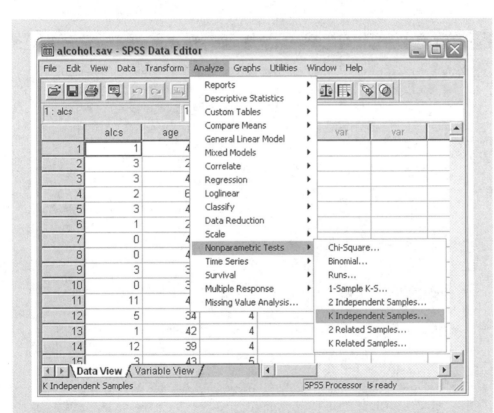

FIGURE 13.19 SPSS nonparametric test menu options. The menu choice for the Kruskal–Wallis test is highlighted. Graph produced with SPSS for Windows, Rel. 11.0.1.2001. Chicago: SPSS Inc. Reprint Courtesy of International Business Machines Corporation.

---

[v] For formulas, see Conover, W. J. (1980). *Practical Nonparametric Statistics* (2nd ed.). New York: John Wiley & Sons, pp. 229–231.
[w] A free online Kruskal–Wallis calculator is available at www.vassarstats.net.

procedure is highlighted ("K Independent Samples"). Check your software's documentation to learn how to access its Kruskal–Wallis procedure.

C. **P-value.** The null hypothesis is rejected when the mean rank in one or more of the groups is too large to fit in with the others. With samples that have at least five individuals in each group, the test is based on a chi-square statistic with $(k - 1)$ degrees of freedom.[x] With smaller samples, Kruskal–Wallis tables based on exact probabilities are needed.

D. **Significance level.** The P-value is compared to various α-levels to gauge the strength of evidence against the null hypothesis.

E. **Conclusion.** A conclusion is derived in the context of the data and research question.

ILLUSTRATIVE EXAMPLE

Kruskal–Wallis Test (Alcohol and income). Recall the illustrative example on the topic of alcohol consumption and income level. Evidence of distributional asymmetry and unequal variances was demonstrated (Figures 13.17 and 13.18 and Table 13.11). Although ANOVA and other Normal techniques are precluded by these facts, we may still apply a Kruskal–Wallis test.

A. **Hypotheses.** H_0: alcohol consumption distributions in the five income groups are the same against H_a: the distributions differ.

B. **Test statistic.** Figure 13.20 shows the output from SPSS for this problem. Group 3 has the largest mean rank and group 1 the lowest. The Kruskal–Wallis chi-square statistic is 7.793 with 4 degrees of freedom.

C. **P-value.** The output reports $P = 0.099$ ("Asymp. Sig").

D. **Significance level.** The evidence against the null hypothesis is significant at the $\alpha = 0.10$ threshold but not at the $\alpha = 0.05$ threshold.

E. **Conclusion.** Alcohol consumption differs by income level ($P = 0.099$).

Note: Figure 13.17 demonstrates that group medians do not differ: all five groups share a median of 3. However, the shapes and the spreads of the distributions appear to be quite different. For example, income group 3 has a much higher Q3 than income group 1, explaining why it had much higher mean alcohol consumption score (4.46 vs. 2.83).

---

[x] Section 18.3 introduces chi-square PDFs.

Kruskal–Wallis Test

Ranks

INC		N	Mean Rank
ALCS	1	46	303.58
	2	88	344.68
	3	140	385.17
	4	250	345.26
	5	189	370.40
	Total	713	

Test Statistics[a,b]

	ALCS
Chi-Square	7.793
df	4
Asymp. Sig.	0.099

a. Kruskal–Wallis test
b. Grouping variable: INC

FIGURE 13.20   Kruskal–Wallis test results, alcohol and income illustrative example. Graph produced with SPSS for Windows, Rel. 11.0.1.2001. Chicago: SPSS Inc. Reprint Courtesy of International Business Machines Corporation.

Exercises

13.7 *Smoking and birth weight, Kruskal–Wallis test.* Exercises 13.1, 13.3, and 13.5 considered birth weight in four groups of mothers. Table 13.3 lists the data. Data are also stored online in the file SMOK_BW.SAV. The ANOVA test found significant evidence against H_0 ($P = 0.042$). Now apply a Kruskal–Wallis test. Show all hypothesis-testing steps. Are the Kruskal–Wallis results consistent with the ANOVA results?

13.8 *Laboratory experiment, Kruskal–Wallis test.* Recall the experiment concerning weight gain in lab animals fed different diets (Exercises 15.2, 15.4, and 15.6). Table 13.4 lists the data. Conduct a Kruskal–Wallis test for these data, showing all hypothesis-testing steps.

Summary Points (One-Way ANOVA)

1. **Analysis of variance (ANOVA)** is used to test for differences among means from k independent populations, where $k \geq 2$. Data are presented in the form of a categorical explanatory variable and quantitative response variable.

2. As is our habit, we begin by **exploring and comparing** the distributions with graphical techniques (e.g., side-by-side boxplots) and group summary statistics (means, standard deviations, and sample sizes).

3. **The problem of multiple comparisons** refers to the elevated family-wise type I error rate that is determined by making many pair-wise comparisons. ANOVA averts this problem by reducing the assessment to a single test.

4. ANOVA separates out the **sources of variability** in the data. If the variance within groups greatly exceeds the variance between groups (large signal-to-noise ratio and large F-statistic), evidence of nonrandom differences is supported.

5. The ANOVA tests $H_0: \mu_1 = \mu_2 = \cdots = \mu_k$ versus H_a: at least one of the μ_is differs from the others via $F_{\text{stat}} = \dfrac{\text{MSB}}{\text{MSW}}$ according to the results in the following table:

Source of Variation	Sum of Squares	df	Mean Square
Between groups	$SS_B = \sum\limits_{i=1}^{k} n_i(\bar{x}_i - \bar{x})^2$	$df_B = k - 1$	$MSB = \dfrac{SS_B}{df_B}$
Within groups	$SS_W = \sum\limits_{i=1}^{k} (n_i - 1)S_i^2$	$df_W = N - k$	$MSW = \dfrac{SS_W}{df_W}$
Total	$SS_T = SS_B + SS_W$	$df = df_B + df_W$	

6. Rejection of the ANOVA does not identify which means differ from each other. To isolate group differences, **post hoc tests** are conducted. For research questions that were formatted before data were collected ("planned comparisons"), use the least squares distance method. For unplanned comparisons, a method that maintains the family-wise error rate (e.g., Bonferroni's method) should be used.

7. Valid ANOVA requires <u>I</u>ndependence within and between groups, <u>N</u>ormal sampling distributions, and <u>E</u>qual variances in the populations. These assumptions form the mnemonic "INE" ("LINE" without the "L").

8. The **equal variance assumption** permits the pooling of group variances to derive a single estimate for the variance within groups. This assumption can be assessed with exploratory methods (e.g., side-by-side boxplots) or with a formal test of $H_0: \sigma_1^2 = \sigma_2^2 = \cdots = \sigma_k^2$ with Levene's test statistic. Informally, if the ratio of largest and smallest standard deviations exceeds 2, unequal variance (heteroscedasticity) may be present.

Vocabulary

Bonferroni's method	Kruskal–Wallis test
Degrees of freedom between	Least squares difference (LSD) method
Degrees of freedom within	Levene's test
F-distributions	Mean square between (MSB)
F-statistic	Mean square within (MSW)
Family-wise type I error rate	Nonparametric tests
Grand mean	One-way analysis of variance (ANOVA)
Heteroscedasticity	Post hoc comparisons
Homoscedasticity	

Review Questions

13.1 What does ANOVA stand for?

13.2 Select the best response: This chapter compares the results from two or more groups when the response variable is _____.

(a) quantitative

(b) ordinal

(c) categorical

13.3 List two graphical techniques that can be used to compare the shapes, locations, and spreads of a data from two or more independent groups.

13.4 Identify two indicators of spread that are visible on boxplots.

13.5 Fill in the blank: When reporting group summary statistics for quantitative response variables, you should always report group sample sizes, means, and _____.

13.6 Select the best response: The Problem of Multiple Comparisons occurs when you identify too many _____ differences as a result of making too many comparisons.

(a) true

(b) systematic

(c) random

13.7 Completing multiple hypothesis tests undermines the intention of restricting the number of type I errors advertised by the nominal _____ level.

(a) α

(b) β

(c) power

13.8 How many different possible pair-wise comparisons are there in testing 10 groups?

13.9 In testing 45 true null hypotheses using an α threshold of 0.05, what is the probability that you will retain all 45 true null hypothesis? What is the probability that you will reject at least one (true) null hypothesis?

13.10 In testing 45 true null hypotheses using an α threshold of 0.01, what is the probability that you will retain all 45 true null hypothesis? What is the probability that you will reject at least one (true) null hypothesis?

13.11 Fill in the blank: The null hypothesis in testing the means from four groups with one-way ANOVA is _____.

13.12 Fill in the blank: The alternative hypothesis in testing the means from four groups with one-way ANOVA is _____.

13.13 Select the best response: The variance is equal to the _____ standard deviation.

(a) logarithm

(b) square root

(c) squared

13.14 Select the best response: The variance is a measure of _____.

(a) shape

(b) spread

(c) location

13.15 Select the best response: Evidence of nonrandom differences in group means occurs when the variance between groups is _____ the variance within groups.

(a) much greater than

(b) much less than

(c) about equal to

13.16 Select the best response: The variance within groups is estimated by the

(a) mean difference

(b) mean square between

(c) mean square within

13.17 Select the best response: The mean of all the observations combined in the data set is called the

(a) mean difference

(b) grand mean

(c) group mean

13.18 Select the best response: This statistic quantifies the variability of the group means around the grand mean.

(a) mean difference

(b) mean square between

(c) mean square within

13.19 The means square within (MSW) is equal to _____ divided by df_W.

13.20 The F-statistic is equal to the mean square between (MSB) divided by the _____ .

13.21 A large F-statistics indicates a large signal to _____ ratio, suggesting that the observed difference is not random.

13.22 F-probability density function curves have long right _____ .

13.23 Select the best response: One-way ANOVA when restricted to the comparison of means from two groups is equivalent to a(n) _____ test.

(a) z

(b) unequal variance independent t-test

(c) equal variance independent t-test

13.24 One procedure for making post hoc comparisons following ANOVA procedure is referred to as the LSD method. What does LSD stand for?

13.25 True or false? Post hoc comparisons made with LSD procedures are usually reserved for unplanned comparisons.

13.26 Select the best response: A Bonferroni-adjusted P-value compensates _____ .

(a) invalid data

(b) the problem of multiple comparisons

(c) violations in the conditions for the ANOVA

13.27 Complete this sentence: The Bonferroni adjustment for multiple comparisons avoids more type I errors than the LSD method, but makes more _____ errors.

13.28 List the three statistical conditions required for ANOVA.

13.29 Change one word in this sentence to make it correct: The conditions required for valid ANOVA are the same as those for an unequal variance independent t-test.

13.30 What does *homoscedastic* mean?

13.31 List three methods to assess whether population variances are equal.

13.32 Write the null hypothesis addressed by Levene's test of three groups.

13.33 List options for dealing with heteroscedasticity.

13.34 What makes a hypothesis test "robustness"?

13.35 What is the name of the nonparametric test that is functionally equivalent to one-way ANOVA?

13.36 State the null hypothesis tested by the Kruskal–Wallis procedure.

Exercise

13.9 *Antipyretics trial.* A trial evaluated the fever-reducing effects of three substances. Study subjects were adults seen in an emergency room with diagnoses of flu with body temperatures between 100.0 and 100.9°F. The three treatments (aspirin, ibuprofen, and acetaminophen) were assigned randomly to study subjects. Body temperatures were reevaluated 2 hours after administration of treatments. Table 13.14 lists the data.

(a) Explore these data with side-by-side boxplots. Discuss your findings.

(b) Calculate the mean and standard deviation of each group.

TABLE 13.14   Data for Exercise 13.9. Decreases in body temperatures (degrees Fahrenheit).

Group 1 (aspirin):	0.95	1.48	1.33	1.28		
Group 2 (ibuprofen):	0.39	0.44	1.31	2.48	1.39	
Group 3 (acetaminophen):	0.19	1.02	0.07	0.01	0.62	−0.39

Data are also stored online in the file ANTIPYRE.SAV.

(c) Complete an ANOVA for the problem. What do you conclude?

(d) Conduct post hoc comparisons with the LSD method. Which groups differ significantly at $\alpha = 0.05$?

13.10 *Melanoma treatment.* An experiment evaluated the use of genetically engineered white blood cells in the treatment of patients with melanoma.[y] Patients were divided into three cohorts. Patient cohort 1 received genetically engineered lymphocytes that were cultured *ex vivo* for 19 days. Cohort 2 received cells that were cultured for between 6 and 9 days. Cohort 3 received cells that were generated by a second rapid expansion performing after 8 to 9 days. Cell doubling times were as follows.

Cohort 1	8.7	11.9	10.0							
Cohort 2	1.4	1.0	1.3	1.0	1.3	2.0	0.6	0.8	0.7	0.9	1.9
Cohort 3	0.9	3.3	1.2	1.1						

---

[y] Morgan, R. A., Dudley, M. E., Wunderlich, J. R., Hughes, M. S., Yang, J. C., Sherry, R. M., et al. (2006). Cancer regression in patients after transfer of genetically engineered lymphocytes. *Science, 314*(5796), 126–129.

Either create a file with these data or download the file MORGAN2006.[*] from the companion website. Most statistical programs require you to enter that data into two columns: one column for the explanatory variable (COHORT) and the other for the response variable (DOUBLING). Your final data table should have 2 columns and 18 rows.

If you do not have access to a statistical program, consider using BrightStat. com.[z] Use of BrightStat.com is free. However, you must register as a user in order to access its ANOVA features.

(a) Calculate the means, standard deviations, and sample sizes of each group. Use the "double the standard deviation" benchmark to assess whether group variances are significantly different. Is there evidence of heteroscedasticity?

(b) Conduct Levene's test for unequal variance. Show all hypothesis testing steps. What do you conclude?

---

[z] Stricker, D. (2008). BrightStat.com: Free statistics online. *Computer Methods and Programs in Biomedicine, 92*, 135–143.

14 | Correlation and Regression

14.1 Data

This chapter considers methods used to assess the relationship between two quantitative variables. One variable is the **explanatory variable**, and the other is the **response variable**. Table 14.1 lists synonyms for these terms.

TABLE 14.1 Synonyms for *explanatory variable* and *response variable*.

Explanatory Variable	→	Response Variable
X	→	Y
independent variable	→	dependent variable
factor	→	outcome
treatment	→	response
exposure	→	disease

ILLUSTRATIVE EXAMPLE

Data (Doll's ecological study of smoking and lung cancer). Data from a historically important study by Sir Richard Doll[a] published in 1955 are used to illustrate techniques in this chapter. This study looked at lung cancer mortality rates by region (response variable) according to per capita cigarette consumption (explanatory variable). Table 14.2 lists the data.

continues

---

[a] Richard Doll (1912–2005) was a British epidemiologist well known for his studies linking smoking to various health problems.

continued

TABLE 14.2   Data used for chapter illustrations. Per capita cigarette consumption in 1930 (cig1930) and lung cancer cases per 100,000 in 1950 (LUNGCA) in 11 countries.

COUNTRY	CIG1930	LUNGCA
USA	1300	20
Great Britain	1100	46
Finland	1100	35
Switzerland	510	25
Canada	500	15
Holland	490	24
Australia	480	18
Denmark	380	17
Sweden	300	11
Norway	250	9
Iceland	230	6

Data from Doll, R. (1955). Etiology of lung cancer. *Advances Cancer Research, 3,* 1–50. Data stored online in the file DOLL-ECOL.SAV.

We will inspect the relationship between cigarette consumption (X) and lung cancer mortality (Y) with graphical explorations, numerical summaries, and inferential techniques.

■ 14.2 Scatterplot

The first step of our analysis is to create a **scatterplot** of the relationship between the explanatory variable (horizontal axis) and response variable (vertical axis). After creating the scatterplot, we inspect its

- **form** (e.g., straight line = **linear**; curved; random),
- **direction** (upward trend = positive association; downward trend = negative association; flat trend = no association), and
- **strength** (how closely data points adhere to an imaginary trend line).

We also check for the presence of **outliers** (striking deviations from the overall pattern).

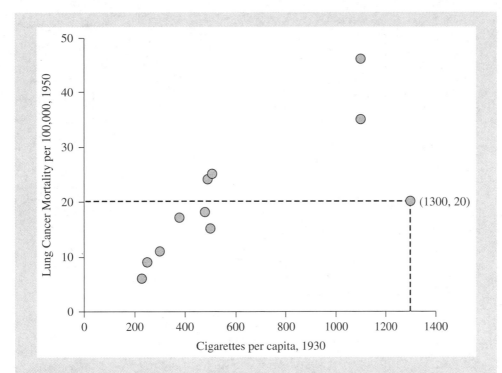

FIGURE 14.1 Scatterplot of Doll's illustration of the correlation between smoking and lung cancer rates. (Data listed in Table 14.2.) The data point for the United States is highlighted. Data from Doll, R. (1955). Etiology of lung cancer. *Advances Cancer Research, 3,* 1–50.

Figure 14.1 is a scatterplot of the illustrative data. The data point for the first observation (United States) has been highlighted to show an example of how values are plotted. This figure reveals a linear (straight line) form, positive association, and no apparent outliers.[b] Notice that as the values of cig1930 go up, the values for LUNGCA also tend to go up. The strength of the association is difficult to assess, but appears to be moderately strong.

Figure 14.2 demonstrates that it is difficult to judge the *strength* of a relationship based on visual clues alone. The data in these two plots are identical, yet the bottom plot appears to show a stronger relationship than the top plot. This is an artifact of the way axes have been scaled; the large amount of space between points in the top plot makes it appear to have a less-strong correlation. The eye is *not* a good judge of correlational strength; we need an objective way to make this type of assessment. The correlation coefficient will serve this purpose.

---

[b] The U.S. data point is lower than expected. Whether it is strikingly low or just a random fluctuation is unclear.

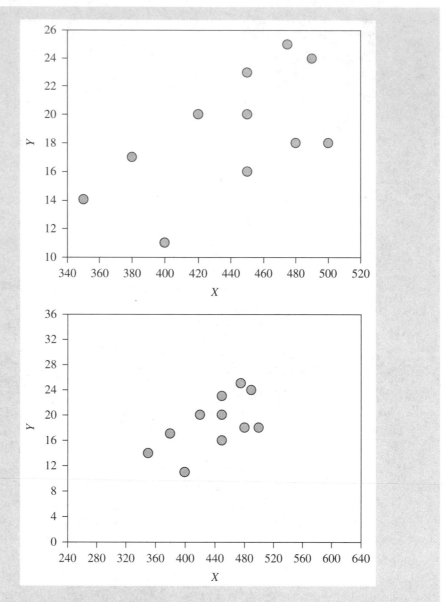

FIGURE 14.2 Scatterplots of the same data with different axis scalings. It is difficult to determine correlational strength visually.

■ 14.3 Correlation

The strength of the linear relationship between two quantitative variables is calculated with Pearson's **correlation coefficient (*r*)**. When all data points fall directly on a line with an upward slope, $r = 1$. When all points fall on a trend line with a negative slope, $r = -1$. Less-perfect correlations fall between these extremes. Lack of linear correlation is indicted by $r \approx 0$.

Figure 14.3 demonstrates correlations of various directions and strengths. The sign of *r* indicates the *direction* of the relationship (positive or negative). The absolute value of *r* indicates the *strength* of the relationship. The closer |*r*| gets to 1, the stronger the *linear* correlation.

This formula will provide insights into how the correlation coefficient does its job:

$$r = \frac{1}{n-1}\Sigma z_X z_Y$$

where $z_X = \dfrac{x_i - \bar{x}}{s_X}$ and $z_Y = \dfrac{y_i - \bar{y}}{s_Y}$.

Recall that *z*-scores quantify how many standard deviations a value lies above or below its mean (Section 7.2). This formula for *r* shows that the correlation coefficient is the average *product* of *z*-score for *X* and *Y*. When *X* and *Y* values are both above their averages, both *z*-scores are positive and the product of the *z*-scores is positive. When *X* and *Y* values are both below their averages, their *z*-scores will both be negative, so the product of *z*-scores is positive. When *X* and *Y* track in opposite directions (higher than average *X* values associated with lower than average *Y* values), one of the *z*-scores will be positive and the other will be negative, resulting in a negative product. The values of $z_X \cdot z_Y$ are summed and divided by $(n-1)$ to determine *r*.[c]

ILLUSTRATIVE EXAMPLE

Correlation coefficient. Table 14.3 shows the calculation of the correlation coefficient for the illustrative data. The correlation coefficient $r = 0.737$ represents a strong positive correlation.

continues

---

[c] The "minus 1" reflects a loss of one degree of freedom.

continued

TABLE 14.3   Calculation of correlation coefficient r, illustrative data.

$$\bar{x} = \frac{6640}{11} = 603.6364 \qquad s_X = 378.4514$$

$$\bar{y} = \frac{226}{11} = 20.54545 \qquad s_Y = 11.725$$

i	Country	X	$z_X = \dfrac{x_i - \bar{x}}{s_X}$	Y	$z_Y = \dfrac{y_i - \bar{y}}{s_Y}$	$z_X \cdot z_Y$
1	USA	1300	1.840	20	−0.047	− 0.086
2	Great Britain	1100	1.312	46	2.171	2.848
3	Finland	1100	1.312	35	1.233	1.618
4	Switzerland	510	−0.247	25	0.380	−0.094
5	Canada	500	−0.274	15	−0.473	0.130
6	Holland	490	−0.300	24	0.295	−0.089
7	Australia	480	−0.327	18	−0.217	0.071
8	Denmark	380	−0.591	17	−0.302	0.178
9	Sweden	300	−0.802	11	−0.814	0.653
10	Norway	250	−0.934	9	−0.985	0.920
11	Iceland	230	− 0.987	6	−1.241	1.225
	Sums →	6640	0.002	226	0	7.374

$$r = \frac{1}{n-1}\Sigma z_X z_Y = \frac{1}{11-1} \cdot 7.374 = 0.737$$

Notes

1. **Direction.** The sign of correlation coefficient r indicates whether there is a positive or negative linear correlation between X and Y. Correlation coefficients close to 0 suggest little or no linear correlation.

2. **Strength.** The correlation coefficient falls on a continuum from −1 to 1. The closer it comes to one of these extremes, the stronger the correlation. Although there are no hard-and-fast rules for what constitutes "strong" and "weak" correlations, here are some rough guidelines.

   $|r| \geq 0.7$ indicate strong associations

   $0.3 \leq |r| < 0.7$ indicate moderate associations

   $|r| < 0.3$ indicate weak associations

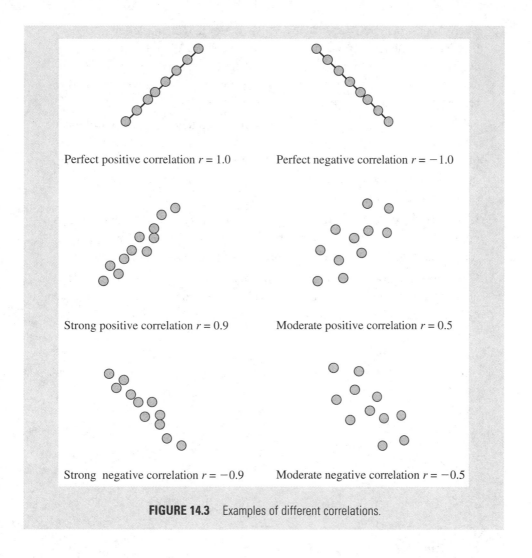

FIGURE 14.3 Examples of different correlations.

3. **Coefficient of determination.** Squaring r derives a statistic called the **coefficient of determination** (r^2). This statistic quantifies the proportion of the variance in Y explained by X. For the illustrative data, $r^2 = 0.737^2 = 0.54$, suggesting that 54% of the variance in lung cancer mortality is explained by per capita cigarette consumption.

4. **Reversible relationship.** Correlations can be calculated without first specifying which variable is the explanatory variable and which is the response variable; the calculation technique makes no assumption about the functional dependency between X and Y. For example, in studying the correlation between, say, arm length and leg length, we need not assume that altering arm length will have an effect on the leg length in order to interpret the correlation between two.

There is no functional dependency between arm length and leg length. It is merely enough to note the strong positive correlation. Correlation coefficient r will do its job whether you denote arm length as variable X or as variable Y. Note, however, that this reversibility property will *not* hold for regression, the technique covered later in this chapter.

5. **Not robust in face of outliers.** Correlation coefficient r is readily influenced by outliers. Figure 14.4 depicts a data set in which $r = 0.82$. The entire correlation in this data set is due to the one "wild-shot" observation. Outliers that lie far to the right or left in the horizontal direction can be especially influential (**influential observations**).

6. **Linear relations only.** Correlation coefficients describe linear relationships only. They do not apply to other forms. Figure 14.5 depicts a strong curved relationship, yet $r = 0$ because the relationship is not linear.

7. **Correlation is not necessarily causation.** Statistical correlations are not always causal. An observed relationship between X and Y may be an artifact of lurking variables (Section 2.2), for example.

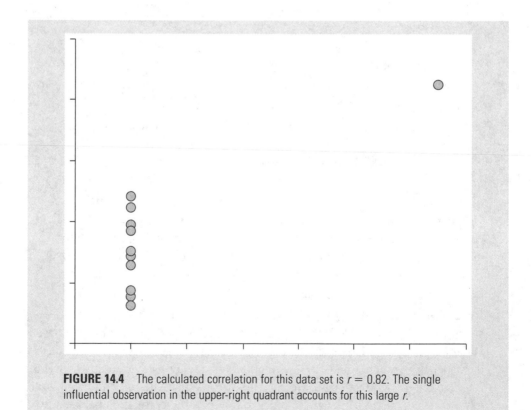

FIGURE 14.4   The calculated correlation for this data set is $r = 0.82$. The single influential observation in the upper-right quadrant accounts for this large r.

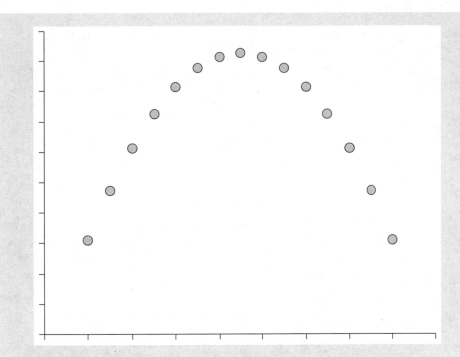

FIGURE 14.5   Correlation and regression apply only to linear relationships. This figure shows a perfect curved relationship, yet $r = 0.00$.

ILLUSTRATIVE EXAMPLE

Confounded correlation (William Farr's analysis of cholera mortality and elevation). This historical illustration shows how correlation does not always relate to causation. It was completed by a famous figure in the history of public health statistics named William Farr.[d] Like many of his contemporaries, Farr erroneously believed that infectious diseases like cholera were caused by unfavorable atmospheric conditions ("miasmas") that originated in suitable low-lying environments. In 1852, Farr reported on the association between elevation and cholera mortality in an article published in the *Journal of the Statistical Society of London.*[e] Data from this report are listed in Table 14.4 and are plotted (on logarithmic axes) in Figure 14.6. This scatter plot reveals a strong negative correlation. With both variables on natural log scales, $r = -0.987$. Farr used

continues

[d] William Farr (1807–1883)—one of the founders of modern epidemiology; first registrar of vital statistics for a nation (England); known as one of the first scientists to collect, tabulate, and analyze surveillance data; recognized the need for standardized nomenclatures of diseases; one of the first to apply actuarial methods to vital statistics.
[e] Farr, W. (1852). Influence of elevation on the fatality of cholera. *Journal of the Statistical Society of London, 15*(2), 155–183. Data stored in FARR1854.SAV.

this correlation to support the theory that "bad air" (miasma) settled into low-lying areas, causing outbreaks of cholera. We now know that this is ridiculous. Cholera is a bacterial disease caused by the waterborne transmission of *Vibrio cholera* and that outbreaks are not influenced by atmospheric conditions.

TABLE 14.4   William Farr's analysis of cholera mortality and elevation above sea level.

i	Mean elevation above sea level (feet)	Cholera mortality per 10,000
1	10	102
2	30	65
3	50	34
4	70	27
5	90	22
6	100	17
7	350	8

Data from Farr, W. (1852). Influence of elevation on the fatality of cholera. *Journal of the Statistical Society of London*, Table on p. 161 15(2), 155–183. Data stored in FARR1854.SAV.

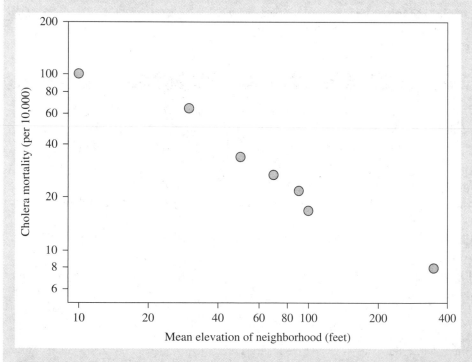

FIGURE 14.6   Cholera mortality and elevation above sea level were strongly correlated in the 1850s ($r = -0.987$), but this correlation was an artifact of confounding by the extraneous factor of "water source."

Farr made this blatant error because he failed to account for the confounding variable of "water source." People who lived in low-lying areas derived their water from nearby rivers and streams contaminated by human waste. The lurking variable "water source" confounded the relationship between elevation and cholera. Therefore, the observed correlation was entirely noncausal.

Statistical Inference About Population Correlation Coefficient ρ

Sample correlation coefficient r is the estimator of **population correlation coefficient ρ** ("rho"). Any observed r must be viewed as an example of an r that could have been derived by a different sample from the same population. Therefore, a positive or negative r could merely reflect sampling chance. For example, Figure 14.7 depicts a situation in which the population of bivariate points has no correlation while the sampled points (circled) have a perfect positive correlation. A hypothesis test is used to decrease the chance of concluding that a correlation exists when no such association is in the population.

Hypothesis Test

Here are the steps for testing a correlation coefficient for statistical significance.

A. **Hypotheses.** We test H_0: $\rho = 0$ against either H_a: $\rho \neq 0$ (two sided), H_a: $\rho > 0$ (one sided to the right) or H_a: $\rho < 0$ (one sided to the left).[f]

B. **Test statistic.** The test statistic is

$$t_{\text{stat}} = \frac{r}{\text{SE}_r}$$

where $\text{SE}_r = \sqrt{\frac{1 - r^2}{n - 2}}$. This test statistic has $n - 2$ degrees of freedom.[g]

C. **P-value.** The t_{stat} is converted to a P-value in the usual manner, using Table C or a software utility. As before, smaller and smaller P-values provide stronger and stronger evidence against the null hypothesis.

---

[f] The procedure described in this section addresses H_0: $\rho = 0$ only. Testing other values of ρ (i.e., H_0: $\rho =$ some value other than 0) requires a different approach. See Fisher, R.A. (1921). On the "probable error" of a coefficient of correlation deduced from a small sample. *Metron, 1,* 3–32.

[g] The loss of 2 degrees of freedom can be traced to using $\bar{x}$ and $\bar{y}$ as estimates of μ_X and μ_Y.

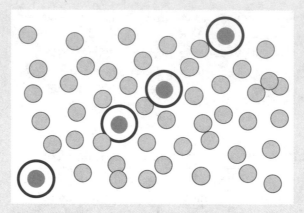

FIGURE 14.7   The encircled sample ($n = 4$) shows a strong correlation, while the population has none.

D. Significance level. The P-value is compared to various α levels to quantify the level of significance. When $P \leq \alpha$, the correlation is said to be significant at the α level.

E. Conclusion. The test results are discussed in the context of the observed correlation and research question.

ILLUSTRATIVE EXAMPLE

Hypothesis test of ρ. We have calculated a correlation coefficient of 0.737 for the ecological association between smoking and lung cancer mortality using the data in Table 14.1. We now test this correlation for statistical significance.

A. Hypotheses. H_0: $\rho = 0$ versus H_a: $\rho \neq 0$

B. Test statistics. The standard error of the correlation coefficient

$$\text{SE}_r = \sqrt{\frac{1 - r^2}{n - 2}} = \sqrt{\frac{1 - 0.737^2}{11 - 2}} = 0.2253.$$ The test statistic

$$t_{\text{stat}} = \frac{r}{\text{SE}_r} = \frac{0.737}{0.2253} = 3.27 \text{ with df} = n - 2 = 11 - 2 = 9.$$

C. P-value. Using Table C, $0.005 < P < 0.01$. Using statistical software, $P = 0.0097$. The evidence against H_0 is strong.

D. Significance level. The correlation coefficient is significant at $\alpha = 0.01$ (reject H_0 at $\alpha = 0.01$).

E. Conclusion. Data demonstrate a strong positive correlation ($r = 0.74$) between per capita cigarette consumption and lung cancer mortality that is statistically significant ($P = 0.0097$).

Confidence Interval for ρ

The **lower confidence limit (LCL)** and **upper confidence limit (UCL)** for population correlation coefficient ρ are given by

$$LCL = \frac{r - \overline{\omega}}{1 - r\overline{\omega}}$$

$$UCL = \frac{r + \overline{\omega}}{1 + r\overline{\omega}}$$

where $\overline{\omega} = \sqrt{\dfrac{t^2_{df,1-\frac{\alpha}{2}}}{t^2_{df,1-\frac{\alpha}{2}} + df}}$ and $df = n - 2$.[h]

ILLUSTRATIVE EXAMPLE

Confidence interval for ρ. We have established the following statistics for the illustrative example: $r = 0.737$, $n = 11$, and $df = 9$. The 95% confidence interval for ρ based on this information uses the following information:

- For 95% confidence, use $t_{9,0.975} = 2.262$; therefore $t^2_{9,0.975} = 2.262^2 = 5.117$

- $\overline{\omega} = \sqrt{\dfrac{2.262^2}{2.262^2 + 9}} = 0.602$

- $LCL = \dfrac{0.737 - 0.602}{1 - (0.737)(0.602)} = 0.243$

- $UCL = \dfrac{0.737 + 0.602}{1 + (0.737)(0.602)} = 0.927$

This allows us to say with 95% confidence that population correlation ρ for per capita cigarette consumption and lung cancer mortality is between 0.243 and 0.927.

Conditions for Inference

Be aware that correlation applies to linear relationships only. It does not apply to curved and other nonlinear relationships. The hypothesis testing and confidence

---

[h] Jeyaratnam, S. (1992). Confidence intervals for the correlation coefficient. *Statistics & Probability Letters, 15*, 389–393.

interval techniques for ρ assume sampling independence from a population in which X and Y have bivariate Normal distributions. Figure 14.8 depicts **bivariate Normality**.

When the correlation between X and Y is weak, deviations from bivariate Normality will be relatively unimportant. However, when the correlation is strong, inferences will be adversely affected by the absence of bivariate Normality. This problem is *not* diminished in larger samples—the central limit theorem does not apply.[i] However, mathematical transformation of variables can be used to impart Normality under some conditions. In addition, it is worth noting that r is still a valid point estimate of ρ even if the population is not bivariate Normal.

Exercises

14.1 *Bicycle helmet use.* Table 14.5 lists data from a cross-sectional survey of bicycle safety. The explanatory variable is a measure of neighborhood socioeconomic status (variable P_RFM). The response variable is "percent of bicycle riders wearing a helmet" (P_HELM).

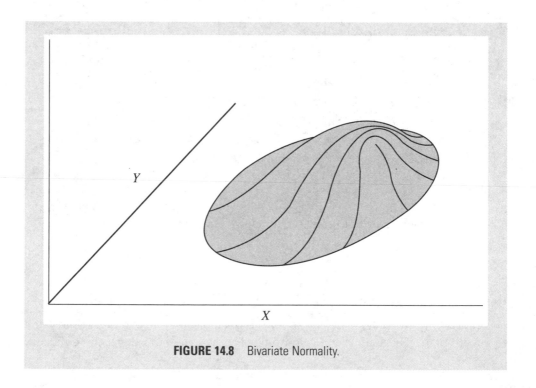

FIGURE 14.8   Bivariate Normality.

[i] Norris, R.C., Hjelm, H.F. (1961). Non-normality and product moment correlation. *The Journal of Experimental Education, 29,* 261–270.

TABLE 14.5 Data for Exercise 14.1. Percent of school children receiving free or reduced-fee lunches at school (variable P_RFM) and percent of bicycle riders wearing a helmet (variable P_HELM). Data for this study was recorded by field observers in October of 1994.

i	SCHOOL	P_RFM	P_HELM
1	Fair Oaks	50	22.1
2	Strandwood	11	35.9
3	Walnut Acres	2	57.9
4	Discov. Bay	19	22.2
5	Belshaw	26	42.4
6	Kennedy	73	5.8
7	Cassell	81	3.6
8	Miner	51	21.4
9	Sedgewick	11	55.2
10	Sakamoto	2	33.3
11	Toyon	19	32.4
12	Lietz	25	38.4
13	Los Arboles	84	46.6

Data from Perales, D., & Gerstman, B. B. (March 27–29, 1995). *A Bi-County Comparative Study of Bicycle Helmet Knowledge and Use by California Elementary School Children.* Paper presented at the Ninth Annual California Conference on Childhood Injury Control, San Diego, CA. Data are stored online in the file BICYCLE.SAV.

(a) Construct a scatterplot of P_REM and P_HELM. If drawing the plot by hand, use graph paper to ensure accuracy. Make sure you label the axes. After you have constructed the scatterplot, consider its form and direction. Identify outliers, if any.

(b) Calculate r for all 13 data points. Describe the correlational strength.

(c) A good case can be made that observation 13 (Los Arboles) is an outlier. Discuss what this means in plain terms.

(d) In practice, the next step in the analysis would be to identify the cause of the outlier. Suppose we determine that Los Arboles had a special program in place to encourage helmet use. In this sense, it is from a different population, so we decide to exclude it from further analyses. Remove this outlier and recalculate r. To what extent did removal of the outlier improve the fit of the correlation line?

(e) Test H_0: $\rho = 0$ (excluding outlying observation 13).

14.2 *Mental health care.* This exercise uses data from a historical study on mental health service utilization.[j] Fourteen Massachusetts counties are considered. The explanatory variable is the reciprocal of the distance to the nearest mental healthcare center (miles^{-1}, REC_DIST). The response variable is the percent of patients cared for in the home (PHOME). Table 14.6 lists the data.

(a) Construct a scatterplot of REC_DIST versus PHOME. Describe the pattern of the plot. Would correlation be appropriate here?

(b) Calculate the correlation coefficient for REC_DIST and PHOME. Interpret this statistic.

(c) Observation 13 (Nantucket) seems to be an outlier. Remove this data point from the data set and recalculate the correlation coefficient. (The variable PHOME2 in the data set LUNATICS.SAV has already removed this observation for you.) Did removing the outlier improve the fit of the linear model?

(d) This exercise plotted the reciprocal of distance to the nearest healthcare center and patient care at home. Now plot direct distance from the healthcare center (variable DIST) versus PHOME2 (percent cared for at home with the outlier removed). Why should we avoid calculating the correlation coefficient for the variables DIST and PHOMEZ?

TABLE 14.6   Data for Exercise 14.2. Percent of mental healthcare patients cared for in the home (PHOME), distance to the nearest mental healthcare center (DIST; miles), and reciprocal of the distance to the nearest mental healthcare center (REC_DIST; miles^{-1}).

i	COUNTY	PHOME	DIST	REC_DIST
1	Berkshire	77	97	0.01031
2	Franklin	81	62	0.01613
3	Hampshire	75	54	0.01852
4	Hampden	69	52	0.01923
5	Worcester	64	20	0.05000
6	Middlesex	47	14	0.07143
7	Essex	47	10	0.10000
8	Suffolk	6	4	0.25000
9	Norfolk	49	14	0.07143
10	Bristol	60	14	0.07143
11	Plymouth	68	16	0.06250

continues

---

[j]This study still has repercussions today. The relation between patient care and distance to the nearest health center remains an important consideration; numerous small hospitals scattered locally are preferable to a large central facility.

continued

i	COUNTY	PHOME	DIST	REC_DIST
12	Barnstable	76	44	0.02273
13	Nantucket	25	77	0.01299
14	Dukes	79	52	0.01923

Data are from a historically important study conducted by the pioneering public health physician and social statistician Edward Jarvis (1803–1884). Jarvis was an early advocate for the humane treatment of mental illness. The source of these data is *The Data and Story Library (DASL)* story titled "Massachusetts Lunatics" (http://lib.stat.cmu.edu/DASL/Stories/lunatics.html). Data are stored online in the file JARVIS.*

■ 14.4 Regression

The Regression Line

Regression, like correlation, is used to quantify the relationship between two quantitative variables. However, unlike correlations, regression can be used to express the functional relationship between X and Y. It does this by fitting a line to the observed bivariate data points. One challenge of this process is to find the best-fitting line to describe the relationship. If all the data points were to fall on a line, this would be a trivial matter. However, with statistical relationships, this will seldom be the case.

We seek the best-fitting line by breaking each observed Y value into two parts—the part predicted by the regression model (**predicted Y**) and the **residual** that is unaccounted for by the regression model:

$$\text{Observed } y = \text{predicted } y + \text{residual}$$

This equation can be reexpressed:

$$\text{Residual} = \text{observed } y - \text{predicted } y$$

Figure 14.9 shows residuals for the illustrative example with dotted lines. The regression line (solid) has been drawn to minimize the sum of the squared residuals. This technique of fitting the line is known as the least squares method and the line itself is the **least squares regression line**.

Notation: Let $\hat{y}$ denote the value of Y predicted by the regression model, a denote the intercept of the regression line, and b denote the slope of the line. The least squares regression line is:

$$\hat{y} = a + bx$$

Intercept a identifies where the regression line crosses the Y axis; slope coefficient b reflects the change in Y per unit X. Figure 14.10 shows how to interpret a and b.

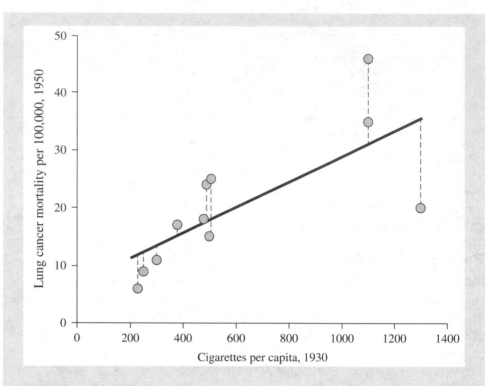

FIGURE 14.9   Fitted regression line and residuals, smoking and lung cancer illustrative data.

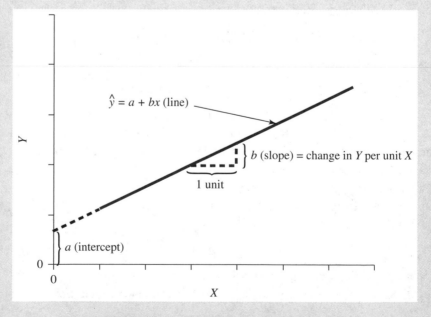

FIGURE 14.10   Components of a regression model.

The **slope** of the least squares regression line is calculated with this equation:

$$b = r\frac{s_Y}{s_X}$$

and the **intercept** is given by

$$a = \bar{y} - b\bar{x}$$

where $\bar{x}$ and $\bar{y}$ are the means of X and Y, s_X and s_Y are their sample standard deviations, and r is the correlation coefficient.

ILLUSTRATIVE EXAMPLE

Regression coefficients. For the illustrative data, we have established $\bar{x} = 603.64$, $\bar{y} = 20.55$, $s_X = 378.451$, $s_Y = 11.725$, and $r = 0.737$.

The slope coefficient $b = r\dfrac{s_Y}{s_X} = 0.737 \cdot \dfrac{11.725}{378.451} = 0.02283$.

The intercept coefficient $a = \bar{y} - b\bar{x} = 20.55 - (0.02283)(603.64) = 6.77$.
The regression model is $\hat{y} = 6.77 + 0.0228x$.

Notes
1. **Interpretation of the slope.** The slope in the illustrative example predicts an increase of 0.0228 lung cancer deaths (per 100,000 individuals per year) for each additional cigarette smoked per capita. Because the relationship is linear, we also can say that an increase of 100 cigarettes per capita predicts an increase of $100 \times 0.0228 = 2.28$ lung cancer deaths (per 100,000). It works the other way, too; a decrease of 100 cigarettes per capita is expected to decrease lung cancer mortality by 2.28 per 100,000.

2. **Predicting Y given x.** A regression model can be used to predict the value of Y for a given value of x. For example, we can ask "What is the predicted lung cancer rate in a country with a per capita cigarette consumption of 800?" According to our model, this predicted value $\hat{y} = 6.77 + (0.0228)(800) = 25.01$ (per 100,000).

3. **Avoid extrapolation.** Extrapolation beyond the observed range of X is *not* recommended.[k] The linear relationship should be applied to the observed range only.

4. **Specification of explanatory and response variables.** In calculating b, it is important to specify which variable is explanatory (X) and which variable is the response (Y). These cannot be switched around in a regression model.

5. **Technology.** We routinely use statistical packages for calculations. Figure 14.11 is a screenshot of computer output for the illustrative example. The intercept and slope coefficients are listed under the column labeled "Unstandardized Coefficient B." The intercept is listed as the model "(Constant)." The slope is listed as the unstandardized coefficient for explanatory variable CIG1930.[l] Notice that the slope is listed as 2.284E–02, which is computerese for $2.284 \times 10^{-02} = 0.02284$.

		Coefficients[a]				
		Unstandardized Coefficients		Standardized Coefficients		
Model		B	Std. Error	β	t	Sig.
1	(Constant)	6.756	4.906		1.377	0.202
	CIG1930	2.284E–02	0.007	0.737	3.275	0.010

[a]Dependent variable: MORTALIT

FIGURE 14.11   SPSSS regression output for the smoking and lung cancer illustrative data. (Graph produced with SPSS for Windows, Rel. 11.0.1.2001. Chicago: SPSS Inc. Reprint Courtesy of International Business Machines Corporation.)

---

[k] "Now, if I wanted to be one of those ponderous scientific people, and 'let on' to prove what had occurred in the remote past by what had occurred in a given time in the recent past, or what will occur in the far future by what has occurred in late years, what an opportunity is here! Geology never had such a chance, nor such exact data to argue from! Nor 'development of species', either! Glacial epochs are great things, but they are vague—vague. Please observe. In the space of one hundred and seventy-six years the Lower Mississippi has shortened itself two hundred and forty-two miles. This is an average of a trifle over one mile and a third per year. Therefore, any calm person, who is not blind or idiotic, can see that in the Old Oolitic Silurian Period, just a million years ago next November, the Lower Mississippi River was upward of one million three hundred thousand miles long, and stuck out over the Gulf of Mexico like a fishing-rod. And by the same token any person can see that seven hundred and forty-two years from now the Lower Mississippi will be only a mile and three-quarters long, and Cairo and New Orleans will have joined their streets together, and be plodding comfortably along under a single mayor and a mutual board of aldermen. There is something fascinating about science. One gets such wholesale returns of conjecture out of such a trifling investment of fact." (Mark Twain, *Life on the Mississippi*, 1883, pp. 173–176).

[l]Output from other statistical packages will have different labels to denote the slope and intercept.

6. **Relationship between the slope coefficient and correlation coefficient.** There is a close connection between correlation and regression.

   This is seen in the formula $b = r\dfrac{s_Y}{s_X}$. A change in one standard deviation of X is associated with a change of r standard deviations in Y. The least squares regression line will always pass through the point $(\bar{x}, \bar{y})$ with a slope of $r \cdot s_Y/s_X$.

7. **b versus r.** Both b and r quantify the relationship between X and Y. How do they differ? Slope b reflects the statistical relationship between X and Y in the same units as the data. For example, the slope for the illustrative data predicts that a decrease in 100 *cigarettes per capita* will be accompanied by an average decrease of 2.28 *lung cancer cases per 100,000* people per year. In contrast, correlation coefficient r provides only a unit-free measure of statistical strength (e.g., $r = 0.74$).

8. **Regression is not robust.** Like correlation, regression is strongly influenced by outliers. Outliers in the Y direction have large residuals. Outliers in the X direction have a large influence on leveraging calculated estimates.

9. **Linear relations only.** Regression describes linear relationships only. It does not apply to other functional forms.

10. **"Association" does not always mean "causation."** As discussed in the section on correlation, statistical associations are not always causal. Take care to consider lurking variables that have the potential to confound results, especially when considering nonexperimental data.

Inferential Methods

Population Regression Model and Standard Error of the Regression

The **population regression model** is:

$$y_i = \alpha + \beta x_i + \epsilon_i$$

where y_i is the value of the response variable for the ith observations, α is the parameter representing the intercept of the model, β is the parameter representing the slope of the model, x_i is the value of the explanatory variable for the ith observations, and ϵ_i is the "error term" or residual for that point. We make the simplifying assumption that residual term ϵ_i varies according to a Normal distribution with mean 0 and uniform standard deviation σ:

$$\epsilon_i \sim N(0, \sigma)$$

The σ in this expression quantifies the random scatter around the regression line. This quantity is assumed to be the same all levels of X. We have thus imposed Normality and equal variance conditions on the model. Figure 14.12 shows this schematically.

We estimate σ with a statistic called the **standard error of the regression**[m] **(denoted $s_{Y|X}$)**, which is calculated

$$s_{Y|X} = \sqrt{\frac{1}{n-2}\Sigma\text{residuals}^2}$$

Recall that a residual is the difference between an observed value of y minus the value of Y predicted by the regression model ($\hat{y}$):

$$\text{Residual} = y - \hat{y}$$

As an example, the point (1300, 20) in the illustrative data set has $\hat{y} = a + bx = 6.77 + (0.0228)(1300) = 36.4$. Therefore, its residual $= y - \hat{y} = 20 - 36.4 = -16.4$. The dotted line in Figure 14.13 depicts this residual.

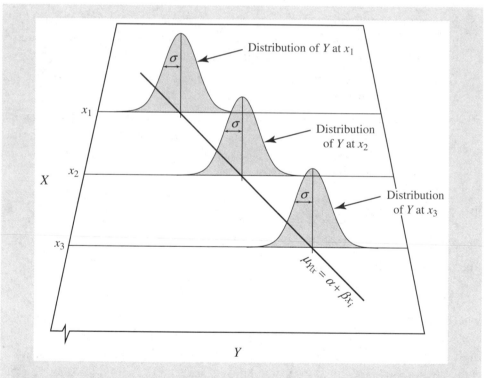

FIGURE 14.12   Population regression model showing Normality and homoscedasticity conditions.

---

[m] It may be easier to think of this statistic as the standard deviation of the scatter around the regression line (i.e., standard deviation of Y at each given level X).

Table 14.7 lists observed values, predicted values, and residuals for each data point in the illustrative data set. The standard error of the regression is calculated as follows:

$$s_{Y|X} = \sqrt{\frac{1}{n-2} \Sigma \text{residuals}^2}$$

$$= \sqrt{\frac{1}{11-2}(-16.4)^2 + 14.1^2 + \cdots + (-6.0)^2}$$

$$= \sqrt{\frac{1}{9} \cdot 626.05} = \sqrt{69.561} = 8.340$$

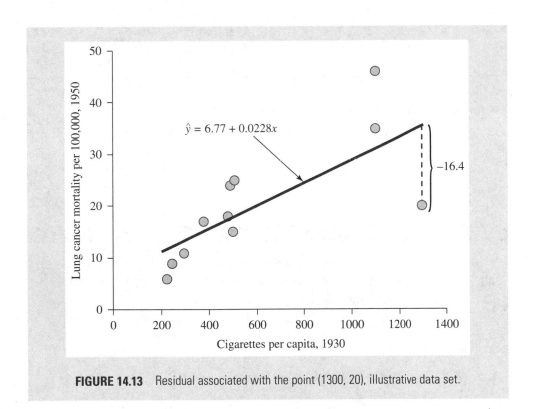

FIGURE 14.13 Residual associated with the point (1300, 20), illustrative data set.

TABLE 14.7   Residuals in smoking and lung cancer illustrate data set.

COUNTRY	x CIG1930	y MORTALITY	$\hat{y} = a + bx$ predicted	$y - \hat{y}$ residual
USA	1300	20	36.4	−16.4
Great Britain	1100	46	31.9	14.1
Finland	1100	35	31.9	3.1
Switzerland	510	25	18.4	6.6
Canada	500	15	18.2	−3.2
Holland	490	24	17.9	6.1
Australia	480	18	17.7	.3
Denmark	380	17	15.4	1.6
Sweden	300	11	13.6	−2.6
Norway	250	9	12.5	−3.5
Iceland	230	6	12.0	−6.0

Confidence Interval for the Population Slope

The confidence interval for slope parameter β can now be estimated with this formula:[n]

$$b \pm t_{n-2,1-\frac{\alpha}{2}} \cdot SE_b, \text{ where } SE_b = \frac{s_{Y|X}}{\sqrt{n-2} \cdot s_X}$$

ILLUSTRATIVE EXAMPLE

Confidence interval for β. We have established for the smoking and lung cancer illustrative example that $n = 11$, $b = 0.02284$, $s_{Y|X} = 8.340$, and $s_X = 378.451$. Let us calculate a 95% confidence for slope parameter β.

- $SE_b = \dfrac{s_{Y|X}}{\sqrt{n-1} \cdot s_X} = \dfrac{8.340}{\sqrt{11-1} \cdot 378.451} = 0.006969$

- For 95% confidence, use $t_{n-2,1-(\alpha/2)} = t_{9,0.975} = 2.262$
- The 95% confidence interval for $\beta = b \pm t_{n-1,1-\frac{\alpha}{2}} \cdot SE_b = 0.02284 \pm$ $(2.262)(0.006969) = 0.02284 \pm 0.01576 = (0.00708 \text{ to } 0.0386)$.

---

[n] An equivalent formula for the standard of error of the slope is $SE_b = \dfrac{s_{Y|X}}{\sqrt{\Sigma(x_i - \bar{x})^2}}$.

t-Test of Slope Coefficient

A *t*-statistic can be used to test the slope coefficient for statistical significance. Under the null hypothesis, there is no linear relationship between *X* and *Y* in the population, in which case, the slope parameter β would be 0. Here are the steps of the testing procedure:

A. **Hypotheses.** H_0: $\beta = 0$ against either H_a: $\beta \neq 0$ (two sided), H_a: $\beta < 0$ (one sided to the left) or H_a: $\beta > 0$ (one sided to the right). The two-sided alternative shall be our default.

B. **Test statistic.** The test statistic is

$$t_{\text{stat}} = \frac{b}{\text{SE}_b}$$

where $\text{SE}_b = \frac{s_{Y|X}}{\sqrt{\Sigma(x - \bar{x})^2}}$. This test statistic has $n - 2$ degrees of freedom.

C. **P-value.** The one-tailed *P*-value $= \Pr(t \geq |t_{\text{stat}}|)$. Use Table C or a software utility such as *StaTable*[o] to determine this probability.

D. **Significance level.** The test is said to be significant at the α-level of significance when $P \leq \alpha$.

E. **Conclusion.** The test results are discussed in the context of the regression model and research question.

ILLUSTRATIVE EXAMPLE

t-statistic. Let us test the slope in the smoking and lung cancer illustrative data for statistical significance. There are 11 bivariate observations. We have established $b = 0.02284$ and $\text{SE}_b = 0.006969$.

A. **Hypotheses.** H_0: $\beta = 0$ versus H_a: $\beta \neq 0$

B. **Test statistic.** $t_{\text{stat}} = \frac{b}{\text{SE}_b} = \frac{0.02284}{0.006969} = 3.28$ with df $= n - 2 = 11 - 2 = 9$

C. **P-value.** $P = 0.0096$, providing good evidence against H_0.

D. **Significance level.** The association is significant at $\alpha = 0.01$.

continues

---

[o]Cytel Software Corp. (1990–1996). *StaTable: Electronic Tables for Statisticians and Engineers.* www.cytel.com/Products/StaTable/.

continued

E. Conclusion. The observed linear association between per capita cigarette consumption and lung cancer mortality is not likely due to chance ($P = 0.0096$).

Note: The hypothesis test does not quantify the extent to which cigarette consumption increases lung cancer mortality. To quantify this relationship, we must return to slope estimate, which suggests an additional 0.0023 lung cancer deaths per 100,000 person-years with each additional cigarette consumed.

Notes:

1. ***t*-tests for correlation and for slope.** The test for H_0: $\rho = 0$ and H_0: $\beta = 0$ produce identical *t*-statistics: $t_{\text{stat}} = \dfrac{r}{\text{SE}_r} = \dfrac{b}{\text{SE}_b}$ with df $= n - 2$.

2. **Relation between confidence interval and test of H_0.** You can use the $(1 - \alpha)100\%$ confidence interval for β to see if results are significant at the α level of significance. When "0" is captured in the $(1 - \alpha)100\%$ confidence interval for the population slope, the data are not significant at that α level. When the value 0 is not captured within the confidence interval, we can say that the slope is significantly different from 0 at that α level. The 95% confidence interval for β for the illustrative data is (0.00706 to 0.03862). Therefore, this slope is significant at $\alpha = 0.05$.

3. **Testing for population slopes other than 0.** The test statistic can be adapted to address population slopes other than 0. Let β_0 represent the population slope posited by the null hypothesis. To test H_0: $\beta = \beta_0$, $t_{\text{stat}} = \dfrac{b - \beta_0}{\text{SE}_b}$ with df $= n - 2$. For example, to test whether the slope in the smoking and lung cancer illustrative data is significantly different from 0.01, the null hypothesis is H_0: $\beta = 0.01$. The test statistic is $t_{\text{stat}} = \dfrac{b - \beta_0}{\text{SE}_b} = \dfrac{0.02284 - 0.01}{0.006976} = 1.84$ with df $= 9$ (two-sided $P = 0.099$). Therefore, the difference is marginally significant (i.e., significant at $\alpha = 0.10$ but not at $\alpha = 0.05$).

Analysis of Variance

An analysis of variance (ANOVA) procedure can be used to test the regression model. Results will be equivalent to the *t*-test. The ANOVA method is presented as a matter of completeness and because it leads to methods useful in multiple regression (Chapter 15).

A. **Hypotheses.** The null hypothesis is H_0: the regression model does not fit in the population. The alternative hypothesis is H_a: the regression model does fit in the population. For simple regression models, these statements are functionally equivalent to $H_0: \beta = 0$ and $H_a: \beta \neq 0$, respectively.

B. **Test statistics.** Variability in the data set is split into regression and residual components. The **regression sum of squares** is analogous to the sum of squares between groups in one-way ANOVA:

$$\text{Regression SS} = \Sigma \, (\hat{y}_i - \bar{y})^2$$

where $\hat{y}_i$ is the predicted value of Y for observation i and $\bar{y}$ is the grand mean of Y.

The **residual sum of squares** is analogous to the sum of squares within groups in one-way ANOVA:

$$\text{Residual SS} = \Sigma \, (y_i - \hat{y}_i)^2$$

where y_i is an observed value of Y for observation i and $\hat{y}_i$ is its predicted value.

Table 14.8 shows how we calculate the mean square regression and mean square residual from their associated sums of squares and degrees of freedom. These mean squares are used to calculate this F statistic:

$$F_{\text{stat}} = \frac{\text{regression MS}}{\text{residual MS}}$$

TABLE 14.8 Calculating mean squares.

	Sum of Squares	df	Mean Square (MS)
Regression	$\Sigma(\hat{y}_i - \bar{y})^2$	1	$\dfrac{\text{Regression SS}}{\text{Regression df}}$
Residual	$\Sigma(y_i - \hat{y}_i)^2$	$n - 2$	$\dfrac{\text{Residual SS}}{\text{Residual df}}$
Total	$\Sigma(y_i - \bar{y})^2$	$n - 1$	

For simple regression models, the F_{stat} has 1 degree of freedom in its numerator and $n - 2$ degrees of freedom in its denominator.

C. **P-value.** The F_{stat} is converted to a P-value with Table D or a software utility (Section 13.3).

D. **Significance level.** The P-value is compared to various α levels to quantify the level of evidence against the null hypothesis.

E. **Conclusion.** The test results are addressed in the context of the data and research question.

ILLUSTRATIVE EXAMPLE

ANOVA for regression. Let us submit the illustrative data (Table 14.2) to an ANOVA test.

A. **Hypotheses.** H_0: the regression model of per capita smoking and lung cancer mortality does not fit in the population against H_a: the null hypothesis is incorrect.

B. **Test statistic.** Table 14.9 demonstrates calculations for sums of squares, mean squares, and the F-statistic. Figure 14.14 displays the SPSS output for the problem. Both show an F_{stat} of 10.723 with 1 and 9 df.

C. **P-value:** The P-value $= 0.010$, which is identical to the two-sided P-value derived by the t-test.

D. **Significance level.** The evidence against the null hypothesis is significant at $\alpha = 0.01$.

E. **Conclusion.** The linear association between lung cancer mortality and per capita cigarette consumption is statistically significant ($P = 0.010$).

ANOVA[b]

Model		Sum of Squares	df	Mean Square	F	Sig.
1	Regression	747.409	1	747.409	10.723	0.010[a]
	Residual	627.319	9	69.702		
	Total	1374.728	10			

[a]Predictors: (constant), CIG1930
[b]Dependent variable: LUNGCA

FIGURE 14.14   ANOVA output for illustrative data set. Graph produced with SPSS for Windows, Rel. 11.0.1.2001. Chicago: SPSS Inc. Reprint Courtesy of International Business Machines Corporation.

TABLE 14.9   Calculating sums of square for regression and residuals components for the smoking and lung cancer illustrative data.

i	COUNTRY	X	Y	Predicted $\hat{y}$	Residual $y_i - \hat{y}_i$	Residual^2 $(y_i - \hat{y}_i)^2$	Regression $\hat{y}_i - \bar{y}$	Regression^2 $(\hat{y}_i - \bar{y})^2$
1	USA	1300	20	36.453	−16.453	270.704	15.908	253.052
2	GrBritain	1100	46	31.884	14.116	199.253	11.339	128.569
3	Finland	1100	35	31.884	3.116	9.708	11.339	128.569
4	Switzerland	510	25	18.406	6.594	43.475	−2.139	4.575
5	Canada	500	15	18.178	−3.178	10.100	−2.367	5.605
6	Holland	490	24	17.950	6.050	36.608	−2.596	6.739

continues

continued

i	COUNTRY	X	Y	Predicted $\hat{y}$	Residual $y_i - \hat{y}_i$	Residual^2 $(y_i - \hat{y}_i)^2$	Regression $\hat{y}_i - \bar{y}$	Regression^2 $(\hat{y}_i - \bar{y})^2$
7	Australia	480	18	17.721	0.279	0.078	−2.824	7.977
8	Denmark	380	17	15.437	1.563	2.444	−5.109	26.099
9	Sweden	300	11	13.609	−2.609	6.808	−6.936	48.111
10	Norway	250	9	12.467	−3.467	12.020	−8.078	65.261
11	Iceland	230	6	12.010	−6.010	36.122	−8.535	72.851

Regression components

- Regression sum of squares $\cong \Sigma\,(\hat{y}_i - \bar{y})^2 = 253.052 + 128.569 + \cdots + 72.851 = 747.409$
- Regression degrees of freedom $= 1$
- Mean square regression $= \dfrac{\text{regression SS}}{\text{regression df}} = \dfrac{747.409}{1} = 747.409$

Residual components

- Residual sum of squares $= \Sigma(y_i - \hat{y}_i)^2 = 270.704 + 199.253 + \cdots + 36.122 = 627.319.$
- Residual degrees of freedom $= n - 2 = 11 - 2 = 9$
- Mean square residual^a $= \dfrac{\text{residual}}{\text{residual df}} = \dfrac{627.319}{9} = 69.702$

$$F_{\text{stat}} = \frac{\text{regression MS}}{\text{residual MS}} = \frac{747.409}{69.702} = 10.723 \text{ with 1 and 9 df.}$$

^{a}This is the variance of the regression model ($s^2_{Y|X}$).

Notes

1. **Coefficient of determination.** The coefficient of determination r^2 (Section 14.3) can be calculated as follows:

$$r^2 = \frac{\text{regression SS}}{\text{total SS}}$$

This is the same coefficient of determination presented in Section 14.1 (i.e., the square of the correlation coefficient). In this form, it is easy to recognize r^2 as the proportion of the total variation in Y accounted for by the regression line.

For the smoking and lung cancer data, $r^2 = \dfrac{\text{regression SS}}{\text{total SS}} = \dfrac{747.409}{1374.728} = 0.544,$

indicating that 54.4% of the variation in the response variable is accounted for by the explanatory variable.

2. **Root mean square residual** = standard error of the regression. The square root of the Mean Square Residual in the ANOVA table is the standard error of the regression model ($s_{Y|X}$). For the illustrative data, $s_{Y|X} = \sqrt{\text{mean square residual}}$ = $\sqrt{69.561} = 8.340.$

Conditions for Inference

Regression inferential procedures require conditions of linearity, sampling independence, Normality, and equal variance. These conditions conveniently form the mnemonic "line."

Linearity refers to the straight line form of X and Y. We can judge linearity by looking directly at a scatterplot or looking at a **residual plot**. This type of graph plots residuals against X values for the data set. Figure 14.15 is a residual plot for the illustrative data set. The horizontal line at 0 makes it easier to judge the variability of the response. Figure 14.15 is difficult to judge because of the sparseness of data points.

Figure 14.16 depicts three different patterns we might see in residual plots. Plots A and B depict linear relationships; there are an equal number of points above and below the 0-reference line throughout the extent of X. Plot C shows a nonlinear pattern.

Sampling **independence** relates to the sampling of bivariate observations. Bivariate data points should represent an SRS of a defined population. There should be no pairing, matching, or repeated measurements of individuals.

Normality refers to the distribution of residuals. Figure 14.12 shows an idealized depiction of this phenomenon. With small data sets, a stemplot of the residuals may be helpful for assessing departures from Normality. Here is the stemplot of the residuals for the smoking and lung cancer illustrative data.[p]

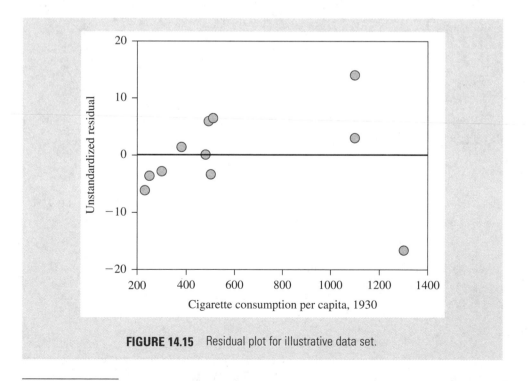

FIGURE 14.15   Residual plot for illustrative data set.

---

[p] See Table 14.7 for a listing that includes residual values.

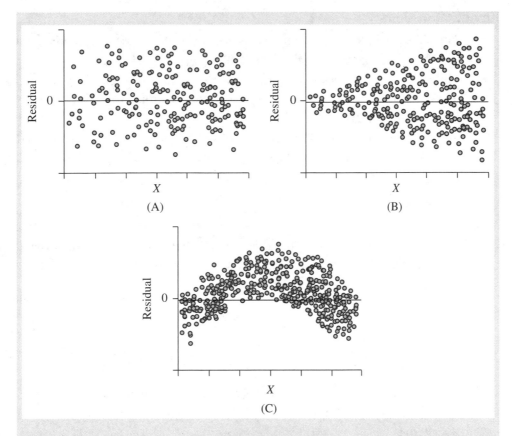

FIGURE 14.16 Residual plots demonstrating (A) linearity with equal variance, (B) linearity with unequal variance, and (C) nonlinearity.

```
−1|6
−0|2336
 0|01366
 1|4
×10
```

This stemplot shows no major departures from Normality.

The **equal variance (homoscedasticity)** condition also relates to the residuals. The spread of residuals should be homogenous at all levels of X (Figure 14.12). Unequal variance is evident when the magnitude of residual scatter changes with levels of X, as demonstrated in Figure 14.16B.

Review Questions

14.1 Fill in the blank: In regression problems, the explanatory variable is represented by X and the response variable is represented by _____.

14.2 Fill in the blank: In regression problems, the explanatory variable is referred to as the independent variable and the response variable is referred to as the _____ variable.

14.3 Select the best response: The explanatory variable in a linear correlation and regression problem is recorded on this measurement scale.

(a) categorical

(b) ordinal

(c) quantitative

14.4 Select the better response: The response variable in a linear correlation and regression problem is recorded on this measurement scale.

(a) categorical

(b) ordinal

(c) quantitative

14.5 Why are scatterplots necessary when investigating the relationship between quantitative variables?

14.6 What four elements do we look for when viewing a scatterplot?

14.7 Why is it difficult to judge the strength of an association by eye when viewing a scatterplot?

14.8 What symbol represents Pearson'e correlation coefficient statistic?

14.9 What symbol represents Pearson's correlation coefficient parameter?

14.10 Fill in the blanks: r is always greater than or equal to _____ and less than or equal to _____

14.11 Which r indicates a stronger association, -0.56 or $+0.48$?

14.12 When there is no linear correction between X and Y, $\rho =$ _____.

14.13 Select the best response: Perfect negative association is present when $r =$ _____.

(a) 1

(b) 0

(c) -1

14.14 Select the best response: According to our benchmarks, a correlation is considered strong when

(a) $r > 0.3$

(b) $r > 0.7$

(c) $|r| > 0.7$

14.15 Complete this statement: H_0: $\rho =$ _____.

14.16 Fill in the blanks: The least squares regression line is given by $\hat{y} =$ _____ + _____ x.

14.17 In the above statement, what does $\hat{y}$ represent?

14.18 What symbol is used to denote the intercept of a regression line?

14.19 What represents the predicted change in Y unit X?

14.20 What symbol represents the slope estimate?

14.21 What symbol represents the slope parameter?

14.22 What is wrong with this statement? "A 95% confidence interval for b is -0.91 to -0.42."

14.23 What is the distance of a data point from the regression line called?

14.24 Besides linearity, what conditions are needed to infer population slope β?

14.25 Besides linearity, what conditions are needed to infer population correlation coefficient ρ?

14.26 Complete this statement: (residual i) = (observed Y value of i) − _____.

14.27 In symbols, residual$_i$ = $y_i -$ _____.

14.28 What is wrong with this statement? "H_0: $r = 0$."

14.29 Fix this statement: "H_a: $\beta = 0$". (Four fixes are possible.)

Exercises

14.3 *Bicycle helmet use, n = 12*. Exercise 14.1 introduced data for a cross-sectional survey of bicycle helmet use in Northern California. Table 14.5 lists the data. Exercise 14.1(a) revealed that observation 13 (Los Arboles) was an outlier.

(a) After eliminating the outlier from the data set (n is now equal to 12), calculate least squares regression coefficients a and b. Interpret these statistics.

(b) Calculate the 95% confidence interval for slope parameter β.

(c) Use the 95% confidence interval to predict whether the slope is significant at $\alpha = 0.05$.

(d) *Optional:* Determine the residuals for each of the 12 data points in the analysis. Plot these residuals as a stemplot. Check for major departures from Normality.

14.4 *Mental health care.* Exercise 14.2 introduced historical data in which the explanatory variable was the reciprocal of the distance to the nearest healthcare facility (miles^{-1}, variable name REC_DIST) and the response variable was the percent of patients cared for at home (variable name PHOME2). Table 14.6 lists the data. Eliminate observation 13 (Nantucket) and then determine the least square regression coefficients for the data.

14.5 *Anscombe's quartet.* "Graphs are essential to good statistical analysis," so starts a 1973 article by Anscombe.[q] This article demonstrates why it is important to *look* at the data before analyzing it numerically. Table 14.10 contains four different data sets. Each of the data sets produces these identical numerical results:

$$n = 11 \quad \bar{x} = 9.0 \quad \bar{y} = 7.5 \quad r = 0.82 \quad \hat{y} = 3 + 0.5X \quad P = 0.0022$$

Figure 14.17 shows scatterplots for each of the data sets. Consider the relevance of the numerical statistics in light of the scatterplots. Would you use correlation or regression to analyze any of these data sets? Explain your reasoning in each instance.

14.6 *Domestic water and dental cavities.* Table 14.11 contains data from a historically important study of water fluoridation and dental caries in 21 North American cities.

(a) Construct a scatterplot of FLUORIDE and CARIES. Discuss the plot. Are there any outliers? Is the relationship linear? If the relationship is not linear, what type of relation *is* evident? Would linear regression be warranted under these circumstances?

(b) Although a single straight line does not fit these data, we may build a valid linear model by reexpressing the data through a mathematical transformation. Apply logarithmic transforms (base *e*) to both FLUORIDE and CARIES. Create a new plot with the transformed data. Discuss the results.

(c) Calculate the coefficients for a least square regression line for the ln–ln transformed data. Interpret the slope estimate.

(d) Calculate r^2 for ln–ln transformed data. Interpret the result.

---

[q]Anscombe, F. J. (1973). Graphs in statistical analysis. *The American Statistician, 27,* 17–21. Data are stored in ANSCOMB.*. Anscombe (1973) states "Most textbooks on statistical methods pay too little attention to graphs. Few of us escape being indoctrinated with these *notions:* (1) Numerical calculations are exact, but graphs are rough; (2) For any particular kind of statistical data there is just one set of calculations constituting a correct statistical analysis; (3) Performing intricate calculations is virtuous, whereas actually looking at data is cheating." This text attempts to avoid these false notions.

TABLE 14.10   Anscombe's quartet, Exercise 14.5.

Data set I		Data set II		Data set III		Data set IV	
X_1	Y_1	X_2	Y_2	X_3	Y_3	X_4	Y_4
10.0	8.04	10.0	9.14	10.0	7.46	8.0	6.58
8.0	6.95	8.0	8.14	8.0	6.77	8.0	5.76
13.0	7.58	13.0	8.74	13.0	12.74	8.0	7.71
9.0	8.81	9.0	8.77	9.0	7.11	8.0	8.84
11.0	8.33	11.0	9.26	11.0	7.81	8.0	8.47
14.0	9.96	14.0	8.10	14.0	8.84	8.0	7.04
6.0	7.24	6.0	6.13	6.0	6.08	8.0	5.25
4.0	4.26	4.0	3.10	4.0	5.39	19.0	12.50
12.0	10.84	12.0	9.13	12.0	8.15	8.0	5.56
7.0	4.82	7.0	7.26	7.0	6.42	8.0	7.91
5.0	5.68	5.0	4.74	5.0	5.73	8.0	6.89

Data from Anscombe, F. J. (1973). Graphs in statistical analysis. *The American Statistician, 27*, 17–21.

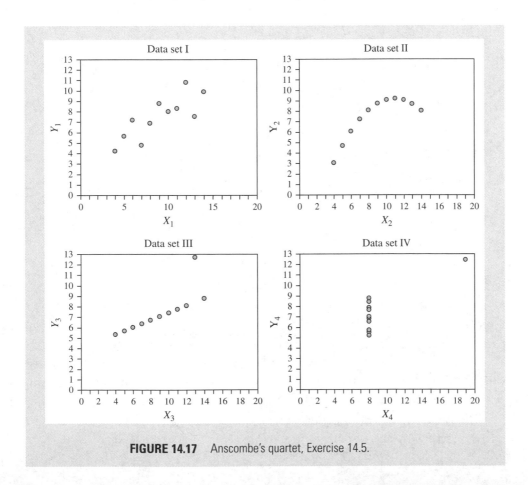

FIGURE 14.17   Anscombe's quartet, Exercise 14.5.

14.7 *Domestic water and dental cavities, range restriction.* Another way to look at the data presented in Exercise 14.6 is to restrict the analysis to a range that can be described with a straight line. This is called a range restriction.

(a) Use the scatterplot you drew in Exercise 14.6(a) to determine whether there is a range of FLUORIDE in which the relationship between FLUORIDE and CARIES is fairly straight. Restrict the data to this range. Then determine the least squares line for this segment of the data. Interpret b.

TABLE 14.11   Fluoride in public water (parts per million, variable name FLUORIDE) and dental caries experience in permanent teeth per 100 children (variable CARIES).

CITYID	FLUORIDE	CARIES
1	1.9	236
2	2.6	246
3	1.8	252
4	1.2	258
5	1.2	281
6	1.2	303
7	1.3	323
8	0.9	343
9	0.6	412
10	0.5	444
11	0.4	556
12	0.3	652
13	0.0	673
14	0.2	703
15	0.1	706
16	0.0	722
17	0.2	733
18	0.1	772
19	0.0	810
20	0.1	823
21	0.1	1037

Data from Dean, H. T., Arnold, F. A., Jr., & Elvove, E. (1942). Domestic water and dental caries. *Public Health Reports, 57,* 1155–1179. Data are stored online in the file DEAN1942.*.

(b) Calculate r^2 for this range-restricted model. Which model has a better fit, this model or the ln–ln transformed model from the previous exercise?

(c) Which model do you prefer, this model or the one created in Exercise 14.6? Explain your reasoning.

14.8 *Correlation matrix.* Statistical packages can easily calculate correlation coefficients for multiple pairings of variables. Results are often reported in the form of a **correlation matrix**. Figure 14.18 displays the correlation matrix for data from a study of geographic variation in cancer rates.[r] The variables are:

CIG   cigarettes sold per capita
BLAD   bladder cancer deaths per 100,000
LUNG   lung cancer deaths per 100,000
KID   kidney cancer deaths per 100,000
LEUK   leukemia cancer deaths per 100,000

Notice that the values for each correlation coefficient appear twice in the matrix, each time the variables intersect in either row or column order. For example, the value $r = 0.704$ occurs for CIG and BLAD and for BLAD and CIG. The correlation of 1 across the diagonal reflects the trivial fact that each variable is perfectly correlated with itself. Review this correlation matrix and interpret its results.

14.9   *True or false?* Identify which of these statements are true and which are false.

(a) Correlation coefficient r quantifies the relationship between quantitative variables X and Y.

(b) Correlation coefficient r quantifies the linear relation between quantitative variables X and Y.

(c) The closer r is to 1, the stronger is the linear relation between X and Y.

(d) The closer r is to -1 or 1, the stronger is the linear relation between X and Y.

(e) If r is close to zero, X and Y are unrelated.

(f) If r is close to zero, X and Y are not related in a linear way.

(g) The value of r changes when the units of measure are changed.

(h) The value of b changes when the units of measure are changed.

14.10   *Memory of food intake.* Retrospective studies of diet and health often rely on recall of distant dietary histories. The validity and reliability of such information is often suspected. An epidemiologic study asked middle-aged adults (median age 50) to recall food intake at ages 6, 18, and 30 years. Recall was validated by comparing recalled results to historical information collected during earlier time periods. Correlations rarely exceeded $r = 0.3$.[s] What do you conclude from this result?

---

[r] Fraumeni, J. F., Jr. (1968). Cigarette smoking and cancers of the urinary tract: Geographic variation in the United States. *Journal of the National Cancer Institute, 41*(5), 1205–1211.
[s] Dwyer, J. T., Gardner, J., Halvorsen, K., Krall, E. A., Cohen, A., & Valadian, I. (1989). Memory of food intake in the distant past. *American Journal of Epidemiology, 130*(5), 1033–1046.

Correlations

		CIG	BLAD	LUNG	KID	LEUK
CIG	Pearson correlation	1	0.704**	0.697**	0.487**	−0.068
	Sig. (two-tailed)	.	0.000	0.000	0.001	0.659
	N	44	44	44	44	44
BLAD	Pearson correlation	0.704**	1	0.659**	0.359*	0.162
	Sig. (two-tailed)	0.000	.	0.000	0.017	0.293
	N	44	44	44	44	44
LUNG	Pearson correlation	0.697**	0.659**	1	0.283	−0.152
	Sig. (two-tailed)	0.000	0.000	.	0.063	0.326
	N	44	44	44	44	44
KID	Pearson correlation	0.487**	0.359*	0.283	1	0.189
	Sig. (two-tailed)	0.001	0.017	0.063	.	0.220
	N	44	44	44	44	44
LEUK	Pearson correlation	−0.068	0.162	−0.152	0.189	1
	Sig. (two-tailed)	0.659	0.293	0.326	0.220	.
	N	44	44	44	44	44

** Correlation is significant at the 0.01 level (two-tailed).
* Correlation is significant at the 0.05 level (two-tailed).

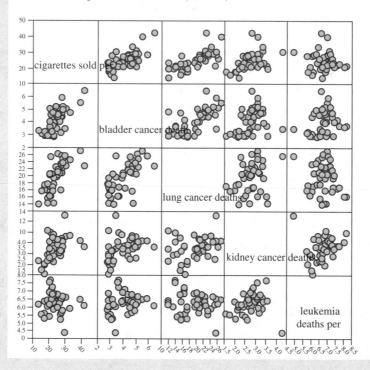

FIGURE 14.18   Correlation matrix and scatterplot matrix for Exercise 14.8.

Summary Points (Correlation & Regression)

1. **Correlation and regression** are used to quantify linear relationships between quantitative variables X and Y.

2. The first step in correlation and regression is to inspect a **scatterplot** in terms of its:

   (a) Form—linear, curved, or nonexistent

   (b) Direction—positive association, negative association, or no association

   (c) Strength—extent to how closely the points on the plot adhere to an imaginary trend line (difficult to judge by eye alone)

   (d) Potential outliers—deviations from the observed overall pattern

3. Use **Pearson's correlation coefficient r** to determine the direction and quantify the strength of linear relationships.

   (a) The closer r is to -1 or 1, the stronger the association ($r = -1$ indicates perfect negative association, $r = 1$ indicates a perfect positive association, and $r = 0$ indicates no association).

   (b) The correlation coefficient is calculated with a statistical calculator, statistical software, or the formula $r = \dfrac{1}{n-1}\Sigma z_X z_Y$, where z_X and z_Y are standardized values for the X and Y variables, respectively.

   (c) Squaring r derives a statistic called the coefficient of determination, which quantifies proportion of the variance in Y mathematically explained by X.

   (d) The null hypothesis $H_0: \rho = 0$ ("no correlation in the population") is tested with $t_{stat} = \dfrac{r}{SE_r}$, where $SE_r = \sqrt{\dfrac{1-r^2}{n-2}}$ and df $= n - 2$.

   (e) A $(1 - \alpha)100\%$ confidence interval for ρ is given by $\left(\dfrac{r - \omega}{1 - r\omega}, \dfrac{1 + \omega}{1 + r\omega} \right)$, where $\omega = \sqrt{\dfrac{t^2_{df, 1-\frac{\alpha}{2}}}{t^2_{df, 1-\frac{\alpha}{2}} + df}}$ and df $= n - 2$.

4. **Linear regression** finds the best fitting line for the data using a least squares method. The relationship is described with the equation $\hat{y} = a + bx$, where $\hat{y}$ denotes the value of Y predicted by the regression model, a denotes the intercept estimate of the regression line, and b denotes the slope estimate of the line.

   (a) Slope b predicts the change in Y per unit increase in X, thus providing a measure of the effect of X on Y.

   (b) Intercept a anchors the line but is of interpretative use only when an X value of 0 makes sense.

(c) These least squares regression coefficients are calculated with a statistical calculator, statistical software, or with the formulas $b = r\dfrac{s_Y}{s_X}$ and $a = \bar{y} - b\bar{x}$.

(d) The null hypothesis H_0: $\beta = 0$ ("zero slope in the population") is tested with $t_{\text{stat}} = \dfrac{b}{SE_b}$, where $SE_b = \dfrac{1}{\sqrt{n-1} \cdot s_X}$, $S_{Y|X} = \sqrt{\dfrac{1}{n-2} \Sigma \text{residuals}^2}$, and df $= n - 2$, or with an ANOVA method.

(e) The $(1 - \alpha)100\%$ confidence interval for $\beta = b \pm t_{\text{df}, 1 - \frac{\alpha}{2}} \cdot SE_b$.

5. Selected **cautions**.

(a) Linear correlation and regression should be used only when the relationship between X and Y can be described by a straight line. (Some nonlinear relationships can be straightened by mathematically transforming one or both variables.)

(b) Neither correlation nor regression is robust: both are susceptible to the influence of outliers.

(c) Do not extrapolate beyond the range of X.

(d) The inferential methods presented in this chapter for ρ require linearity, independence, and bivariate normality. The inferential methods for β require Linearity, Independence, Normality, and Equal variance (mnemonic "LINE").

Vocabulary

Bivariate Normality

Coefficient of determination (r^2)

Correlation coefficient (r denotes the sample correlation and denotes the population correlation coefficient)

Correlation matrix

Direction

Explanatory variable

Form

Influential observations

Intercept

Least squares regression line

Linear

Outliers

Predicted Y ($\hat{y}$)

Regression sum of squares

Residual ($y - \hat{y}$)

Residual plot

Residual sum of squares

Response variable

Scatterplot

Slope (b denotes the sample slope and β denotes the population slope)

Standard error of the regression ($s_{Y|X}$)

Standard error of the slope (SE_b)

Strength

14.11 $(\bar{x}, \bar{y})$ *is always on the least squares regression line.* The least square regression line will always go through the point $(\bar{x}, \bar{y})$. Prove that when $X = \bar{x}$, $\hat{y} = \bar{y}$ (Hint: Recall that $\hat{y} = a + bx$ and $a = \bar{y} - b\bar{x}$; replace a in the first equation for the equivalent provided in the second equation and voilà!)

14.12 *Historically important coronary heart disease study.* Following World War II, many northern European countries experienced notable increases in what was then called degenerative heart disease and is now called coronary heart disease. In 1953, an investigator by the name of Ancel Keys reported the data shown in the following table. In this table, the variable FAT represents "% of total dietary calories consumed as fat" in the six western democracies in question and the variable CHD represents "coronary heart disease mortality per 1000 in 50- to 59-year-old males."

COUNT	FAT	CHD
Japan	8	0.5
Italy	20	1.4
England	33	3.8
Australia	36	5.5
Canada	37	5.7
USA	39	7.1

Data from Keys, A. (1953). Atherosclerosis: a problem in newer public health, Figure 2. *Journal of the Mount Sinai Hospital, 20*, 118–139. Data file on companion website is KEYS1953.*.

(a) Explore the relationship between FAT and CHD with a scatterplot. Interpret this plot (form, direction, strength, and potential outliers). Can linear correlation and regression methods be used on these data as is?

(b) Apply a natural logarithmic transformation to the response variable. Call this new variable LN_CHD. Then generate a plot of FAT versus LN_CHD. Can linear correlation and regression now be used?

(c) Use correlation to determine the proportion of the variance of LN_CHD explained by FAT.

(d) As noted many times in this text, "correlation" does not always equate with "causation." One reason for this nonequivalence is due to the phenomena of confounding, in which the observed relationship between X and Y is explained by their mutual relationship with a third variable Z lurking in the background. Proffer an explanation as to how the lurking variable "obesity" could confound the observed association between fat consumption and coronary heart disease.

14.13 *Nonexercise activity thermogenesis (NEAT).* We routinely burn calories through the nonexercise activities of daily life. In this exercise, we will use the acronym NEAT to refer to "nonexercise activity thermogenesis." NEAT includes such activities of daily living, fidgeting, maintenance of posture, spontaneous muscle contraction, and maintaining posture when not recumbent. It has been hypothesized that individual differences in NEAT may explain why some people tend to gain more weight than others. To study this phenomenon, Levine and coworkers (1999) deliberately overfed 16 young healthy volunteers for an 8-week period. The amount of fat the volunteers gained (FATGAIN, kg) and change in NEAT activity (kcal/day) during this period are listed in the following table.

Observation	NEAT	FATGAIN
1	−100	4.2
2	−60	3.0
3	−20	3.8
4	140	2.6
5	140	3.2
6	150	3.5
7	250	2.4
8	350	1.3
9	400	3.9
10	480	1.7
11	500	1.6
12	540	2.2
13	570	1.0
14	580	0.4
15	620	2.4
16	690	1.1

Data from Levine, J. A., Eberhardt, N. L., & Jensen, M. D. (1999). Role of nonexercise activity thermogenesis in resistance to fat gain in humans, Figure 1C. *Science, 283*(5399), 212–214. Data file on the companion website is LEVINE1999.*.

(a) Use regression methods to address whether the amount of fat gained is related to the change in NEAT. (Start your analysis with a scatterplot.) When you have completed your analyses, return to the practical question and discuss your findings.

(b) Predict the amount of fat gain in this experiment by a person who has a change of 600 calories in NEAT.

(c) Identify the point $(\bar{x}, \bar{y})$ on your graph. The regression line always goes through this point (see Exercise 14.11). Use $(\bar{x}, \bar{y})$ and the point predicted in part (b) of this exercise to draw the least squares regression line on the scatterplot you produced in part (a).

(d) Determine the residuals associated with the first three observations in the data set. Show these on the scatterplot.

14.14 *Neuroimaging social rejection*. A study examined the correlates of social exclusion (a social stressor) by neuroimaging an activity in the anterior cingulate cortex (a part of the brain whose activity is associated with physical pain). Study subjects filled out a questionnaire that assessed the degree to which they felt excluded from a social activity (DISTRESS). A functional MRI was then used to measure activity in the anterior cingulate cortex (fMRI_ACC). Data are:

ID	DISTRESS	fMRI_ACC
1	1.2	−0.05
2	1.9	−0.04
3	1.1	−0.03
4	2.5	−0.02
5	2.2	−0.02
6	2.7	0.02
7	2.0	0.02
8	2.2	0.03
9	2.6	0.03
10	2.8	0.03
11	2.8	0.06
12	3.3	0.08
13	3.7	0.12

Data from Eisenberger, N. I., Lieberman, M. D., & Williams, K. D. (2003). Does rejection hurt? An FMRI study of social exclusion, Figure 2A. *Science, 302*(5643), 290–292. Datafile: EISENBERGER2003.*.

(a) Create a scatterplot of the data. Interpret what you see.

(b) Calculate correlation coefficient *r*. Interpret this result. Test the correlation for statistical significance. Show all hypothesis testing steps.

14.15. *Gorilla ebola.* An ebola virus outbreak in gorillas in the Congo from 2002 to 2003 killed 91 gorillas in 7 ranges. This table list ONSET date relative to the initiation of the outbreak and the number of home ranges separating the gorilla band from the initially infected band (DISTANCE):

DISTANCE	ONSET
1	4
3	21
4	33
4	41
4	43
5	46

Data from Bermejo, M., Rodriguez-Teijeiro, J. D., Illera, G., Barroso, A., Vila, C., & Walsh, P. D. (2006). Ebola outbreak killed 5000 gorillas, Figure 1A. *Science, 314*(5805), 1564. Data file: EBOLA.*.

(a) Which variable, DISTANCE or ONSET, is the independent variable in this analysis? Which is the dependent variable?

(b) Create a scatterplot of the data. Interpret what you see.

(c) Calculate correlation coefficient r. Then describe the strength of the association.

(d) What percent of the variance in ONSET is explained by DISTANCE.

(e) Use a regression model to predict how long on average it takes the outbreak to move from one gorilla band to an adjacent band.

14.16 *Sodium and systolic blood pressure*. Data from 10 individuals in renal failure with high blood pressure are shown in the following table. The variable SODIUM represents salt consumption (in milligrams) and the variable BP represents systolic blood pressure (in mmHg).

SODIUM (mg)	7.0	7.2	7.1	7.5	7.2	7.2	7.5	7.3	6.7	6.6
BP (mmHg):	156	169	164	177	192	160	197	191	170	150

Fictitious data. Datafile: SODIUM-BP.*.

(a) Demonstrate the relationship between sodium intake and systolic blood pressure with a scatterplot. Describe the observed relationship (form, direction, outliers, and strength of the relationship using r as a guide).

(b) Can the observed relationship be easily explained by "chance"? Justify your response with a single statistic.

(c) Determine the least squares regression line for the data. What does the slope coefficient tell you about the relationship between SODIUM and BP.

(d) What is the predicted BP for a person from this population that consumes 7.0 mg of sodium per day?

(e) What is the value of the residual associated with the first observation? The observed value for the first observation is (7.0, 156).

(f) Draw the regression line on the scatter plot you produced in part (a) of this exercise. Then, show the residual associated with the first data point on this graph.

15 | Multiple Linear Regression

■ 15.1 The General Idea

Simple regression addresses a single explanatory variable (X) and response variable (Y):

$$X \to Y$$

Chapter 14 considered simple regression.

Multiple regression is an extension of simple linear regression that addresses multiple explanatory variables (X_1, X_2, ..., X_k) in relation to a response variable (Y):

To start our discussion of multiple regression, let explanatory variable X_1 represent the explanatory variable of prime interest. All other explanatory variables in the model (X_2, ..., X_k) will be considered to be "extraneous" for now. The multiple regression model holds constant the influence of the extraneous variables, allowing us to more readily isolate the effects of the primary explanatory variable. The intention is to "adjust out" confounding effects, leaving the regression coefficient associated with the primary explanatory variable relatively unconfounded. When interest later shifts to (say) explanatory variable X_2, it then becomes the primary explanatory variable and all others are extraneous.

It must be pointed out that the causal nature of an explanatory variable from a nonexperimental study is not tested in a strong way even with multiple regression models. Adjustments imposed by multiple regression are passive. To get a more robust test of cause, active experimentation is needed. "To find out what happens to a system when you interfere with it, you have to interfere with it (not just passively observe it)."[a] Therefore, as was the case with simple regression, it should not be taken for granted that statistics from multiple regression models reflect actual causal relations. An observed association may be causal or non-causal depending on nature, not on the model.

Finally, we will *not* consider all the mathematics behind the multiple regression model. Instead, we will rely on statistical software for computations. This will allow us to introduce the model without getting bogged down in its mathematical complexities.[b]

■ 15.2 The Multiple Linear Regression Model

Simple linear regression and multiple regression are based on similar lines of reasoning.[c] In a simple regression population model, the expected value of Y at a given level of X is based on the linear equation:

$$\mu_{Y|x} = \alpha + \beta x$$

where $\mu_{Y|x}$ is the expected value of Y given x, α is the parameter indicating the model's intercept, and β is the parameter indicating the model's slope. We do not know the true values of model parameters, so we estimate them with a least squares regression line that minimizes the Σresiduals^2. This derives estimates for regression parameters that predict values of Y ($\hat{y}$) with the equation.

$$\hat{y} = a + bx$$

where a is the intercept estimate and b is the slope estimate.

The standard error of the regression is the root mean square of the residuals:

$$S_{Y|X} = \sqrt{\Sigma\text{residuals}^2 / df_{res}}$$

Inferential techniques use this standard error estimate in their application (Section 14.4).

Now consider a regression model with two explanatory variables. The expected value of Y at given levels of X_1 and X_2 is:

$$\mu_{Y|x_1,x_2} = \alpha + \beta_1 x_1 + \beta_2 x_2$$

---

[a] George Box cited in Gilbert, J. P., & Mosteller, F. (1972). The urgent need for experimentation. In F. Mosteller & D. P. Moynihan (Eds.), *On Equality of Educational Opportunity*. New York: Vintage, p. 372.

[b] For discussions of the mathematics of multiple regression model fitting, see Kutner, M. H., Nachtsheim, C. J., Neter J. (2004). *Applied Linear Regression Models* (4th ed.). New York: McGraw-Hill/Irwin.

[c] Chapter 14 covers basic concepts for simple linear regression.

As was the case with simple linear regression, a least squared method minimizing Σresiduals^2 is used to fit the model:

$$\hat{y} = a + b_1 x_1 + b_2 x_2$$

where a is the intercept estimate, b_1 is the slope estimate for X_1, and b_2 is the slope estimate for X_2.

While a simple regression model with one independent variable fits a **response line** in two-dimensional space, a multiple regression model with two independent variables fits a **response plane** in three-dimensional space. Figure 15.1 depicts a response plane for a multiple regression model with two explanatory variables.

Multiple regression models can accommodate more than two explanatory variables by fitting a **response surface** in $k + 1$ dimensional space, where k represents the number of explanatory variables in the model. For example, a model with three explanatory variables uses a four-dimensional response plane. Although we cannot visualize four dimensions, we can still use linear algebra to find the best-fitting surface for the response plane. As before, this is achieved by minimizing the Σresiduals^2 for the plane. The result is a multiple regression model with k explanatory variables in which:

$$\hat{y} = a + b_1 x_1 + b_2 x_2 + \cdots + b_k x_k$$

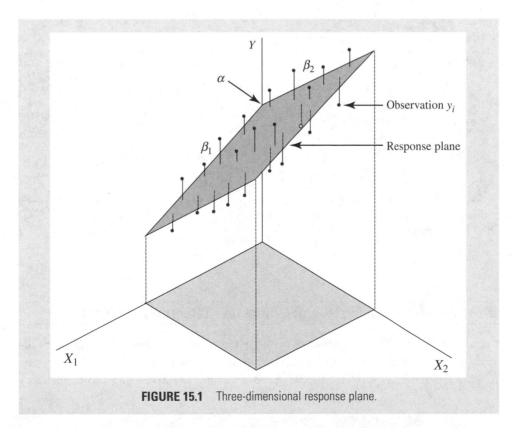

FIGURE 15.1   Three-dimensional response plane.

The coefficients derived by this model have similar interpretations as those derived by simpler models. Intercept estimate (a) is where the regression surface crosses the Y axis, the slope estimate associated with X_1 (b_1) predicts the amount of change in Y per unit increase in X_1 while holding the other variables in the model constant, the slope estimate associated with X_2 (b_2) predicts the amount of change in Y per unit increase in X_2 while holding the other variables in the model constant, and so on.

ILLUSTRATIVE EXAMPLE

Data (Forced expiratory volumes). Our illustrative data comes from a health survey done in adolescents. The illustrative data set includes information on 654 individuals between 3 and 19 years of age. Table 15.1 is a code book for the variables in the data set.

The response variable is a measure of respiratory function called *forced expiratory volume* (variable name FEV) that measures respiratory capacity in liters/second units. High values represent high respiratory capacity. We will consider the effects of two explanatory variables: SMOKE (coded 0 for "nonsmoker" and 1 for "smoker") and AGE (years). Individuals were classified as ever-smokers if they currently smoked or had at some time smoked as much as one cigarette per week. We wish to quantify the effects of SMOKE on FEV while adjusting for AGE.

TABLE 15.1   Summary statistics from illustrative data set, $n = 654$.

Variable	Description	Minimum	Maximum	Mean	Std Deviation
AGE	Years	3	19	9.93	2.954
FEV	Forced expiratory volume (L/sec)	0.7910	5.7930	2.6368	0.86706
HEIGHT	Height in inches	46.00	74.00	61.14	5.704
SEX	0 = female; 1 = male	48.6% female; 51.4% male; none missing			
SMOKE	0 = no; 1 = yes	90.1% no; 9.9% yes; none missing			

Data from Rosner, B. (1990). *Fundamentals of Biostatistics* (Third ed.). Belmont, CA: Duxbury Press. Data are stored online in the file FEV.*.

■ 15.3 Categorical Explanatory Variables in Regression Models

Let us start by looking at the effect of smoking (variable name SMOKE) on forced expiratory volume (FEV) while ignoring AGE for now. The response variables in regression models are sometimes referred to as the **dependent variables** and the explanatory variables are referred to as **independent variables**. In this example, FEV is the dependent variable and SMOKE is the independent variable. Notice that SMOKE is categorical. Until this point, regression models we've considered have addressed quantitative independent (explanatory) variables only. Fortunately, regression models can accommodate binary explanatory variables as long as they are coded 0 for "absence of the attribute" and 1 for "presence of the attribute." Variables of this sort are called **indicator variables** or **dummy variables**.

Regression models can handle categorical explanatory variables with more than two levels of the attribute by reexpressing the variable with multiple indicator variables. Table 15.2 describes how this is done. For now, we will address only binary explanatory variables.

Figure 15.2 is a scatterplot of FEV by SMOKE. The group means are 2.566 L/sec for SMOKE = 0 and 3.277 L/sec for SMOKE = 1. The line connecting the two means on the plot is the regression line for the relationship.

TABLE 15.2 Accommodating a categorical variable with more than two levels of the attribute. When the explanatory variable has k levels of the attribute, it is reexpressed with $k - 1$ dummy variables. For example, an attribute classified into three levels is reexpressed with two dummy variables. As an example, the variable SMOKE2 classifies individuals into three levels of smoking (0 = nonsmoker, 1 = former smoker, 2 = current smoker). To incorporate this information into a regression model, it is recoded as follows:

SMOKE2	DUMMY1	DUMMY2
0	0	0
1	1	0
2	0	1

Here's an example of programming code that can be used for this purpose:

IF SMOKE = 1 THEN DUMMY1 = 1 ELSE DUMMY1 = 0
IF SMOKE = 2 THEN DUMMY2 = 1 ELSE DUMMY2 = 0

The variable DUMMY1 is coded so that 0 = not a former smoker, 1 = former smoker. The variable DUMMY2 is coded so that 0 = not a current smoker, 1 = current smoker. When DUMMY1 and DUMMY2 both equal 0, the individual is identified as a nonsmoker.

The regression line $\hat{y} = a + bx$ is linked to the means as follows:

- The mean response in group 0 $\hat{y}_{x=0} = a + b(0) = a$.
- The mean response in group 1 $\hat{y}_{x=1} = a + b(1) = a + b$.
- The difference between mean responses = $\hat{y}_1 - \hat{y}_0 = (a + b) - a = b$.

Figure 15.3 contains SPSS output for this regression model. FEV is the dependent variable and SMOKE is the independent variable. This output indicates an intercept (a) of 2.566 and slope (b) of 0.711. Therefore, the regression model is $\hat{y} = 2.566 + 0.711x$.

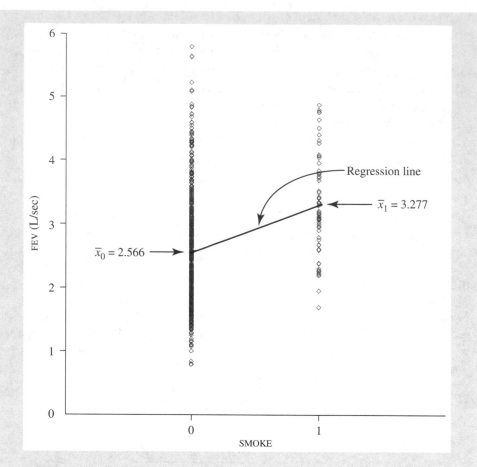

FIGURE 15.2   Scatterplot and regression line for forced expiratory volume in nonsmokers (SMOKE = 0) and smokers (SMOKE = 1), illustrative data set.

- The mean response in group 0 (a) is 2.566.
- The mean response in group 1 ($a + b$) is $2.566 + 0.711 = 3.277$.
- The difference between mean responses (b) is 0.711.

Analogous statements can be made about the population regression model $\mu_{Y|x} = \alpha + \beta x$:

- The expected response in group 0 is $\mu_{Y|x=0} = \alpha + \beta(0) = \alpha$.
- The expected response in group 1 is $\mu_{Y|x=1} = \alpha + \beta(1) = \alpha + \beta$.
- The difference in expectations $\mu_{Y|x=1} - \mu_{Y|x=0} = (\alpha + \beta) - \alpha = \beta$.

Thus, testing $H_0: \mu_1 - \mu_0 = 0$ is equivalent to testing $H_0: \beta = 0$. In addition, a confidence interval for β is equivalent to a confidence interval for $\mu_1 - \mu_0$.

Coefficients[a]

Model	Unstandardized Coefficients		Standardized Coefficients			95% Confidence Interval for β	
	B	Std. Error	β	t	Sig.	Lower Bound	Upper Bound
1 (Constant)	2.566	0.035		74.037	0.000	2.498	2.634
SMOKE	0.711	0.110	0.245	6.464	0.000	0.495	0.927

[a]Dependent Variable: FEV

FIGURE 15.3 Screenshot of SPSS output for the illustrative example in which the response variable (FEV) is regressed on the binary explanatory variable (SMOKE). Graph produced with SPSS for Windows, Rel. 11.0.1.2001. Chicago: SPSS Inc. Reprint Courtesy of International Business Machines Corporation.

The output in Figure 15.3 reveals $t_{stat} = 6.464$ with 652 df and $P \approx 0.000$. It also reveals a confidence interval for β of (0.495 to 0.927) L/sec. These results suggest that the children who smoked had substantially *greater* lung capacity than nonsmokers— but how could this be true given what we know about the deleterious effects of smoking? The answer lies in the fact that the children in the sample who smoked were older than those who did not smoke.[d] AGE confounded the observed relationship between SMOKE and FEV. Fortunately, a multiple regression model can be used to adjust for AGE while comparing FEV levels in smokers and nonsmokers.

■ 15.4 Regression Coefficients

For multiple regression models, we rely on statistical packages to calculate the intercept and slope coefficients. Figure 15.4 depicts the SPSS dialog box needed to estimate the effect of SMOKE on FEV while adjusting for AGE.[e] Figure 15.5 is a screenshot of output from the procedure. The intercept coefficient (a) is 0.367,[f] the slope coefficient for SMOKE (b_1) is -0.209, and the slope coefficient for AGE (b_2) is 0.231. The multiple regression model is:

$$\hat{y}_i = 0.367 + (-0.209) \cdot x_{1i} + 0.231 \cdot x_{2i}$$

where $\hat{y}_i$ is the predicted value of the ith observation, x_{1i} is the value of SMOKE for the ith observation, and x_{2i} is the value of the AGE variable of the ith observation.

[d]The 589 nonsmokers have a mean age of 9.5 years (standard deviation 2.7 years). The 65 smokers have a mean age of 13.5 (standard deviation 2.3 years).
[e]The illustration uses SPSS, but any reliable statistical package would do.
[f]This is labeled (constant).

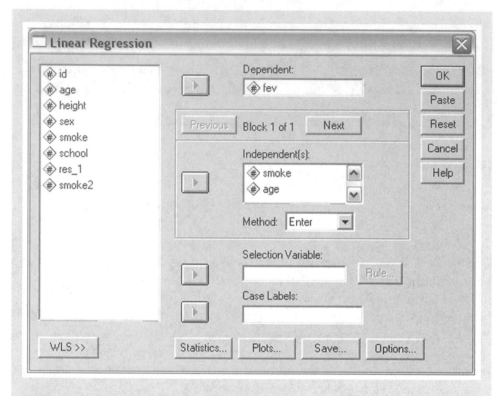

FIGURE 15.4   Screenshot of SPSS dialog box for setting up multiple regression. Graph produced with SPSS for Windows, Rel. 11.0.1.2001. Chicago: SPSS Inc. Reprint Courtesy of International Business Machines Corporation.

- The intercept in this model (0.367) is an extrapolation of where the regression plane would slice through the Y axis. This has no practical application.

- The slope for SMOKE (-0.209) predicts that going from SMOKE $= 0$ (nonsmoker) to SMOKE $= 1$ (smoker) is associated with a mean 0.209 *decline* in FEV.

- The slope for AGE (0.231) predicts that each additional year of age is associated with a 0.231 increase in FEV.

This multiple regression model has adjusted for AGE to provide a more meaningful prediction of the effects of smoking.

Figure 15.5 includes the following results:

- The **standard error of the slope** (SE_{b_i}) for each of independent variables $i = 1$ through k is listed in the column labeled "Std. Error." (Formula not presented; we rely instead on statistical software.) Based on the output in Figure 15.5, the standard error of the slope associated with SMOKE (SE_{b_1}) is 0.081, and the standard error of the slope associated with AGE (SE_{b_2}) is 0.008. As is the case with all standard errors, these statistics quantify the precision of their associated estimate.

Coefficientsa

Model	Unstandardized Coefficients		Standardized Coefficients	t	Sig.	95.0% Confidence Interval for B	
	B	Std. Error	β			Lower Bound	Upper Bound
1 (Constant)	0.367	0.081		4.511	0.000	0.207	0.527
SMOKE	−0.209	0.081	−0.072	−2.588	0.010	−0.368	−0.050
AGE	0.231	0.008	0.786	28.176	0.000	0.215	0.247

aDependent Variable: FEV

FIGURE 15.5 Screenshot of multiple linear regression output, illustrative example . Graph produced with SPSS for Windows, Rel. 11.0.1.2001. Chicago: SPSS Inc. Reprint Courtesy of International Business Machines Corporation.

- The last two columns of Figure 15.5 list **95% confidence intervals** for the regression coefficients. The 95% confidence interval for the slope associated with SMOKE is (−0.368 to −0.050). The 95% confidence interval for slope associated with AGE is (0.215 to 0.247). The interpretation of these confidences is similar to those derived from simple regression models. However, the confidence limits have now been adjusted for the other independent variables in the model. For example, the 95% confidence interval for SMOKE suggests that as we go from 0 (nonsmoker) to 1 (smoker), dependent variable FEV *decreases* (the confidence limits have negative signs) between 0.368 and 0.050 (L/sec) after controlling for AGE.

- The **Standardized Coefficients** represent the predicted change in Y per unit increase in X_i in standard deviation units. The standardized coefficient for the slope of variable i is equal to $b_i \dfrac{S_{X_i}}{S_Y}$.

- The t column provides t-statistics for testing $H_0: \beta_i = 0$. In each case, $t_{stat} = \dfrac{b_i}{SE_{b_i}}$ with $n - k - 1$ degrees of freedom. Each of these test statistics has been adjusted for the contributions of the other independent variables in the model.

- The **Sig.** column provides two-sided P-values for each of the t-tests. For example, Figure 15.5 shows that $P = 0.010$ in testing $H_0: \beta_1 = 0$, where β_1 represents the population slope associated with the SMOKE variable.

■ 15.5 ANOVA for Multiple Linear Regression

Test Procedure

Analysis of variance can be used to test the overall fit of a multiple regression model. Here is a step-by-step procedure for the test.

A. Hypotheses. The null hypothesis is H_0: the multiple regression model does not fit in the population. The alternative hypothesis is H_a: the multiple regression model does fit in the population. These statements are functionally equivalent to H_0: all $\beta_i = 0$ versus H_a: at least one of the $\beta_i \neq 0$. Note that some of the population slopes can be 0 under the alternative hypothesis.

B. Test statistics. As discussed in Section 14.4, two components of variance are analyzed: the mean square of the residuals and the mean square of the regression.

Mean square residual: Recall that a **residual** is the difference between an observed response and the response predicted by the regression model. For the ith observation:

$$\text{Residual}_i = y_i - \hat{y}_i$$

where y_i is the observed value and $\hat{y}_i$ is the predicted value of y_i. The multiple regression model has been constructed so that a, b_1, b_2, ..., b_k minimized Σresiduals^2. The variance of the residuals (call it σ^2) is assumed to be constant at all levels and combinations of the explanatory variables. We estimate this variance with the statistic:

$$\text{Mean square residual} = \frac{\Sigma\text{residuals}^2}{\text{df}_{\text{residuals}}}$$

where Σresiduals^2 is the sum of squared residuals and df$_{\text{residuals}} = n - k - 1$ and k represents the number of explanatory variables in the model. You lose $(k + 1)$ degrees of freedom in this variance estimate because $k + 1$ parameters are estimated—the slopes for each explanatory variable plus the intercept. For the illustrative data, you lose 3 degrees of freedom in estimating α, β_1, and β_2. Therefore, df$_{\text{residuals}} = 654 - 2 - 1 = 651$.

The symbol $s^2_{Y|x_1,x_2}$ can be used to represent the mean square residual. This is the estimated variance of Y given X_1 and X_2 on the regression plane. The square root of this mean square residual is the **standard error of the multiple regression**:

$$S_{Y|x_1 x_2,\ldots,x_k} = \sqrt{S^2_{Y|x_1 x_2,\ldots,x_k}}.$$

Mean square regression: The mean square regression quantifies the extent to which the response surface deviates from the grand mean of Y:

$$\text{Mean square regression} = \frac{\Sigma\text{regressions}^2}{\text{df}_{\text{regression}}}$$

where Σregressions^2 is the sum of squares of the regression components and df$_{\text{regression}} = k$. The regression component of observation i is given by $\hat{y}_i - \bar{y}$ where $\hat{y}_i$ is the predicted value of Y and $\bar{y}$ is the grand mean of Y.

ANOVA table and F-statistic: Sums of squares, mean squares, and degrees of freedom are organized to form an ANOVA table (see Table 15.3).

TABLE 15.3 ANOVA table.

	Sum of Squares	df	Mean Square
Regression[a]	$\Sigma\,(\hat{y}_i - \bar{y})^2$	k	$\dfrac{\text{SS regression}}{\text{df regression}}$
Residual[b]	$\Sigma\,(y_i - \hat{y}_i)^2$	$n - k - 1$	$\dfrac{\text{SS residual}}{\text{df residual}}$
Total	$\Sigma\,(y_i - \bar{y})^2$	$n - 1$	

[a]Some statistical packages will indicate this line of output with the label "Model" (i.e., regression model).
[b]Some statistical packages will indicate this line of output with the label "Error" (i.e., residual error).

The *F*-statistic is:

$$F_{\text{stat}} = \frac{\text{MS regression}}{\text{MS residual}}$$

Figure 15.6 shows computer output of the ANOVA table for the illustrative example. This reveals $F_{\text{stat}} = 443.25$ with 2 and 651 degrees of freedom.

1. **P-value.** A conservative estimate for *P*-value can be derived from Appendix Table D by looking up the tail region for an *F*-distribution with 2 numerator df and 100 denominator df. (Always go down to the next available degree of freedom for a conservative *P*-value.) This lets us know that an $F_{2,100}$ critical value of 7.41 has a right tail area of 0.001. The observed *F*-statistic falls further into this tail. Therefore, $P < 0.001$.

2. **Significance level.** The evidence against the null hypothesis is significant at the $\alpha = 0.001$ level.

3. **Conclusion.** The *F*-test is directed toward H_0: $\beta_1 = \beta_2 = 0$ versus H_a: "H_0 is false." Therefore, this analysis merely tells us that age and/or smoking are significantly associated with forced expiratory volume in this adolescent population

ANOVA[b]

Model		Sum of Squares	df	Mean Square	F	Sig.
1	Regression	283.058	2	141.529	443.254	0.000[a]
	Residual	207.862	651	0.319		
	Total	490.920	653			

[a]Predictors: (constant), AGE, SMOKE
[b]Dependent variable: FEV

FIGURE 15.6 Screenshot of computer output, multiple linear regression ANOVA statistics, forced expiratory volume illustrative example. Graph produced with SPSS for Windows, Rel. 11.0.1.2001. Chicago: SPSS Inc. Reprint Courtesy of International Business Machines Corporation.

($P < 0.001$). To determine which slopes are significant, we must look at the individual contributions of the variables as addressed in Section 15.4.

Model Fit

The ratio of the sum of squares of the regression and total sum of squares forms the **multiple coefficient of determination (R^2):**

$$R^2 = \frac{\text{sum of squares regression}}{\text{sum of squares total}}$$

This statistic quantifies the proportion of the variance in Y that is explained by the regression model, thus providing a statistic of overall model fit.

The output in Figure 15.6 shows that $SS_{regression} = 283.058$ and $SS_{Total} = 490.920$. Therefore, $R^2 = 283.058/490.920 = 0.577$. This suggests that about 58% of the variability of the response is explained by the independent variables in the model.

The square root of the multiple coefficient of determination is the **multiple correlation coefficient (R): $R = \sqrt{R^2}$.** This statistic is analogous to Pearson's correlation coefficient (r) except that it considers the contribution of multiple explanatory factors. This statistic will always be a positive number between 0 and 1. The multiple correlation coefficient for the illustrative example $R = \sqrt{0.577} = 0.76$.

Figure 15.7 is a screenshot of output of model summary statistics. Besides R and R^2, this output includes two additional statistics: the **Adjusted R Square**, which is equal to $1 - (1 - R^2)\dfrac{n - 1}{n - k - 1}$ and the **Std. Error of the Estimate**, which is equal to $\sqrt{MS \text{ residual}}$. The Adjusted R Square more closely than R^2 reflects the goodness of fit of the model in the population.[g] The standard error of the estimate is the standard error of the multiple regression, which is equal to $\sqrt{\text{mean square residual}}$.

Model Summary

Model	R	R Square	Adjusted R Square	Std. Error of the Estimate
1	0.759a	0.577	0.575	0.5650627

aPredictors: (constant), AGE, SMOKE

FIGURE 15.7   Screenshot of computer output, multiple linear regression model summary, forced expiratory volume illustrative example. Graph produced with SPSS for Windows, Rel. 11.0.1.2001. Chicago: SPSS Inc. Reprint Courtesy of International Business Machines Corporation.

---

[g] This adjustment is needed because as predictors are added to the model, some of the variance in Y is explained simply by chance.

■ 15.6 Examining Multiple Regression Conditions

The conditions needed for multiple linear regression mirror those of simple linear regression. Recall the mnemonic "line," which stands for linearity, independence, normality, and equal variance (Section 14.4). Statistical packages often incorporate tools to help examine the multiple regression response surface for these conditions. Although thorough consideration of these tools is beyond the scope of this general text, we examine two such tools by way of introduction.

Figure 15.8 is a plot of standardized residuals against standardized predicted values for the illustrative example. The standardized residual of observation i is $\dfrac{y_i - \hat{y}_i}{s_{Y|x_1 x_2 \ldots x_k}}$, and its standardized predicted value is $\dfrac{\hat{y}_i}{s_{Y|x_1 x_2 \ldots x_k}}$, where $s_{Y|x_1 x_2, \ldots, x_k} = \sqrt{\text{MS residual}}$.

This plot shows less variation at the lower end of predicted values than at the higher end. This is not unexpected with these data; the lower FEV values are likely to represent younger subjects where respiratory function is expected to be less variable. The extent to which this undermines the equal variance assumption and utility of the inferential model is a matter of judgment and may warrant further consideration by a specialist.

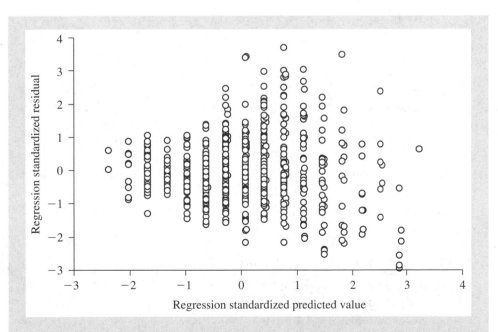

FIGURE 15.8   Standardized residuals plotted against standardized predicted values, multiple linear regression model. Graph produced with SPSS for Windows, Rel. 11.0.1.2001. Chicago: SPSS Inc. Reprint Courtesy of International Business Machines Corporation.

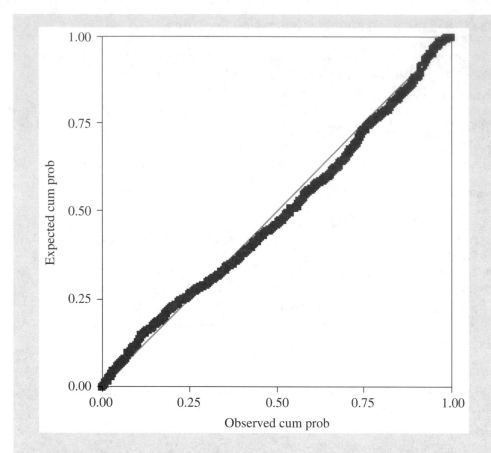

FIGURE 15.9   Normal Q-Q plot, multiple linear regression illustration. Graph produced with SPSS for Windows, Rel. 11.0.1.2001. Chicago: SPSS Inc. Reprint Courtesy of International Business Machines Corporation.

Figure 15.9 is a Normal Q-Q plot of the standardized residuals. In this plot, we look for deviations from the diagonal line as evidence of non-Normality (Section 7.4). The data in this plot does a fairly good job adhering to the diagonal line, suggesting no major departures from Normality.

Summary Points (Multiple Regression)

1. **Multiple regression** is used to quantify linear relationship between explanatory variable X_1 and quantitative response variable Y while adjusting for potential confounding variables X_2, X_3, …, X_k.

2. **Categorical explanatory variables** can be incorporated into a regression model by coding them as 0/1 dummy variables.

3. The **population model** for multiple regression is $\mu_{Y|x_1,x_2,\ldots,x_k} = \alpha + \beta_1 x_1 + \beta_2 x_2 + \cdots + \beta_k x_k$, where $\mu_{Y|x_1,x_2,\ldots,x_k}$ represents the expected value of response variable Y given specific values for explanatory variables $X_1, X_2, \ldots, X_k$. $\beta_1, \beta_2, \ldots, \beta_k$ are the population slopes for the explanatory variables $X_1, X_2, \ldots, X_k$; α is the population intercept.

4. We rely on **statistical software** to provide estimates for each regression coefficient. These estimates are called $a, b_1, b_2, \ldots, b_k$. The regression equation for the data is $\hat{y} = a + b_1 x_1 + b_2 x_2 + \cdots + b_k x_k$, where $\hat{y}$ represents the predicted value of Y given values $x_1, x_2, \ldots, x_k$.

   (a) Each slope $b_1, b_2, \ldots, b_k$ is interpreted as follows: *if all the X variables except X_i are held constant, the predicted change in Y per unit change in X_i is equal to b_i.*

   (b) Each slope estimate can be tested for statistical significance with t-test $H_0: \beta_i = 0$. We rely on statistical software to compute the statistics.

   (c) 95% confidence intervals for each β_i are provided by the software.

5. The **residual** associated with observation $i(\epsilon_i)$ is the distance of data point from regression plane fit by the model. It is assumed that these residuals show constant variance at all points along the plane and are Normally distributed with a mean of 0: $\epsilon_i \sim N(0, \sigma)$. These assumptions can be examined with a residual plot.

6. The **variance of the regression** σ^2 is estimated by the mean square residual $\dfrac{\Sigma\text{residuals}^2}{\text{df}_{\text{residuals}}}$, where $\Sigma\text{residuals}^2$ is the sum of squared residuals and $\text{df}_{\text{residuals}} = n - k - 1$. The mean square residual is also called the mean square error.

Vocabulary

Adjusted R square	Response line
Dependent variables	Response plane
Dummy variables	Response surface
Independent variables	Simple regression
Indicator variables	Standard error of the multiple regression
Multiple correlation coefficient (R)	$\quad(S_{Y\,|\,x_1 x_2,\ldots,x_k})$
Multiple regression	
Multiple regression of determination (R^2)	

Exercises

15.1 *The relation between FEV and SEX in the illustrative data set.* Download the illustrative data set (FEV.*) used in this chapter. See Table 15.1 for a codebook of variables. Examine the relationship between FEV and SEX. Then, examine the relationship between FEV and SEX while adjusting for AGE. Did AGE confound the relationship?

15.2 *Cognitive function in centenarians.* A geriatric researcher looked at the effect of AGE (years), EDUCATION level (years of schooling), and SEX (0 = female, 1 = male) on cognitive function in 162 centenarians (fictitious data). Cognitive function was measured using standardized psychometric and mental functioning tests with higher scores corresponding to better cognitive ability. Using these data, a multiple regression computer run derived the following multiple regression model:

$$\hat{y} = 99.4 + (-0.165) \cdot \text{AGE} + 0.522 \cdot \text{EDUCATION} + 1.112 \cdot \text{SEX}$$

Based on this model, describe the effect of AGE, EDUCATION, and SEX on cognition scores in centenarians.

Part III

Categorical Response Variable

16 | Inference About a Proportion

■ 16.1 Proportions

This chapter considers the analysis of a categorical response derived from a single simple random sample. We consider only categorical variables with two possible responses in this chapter. This type of response is common in public health research, where many health outcomes are binary by nature (e.g., diseased/not diseased) and others are classified into groups based on a cutoff point (e.g., hypertensive/normotensive).

We take a simple random sample (SRS) of n individuals from a population in which $p \times 100\%$ of the individuals are classified as successes. The **proportion** of successes in a sample, denoted $\hat{p}$ ("p hat"), is:

$$\hat{p} = \frac{x}{n}$$

where x represents the number of successes observed in the sample. Sample proportion $\hat{p}$ is an unbiased estimator of population proportion p.

ILLUSTRATIVE EXAMPLE

Data (Prevalence of smoking in a community). A random-digit dialing technique is used to select 57 individuals from a community. Seventeen of 57 individuals in the sample are classified as smokers.

Therefore, the proportion of smokers in the sample $\hat{p} = \frac{x}{n} = \frac{17}{57} = 0.298$. Assuming the sample is an SRS and the information is properly classified, this in an unbiased estimate of the prevalence of smoking in the population.

Notes

1. **Counts and proportions.** While quantitative data are often described with sums and averages, descriptive statistics for categorical outcomes are based on counts and proportions (Figure 16.1).

2. **Incidence and prevalence.** We are particularly interested in two specific types of proportions: prevalence proportions and incidence proportions. **Prevalence proportions** are the proportion of individuals in a population affected by a condition at particular time. **Incidence proportions** are the proportion of individuals at risk who go on to develop a condition over a specified period of time. Synonyms for *incidence proportion* include **cumulative incidence** and average **risk**.

3. **Proportions are a type of average.** Consider a binary outcome in which each success is coded "1" and each failure is coded "0." Here are 10 such observations:

    0 0 0 1 0 0 0 0 1 0

    Notice that $n = 10$, $\Sigma x_i = 2$, and $\bar{x} = \dfrac{2}{10} = 0.2$. Also notice that $\hat{p} = \dfrac{2}{10}$. This shows the equivalency of $\hat{p}$ and $\bar{x}$. A sample proportion is an average of zeros and ones.

4. **Statistics and parameters.** The proportion in the sample (denoted $\hat{p}$) is a statistic. The proportion in the population (denoted p) is a parameter. Principles of statistical inference introduced in Chapters 8 through 10 apply. The population is sampled; statistics are calculated in the sample; sample statistics are then used to help infer population parameters (Figure 16.2). In previous chapters, knowledge of the sampling distribution of $\bar{x}$ was used to help infer population mean μ. In this chapter, we will use knowledge about the sampling distribution of $\hat{p}$ to help infer parameter p.

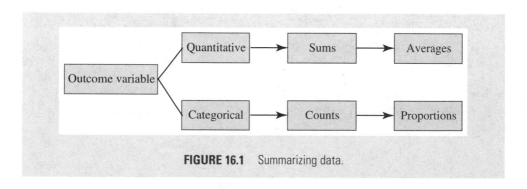

FIGURE 16.1   Summarizing data.

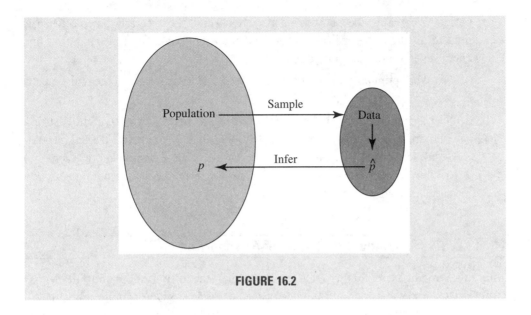

FIGURE 16.2

Exercises

16.1 *AIDS-related risk factor.* A national study of AIDS risk factors used a random-digit dialing technique to contact study participants. Among the 2673 heterosexual adults in the sample, 170 reported two or more sexual partners in the past 12 months.[a]

(a) Describe the population to which inferences will be made. What population parameter will be estimated? Calculate the prevalence of multiple sexual partners in the sample.

(b) Practical problems of bias can pose a threat to the validity of a study of this type. What specific types of selection biases might evolve from the sampling method used by this study?

(c) This study relied on the truthfulness of responses. Although the investigators tried their best to make respondents feel comfortable, they could not be assured that the information was fully accurate. If an information bias existed in responses, hypothesize whether it would tend to overestimate or underestimate the prevalence of the risk factor.

16.2 *Patient preference.* An investigation uses two different methods of nonvolitional muscle function testing to study the effects of parenteral nutrition in debilitated

[a] Catania, J. A., Coates, T. J., Stall, R., Turner, H., Peterson, J., Hearst, N., et al. (1992). Prevalence of AIDS-related risk factors and condom use in the United States. *Science, 258*(5085), 1101–1106.

patients. As part of the study, subjects were asked if they preferred method A or B muscle testing (in terms of comfort). Of the eight patients expressing a preference, seven preferred method A.[b]

(a) To what population will inferences be made? What parameter will be estimated?

(b) Using intuition, do you believe the current evidence seven of eight is strong enough to conclude a preference for method A in the entire patient population? (Later in the chapter, we will address this question more formally.) Either an affirmative or a negative response is acceptable as long as you explain your reasoning.

■ 16.2 The Sampling Distribution of a Proportion

Inferential methods in this chapter rest on using the **binomial probability mass function** to model the random number of successes in a given sample (Chapter 6). The **Normal approximation to the binomial** can be used when the sample is large (Section 8.3). Because these distributions were covered elsewhere, only a brief review is presented here.

Binomial random variables are based on counting the number of successes in n independent Bernoulli trials. The probability of success in each trial is assumed to be a constant p. The random number of successes (X) will vary according to a binomial distribution with parameters n and p: $X \sim b(n,p)$. Random variable X has expected value $\mu = np$ and standard deviation $\sigma = \sqrt{npq}$, where $q = 1 - p$. Probabilities are calculated with this formula:

$$\Pr(X = x) = {}_nC_x p^x q^{n-x}$$

where ${}_nC_x = \dfrac{n!}{x!(n-x)!}$, n represents the number of independent trials, and p represents the probability of success for each trial, and $q = 1 - p$.

Because calculation of binomial probabilities can be tedious, a Normal approximation to the binomial can be used in large samples. This will simplify calculations with only a minimal and inconsequential loss of accuracy. For our purposes, the sample is considered sufficiently large to use this Normal approximation when $npq \geq 5$.

---

[b] Brooks, S. D., Gerstman, B. B., Sucher, K. P., & Kearns, P. J. (1998). The reliability of muscle function analysis using different methods of stimulation. *Journal of Parenteral and Enteral Nutrition, 22*(5), 331–334.

The Normal approximation to the binomial states that, in large samples, binomial random variable X varies according to a Normal distribution with $\mu = np$ and $\sigma = \sqrt{npq}$. This is equivalent to saying that sample proportion $\hat{p}$ varies according to a Normal distribution with $\mu = p$ and standard deviation $\sigma = \sqrt{\frac{pq}{n}}$.

Figure 16.3 depicts a simulation of a **sampling distribution of a proportion.** Repeated samples of size n are taken from a population in which $p \times 100\%$ of the individuals are classified as successes. Some of the samples produce $\hat{p}$s that are less than the actual value of p, and some produce $\hat{p}$s that are larger than p. The distribution of sample proportions is approximately Normal with $\mu = p$ and $\sigma = \sqrt{\frac{pq}{n}}$. This sampling distribution model will be used to help draw inferences about the unknown value of population parameter p.

Exercises

16.3 *AIDS-related risk factor.* Exercise 16.1 introduced a problem in which a random sample of 2673 adult heterosexuals were asked about presence of an HIV risk factor.

    (a) Describe the sampling distribution of the number of individuals who are positive for this risk factor.

    (b) Do you think a Normal approximation can be used to characterize the sampling distribution? Explain your reasoning.

16.4 *Patient preference.* For the problem about patient preference described in Exercise 16.2, characterize the sampling distribution of the number of patients preferring method A. Do you think a Normal approximation could be used to describe this sampling distribution?

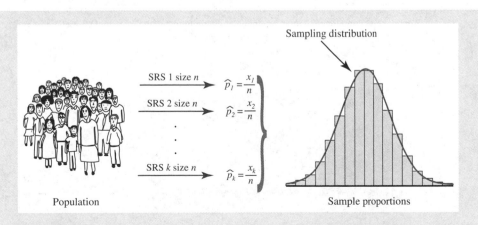

FIGURE 16.3   Simulation of a sampling distribution of proportions showing superimposed Normal approximation.

■ 16.3 Hypothesis Test, Normal Approximation Method

With large samples, a Normal approximation to the binomial distribution can be used to test a proportion for statistical significance. Again, we use the *npq* rule as a rough guide to determine whether the sample is large enough to use a Normal approximation (Section 8.3). This rule states that it is acceptable to use the Normal approximation to the binomial when $npq \geq 5$, where $q = 1 - p$. Under these conditions, the sampling distribution of $\hat{p}$ is approximately Normal with mean p and standard deviation $\sigma = \sqrt{\frac{pq}{n}}$, as depicted in Figure 16.4.

The testing procedure is based on using $\dfrac{\hat{p} - p}{\sqrt{pq/n}}$ as a Standard Normal random variable. Here is step-by-step guide to the procedure.

A. **Hypotheses.** The null hypothesis is $H_0\colon p = p_0$, where p_0 represents the proportion under the null hypothesis. The value of p_0 comes from the research question itself, not from the data. The alternative hypothesis is either $H_a\colon p > p_0$ (one sided to the right), $H_a\colon p < p_0$ (one sided to the left), or $H_a\colon p \neq p_0$ (two sided).

B. **Test statistic.** After confirming that a Normal approximation to the binomial can be used (by checking whether $np_0q_0 \geq 5$), calculate:

$$z_{\text{stat}} = \frac{\hat{p} - p_0}{\text{SE}_{\hat{p}}}$$

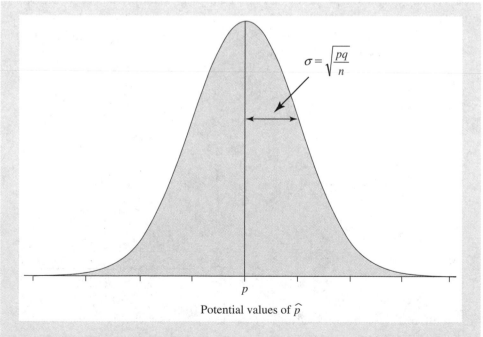

FIGURE 16.4 Sampling distribution of a proportion, Normal approximation.

where $\hat{p}$ represents the sample proportion, p_0 represents the value of p under the null hypothesis, and $SE_{\hat{p}} = \sqrt{\frac{p_0 q_0}{n}}$ where $q_0 = 1 - p_0$.

Optional: A **continuity-correction**[c] can be incorporated into the test statistics as follows:

$$z_{stat,\, c} = \frac{|\hat{p} - p_0| - \dfrac{1}{2n}}{SE_{\hat{p}}}$$

This continuity correction is introduced because the *P*-valve produced by the continuity z method better approximates the probability that would be derived by an exact binomial procedure (Section 16.4) had one been pursued. This is especially true in small samples.

C. **P-value.** Convert the z-statistic to a *P*-value in the usual fashion (with either Table B or a software utility). An additional z table, Appendix Table F, is included in the back of the book to make it easier to look up two-tailed *P*-values directly from the $|z_{stat}|$. If a one-sided *P*-value is needed when using Table F, divide the table entry by 2.

D. **Significance level.** The difference is said to be statistically significant at the α-level of significance when $P \leq \alpha$, in which case H_0 is rejected. Keep in mind that failure to reject the null hypothesis should *not* be construed as its acceptance.

E. **Conclusion.** The test results are addressed in the context of the data and research question.

ILLUSTRATIVE EXAMPLE

z-test of a proportion. We propose to test whether the prevalence of smoking considered in the prior illustration is significantly higher than that of the United States as a whole. The prevalence of smoking in U.S. adults according to an NCHS report is 0.21.[d] An SRS of 57 individuals from a particular community reveals 17 smokers. Thus, $\hat{p} = 17/57 = 0.298$. A two-sided test is used to determine whether this is significantly different than 0.21.

continues

---

[c] The continuity correction adjusts for the fit of the smooth Normal curve to the chunky binomial function. The continuity-correct statistic produces larger *P*-values than the regular test statistic.
[d] National Center for Health Statistics. (2006). *Early Release of Selected Estimates Based on Data from the January–September 2005 National Health Interview Survey.* Retrieved November 2006 from www.cdc.gov/nchs/data/nhis/earlyrelease/200603_08.pdf.

continued

Solution:

A. Hypotheses. H_0: $p = 0.21$ against H_a: $p \neq 0.21$

B. Test statistic. Before calculating the z_{stat}, we confirm that a Normal approximation to the binomial can be used. Notice that $n = 57$, $p_0 = 0.21$, so $q_0 = 1 - 0.21 = 0.79$. Therefore, $np_0q_0 = 57 \cdot 0.21 \cdot 0.79 = 9.5$, showing the sample to be large enough to apply the Normal approximation test.

(a) $SE_{\hat{p}} = \sqrt{\dfrac{p_0 q_0}{n}} = \sqrt{\dfrac{0.21 \cdot 0.79}{57}} = 0.05395$

(b) $z_{stat} = \dfrac{\hat{p} - p_0}{SE_{\hat{p}}} = \dfrac{0.298 - 0.21}{0.05395} = 1.63$

C. *P*-value. Figure 16.5 depicts the sampling distribution of $\hat{p}$ under the null hypothesis. Table F is used to convert the z_{stat} to $P = 0.1031$.

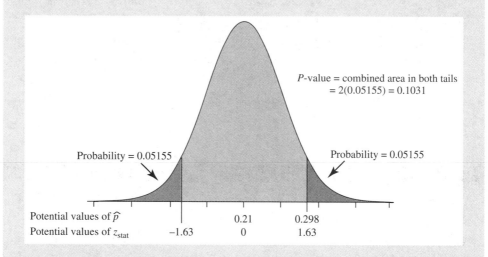

FIGURE 16.5 *P*-value, smoking prevalence illustrative example.

D. Significance level. The evidence against H_0 is not sufficient to warrant its rejection at $\alpha = 0.10$.

E. Conclusion. We do not have enough evidence here to conclude that the prevalence of smoking in this community is any different than the nationally reported average ($P = 0.103$).

Use of the continuity-correction results in $z_{\text{stat}, c} = \dfrac{|\hat{p} - p_0| - \dfrac{1}{2n}}{SE_{\hat{p}}} =$
$\dfrac{|0.298 - 0.21| - \frac{1}{2 \times 57}}{0.05395} = 1.47$, P (two sided) $= 0.1416$. This does not materially change our conclusion.

Exercises

16.5 *AIDS-related risk factor.* Exercises 16.1 and 16.3 considered a survey in which 170 of 2673 individuals (6.4%) reported having two or more sexual partners in the prior 12 months. This study was completed in the early 1990s. Suppose an earlier study (completed in the 1970s) suggested that the prevalence of this attribute in the population was 7.5%. Is the observed proportion significantly different from the prior prevalence? Use a two-sided alternative hypothesis. Show all hypothesis-testing steps.

16.6 *AIDS-related risk factor* (*Continuity-corrected z-statistic*). Recalculate the P-value for Exercise 16.5 using the continuity-corrected z-statistic. Is the P-value from the continuity-corrected z-statistic larger or smaller than that from the non-corrected z-statistic?

■ 16.4 Hypothesis Test, Exact Binomial Method

The z-test for proportions is accurate only in large samples. When working with small samples, an **exact test** based on binomial calculations is required. Here are the steps of **Fisher's exact test**.

A. Hypotheses. The hypothesis statements are the same as used in the prior section (i.e., H_0: $p = p_0$).[e]

B. Test statistic. The test statistic is the observed count of successes in the sample, denoted x.

---

[e] The hypotheses can also be stated in terms of expected number of successes, that is, H_0: $\mu = np_0$.

C. **P-value.** The *P*-value is the probability of observing *x* successes or a value more extreme than *x* under the conditions set forth by the null hypothesis.

   (a) For one-sided alterative hypotheses to the right, $P = \Pr(X \geq x) = \Pr(X = x) + \Pr(X = x + 1) + \cdots + \Pr(X = n)$, assuming $X \sim b(n, p_0)$.

   (b) For one-sided alterative hypotheses to the left, $P = \Pr(X \leq x) = \Pr(X = x) + \Pr(X = x - 1) + \cdots + \Pr(X = 0)$, assuming $X \sim b(n, p_0)$.

   (c) For two-sided alterative hypotheses, $P = \Pr(X \leq x_1) + \Pr(X \geq x_2)$, where x_1 or x_2 is equal to *x* and the other x_i is a value as extreme in the opposite direction.[f]

D. **Significance level.** The *P*-value is compared to various type I error thresholds (α) to gauge the strength of evidence against the null hypothesis.

E. **Conclusion.** The test results are addressed in the context of the data and research question.

ILLUSTRATIVE EXAMPLE

Exact binomial test (Fisher's tea challenge). A memorable story in statistical lore has R. A. Fisher drinking tea with an heiress when the heiress claims she can discern by taste alone whether milk is added to the cup before or after the tea is poured.[g] Fisher challenges the heiress to a test in which he gives her eight cups of tea in random order and tells her that four had the tea added first and the remaining four had the milk added first. She correctly identifies the order of adding the milk in six of the eight attempts. Can we say from this test that the heiress is doing better than random guessing?

A. **Hypotheses.** With random guessing, there is a 50/50 chance of guessing correctly for each trial in this experiment. Therefore, H_0: $p = 0.5$. With better-than-random guessing, the taster has better than a 50% chance of guessing correctly, H_a: $p > 0.5$.

---

[f] Some statisticians merely double the one-sided *P*-value for a two-sided test. For justification, see Dupont, W. D. (1986). Sensitivity of Fisher's exact test to minor perturbations in 2 × 2 contingency tables. *Statistics in Medicine, 5*(6), 629–635.

[g] Even this simple problem raises interesting statistical questions such as "How many cups should be tested? Should the cups be paired? In what order should the cups be presented? What should be done about chance variation in temperature, sweetness, and so on?" What conclusions could be drawn from a perfect score? *Source:* Box, J. F. (1978). *R. A. Fisher, the Life of a Scientist.* New York: John Wiley.

B. Test statistic. Note that the *npq* rule derives $8 \cdot 0.5 \cdot 0.5 = 2$, confirming the need for an exact binomial procedure. The test statistic for the exact test is observed number of successes $x = 6$.

C. *P*-value. *P*-value $= \Pr(X \geq 6)$ while assuming X is a binomial random variable with $n = 8$ and $p = 0.5$. Therefore, we calculate:

- $\Pr(X = 6) = (_8C_6)(0.5)^6(1 - 0.5)^{8-6} = (28)(0.0156)(0.25) = 0.1094$
- $\Pr(X = 7) = (_8C_7)(0.5)^7(1 - 0.5)^{8-7} = (8)(0.0078)(0.5) = 0.0313$
- $\Pr(X = 8) = (_8C_8)(0.5)^8(1 - 0.5)^{8-8} = (1)(0.0039)(1) = 0.0039$

The (one-sided) *P*-value $= \Pr(X \geq 6) = 0.1094 + 0.0313 + 0.0039 = 0.1446$. Figure 16.6 depicts this graphically.

D. Significance level. The evidence against the null hypothesis is not significant at the $\alpha = 0.10$ level.

E. Conclusion. Selecting six of the eight "milk-in-tea sequences" correctly under the stated conditions is not significantly better than mere "50/50" guessing ($P = 0.1446$).

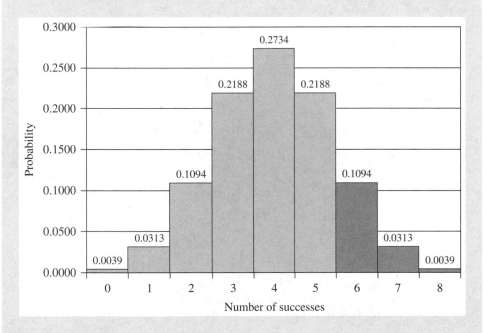

FIGURE 16.6 Exact *P*-value (shaded), tea challenge illustrative example.

Notes

1. **How many correct guesses would provide good evidence against H_0?**
   In the tea challenge illustrative example, we may ask, "How many correct
   identifications would provide good evidence against guessing?" Because eight
   of eight would occur only 0.0039 of the time with random guessing, this
   would provide strong evidence against H_0. Seven of eight correct guesses
   would derive a one-sided P-value = $\Pr(X = 7) + \Pr(X = 8) = 0.0313 +$
   $0.0039 = 0.0352$, which is also good evidence against H_0.

2. **Mid-P correction.** The discrete nature of the binomial distribution causes
   Fisher's test to produce P-values that are a bit too large; a P-value from Fisher's
   test will fail to reject H_0 more than $\alpha \times 100\%$ of the time. A better P-value
   can be achieved by including only half of $\Pr(X = x)$ into the calculation of the
   P-value. This method is called the **Mid-P test.** For the tea challenge illustrative
   example, the Mid-P procedure uses $\frac{1}{2} \cdot \Pr(X = 6) = \frac{1}{2} \cdot 0.1094 = 0.0547$ in
   its calculation of the P-value. Thus, $P_{\text{Mid-P}} = \frac{1}{2} \cdot \Pr(X = 6) + \Pr(X = 7) +$
   $\Pr(X = 8) = 0.0547 + 0.0313 + 0.0039 = 0.0899$. Figure 16.7 depicts this
   graphically.

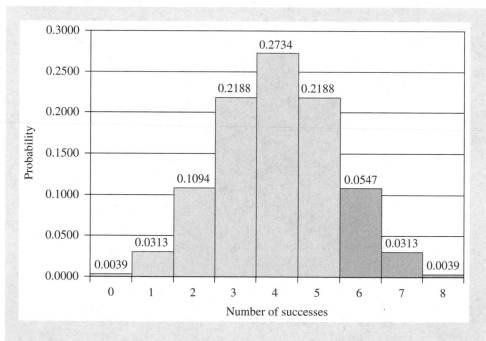

FIGURE 16.7   Mid-P P-value, tea challenge illustrative example.

3. **Software utilities.** In practice, exact tests are usually calculated with statistical packages and software utilities. Figure 16.8 is a screenshot from *WinPepi*'s Describe program (option A)[h] with fields filled in for testing the tea-challenge data. Figure 16.9 is the output from the program. Results replicate those of our prior calculations (one-tailed Fisher's $P = 0.145$ and one-tailed $P_{\text{Mid-P}} = 0.090$).

4. **OK to use exact tests in large samples.** Exact procedures are necessary when testing small samples and also may be used in large samples when computational software is available.

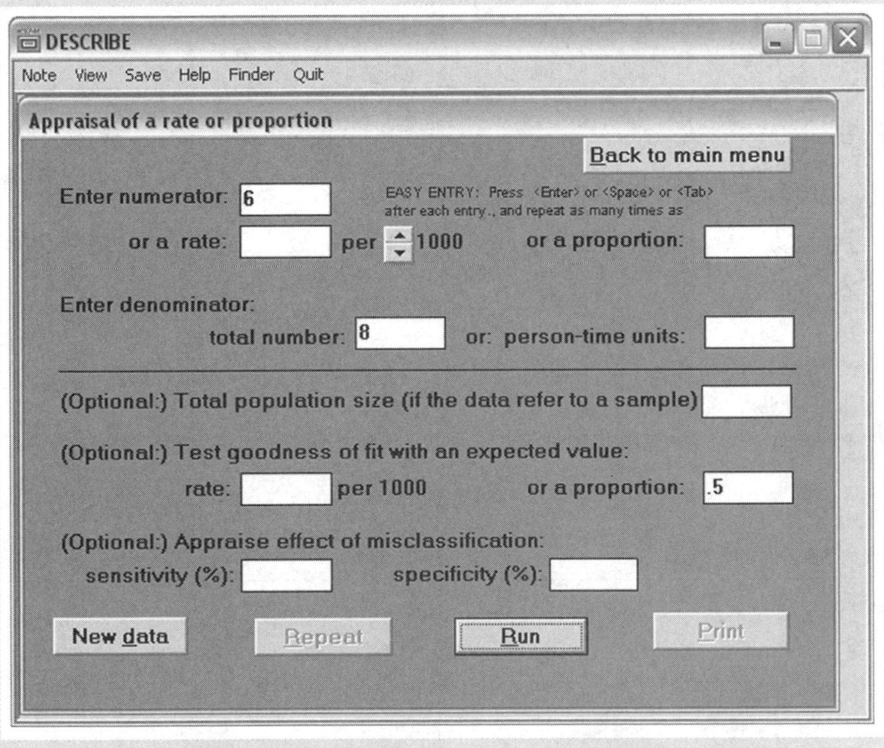

FIGURE 16.8 *WinPepi*'s data-entry screen for exact binomial test, tea challenge illustrative example. Abramson J.H. (2011). WINPEPI updated: computer programs for epidemiologists, and their teaching potential. *Epidemiologic Perspectives & Innovations, 8*(1), 1.

---

[h] Abramson, J. H. (2004). WINPEPI (PEPI-for-Windows): Computer programs for epidemiologists. *Epidemiologic Perspectives & Innovations, 1*(1), 6.

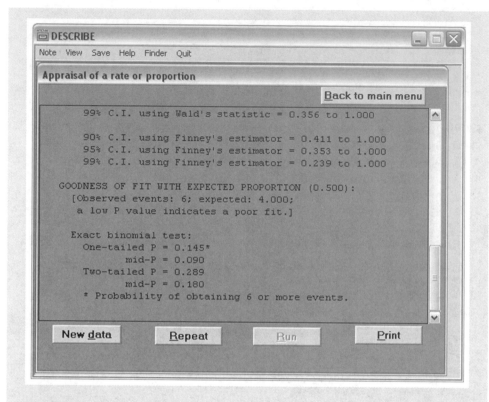

FIGURE 16.9   *WinPepi*'s output for tea challenge illustrative example. Abramson J.H. (2011). WINPEPI updated: computer programs for epidemiologists, and their teaching potential. *Epidemiologic Perspectives & Innovations, 8*(1), 1.

Exercises

16.7 *Patient preference, Fisher's method.* Exercises 16.2 and 16.4 considered a problem in which seven of eight patients expressed a preference for medical procedure A compared to medical procedure B. A Normal approximation test was precluded because of the small sample size. Test the hypothesis of equal preference for medical procedure A and medical procedure B with Fisher's procedure.

16.8 *Patient preference, exact binomial test, Mid-P method.* What is the exact Mid-P *P*-value for the problem in Exercise 16.7?

■ 16.5 Confidence Interval for a Population Proportion

Plus-Four Method

We will use a **plus-four method** as our primary method for calculating confidence intervals for proportions. This method is based on the **Wilson score test,**[i] is simple to calculate, is more accurate than standard Normal approximation methods, and can be used in samples as small as $n = 10$.[j]

We start by adding four imaginary observations to the data set before calculating the confidence interval; the imaginary sample size $\tilde{n} = n + 4$. Half of these imaginary observations go into the numerator of the proportion, so the imaginary number of success $\tilde{x} = x + 2$. The plus-four proportion $\tilde{p} = \dfrac{\tilde{x}}{\tilde{n}}$. Now use $\tilde{p}$ as the center of the confidence interval,[k] so $(1-\alpha)100\%$ confidence interval for p is:

$$\tilde{p} \pm z_{1-\frac{\alpha}{2}} \cdot \text{SE}_{\tilde{p}}$$

where $\text{SE}_{\tilde{p}} = \sqrt{\dfrac{\tilde{p}\,\tilde{q}}{\tilde{n}}}$. Use $z_{1-(\alpha/2)} = 1.645$ for 90% confidence, $z_{1-(\alpha/2)} = 1.96$ for 95% confidence, and $z_{1-(\alpha/2)} = 2.576$ for 99% confidence.

After calculating the confidence interval for population proportion p, interpret the results in the context of the data and research question.

ILLUSTRATIVE EXAMPLE

Confidence interval for population proportion, plus-four method (Smoking prevalence). Recall the survey that found 17 smokers in an SRS on $n = 57$ in a particular community. What is the 95% confidence interval for the prevalence of smoking in this population?

continues

---

[i] Wilson, E. B. (1927). Probable inference, the law of succession, and statistical inference. *Journal of the American Statistical Association, 22,* 209–212. Also see page 14 in Fleiss, J. L. (1981). *Statistical Methods for Rates and Proportions.* (2nd ed.). New York: John Wiley & Sons.

[j] Agresti, A., & Coull, B. A. (1998). Approximate is better than "exact" for interval estimation of binomial proportions. *The American Statistician, 52*(2), 119–126. Agresti, A., & Caffo, B. (2000). Simple and effective confidence intervals for proportions and differences of proportions result from adding two successes and two failures. *The American Statistician, 54*(4), 280–288. Brown, L. D., Cair, T. T., & DasGupta, A. (2001). Interval estimation for a binomial proportion. *Statistical Science, 16*(2), 101–117.

[k] When n is large, $\tilde{p} \approx \hat{p}$.

continued

Solution:

- $\widetilde{n} = n + 4 = 57 + 4 = 61$
- $\widetilde{x} = x + 2 = 17 + 2 = 19$
- $\widetilde{p} = \dfrac{\widetilde{x}}{\widetilde{n}} = \dfrac{19}{61} = 0.3115$
- $\widetilde{q} = 1 - \widetilde{p} = 1 - 0.3115 = 0.6885$
- $SE_{\widetilde{p}} = \sqrt{\dfrac{(0.3115)(0.6885)}{61}} = 0.0593$
- The 95% confidence interval for $p = 0.3115 \pm (1.96)(0.0593) = 0.3115 \pm 0.1162 = (0.1953 \text{ to } 0.4277)$ or about (20% to 43%).

We can conclude with 95% confidence that the prevalence of smoking in this community is between 20% and 43%.

Note: Keep in mind that confidence intervals and *P*-values address *random* sampling errors only and do not protect us from *systematic* forms of errors as might arise from selection bias and information bias.

Exact Confidence Intervals

With very small samples ($n < 10$), the confidence interval for population proportion p should be based directly on the binomial distribution. These methods are analogous to the exact binomial tests presented in Section 16.4.[1] Calculations can be based on the relation between the *F*-distribution and binomial distribution by using the following formulas for the lower confidence limit (LCL) and upper confidence limit (UCL)[m]:

$$LCL = \frac{x}{x + (n - x + 1)F_{\frac{\alpha}{2}, df_1, df_2}}$$

where $df_1 = 2(n - x + 1)$ and $df_2 = 2x$ and

$$UCL = \frac{(x + 1)F_{\frac{\alpha}{2}, df_1', df_2'}}{n - x + (x + 1)F_{\frac{\alpha}{2}, df_1', df_2'}}$$

where $df_1' = 2(x + 1)$ and $df_2' = 2(n - x)$.

---

[1] Clopper, C. J., & Pearson, E. A. (1934). The use of confidence or fiducial limits illustrated in the case of the binomial. *Biometrika, 26*(4), 404–413.

[m] Rothman, K. J., & Boyce, J. D. J. (1979). *Epidemiologic Analysis with a Programmable Calculator.* Washington, D.C.: U.S. Government Printing Office. p. 31.

A Mid-P adjustment analogous to the one reported in the prior section can be incorporated in the confidence interval.[n] We will use the *WinPepi* Describe.exe program to perform our calculations.

Confidence interval for proportion, exact method (Tea-challenge). Recall the tea-tasting challenge illustration based on whether a person could predict whether milk was added to a cup before or after tea is poured. The taster proved right in six of eight trials. Based on this finding, determine the proportion that the tester can correctly identify in the long run.

Data are entered into *WinPepi*'s Describe program (option A), as shown in Figure 16.8. Figure 16.10 contains output from the program showing confidence

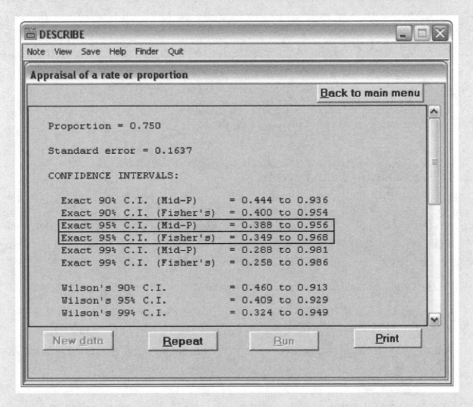

FIGURE 16.10 *WinPepi* output showing confidence intervals for *p*, "tea challenge" illustrative data. Abramson J.H. (2011). WINPEPI updated: computer programs for epidemiologists, and their teaching potential. *Epidemiologic Perspectives & Innovations, 8*(1), 1.

continues

[n] Berry, G., & Armitage, P. (1995). Mid-P confidence intervals: A brief review. *The Statistician, 44*(4), 417–423.

continued

intervals calculated at three levels of confidence (90%, 95%, and 99%) according to three different methods (exact Mid-P, exact Fisher's, and Wilson's). The formulas presented previously correspond to the output labeled "(Fisher's)." The 95% confidence interval by this method is 0.349 to 0.968. The 95% confidence interval by the Mid-P method is 0.388 to 0.956, reflecting its less-conservative approach.

Exercises

16.9   *AIDS-related risk factor.* Use the plus-four method to calculate a 95% confidence interval for the prevalence of multiple sexual partners in adult heterosexuals using the information in Exercise 16.1. Recall that 170 of the 2673 subjects reported this behavior.

16.10 *AIDS-related risk factor, 90% confidence interval.* Calculate a 90% confidence interval for the problem presented in Exercise 16.9.

16.11 *Patient preference.* Exercise 16.2 stated that seven of eight patients expressed a preference for a particular medical procedure. Use an exact procedure to calculate a 95% confidence interval for p.

■ 16.6 Sample Size and Power

Concepts and methods needed to determine the sample size requirements for estimating and testing means were introduced in Sections 9.6, 10.3, and 11.7. Similar methods apply when determining the sample size requirements for estimating and testing proportions. We will approach this problem from three interrelated angles. We ask:

- What is the sample size needed to estimate p with a given margin of error?
- What is the sample size needed to test a proportion for significance at a stated α-level with given power?
- What is the power of a significance test of a proportion given a stated sample size and α-level?

Sample Size Requirements for Estimating p with Margin of Error m

Define **margin of error m** as half the confidence interval width. For large samples,

$$m = z_{1-\frac{\alpha}{2}} \cdot \sqrt{\frac{pq}{n}}$$

where $q = 1 - p$ and z is the value of a Standard Normal variable with cumulative probabililty $1 - \dfrac{\alpha}{2}$. Because the value of m depends on p, we must make an educated guess of its value before calculating the sample size requirement. Let p^* represent this educated guess. Rearranging the formula $m = z_{1-\frac{\alpha}{2}} \cdot \sqrt{\frac{p^* q^*}{n}}$ to solve for n derives the following sample size requirement:

$$ n = \frac{z_{1-\frac{\alpha}{2}}^2 p^* q^*}{m^2} $$

where $q^* = 1 - p^*$. Round results up to the next integer to make sure the margin of error is no greater than m.

ILLUSTRATIVE EXAMPLE

Sample-size requirement, confidence interval. The smoking prevalence example in this chapter ($n = 57$, $x = 17$) estimated a population proportion of 0.30 with margin of error (assuming 95% confidence) equal to ± 0.12. How large a sample would be needed to derive an estimate with a margin of error of ± 0.05?

Solution: $n = \dfrac{z_{1-\frac{\alpha}{2}}^2 p^* q^*}{m^2} = \dfrac{1.96^2 \cdot 0.30 \cdot 0.70}{0.05^2} = 322.7$. Use a sample of 323 observations.

How large a sample would be needed to shrink this margin of error to ± 0.03?

Solution: $n = \dfrac{1.96^2 \cdot 0.30 \cdot 0.70}{0.03^2} = 896.4$. Use a sample of 897 observations.

When no educated guess of p is available, use $p^* = 0.50$ to provide a "worst-case scenario" estimate. This will ensure that enough data are collected to achieve no less than the required margin of error. Figure 16.11 shows the relation between the sample size requirements to achieve a margin of error of 0.05 at various assumed values of p^*, demonstrating that sample size requirements are maximum for $p = 0.5$.

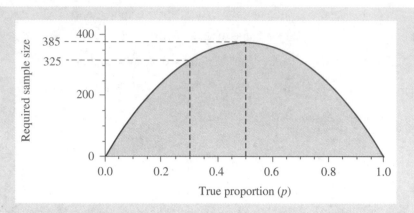

FIGURE 16.11   Sample size requirements for a study to estimate a proportion with margin of error 0.05. The required sample size has its maximum when $p = 0.5$.

Sample Size Requirement for Testing a Proportion

In testing a proportion from a single sample, let p_0 represent the population proportion under the null hypothesis and p_1 represent the population proportion under the alternative hypothesis. To test $H_0: p = p_0$ versus $H_a: p = p_1$ with $1 - \beta$ power at a two-sided α-level of significance, use a sample of size:

$$n = \left(\frac{z_{1-\frac{\alpha}{2}}\sqrt{p_0 q_0} + z_{1-\beta}\sqrt{p_1 q_1}}{p_1 - p_0} \right)^2$$

Round results derived from this formula up to the next integer to ensure the required power.

ILLUSTRATIVE EXAMPLE

Sample-size requirement, hypothesis test. A prior illustrative example in this chapter had us conduct a two-sided test of $H_0: p = 0.21$ using a sample of $n = 57$. How large a sample is needed to detect a proportion of 0.31 in the population (i.e., $p_1 = 0.31$) at $\alpha = 0.05$ (two sided) with 90% power?

Solution: $n = \left(\dfrac{z_{1-\frac{\alpha}{2}}\sqrt{p_0 q_0} + z_{1-\beta}\sqrt{p_1 q_1}}{p_1 - p_0} \right)^2$

$$\left(\frac{1.96\sqrt{(0.21)(0.79)} + 1.28\sqrt{(0.31)(0.69)}}{0.31 - 0.21} \right)^2 = 193.3.$$

The sample should include 194 observations.

Power

The sample-size formula is rearranged to determine the power of a test of H_0: $p = p_0$ as follows:

$$1 - \beta = \phi\left(\frac{|p_1 - p_0|\sqrt{n} - z_{1-\frac{\alpha}{2}}\sqrt{p_0 q_0}}{\sqrt{p_1 q_1}}\right)$$

where p_0 is the value of the population proportion under the null hypothesis, p_1 is the value of the population proportion under the alternative hypothesis, n is the sample size, α is the significance level of the test, and $\Phi(z)$ represents the cumulative probability of a Standard Normal random variable.[o] For a one-sided alternative hypothesis, replace $z_{1-(\alpha/2)}$ with $z_{1-\alpha}$ in this formula.

ILLUSTRATIVE EXAMPLE

Power. A test of H_0: $p = 0.21$ using a sample of $n = 57$ was not significant ($P = 0.103$). Suppose the true prevalence in the population p_1 was 0.31. What was the power of the test assuming a two-sided α-level of 0.05?

Solution: $1 - \beta = \phi\left(\dfrac{|p_1 - p_0|\sqrt{n} - z_{1-\frac{\alpha}{2}}\sqrt{p_0 q_0}}{\sqrt{p_1 q_1}}\right) =$

$\phi\left(\dfrac{|0.31 - 0.21|\sqrt{57} - 1.96\sqrt{(0.21)(0.79)}}{\sqrt{(0.31)(0.69)}}\right) = \Phi(-0.09) = 0.4641$

(from Table B). The power of the test under the stated conditions was about 46%.

Summary Points (Inference About a Proportion)

1. This chapter addresses the analysis of a **binary categorical response**.

2. **Descriptive statistics** are limited to counts and proportions.

3. The proportion of individuals at risk who develop an outcome over a specified period of time is the **incidence proportion** or, more informally, "risk." The proportion of individuals with a particular condition at a given point in time is the **prevalence proportion**.

---

[o] Use Table B or a statistical utility to look up this cumulative probability.

4. The **parameter** of interest is population proportion p. When data are based on an SRS of size n, population proportion p is synonymous with binomial parameter p (Chapter 6).

5. Sample proportion $\hat{p}$ is the **point estimator** of parameter p.

6. A **$(1 - \alpha)100\%$ confidence interval for p** $= \tilde{p} + z_{1-\frac{\alpha}{2}} \cdot \text{SE}_{\tilde{p}}$, where, $\tilde{p} = \frac{\tilde{x}}{\tilde{n}}$, $\tilde{x} = x + 2$, $\tilde{n} = n + 4$, $\text{SE}_{\tilde{p}} = \sqrt{\frac{\tilde{p}\tilde{q}}{\tilde{n}}}$, $\tilde{q} = 1 - \tilde{p}$. This method is referred to as the "plus-four method" and is reliable in samples as small as $n = 10$. Smaller samples require exact binomial procedures.

7. A **test of $H_0\colon p = p_0$** (where the value of p_0 is derived from the research question) is conducted with $z_{\text{stat}} = \dfrac{\hat{p} - p_0}{\sqrt{p_0 q_0 / n}}$ when the sample is large. Small samples require exact binomial procedures.

8. **Sample size requirements and power**

   (a) To derive a confidence interval with a margin of error m, the sample size requirement is $n = \dfrac{z_{1-\frac{\alpha}{2}}^2 p^* q^*}{m^2}$, where p^* represents an educated guess for p.

   (b) To conduct a two-sided test of $H_0\colon p = p_0$ with $1 - \beta$ power at a given α-level, the sample size requirement is.

   $$n = \left(\frac{z_{1-\frac{\alpha}{2}}\sqrt{p_0 q_0} + z_{1-\beta}\sqrt{p_1 q_1}}{p_1 - p_0} \right)^2.$$

   (c) The power when testing $H_0\colon p = p_0$ is

   $$1 - \beta = \phi\left(\frac{|p_1 - p_0| \sqrt{n} - z_{1-\frac{\alpha}{2}}\sqrt{p_0 q_0}}{\sqrt{p_1 q_1}} \right).$$

Vocabulary

Cumulative incidence

Exact test

Fisher's exact test

Incidence proportions (cumulative incidence or average risk)

Mid-P test

Normal approximation to the binomial

Plus-four method

Prevalence proportions

Proportion (the sample proportion is denoted $\hat{p}$; the population proportion is denoted p)

Risk

Sampling distribution of a proportion

Review Questions

16.1 What symbol denotes the sample proportion? $\hat{p}$

16.2 What symbol denotes the population proportion? It also denotes the binomial parameter for the probability of success. p

16.3 What is the *complement* of p? $1-p$

16.4 Fill in the blank: While quantitative data are described with sums and averages, categorical data are described with counts and ___ % Proportions

16.5 What term refers to the proportion of individuals at risk of developing a disease who develop the condition over a specified period of time? incidence proportion

16.6 What term refers to the proportion of individuals who have a condition at a particular point in time?

16.7 Fill in the blank: A proportion is an _____ of ones and zeros, where "one" represents presence or the condition and 0 represents its absence.

16.8 Fill in the blanks: The random number of successes in n independent Bernoulli trials has a binomial distribution with parameters n and p. When the sample is large, the random number of successes can also be described by a _____ distribution with $\mu =$ ___ and $\sigma =$ ___.

16.9 What does the symbol p_0 represent in the expression H_0: $p = p_0$?

16.10 Where does the value of p_0 come from when stating H_0?

16.11 Exact inferential procedures for counts and proportions are based on what probability mass function?

16.12 Select the best response: The plus-four method of calculating a confidence interval for p is based on _____ score method.

    (a) Student's

    (b) Fisher's

    (c) Wilson's

16.13 The plus-four confidence interval for p adds ___ (a number) imaginary successes to the *numerator* of the proportion estimate and ___ (a number) imaginary successes to the *denominator* to derive $\hat{p}$.

16.14 Fill in the blanks: For 95% confidence intervals, the margin of error m is equal to approximately _____ the confidence interval width.

16.15 In determining the sample size requirements for estimating p, we specify p^* as an educated guess for population proportion p. When no such educated guess is available, we let $p^* = 0.5$. What is the justification for doing this?

16.16 What factors determine the sample size requirements for estimating population proportion p?

16.17 What is the distinction between selection bias and information bias?

16.18 True or false? Confidence intervals compensate for selection bias.

16.19 True or false? Confidence intervals compensate for random sampling error.

16.20 True or false? *P*-values address information bias.

16.21 True or false? *P*-values address random sampling error.

Exercises

16.12 ***Drove when drinking alcohol.*** The Youth Risk Behavior Surveillance survey for 2005 estimated that, within a 30-day period, 10% of the adolescent population had driven or ridden in a car or other vehicle when the driver had been drinking alcohol.[p]

    (a) The overall response rate was about 67%. How might this bias the results of the survey?

    (b) A separate validation study documented fair to good repeatability of responses when the questionnaire was administered on separate occasions.[q] This does not guarantee validity (responses can be repeatedly inaccurate) but does provide some reassurance. Provide examples of things we need to consider when thinking about the accuracy of responses.

16.13 ***Cerebral tumors and cell phone use.*** In a case-controlled study on cerebral tumors and cell phone use, tumors occurred more frequently on the same side of the head where cellular telephones had been used in 26 of 41 cases.[r] Test the hypothesis that there is an equal distribution of contralateral and ipsilateral tumors in the population. Use a two-sided alternative. Show all hypothesis-testing steps.

16.14 ***BRCA1 mutations in familial breast cancer cases.*** Of 169 women having breast cancer and a familial risk factor, 27 had an inherited *BRCA1* mutation.[s] Based on this information, estimate the prevalence of *BRCA1* mutation in

---

[p] Eaton, D. K., Kann, L., Kinchen, S., Ross, J., Hawkins, J., Harris, W. A., et al. (2006). Youth risk behavior surveillance—United States, 2005. *MMWR Surveillance Summaries, 55*(5), 1–108. The actual survey used a multistage sampling method. Reported results here have been simplified through "reverse engineering" to accommodate a design effect of about 2.

[q] Brener, N. D., Kann, L., McManus, T., Kinchen, S. A., Sundberg, E. C., & Ross, J. G. (2002). Reliability of the 1999 youth risk behavior survey questionnaire. *Journal of Adolescent Health, 31*(4), 336–342.

[r] Muscat, J. E., Malkin, M. G., Thompson, S., Shore, R. E., Stellman, S. D., McRee, D., et al. (2000). Handheld cellular telephone use and risk of brain cancer. *JAMA, 284*(23), 3001–3007.

[s] Couch, F. J., DeShano, M. L., Blackwood, M. A., Calzone, K., Stopfer, J., Campeau, L., et al. (1997). BRCA1 mutations in women attending clinics that evaluate the risk of breast cancer. *New England Journal of Medicine, 336*(20), 1409–1415.

women with familial breast cancer. Include a 95% confidence interval for the prevalence.

16.15 *Insulation workers.* Twenty-six cancer deaths were observed in a cohort of 556 insulation workers. Based on national statistics, a cohort of this size and age distribution was expected to experience 14.4 incident cases during the observation period. Therefore, under the null hypothesis, $p = p_0 = 14.4/556 = 0.02590$. Test whether the observed incidence is significantly greater than expected. Show all hypothesis-testing steps.

16.16 *AIDS-related risk factor.* In a study of AIDS-related risk factors, 5 of 2673 heterosexual respondents reported a history of receiving a blood transfusion or having a sexual partner from a high-risk group.[t] Assume this is an SRS of U.S. heterosexuals. Provide a 95% confidence interval for the prevalence of this combined risk factor.

16.17 *Kidney cancer survival.* An oncologist treats 40 kidney cancer cases. Sixteen of the cases survive at least 5 years. Historically, one in five cases were expected to survive this long. Test whether there has been a significant improvement in survival.

16.18 *Leukemia gender preference.* A simple random sample of 262 leukemia cases consisted of 150 males and 112 females. Does this provide evidence of a gender preference for the disease? (Observed proportion, male $\hat{p} = 150/(150 + 112) = 0.5725$. Test H_0: $p = 0.5$.)

16.19 *Sample-size requirement.* You are planning a study that wants to estimate a population proportion with 95% confidence. How many individuals do you need to study to achieve a margin of error of no greater than 6%? A reasonable estimate for the population proportions is not available before the study, so assume $p* = 0.50$ to ensure adequate precision.

16.20 *Sample-size requirement.* As in the prior exercise, you are planning a study that intends to estimate a population proportion with 95% confidence. How many individuals do you need if you intend to cut your margin of error in half (i.e., $m = 0.03$)? Why is the sample size requirement for this exercise four times that of Exercise 16.19?

16.21 *Alternative medicine.* According to an April 29, 1998, *New York Times* article, a nationwide telephone survey conducted for a managed alternative care company found that of 1500 adults interviewed, 660 said they would use alternative medicine if traditional medical care failed to produce the desired results.[u] Calculate a 95% confidence interval for the population proportion. (Assume data represent an SRS of a defined population.)

---

[t] Catania, J. A., Coates, T. J., Stall, R., Turner, H., Peterson, J., Hearst, N., et al. (1992). Prevalence of AIDS-related risk factors and condom use in the United States. *Science, 258*(5085), 1101–1106.
[u] Brody, J. E. (1998, April 28). Alternative medicine makes inroads, but watch out for curves. *New York Times.*

16.22   *Perinatal growth failure.* Failure to grow normally during the first year of life (perinatal growth failure) was observed in 33 of 249 very-low birth-weight babies.[v] Calculate a 90% confidence interval for p based on this information. Assume data were derived by an SRS.

16.23   *Perinatal growth failure.* Among the 33 perinatal growth failure cases discussed in the prior exercise, eight (24.2%) had very-low intelligence test scores when tested at 8 years of age. In normal birth-weight babies, we'd expect 2.5% of the population to exhibit this trait. Perform a one-sided exact binomial test to address whether the observation is statistically significant. Show all hypothesis-testing steps.

16.24   *Binge drinking in U.S. colleges.* Alcohol abuse is a serious problem on college campuses. A nationwide survey of students at 4-year colleges found that 3314 of the 17,096 student respondents met the criterion for being a "frequent binge drinker" (five or more drinks in a row three or more times in the past 2-week period).[w] Assume data represent an SRS of 4-year colleges.

(a) Calculate the observed prevalence of frequent binge drinking.

(b) Calculate a 95% confidence interval for p.

(c) Data were self-reported. In addition, the response rate was 69%. How might these facts influence the estimates?

16.25   *Incidence of improvement.* Of 75 patients in a clinical study, 20 showed spontaneous improvement within a month. Calculate the 1-month incidence proportion of improvement. Include a 95% confidence interval for the proportion. Assume that the data represent an SRS of a defined clinical population.

16.26   *Sample size requirements.* How large a sample is needed to estimate the incidence of female breast cancer in a population with 95% confidence and a margin of error that is no greater than 1% (0.01)? Assume that the expected incidence proportion in the population is 3% (0.03). How large a sample would be needed if you were willing to settle for 90% confidence? What was the effect of decreasing the required level of confidence?

16.27   *Familial history of breast cancer, sample size requirements.* Suppose we want to test the hypothesis that women with a family history of breast cancer are at a higher risk of developing breast cancer than women who do not have

[v] Hack, M., Breslau, N., Weissman, B., Aram, D., Klein, N., & Borawski, E. (1991). Effect of very-low birth-weight and subnormal head size on cognitive abilities at school age. *New England Journal of Medicine, 325*(4), 231–237.

[w] Wechsler, H., Davenport, A., Dowdall, G., Moeykens, B., & Castillo, S. (1994). Health and behavioral consequences of binge drinking in college. A national survey of students at 140 campuses. *JAMA, 272*(21), 1672–1677.

this family history. Let us assume that 3% (0.03) of women who do not have a family history of breast cancer will develop the disease in their 60s. We take a simple random sample of women entering their 60s who have a family history of breast cancer.

(a) How large a sample would be needed to detect a different breast cancer risk in our study population if women with a family history of the disease actually had a risk of 5% (0.05) compared to the 3% expected in other women? Let $\alpha = 0.05$ (two-sided) while seeking a statistical power of 90% (0.90).

(b) How large a sample would be needed if the true incidence among women with a familial history is actually 6%? Again, let $\alpha = 0.05$ (two-sided) and seek 90% power.

(c) Based on your findings in part (a) and (b) of this exercise, characterize the relationship between the expected differences under the null and alterative hypotheses and the sample size requirements of the study.

(d) Replicate the conditions expressed in part (a) of this exercise, except for insisting on an α-level of 0.01 (two-sided) for your test. How many individuals must now be studied?

(e) Characterize the relationship between the required α-level and the sample size requirement.

16.28 *Power.* Determine the power of a test of H_0: $p = 0.10$ when:

(a) p is actually 0.20, $\alpha = 0.05$ (two-sided), and $n = 25$.

(b) p is actually 0.15, $\alpha = 0.01$ (two-sided), and $n = 100$.

16.29 *Freshman binge drinking.* A survey of drinking in college students found 1802 binge drinkers among 5266 U.S. freshman completing a survey.[x]

(a) The prevalence of binge drinking in this sample is $\hat{p} = \dfrac{1802}{5206} = 0.3422$ or 34.2%. Assume that the data represent an SRS of U.S. freshman. Also assume that the responders provided accurate information. Using this information, determine the prevalence of binge drinking in U.S. freshman with 95% confidence.

---

[x] Stahlbrandt, H., Andersson, C., Johnsson, K. O., Tollison, S. J., Berglund, M., & Larimer, M. E. (2008). Cross-cultural patterns in college student drinking and its consequences—A comparison between the USA and Sweden. *Alcohol and Alcoholism, 43*(6), 698–705.

(b) Calculate the 99% confidence interval for the prevalence. Interpret the result.

(c) A study from the late 1990s found that 20% of freshman engaged in binge drinking. Assume that 20% is an accurate and reliable estimate for the prevalence of binge drinking in the 1990s. Do the current data provide reliable evidence of an increase in the prevalence of binge drinking since the 1990s?

17 | Comparing Two Proportions

■ 17.1 Data

In this chapter, we compare binary responses from two groups. With this type of data, it is common to use the term *exposure* to refer to the explanatory variable and *outcome* to refer to the response variable. The exposure can represent any experimental intervention or nonexperimental factor that distinguishes the groups. We tally the number of "successes" in each group and convert the counts to proportions (as we did in Chapter 16). Table 17.1 shows the notation we will use:

TABLE 17.1 Notation.

Group	Number of successes	Sample size	Sample proportion (Statistic)	Population proportion (Parameter)
1 (exposed)	a_1	n_1	$\hat{p}_1$	p_1
2 (nonexposed)	a_2	n_2	$\hat{p}_2$	p_2

The observed proportion in group 1 is $\hat{p}_1 = \dfrac{a_1}{n_1}$. The observed proportion in group 2 is $\hat{p}_2 = \dfrac{a_2}{n_2}$. These are the estimators of parameters p_1 and p_2, respectively.

ILLUSTRATIVE EXAMPLE

Data (Estrogen trial). One component of the NIH-sponsored Women's Health Initiative study was a trial in which postmenopausal women were randomly assigned either estrogen or an identical-looking placebo. The estrogen-exposed group comprised $n_1 = 8506$ subjects. The placebo group had $n_2 = 8102$. Incidents of a *combined index outcome* (consisting of coronary disease, stroke, pulmonary embolism, breast cancer, endometrial cancer, colorectal cancer, hip

continues

---

continued

fracture, and death due to other causes) were tallied in the groups. After a mean 5.2 years of follow-up, the estrogen-exposed group had 751 incidents and the control group had 623 incidents.[a] Therefore, cumulative incidence proportions were $\hat{p}_1 = \dfrac{a_1}{n_1} = \dfrac{751}{8506} = 0.08829$ and $\hat{p}_2 = \dfrac{a_2}{n_2} = \dfrac{623}{8102} = 0.07689.$

Notes

1. **Prevalence and cumulative incidence.** Recall that **prevalence** is the proportion of individuals with the attribute at a particular time, and **incidence** is the proportion of individuals at risk who develop a condition over a specified period of time. The term **risk** is used synonymously with incidence proportion (Section 16.1).

2. **How many significant digits should I carry?** Students often inquire about the number of digits to carry when working with numbers. The answer to this question depends on the accuracy of the data and its intended use. As a general rule, maintain *at least* four significant digits when performing calculations. Then, before reporting results, round the proportion to three significant digits. For example, 0.08829 is reported with three significant digits as 0.0883, 8.83%, or 88.3 per 1000. (Leading zeros do not count as significant digits.)

■ 17.2 Proportion Difference (Risk Difference)

Point Estimate

One way to compare proportions is in the form of a difference. Comparisons may represent **prevalence differences** or **incidence (risk) differences**, depending on the nature of the underlying proportions. For the sake of economy of language, we will focus on risk differences, keeping in mind that these techniques also apply to prevalence differences.

The risk difference parameter is $p_1 - p_2$. Its estimator $\hat{p}_1 - \hat{p}_2$ measures the *excess risk* associated with an exposure in absolute terms.

[a] Writing Group for the Women's Health Initiative Investigators. (2002). Risks and benefits of estrogen plus progestin in healthy postmenopausal women: Principal results from the Women's Health Initiative randomized controlled trial. *JAMA, 288*(3), 321–333.

ILLUSTRATIVE EXAMPLE

Risk difference. What is the risk difference for the combined outcome in the estrogen trial alluded to earlier? We established 5.2-year risks of $\hat{p}_1 = 0.08829$ in the treatment group and $\hat{p}_2 = 0.07689$ in the control group.

   Solution: $\hat{p}_1 - \hat{p}_2 = 0.08829 - 0.07689 = 0.0114$. This is equivalent to 11.4 additional cases per 1000 exposed individuals over the 5.2-year period.

Sampling Distribution of the $\hat{p}_1 - \hat{p}_2$

To get a better sense of how well $\hat{p}_1 - \hat{p}_2$ reflects $p_1 - p_2$, we need to know something about its sampling variability. We start with these facts:

- In large samples, the sampling distribution of $\hat{p}_1 - \hat{p}_2$ is Normal.

- The sampling distribution of $\hat{p}_1 - \hat{p}_2$ has an expected value (μ) of $p_1 - p_2$. Therefore, the sample risk difference is an unbiased estimator of the population risk difference.

- The standard deviation of the sampling distribution (denoted $\sigma_{\hat{p}_1 - \hat{p}_2}$ is equal to $\sqrt{\frac{p_1 q_1}{n_1} + \frac{p_2 q_2}{n_2}}$, where $q_i = 1 - p_i$.

- Using the notation established earlier, the probability model that predicts the sampling variability of a risk difference in large samples is

$$\hat{p}_1 - \hat{p}_2 \sim N\left(p_1 - p_2, \sqrt{\frac{p_1 q_1}{n_1} + \frac{p_2 q_2}{n_2}}\right).$$

Figure 17.1 depicts these facts.

ILLUSTRATIVE EXAMPLE

Sampling distribution of a risk difference (Hypothetical). One of the difficult parts about statistical inference is using our imagination to understand the theoretical sampling distributions of an estimator. To exercise our imagination, let us consider the sampling distribution of a hypothetical situation in which two groups, each composed of 1000 individuals, come from populations with risks of 10% for a particular outcome. In studying 1000 individuals per group, the sampling distribution of the risk difference is approximately Normal with expected value 0 and standard deviation

continues

continued

$$\sigma_{\hat{p}_1 - \hat{p}_2} = \sqrt{\frac{p_1 q_1}{n_1} + \frac{p_2 q_2}{n_2}} = \sqrt{\frac{(0.1)(1 - 0.1)}{1000} + \frac{(0.1)(1 - 0.1)}{1000}} = 0.0134$$

Using the notation for Normal random variables established in prior chapters, $\hat{p}_1 - \hat{p}_2 \sim N(\mu = 0, \sigma = 0.0134)$.

Application of the 68–95–99.7 rule allows us to say that 68% of the sample risk difference based in these conditions will fall in the range 0 ± 0.0134, 95% will fall in the range $0 \pm (2)(0.0134) = 0 \pm 0.0268$ (and so on). In addition, we can use this sampling distribution to determine the probability of a given range of observations. To do this, standardize the observed risk difference and determine the probability of the range under the assumed sampling distribution. For example, the probability of observing a risk difference of 0.02 or greater under the stated conditions is $Pr(\hat{p}_1 - \hat{p}_2 \geq 0.02) = Pr\left(z \geq \dfrac{0.02 - 0}{0.0134}\right) = Pr(z \geq 1.49) = 0.0681$.

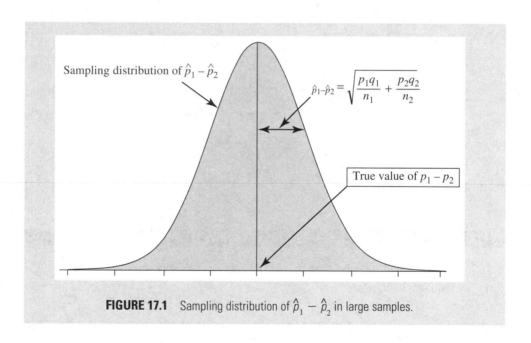

FIGURE 17.1   Sampling distribution of $\hat{p}_1 - \hat{p}_2$ in large samples.

Notice that the formula $\sqrt{\frac{p_1 q_1}{n_1} + \frac{p_2 q_2}{n_2}}$ uses parameters p_1 and p_2. Because these values are seldom known in practice, we replace them with $\hat{p}_1$ and $\hat{p}_2$ to derive the **estimated standard error of the risk difference:**

$$SE_{\hat{p}_1 - \hat{p}_2} = \sqrt{\frac{\hat{p}_1 \hat{q}_1}{n_1} + \frac{\hat{p}_2 \hat{q}_2}{n_2}}$$

This standard error then can be used to quantify the precision of $\hat{p}_1 - \hat{p}_2$ as an estimator of $p_1 - p_2$ for use in inferential procedures.

Exercises

17.1 *Prevalence of cigarette use in two ethnic groups.* Among U.S. adults from 1999 to 2000, American Indians/Alaska Natives had the greatest prevalence of smoking ($p_1 = 0.40$); Chinese-Americans had the lowest prevalence ($p_2 = 0.12$).[b] Suppose we take an SRS of 1000 individuals from each of these groups ($n_1 = n_2 = 1000$).

(a) Describe the sampling distribution of $\hat{p}_1 - \hat{p}_2$.

(b) What is the probability a sample will show a prevalence difference of 0.26 or less?

(c) How likely is it for a sample to show no difference in prevalences?

17.2 *Hypothetical situation.* Consider a fictitious study based in two populations. Twenty-five percent of the individuals in both populations have a particular risk factor ($p_1 = p_2 = 0.25$). We randomly selected 3750 subjects from each population ($n_1 = n_2 = 3750$).

(a) These are large samples, so the sampling distribution of $\hat{p}_1 - \hat{p}_2$ will be approximately Normal. What are the mean and standard deviation of the sampling distribution?

(b) What percent of repeated independent SRSs of the given sizes will have prevalence differences that fall in the range -0.02 to 0.02?

Confidence Interval for $p_1 - p_2$

A plus-four method (similar to the one presented in Section 16.5) is our primary method for determining confidence intervals for proportion differences. This method, which is related to a Wilson score confidence interval, can be used in samples with as few as five observations per group.[c]

When comparing two groups, the plus-four method adds one success and one failure to each group. Let:

$\tilde{a}_i \equiv$ the observed number of successes in group i plus 1 (i.e., $\tilde{a}_i = a_i + 1$)

$\tilde{n}_i \equiv$ the sample size of group i plus 2 ($\tilde{n}_i = n_i + 2$)

$\tilde{p}_i = \dfrac{\tilde{a}_i}{\tilde{n}_i}$ for group i [i: 1,2]

[b] CDC. (2004). Prevalence of cigarette use among 14 racial/ethnic populations—United States, 1999–2001. *MMWR, 53*(3), 49–52.
[c] Agresti, A., & Caffo, B. (2000). Simple and effective confidence intervals for proportions and differences of proportions result from adding two successes and two failures. *The American Statistician, 54*(4), 280–288.

The standard error of the proportion difference is:

$$SE_{\widetilde{p}_1 - \widetilde{p}_2} = \sqrt{\frac{\widetilde{p}_1 \widetilde{q}_1}{\widetilde{n}_1} + \frac{\widetilde{p}_2 \widetilde{q}_2}{\widetilde{n}_2}}$$

The typical "point estimate $\pm z \cdot SE$" formula is applied using $\widetilde{p}_1 - \widetilde{p}_2$ as the point estimate. The $(1 - \alpha)100\%$ confidence interval for $p_1 - p_2$ is:

$$(\widetilde{p}_1 - \widetilde{p}_2) \pm z_{1-\frac{\alpha}{2}} \cdot SE_{\widetilde{p}_1 - \widetilde{p}_2}$$

Use $z_{1-\alpha/2} = 1.645$ for 90% confidence, $z_{1-\alpha/2} = 1.96$ for 95% confidence, and $z_{1-\alpha/2} = 2.576$ for 99% confidence.

ILLUSTRATIVE EXAMPLE

Confidence interval for risk difference. We propose to calculate a 95% confidence interval for the risk difference in the estrogen trial presented earlier. We've established that $a_1 = 751$ and $n_1 = 8506$ in the treatment group. In the control group, $a_2 = 623$ and $n_2 = 8102$.

Solution:

- In the treatment group, $\tilde{a}_1 = a_1 + 1 = 751 + 1 = 752$, $\tilde{n}_1 = n_1 + 2 = 8506 + 2 = 8508$, and $\widetilde{p}_1 = \dfrac{752}{8508} = 0.08839$.

- In the control group, $\tilde{a}_2 = a_2 + 1 = 623 + 1 = 624$, $\tilde{n}_2 = n_2 + 2 = 8102 + 2 = 8104$, and $\widetilde{p}_2 = \dfrac{624}{8104} = 0.07700$.

- $SE_{\widetilde{p}_1 - \widetilde{p}_2} = \sqrt{\frac{\widetilde{p}_1 \widetilde{q}_1}{\widetilde{n}_1} + \frac{\widetilde{p}_2 \widetilde{q}_2}{\widetilde{n}_2}} =$

  $\sqrt{\frac{(0.08839)(0.91161)}{8508} + \frac{(0.07700)(0.92300)}{8104}} = 0.004271$

- The 95% confidence interval for $p_1 - p_2 = \widetilde{p}_1 - \widetilde{p}_2 \pm z_{1-\frac{\alpha}{2}} \cdot SE_{\widetilde{p}_1 - \widetilde{p}_2} =$ $(0.08839 - 0.07700) \pm (1.96)(0.004271) = 0.01139 \pm 0.00837 = (0.00302$ to $0.01976)$, or approximately from 0.3% to 2.0%.

- We conclude with 95% confidence that estrogen increases the risk of the combined index outcome (coronary heart disease, invasive breast cancer, stroke, pulmonary embolism, endometrial cancer, colorectal cancer, hip fracture, and death due to other causes) in the population by between 0.3% and 2.0%.

WinPepi's Compare2.exe program (option B)[d] calculates confidence intervals for proportion difference via three different methods. Our plus-four method corresponds most closely with the noncontinuity-corrected Wilson score method.

Exercises

17.3 *Cytomegalovirus and coronary restenosis.* Each year, cardiologists surgically repair thousands of clogged coronary arteries only to have many of these arteries narrow again (restenose) soon after surgery. A study sponsored by the NIH was conducted to help determine whether infection with a common type of virus, cytomegalovirus (CMV), contributed to coronary restenosis. Forty-nine of the subjects showed serological evidence of CMV infection, while 26 showed no such evidence. In the CMV+ group, 21 individuals restenosed within 6 months of atherectomy. In the CMV− group, 2 individuals restenosed.[e]

(a) Calculate the risk difference of restenosis associated with seropositivity.

(b) Calculate a 95% confidence interval for $p_1 - p_2$. Explain the meaning of this confidence interval in terms a layperson can understand.

(c) *Optional*: Use *WinPepi* Compare2.exe to calculate 95% confidence intervals for $p_1 - p_2$. How do *WinPepi*'s calculations compare with the confidence interval calculated by hand in part (b)?

■ 17.3 Hypothesis Test

An observed difference in sample proportions can reflect either a true difference in the population, random sampling error, or bias. We will ignore the possibility of bias for now and focus instead on how to test against random sampling error. Two methods are considered: a *z*-test (for use in large samples) and an exact binomial procedure.

z-Test

A *z*-test can be used when we expect at least five successes and five failures in each of the groups.[f]

---

[d] Abramson, J. H. (2004). *WINPEPI* (PEPI-for-Windows): Computer programs for epidemiologists. *Epidemiologic Perspectives & Innovations, 1*(1), 6.

[e] Zhou, Y. F., Leon, M. B., Waclawiw, M. A., Popma, J. J., Yu, Z. X., Finkel, T., et al. (1996). Association between prior cytomegalovirus infection and the risk of restenosis after coronary atherectomy. *New England Journal of Medicine, 335*(9), 624–630.

[f] This is a simplification of the rule that applies to expected frequencies, which is covered in Section 18.3.

A. **Hypotheses.** The null hypothesis is $H_0: p_1 - p_2 = 0$ or, equivalently, $H_0: p_1 = p_2$. The two-sided alternative hypothesis is $H_a: p_1 - p_2 \neq 0$ (equivalently, $H_a: p_1 \neq p_2$). One-sided alternatives ($H_a: p_1 > p_2$ or $H_a: p_1 < p_2$) are used occasionally.

B. **Test statistic.** The test statistic is

$$z_{stat} = \frac{\hat{p}_1 - \hat{p}_2}{\sqrt{\bar{p}\bar{q}\left(\frac{1}{n_1} + \frac{1}{n_2}\right)}}$$

where $\bar{p} = \dfrac{\text{number of success, both samples combined}}{\text{total observations, both samples combined}}$ and $\bar{q} = 1 - \bar{p}$.

C. **P-value.** The P-value is the probability of observing a z_{stat} that is equal to or more extreme than the observed z_{stat} under the null hypothesis. The two-tailed P-value is twice the area under the curve beyond the $|z_{stat}|$ on a Standard Normal distribution. Use Appendix Table F or a software utility to convert the test statistic to a P-value.

D. **Significance level.** The P-value is compared to type I error rates (α) to gauge the strength of evidence against the null hypothesis.

E. **Conclusion.** The test results are addressed in the context of the data and research question.

Stay diligent to the fact that statistical hypothesis testing and confidence intervals address only random sampling error and do not control for nonrandom sources of error in the data.

ILLUSTRATIVE EXAMPLE

z-test of independent proportions. We apply a two-sided z-test to the estrogen trial data presented earlier. Recall that $\hat{p}_1 = \dfrac{751}{8506} = 0.08829$ and $\hat{p}_2 = \dfrac{623}{8102} = 0.07689$.

A. **Hypotheses.** $H_0: p_1 = p_2$ versus $H_a: p_1 \neq p_2$
B. **Test statistic.**

    (a) $\bar{p} = \dfrac{(751 + 623)}{(8506 + 8102)} = 0.08273$

    (b) $\bar{q} = 1 - \bar{p} = 1 - 0.08273 = 0.91727$

    (c) $z_{stat} = \dfrac{\hat{p}_1 - \hat{p}_2}{\sqrt{\bar{p}\bar{q}\left(\frac{1}{n_1} + \frac{1}{n_2}\right)}} =$

$$\frac{0.08829 - 0.07689}{\sqrt{(0.08273)(0.91727)\left(\frac{1}{8506} + \frac{1}{8102}\right)}} = 2.67$$

C. P-value. $P = 0.00759$ (from Table F). Figure 17.2 shows the sampling distribution of the test statistic for this problem. To better understand the basis of the P-value, take note of distributional landmarks on the horizontal axis. The small P-value regions indicate that the observed difference in proportions is far enough from zero to suggest that the observed difference is highly significant.

D. Significance level. The evidence against the null hypothesis is significant at the $\alpha = 0.01$ level.

E. Conclusion. The evidence that estrogen increases their risk of the combined index outcome in the source population is statistically significant ($P = 0.0076$).

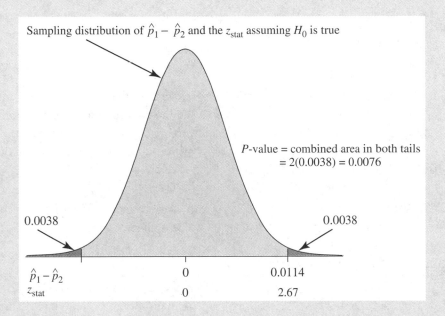

FIGURE 17.2 *P*-value, illustrative example.

Notes

1. **Dissecting the z-statistic.** The numerator of the z_{stat} is the observed risk difference. Its denominator is the standard error of the difference when $p_1 = p_2$.

2. **Continuity-corrected test statistic.** A continuity correction can be applied to the z_{stat} to accommodate the discrete nature of binomial counts. The continuity-corrected z-statistic is:

$$z_{stat,c} = \frac{|\hat{p}_1 - \hat{p}_2| - \left(\frac{1}{2n_1} + \frac{1}{2n_2}\right)}{\sqrt{\bar{p}\bar{q}\left(\frac{1}{n_1} + \frac{1}{n_2}\right)}}$$

For this illustrative example, $z_{stat,c} = \dfrac{|\hat{p}_1 - \hat{p}_2| - \left(\frac{1}{2n_1} + \frac{1}{2n_2}\right)}{\sqrt{\bar{p}\bar{q}\left(\frac{1}{n_1} + \frac{1}{n_2}\right)}} = $

$$\frac{|0.08829 - 0.07689| - \left(\frac{1}{2 \cdot 8506} + \frac{1}{2 \cdot 8102}\right)}{\sqrt{(0.08273)(0.91727)\left(\frac{1}{8506} + \frac{1}{8102}\right)}} = 2.64,$$

corresponding to $P = 0.00829$ (two-sided).

3. **Is the continuity-corrected z-statistic necessary?** Statisticians differ in their opinion in response to this question. Some see the continuity correction as an improvement, while others see it as counterproductive. In large samples, the regular and continuity-corrected z-statistic will produce nearly identical results. In small samples, the continuity-corrected z-statistic will produce a slightly larger P-value. Either statistic is acceptable in practice.

4. **Chi-square test.** The test of H_0: $p_1 = p_2$ can also be performed with a chi-square statistic, as we will discuss in the next chapter.

Exact Tests for Comparing Proportions

The z-test of H_0: $p_1 = p_2$ is based on the Normal approximation to the binomial distribution (see Section 8.3). This Normal approximation is accurate in large samples but is not a good fit in small samples. To determine whether a sample is large enough for this Normal approximation, we first examine the data in the form of a **two-by-two cross tabulation**. Table 17.2 shows the notation we will use in addressing two-by-two tables. In this notation, a_1 represents the number of cases in group 1, b_1 represents the number of noncases in group 1, n_1 represents the number of individuals in group 1, and so on.

TABLE 17.2 Notation for two-by-two cross-tabulations of frequencies.

	Successes	Failures	Total
Group 1	a_1	b_1	n_1
Group 2	a_2	b_2	n_2
Total	m_1	m_2	N

$$\hat{p}_1 = \frac{a_1}{n_1} \qquad \hat{p}_2 = \frac{a_2}{n_2}$$

To demonstrate a two-by-two cross tabulation, let us consider this new illustrative data set.

ILLUSTRATIVE EXAMPLE

Two-by-two Cross Tabulation (*Kayexelate® and colonic necrosis*). Sodium polystyrene (brand name Kayexelate®g) in sorbitol is a drug used to reduce potassium levels in patients with hyperkalemia, commonly in renal failure patients. This drug was suspected of causing an adverse reaction leading to gangrene of the intestine (colonic necrosis). To address this issue, a study compared the incidence of colonic necrosis in 117 Kayexelate-exposed postoperative patients and 862 nonexposed postoperative patients. Two colonic necrosis cases occurred in the Kayexelate-exposed group, while no cases occurred in the nonexposed group. Table 17.3 shows the data in a two-by-two format.

TABLE 17.3 Kayexelate and colonic necrosis, illustrative example of two-by-two cross tabulation.

	Successes	Failures	Total
Exposed (group 1)	2	115	117
Nonexposed (group 2)	0	862	862
Total	2	977	979

Data from Gerstman, B. B., Kirkman, R., & Platt, R. (1992). Intestinal necrosis associated with postoperative orally administered sodium polystyrene sulfonate in sorbitol. *American Journal of Kidney Diseases, 20*(2), 159–161.

Using our established notation, the incidence of colonic necrosis in the exposed group $\hat{p}_1 = \frac{a_1}{n_1} = \frac{2}{117} = 0.017$. The incidence in the nonexposed group $\hat{p}_2 = \frac{a_2}{n_2} = \frac{0}{862} = 0.000$.

---

g Winthrop Pharmaceutical, New York, New York.

To address whether a Normal-approximation (z) test is acceptable for testing proportions, we must first look at the **expected values in each of the table cells**. When an expected value falls below 5 in any of the table cells, the sample is too small to use a Normal-approximation–based test.

Expected cell values are calculated as the product of the row and column marginal totals divided by the table total:

$$E_i = \frac{\text{row total} \times \text{column total}}{\text{table total}}$$

where E_i is the expected value in table cell i. For example, the expected value in table cell a_1 is $E_{a_1} = \dfrac{\text{row 1 total} \times \text{column 1 total}}{\text{table total}}$. Expected values for the colonic necrosis and Kayexelate® illustrative data are shown in Table 17.4.

TABLE 17.4   Kayexelate and colonic necrosis illustration, expected counts if the null hypothesis were true.

	Successes	Failures	Total
Exposed Group 1	$\dfrac{2 \times 117}{979} = 0.24$	$\dfrac{977 \times 117}{979} = 116.76$	117
Nonexposed Group 2	$\dfrac{2 \times 862}{979} = 1.76$	$\dfrac{977 \times 862}{979} = 860.24$	862
Total	2	977	979

Notice that table cells a_1 and a_2 have expected values that are less than 5. Therefore, the Normal-approximation–based z-test of proportions is not appropriate in this situation. Under such circumstances, we use a test based on the calculation of exact binomial probabilities ("exact tests").

Exact tests for two-by-two cross tabulations are based on calculating exact probabilities for all possible outcomes that are equal to or are more extreme than the observed set of counts. Such calculations are tedious. We therefore rely on computer programs or applets for such calculations. Let us demonstrate the application of an exact test of the "colonic necrosis" illustrative data using the free *WinPEPI* application.

ILLUSTRATIVE EXAMPLE

Exact tests (*Kayexelate® and colonic necrosis*). Recall the prior illustrative example on colonic necrosis in patients exposed to Kayexelate® in sorbitol (Table 17.3). The steps involved in testing the proportions with **Fisher's exact test** follow.

A. Hypotheses. H_0: $p_1 = p_2$ against H_a: $p_1 \neq p_2$. Notice that the null hypothesis states no difference in risks in the underlying populations.

B. Test statistic. The test statistic is the observed counts in the two-by-two table (Table 17.4).

C. *P*-value. We use *WinPepi* Compare2.exe to calculate exact test results. Figure 17.3 shows part of the output. Three exact procedures are calculated (Fisher's, Mid-P, Overall's). The two-sided *P*-values by Fisher's method is 0.014.

D. Significance level. The evidence against the null hypothesis is significant at the $\alpha = 0.025$ level but not at the $\alpha = 0.01$ level.

E. Conclusion. Sodium polystyrene increases the postoperative incidence of colonic necrosis ($P = 0.014$).

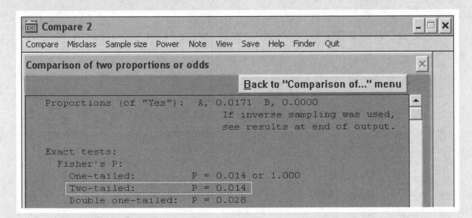

FIGURE 17.3   Screenshot of *WinPepi*'s results for the kayexelate and colonic necrosis illustrative example. Abramson J.H. (2011). WINPEPI updated: computer programs for epidemiologists, and their teaching potential. *Epidemiologic Perspectives & Innovations, 8*(1), 1.

Notes

1. **Which exact procedure is best?** Statisticians differ in their preference for exact tests. Fisher's method[h] produces the largest *P*-values, while Overall's[i] produces the smallest. The mid-P method[j] provides a middle ground and has a firm theoretical basis.[k]

[h] Fisher, R. A. (1934). *Statistical Methods for Research Workers* (5th ed.). Edinburgh, Scotland: Oliver and Boyd.

[i] Overall, J. E. (1990). Comment. *Statistics in Medicine, 9*, 379–382.

[j] Miettinen, O. (1974). Comment. *Journal of the American Statistical Association, 69*, 380–382.

[k] Berry, G., & Armitage, P. (1995). Mid-P confidence intervals: A brief review. *The Statistician, 44*(4), 417–423.

2. **One tailed or two?** The output in Figure 17.3 lists one-tailed, two-tailed, and double one-tailed P-values. The one-tailed and two-tailed P-values are identical for this particular problem because there is no tail showing fewer than 0 cases in the nonexposed group. The doubled one-tailed P-value is acceptable,[1] but is used only rarely in practice.

3. **What exactly does "exact" mean?** Do not confuse "exact test" with "exactly true test." Exact in this sense means that probabilities were calculated with precise binomial formulas. However, the exact statistics, like Normal approximations, do *not* account for nonrandom sources of error. Exact tests are not foolproof.

Calculation of Fisher's and the Mid-P Exact P-Values (Optional)

Exact tests do their job by calculating probabilities for tables more extreme than the observed counts conditional on the table's marginal frequencies. Using the notation established in Table 17.2, the probability of a given cross tabulation conditioned on the observed margins of the table is:

$$\frac{n_1!n_2!m_1!m_2!}{N!a_1!b_1!a_2!b_2!}$$

Here are the exact probabilities for the Kayexelate and colonic necrosis illustrative example (Table 17.5):

TABLE 17.5   Calculation of exact probabilities for the kayexelate illustrative example.

Table 0

0	117	117
2	860	862
2	977	979

The probability of Table 0, $P_0 = \dfrac{117! \cdot 862! \cdot 2! \cdot 977!}{979! \cdot 0! \cdot 117! \cdot 2! \cdot 860!} = \dfrac{862 \cdot 861}{979 \cdot 978} = 0.775156$

Table 1

1	116	117
1	861	862
2	977	979

The probability of Table 1, $P_1 = \dfrac{117! \cdot 862! \cdot 2! \cdot 977!}{979! \cdot 1! \cdot 116! \cdot 1! \cdot 861!} = \dfrac{117 \cdot 2 \cdot 862}{979 \cdot 978} = 0.210669$

---

[1]Dupont, W. D. (1986). Sensitivity of Fisher's exact test to minor perturbations in 2 × 2 contingency tables. *Stat Med,* 5(6), 629–635.

Table 2

2	115	117
0	862	862
2	977	979

The probability of Table 2, $P_2 = \dfrac{117! \cdot 862! \cdot 2! \cdot 977!}{979! \cdot 2! \cdot 115! \cdot 0! \cdot 862!} = \dfrac{117 \cdot 116}{979 \cdot 978} = 0.014175$

With the marginal totals fixed, these tables represent all possible outcomes for the study. Notice that these probabilities sum to 1: $P_0 + P_1 + P_2 = 0.775156 + 0.210669 + 0.014175 = 1.00000$.

The P-value for Fisher's exact test is the sum of probabilities for the observed table and tables more extreme than the observed table. Because there are no outcomes more extreme than the last cross tabulation in Table 17.5, this conditional probability represents both the one-sided and two-sided P-value for the problem by Fisher's method. The mid-P method takes half this probability to derive $P_{\text{mid-P}} = \frac{1}{2}(0.014175) = 0.0070875$.

Exercises

17.4 *Smoking cessation trial.* A randomized controlled, double-blind trial was conducted to see whether sustained-release bupropion (a pharmaceutical normally used to treat depression) provided benefit over use of the nicotine patch alone in helping people stop smoking. The control group (nicotine patch alone) included 244 smokers who wanted to stop smoking. The treatment group consisting of 245 individuals received sustained-release bupropion in combination with a nicotine patch. After 1 year, 40 individuals in the control group remained smoke-free. In contrast, 87 in the treatment group had done the same.[m]

    (a) Calculate the incidence proportions of cessation in the groups.

    (b) Test the difference in proportions for significance. (Show all hypothesis-testing steps.)

    (c) Concisely summarize the results of the study.

17.5 *Joseph Lister and antiseptic surgery.* Joseph Lister (1827–1912) demonstrated that postoperative mortality dropped from 16 fatalities in 35 procedures to 6 fatalities in 40 procedures after adopting antiseptic surgical techniques. Determine the risks of postoperative mortality in each group. Then, test the proportions for inequality.

---

[m] Jorenby, D. E., Leischow, S. J., Nides, M. A., Rennard, S. I., Johnston, J. A., Hughes, A. R., et al. (1999). A controlled trial of sustained-release bupropion, a nicotine patch, or both for smoking cessation. *New England Journal of Medicine, 340*(9), 685–691.

17.6 *Médecine d'observation.* Pierre-Charles Alexandre Louis (1787–1872) is often referred to as the "father of clinical statistics." In 1837, he wrote "I conceive that without the aid of statistics nothing like real medical science is possible."[n] In his most famous study, Louis evaluated bloodletting as a treatment for pneumonia. At the time, bloodletting was an extremely popular form of therapy. Two forms were practiced: by lancet (cutting a vein) and by placement of leeches on specific body parts.[o] Louis called into question the effectiveness of these therapies by carefully monitoring and recording outcomes in treatment groups. In one analysis, he compared patients who received early bloodletting (treatment group) to those who received late treatment (control group). He found that 18 of 41 patients in the early treatment group died. In contrast, 9 of 36 control patients died.[p]

(a) Calculate the risk difference in the groups.

(b) Test the difference for significance. Report a two-tailed P-value for the test. Interpret the results.

17.7 *Induction of labor and meconium staining.* Meconium staining during childbirth may be a sign of fetal distress. A randomized trial was conducted to see whether induction of labor (by pitocin and other drugs delivered near term) would reduce the risk of meconium staining. The treatment group ($n_1 = 111$) had elective inductions at 39 to 40 weeks of pregnancy. One delivery in the treatment group experienced meconium staining. In contrast, 13 of 117 pregnancies in the more conservatively managed group experienced meconium staining.[q]

(a) Determine the risks of meconium staining in each group.

(b) Even though the counts in table are small, can a Normal approximation (z-procedure) still be used to test the data? Show why.

(c) Conduct the hypothesis test using a z-procedure.

(d) *Optional.* Use a computer applet (e.g., www.OpenEpi.com) to calculate a P-value by Fisher's exact test. Compare this P-value to the one derived from your z-procedure. How do they compare?

17.8 *Induction of labor and meconium staining, risk difference estimate.* Calculate a 95% confidence for the risk difference (induced minus noninduced) for the data in Exercise 17.7.

---

[n] Louis, P. C. A. (1837). Louis on the application of statistics to medicine. *American Journal of Medical Science, 21,* 525–528.

[o] In 1833, France imported more than 42 million leeches for this purpose.

[p] Louis, P. C. A. (1836). *Researches on the effects of bloodletting in some inflammatory diseases* (C. G. Putnam, Trans.). Boston: Hilliard, Gray, and Co.

[q] Osborn, J. F. (1979). *Statistical Exercises in Medical Research.* New York: John Wiley & Sons. Page 37 cites "M. Sc. Social Medicine, September 1975" as the data's source.

■ 17.4 Proportion Ratio (Relative Risk)

The ratio of two proportions is a commonly encountered epidemiologic statistic. The ratio of two prevalences is a **prevalence ratio**. The ratio of two incidence proportions is a **risk ratio**. In loose terms, and for economy of language, we call both **relative risks**.[r] The relative risk estimator, denoted $\widehat{RR}$ ("RR hat") is $\dfrac{\hat{p}_1}{\hat{p}_2}$. The relative risk parameter is denoted RR.

ILLUSTRATIVE EXAMPLE

Relative risk (Estrogen trial). The relative risk of the combined index outcome in the estrogen trial illustrative example is $\widehat{RR} = \dfrac{\hat{p}_1}{\hat{p}_2} = \dfrac{0.08829}{0.07689} = 1.1483 \cong 1.15.$

1. **Risk multiplier.** We rearrange the formula $\widehat{RR} = \dfrac{\hat{p}_1}{\hat{p}_2}$ to show that $\hat{p}_1 = \widehat{RR} \times \hat{p}_2$. This reveals that the relative risk is a risk multiplier. Use it to multiply the risk in the nonexposed group ($\hat{p}_2$) to derive the risk in the exposed group ($\hat{p}_1$). The illustrative example's relative risk of 1.15 indicates the risk in the exposed group is 1.15 times that of the nonexposed group.
2. **Risk relative to baseline.** The baseline relative risk is 1, not 0; a relative risk of 1 implies no increase in risk. A relative risk of 1.15 implies that the risk in the exposed group is 15% *above* baseline. By the same token, a relative risk of 0.85 would imply that the risk in the exposed group is 15% below the baseline established by the nonexposed group. You must subtract 1 from the value of $\widehat{RR}$ to determine risk above or below the baseline.
3. **Direction of association.** A risk ratio of 1 indicates that risks are equal in the exposed group and nonexposed group (*no association* between the explanatory variable and response variable). Risk ratios greater than 1 indicate a *positive association*. Risk ratios between 0 and 1 indicate a *negative association*.
4. **Switching group designations.** Switching the designation of "exposure" turns the relative risk into its reciprocal. For example, had we designated lack of estrogen supplementation as the exposure in the illustrative example, the relative risk would have been $1.15^{-1} = 0.87$. This does *not* materially affect the interpretation of the statistic.

---

[r]The prevalence ratio will equal the risk ratio when (1) risks are small (say, less than 5%), (2) the inflow and outflow of cases and noncases in the population is in steady state, (3) the mean duration of disease is the same in the exposed and nonexposed groups, and (4) the outcome does not influence the status of the exposure.

5. **Relative risk versus risk difference.** Both the relative risk $\left(\widehat{RR} = \dfrac{\hat{p}_1}{\hat{p}_2}\right)$ and the risk difference $(\widehat{RD} = \hat{p}_1 - \hat{p}_2)$ can be used to qualify the effect of an exposure. The $\widehat{RR}$ quantifies the effect of exposure in relative terms. The $\widehat{RD}$ quantifies the effect of the exposure in absolute terms. Take, for example, the estrogen trial illustrative example in which $\hat{p}_1 = \dfrac{751}{8506} = 0.0883$ (about 8.8%) and $\hat{p}_2 = \dfrac{623}{8102} = 0.0769$ (about 7.7%). The relative risk for these data is $\widehat{RR} = \dfrac{0.0883}{0.0769} = 1.15$; the risk in the exposed group is 1.15 times that of the nonexposed group. The risk difference for these data is $\widehat{RD} = 0.0883 - 0.0769 = 0.0114$ showing a 1.1% increase in risk in absolute terms. The relationship between the relative risk and risk difference can be expressed as $\widehat{RR} = 1 + \dfrac{\widehat{RD}}{\hat{p}_2}$, where $\dfrac{\widehat{RD}}{\hat{p}_2}$ is the increase or decrease in risk relative to baseline. For the illustrative data $\widehat{RR} = 1 + \dfrac{0.0114}{0.0769} = 1.15$.

Confidence Interval for the RR

Figure 17.4 is a schematic of the sampling distribution of the natural logarithm of the relative risk in large samples. This sampling distribution is Normal with expectation (mean) lnRR and standard error:

$$SE_{\ln\widehat{RR}} = \sqrt{\frac{1}{a_1} - \frac{1}{n_1} + \frac{1}{a_2} - \frac{1}{n_2}}$$

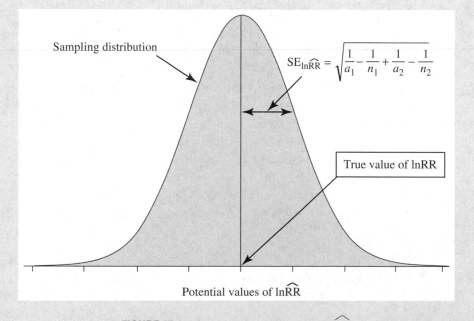

Sampling distribution

$$SE_{\ln\widehat{RR}} = \sqrt{\frac{1}{a_1} - \frac{1}{n_1} + \frac{1}{a_2} - \frac{1}{n_2}}$$

True value of lnRR

Potential values of $\ln\widehat{RR}$

FIGURE 17.4   Sampling distribution of the $\ln\widehat{RR}$.

See Table 17.2 for notation.

The 95% confidence interval for the lnRR is:

$$\widehat{\ln RR} \pm z_{1-\frac{\alpha}{2}} \cdot SE_{\widehat{\ln RR}}$$

where $z_{1-\alpha/2} = 1.645$ for 90% confidence, $z_{1-\alpha/2} = 1.96$ for 95% confidence, and $z_{1-\alpha/2} = 2.576$ for 99% confidence. Take the antilog of these limits to get the confidence interval for RR.[s]

ILLUSTRATIVE EXAMPLE

Confidence interval for the relative risk (Estrogen trial). We calculate 90% and 95% confidence intervals for the RR for the estrogen trial illustrative data. Table 17.3 shows the data. We established that $\widehat{RR} = \dfrac{\hat{p}_1}{\hat{p}_2} = \dfrac{0.08829}{0.07689} = 1.1483$.

Solution:

- $\ln(\widehat{RR}) = \ln(1.1483) = 0.1383$

- $SE_{\widehat{\ln RR}} = \sqrt{\dfrac{1}{a_1} - \dfrac{1}{n_1} + \dfrac{1}{a_2} - \dfrac{1}{n_2}} = \sqrt{\dfrac{1}{751} - \dfrac{1}{8506} + \dfrac{1}{623} - \dfrac{1}{8102}} = 0.051920$

- The 90% confidence interval for lnRR $= \widehat{\ln RR} \pm z_{1-\alpha/2} \cdot SE_{\widehat{\ln RR}} = 0.1383 \pm (1.645)(0.05192) = 0.1383 \pm 0.0854 = (0.0529, 0.2237)$. Therefore, the 90% confidence interval of RR $= e^{0.0527, 0.2239} = (1.05$ to $1.25)$.

- The 95% confidence interval for lnRR $= 0.1383 \pm (1.96)(0.05192) = 0.1383 \pm 0.1018 = (0.0365, 0.2401)$. Therefore, the 95% confidence interval of RR $= e^{0.0365, 0.2401} = (1.04$ to $1.27)$.

We can conclude with 95% confidence that the relative risk associated with estrogen use in this population is between 1.04 and 1.27.

Notes

1. **Confidence intervals address random errors only.** By calculating a confidence interval for the RR, we are acknowledging the imprecision of our estimate. Keep in mind that the confidence intervals address random sources of errors only and do not address systematic errors.

2. **Capturing the RR.** The confidence interval seeks to capture the risk ratio parameter (RR), not the risk ratio estimate (RR).

3. **No guarantees.** Even if we assume that the only sources of error we have to confront are random, there is still an $\alpha \cdot 100\%$ chance that the confidence interval for the RR will miss its true value.

---

[s] Equivalently, the 95% confidence interval for the RR $= e^{\widehat{\ln RR} \pm z_{1-\alpha/2} \cdot SE_{\widehat{\ln RR}}}$.

Exercises

17.9 *Framingham Heart Study.* An early publication from the Framingham Heart Study identified 51 incidences of coronary disease among 424 men with high serum cholesterol (245 mg per 100 mL and above) over a 6-year period. Among 454 men with lower cholesterol (less than 210 mg per 100 ml), there were 16 incident cases.[t] Calculate the relative risk of coronary disease associated with high cholesterol. Include a 95% confidence interval for the RR.

■ 17.5 Systematic Sources of Error

Confidence intervals and hypothesis tests are effective in addressing the **random error** associated with sampling and randomization. These methods, however, do nothing to address **systematic error**. Systematic errors come in these three general forms:

1. confounding
2. information bias
3. selection bias

   Confounding is especially problematic in nonexperimental studies. Recall that confounding derives from the mixing together of the effects of the explanatory variable and extraneous variables lurking in the background (Section 2.2). Experimentation mitigates confounding by balancing the distribution of extraneous variables in the groups being compared. Without randomization, nonexperimental studies are prone to the unbalanced effects of **lurking factors**. Chapter 15 introduced a method for mathematically adjusting for the influence of lurking factors when studying linear relationships. Chapter 19 will introduce a method used for a similar purpose when studying categorical outcomes. Even so, mathematical methods such as these are limited,[u] as this example will illustrate:

ILLUSTRATIVE EXAMPLE

Confounding. The large estrogen trial (experimental study) discussed earlier in this chapter was preceded in time by many nonexperimental studies on the same topic. Whereas the experimental study found that postmenopausal estrogen

---

[t] Kannel, W. B., Dawber, T. R., Kagan, A., Revotskie, N., & Stokes III, J. (1961). Factors of risk in the development of coronary heart disease—Six-year follow-up experience. *Annals of Internal Medicine, 55,* 33–50.
[u] Petitti, D. B., & Freedman, D. A. (2005). Invited commentary: How far can epidemiologists get with statistical adjustment? *American Journal of Epidemiology, 162*(5), 415–418; discussion 419–420.

supplementation increased the risk of several serious adverse health outcomes, the earlier nonexperimental studies found just the opposite[v]. The nonexperimental studies had even adjusted for potentially confounding variables using advanced mathematical methods. How do we explain these discrepant results? The most plausible explanation is that the nonexperimental studies did not sufficiently control for confounding variables.[w]

Information bias arises from defects in measurement. With categorical data, this corresponds to *misclassification* of the explanatory variable or response variable (or both). Such misclassifications may be either differential or nondifferential. **Nondifferential misclassification** occurs to the same extent in the groups being compared. **Differential misclassification** affects some groups more than others. Table 17.6 illustrates examples of nondifferential and differential misclassification.

TABLE 17.6 Examples of nondifferential and differential misclassification, and their effect on relative risk estimates.

Scenario A: Nondifferential misclassification: 10% of the cases in the exposed and nonexposed groups are misclassified as noncases.

	Accurate data				Misclassified data		
	+	−	Total		+	−	Total
Exposed	100	900	1000	→	90	910	1000
Nonexposed	50	450	500		45	455	500

$$\widehat{RR} = \frac{100/1000}{50/500} = 1.00 \qquad \widehat{RR} = \frac{90/1000}{45/500} = 1.00$$

Scenario B: Differential misclassification: none (0%) of the cases are misclassified in the exposed group, while 10% are misclassified in the nonexposed group.

	Accurate data				Misclassified data		
	+	−	Total		+	−	Total
Exposed	100	900	1000	→	100	900	1000
Nonexposed	50	450	500		45	455	500

$$\widehat{RR} = \frac{100/1000}{50/500} = 1.00 \qquad \widehat{RR} = \frac{100/1000}{45/500} = 1.11$$

[v] Grady, D., Rubin, S. M., Petitti, D. B., Fox, C. S., Black, D., Ettinger, B., et al. (1992). Hormone therapy to prevent disease and prolong life in postmenopausal women. *Annals of Internal Medicine, 117*(12), 1016–1037.
[w] See Volume 61, issue 4 of *Biometrics,* 2005.

Nondifferential misclassification biases result either toward the null[x] or not at all,[y] as shown in **scenario A** in the table. Differential misclassification, on the other hand, can lead to false positive or false negative results, as demonstrated in **scenario B** in Table 17.6.

Selection bias is due to the manner in which subjects are selected for study. There are many types of selection biases. Here is an example:

ILLUSTRATIVE EXAMPLE

After-the-fact identification of a cluster. In the spring of 1989, physicians and pharmacists at the University of Wisconsin Hospital clinics became aware of an increase in the frequency an adverse reaction (cerebellar toxicity) to a drug used in bone marrow transplant patients. This association became apparent after the hospital switched from the original formulation of the drug to a generic equivalent. The U.S. Food and Drug Administration was contacted, and a team of investigators was sent to look into the problem. A subsequent record review confirmed the increased frequency of this serious adverse reaction. Of the 25 patients treated with the generic form of the drug, 11 (44%) experienced cerebellar toxicity. In comparison, 3 (8.8%) of 34 individuals treated with the brand-name product experienced toxicity.[z] Thus, the relative risk associated with the generic drug was $\frac{44.0\%}{8.8\%} = 5.0$ ($P = 0.002$). Although this particular hospital experienced much greater toxicity with the generic formulation, there is difficulty in interpreting these results. This hospital was *selected* for study just because of its high rate of occurrence. No other institution reported a similar problem. Somewhere, there might be a hospital where the generic formulation proved safer. Thus, there is no way of knowing whether the identified problem represents a random **cluster** of events or a true increase in occurrence. Identifying random clusters after the fact has been likened to shooting the broad side of a barn and drawing the bull's-eye around where the arrow has already landed.

---

[x] Bross, I. D. J. (1954). Misclassification in 2×2 tables. *Biometrics, 10,* 478–486.

[y] Poole, C. (1985). Exception to the rule about nondifferential misclassification. (Abstract). *American Journal of Epidemiology, 122,* 508.

[z] Jolson, H. M., Bosco, L., Bufton, M. G., Gerstman, B. B., Rinsler, S. S., Williams, E., et al. (1992). Clustering of adverse drug events: Analysis of risk factors for cerebellar toxicity with high-dose cytarabine. *Journal of the National Cancer Institute, 84,* 500–505.

■ 17.6 Power and Sample Size

Sample Size for Testing H_0: $p_1 = p_2$

Equal Sample Sizes

Suppose we want to test H_0: $p_1 = p_2$ at a specified α-level with power equal to $1 - \beta$. In using equal group sample sizes ($n = n_1 = n_2$), use this many observations per group:

$$n = \left(\frac{z_{1-\frac{\alpha}{2}}\sqrt{2\overline{p}\,\overline{q}} + z_{1-\beta}\sqrt{p_1 q_1 + p_2 q_2}}{|p_1 - p_2|} \right)^2$$

where p_1 and p_2 are educated guesses for the proportions in the population, $q_i = 1 - p_i$, and $\overline{p} = \dfrac{p_1 + p_2}{2}$. For one-sided alternatives hypotheses, replace $z_{1-(\alpha/2)}$ with $z_{1-\alpha}$. Round up the results to the next integer to ensure the power does not dip below the stated level.

If you want to incorporate a continuity correction into your test, then take this further step:

$$n' = \frac{n}{4}\left(1 + \sqrt{1 + \frac{4}{n\,|p_1 - p_2|}} \right)^2$$

where n comes from the previous formula.[aa]

ILLUSTRATIVE EXAMPLE

Sample-size requirement, equal sample sizes. Determine the sample size required for conducting a two-sided test of H_0: $p_1 = p_2$ with an equal number of subjects in each group, $\alpha = 0.05$ ($z_{1-(0.05/2)} = 1.96$), $1 - \beta = 0.90$ ($z_{0.90} = 1.282$), $p_1 = 0.10$, and $p_2 = 0.05$.

Solution:

$$\overline{p} = \frac{p_1 + p_2}{2} = \frac{0.10 + 0.05}{2} = 0.075; \overline{q} = 1 - \overline{p} = 1 - 0.075 = 0.925$$

continues

[aa] Fleiss, J. L. (1981). *Statistical Methods for Rates and Proportions* (2nd ed.). New York: John Wiley & Sons, pp. 41–42.

continued

$$n = \left(\frac{z_{1-\frac{\alpha}{2}}\sqrt{2\bar{p}\,\bar{q}} + z_{1-\beta}\sqrt{p_1 q_1 + p_2 q_2}}{|p_1 - p_2|} \right)^2$$

$$= \left(\frac{1.96\sqrt{2(0.075)(0.925)} + 1.282\sqrt{(0.10)(0.90) + (0.05)(0.95)}}{|0.10 - 0.05|} \right)^2$$

$$= 581.26$$

Resolve to use 582 per group.

How many individuals should be studied if we want to incorporate a continuity correction into our test statistic?

$$\text{Solution: } n' = \frac{n}{4}\left(1 + \sqrt{1 + \frac{4}{n|p_1 - p_2|}} \right)^2$$

$$= \frac{581.26}{4}\left(1 + \sqrt{1 + \frac{4}{581.26\,|0.10 - 0.05|}} \right)^2$$

$$= 620.6$$

Therefore, resolve to study 621 observations per group.

Because of the complex nature of these calculations, we often use a software utility for calculations. Figure 17.5 is an annotated screenshot of *WinPepi*'s sample size program for comparisons of proportions. Results for required sample sizes correspond to the hand calculations shown in the previous illustration. The line labeled "expected precision" refers to the margin of error for a risk difference with incorporation of continuity correction calculated as follows:

$$m = 1.96 \times \left(\sqrt{\frac{p_1 q_1}{n_1} + \frac{p_2 q_2}{n_2}} + \frac{1}{n_1} + \frac{1}{n_2} \right).\text{[bb]}$$

[bb] Bristol, D. (1989). Sample sizes for constructing confidence intervals and testing hypotheses. *Statistics in Medicine, 8,* 803–811.

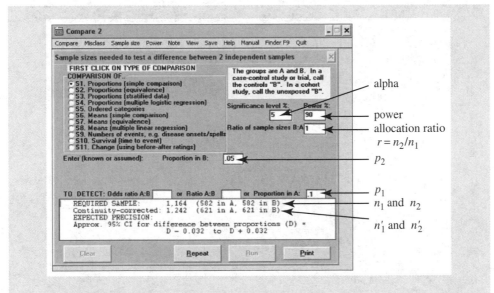

FIGURE 17.5 Annotated output from *WinPepi*. Results correspond to the equal sample sizes illustrative example. Abramson J.H. (2011). WINPEPI updated: computer programs for epidemiologists, and their teaching potential. *Epidemiologic Perspectives & Innovations, 8*(1), 1.

Unequal Group Sample Sizes

Maximum efficiency is gained by using groups of equal size. However, when this is not feasible, group 1 should have:

$$n_1 = \frac{\left(z_{1-\frac{\alpha}{2}}\sqrt{(r+1)\overline{p}'\,\overline{q}'} + z_{1-\beta}\sqrt{rp_1q_1 + p_2q_2}\right)^2}{r(p_1 - p_2)^2}$$

where r is **allocation ratio** $r = \dfrac{n_2}{n_1}$ and $\overline{p}' = \dfrac{p_1 + rp_2}{r+1}$. The size of group 2 is then determined by $n_2 = r \cdot n_1$.[cc]

To incorporate the continuity correction, use

$$n_1' = \frac{n_1}{4}\left(1 + \sqrt{1 + \frac{2(r+1)}{rn_1\,|\,p_1 - p_2\,|}}\right)^2$$

where n_1 comes from the first formula. The size of group 2 is now $n_2' = r \cdot n_1'$.

[cc] Ury, H. K., & Fleiss, J. L. (1980). On approximate sample sizes for comparing two independent proportions with the use of Yates' correction. *Biometrics, 36*(2), 347–351. Fleiss, J. L., Tytun, A., & Ury, H. K. (1980). A simple approximation for calculating sample sizes for comparing independent proportions. *Biometrics, 36*(2), 343–346.

ILLUSTRATIVE EXAMPLE

Sample-size requirement, unequal sample sizes. Determine the sample sizes required for testing H_0: $p_1 = p_2$ with group 2 having twice the number of observations as group 1. Otherwise, assume the same conditions used in the prior illustrative example.

$$\bar{p}' = \frac{p_1 + rp_2}{r + 1} = \frac{0.10 + (2)(0.05)}{2 + 1} = 0.06667$$

$$\bar{q}' = 1 - 0.06667 = 0.93333$$

$$n_1 = \frac{\left(z_{1-\frac{\alpha}{2}}\sqrt{(r + 1)\bar{p}'\bar{q}'} + z_{1-\beta}\sqrt{rp_1q_1 + p_2q_2}\right)^2}{r(p_1 - p_2)^2}$$

$$= \frac{\left(1.96\sqrt{(2 + 1)(0.06667)(0.93333)} + 1.282\sqrt{(2)(0.10)(0.90) + (0.05)(0.95)}\right)^2}{2(0.10 - 0.05)^2}$$

$$= 425.33$$

Therefore, $n_1 = 426$ and $n_2 = r \cdot n_1 = 2 \cdot 426 = 852$.

If you plan on incorporating a continuity correction into your test statistic, then use $n_1' = \frac{n_1}{4}\left(1 + \sqrt{1 + \frac{2(r + 1)}{rn_1 \mid p_1 - p_2 \mid}}\right)^2 =$

$$\frac{425.33}{4}\left(1 + \sqrt{1 + \frac{2(2 + 1)}{2 \cdot 425.33 \cdot \mid 0.10 - 0.05 \mid}}\right)^2 = 454.84 \approx 455.$$

Group 2 will now have $n_2' = r \cdot n_1' = 2 \cdot 455 = 910$ individuals.

Figure 17.6 shows annotated results for these conditions calculated by *WinPepi*.

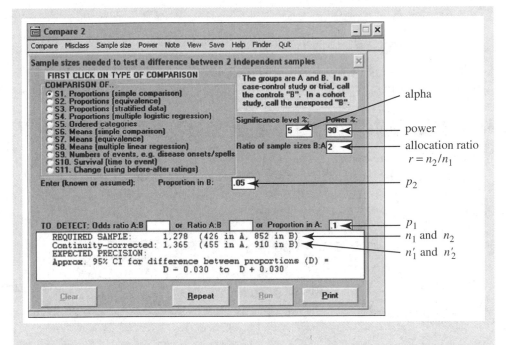

FIGURE 17.6 Annotated output from *WinPepi*'s sample size program for comparing proportions. Results correspond to those of the unequal sample sizes example. Abramson J.H. (2011). WINPEPI updated: computer programs for epidemiologists, and their teaching potential. *Epidemiologic Perspectives & Innovations, 8*(1), 1.

When planning a study, allowances should be made to anticipate for dropouts. To adjust for this, recruit $n_d = \dfrac{n}{(1-R)^2}$ individuals, where n_d is the sample size factoring in **drop-out rate** R and n is the sample size assuming no dropouts.[dd]

Power

The power of the test of proportions is:

$$1 - \beta = \Phi\left(\frac{|p_1 - p_2|}{\sqrt{\frac{p_1 q_1}{n_1} + \frac{p_2 q_2}{n_2}}} - \frac{z_{1-\alpha/2} \sqrt{\overline{p}\,\overline{q}\left(\frac{1}{n_1} + \frac{1}{n_2}\right)}}{\sqrt{\frac{p_1 q_1}{n_1} + \frac{p_2 q_2}{n_2}}} \right)$$

[dd]Lachin, J. M. (1981). Introduction to sample size determination and power analysis for clinical trials. *Controlled Clinical Trials, 2*(2), 93–113, see p. 97.

where p_1 and p_2 are educated guesses for the group proportions, $q_i = 1 - p_i$, $\bar{p} = \dfrac{n_1 p_1 + n_2 p_2}{n_1 + n_2}$, $\bar{q} = 1 - \bar{p}$, and $\Phi(z)$ is the cumulative probability of Standard Normal random variable z.[ee] This formula applies to two-sided tests and does not incorporate a continuity correction.

ILLUSTRATIVE EXAMPLE

Power. We want to test H_0: $p_1 = p_2$ at $\alpha = 0.05$ (two sided) with sample sizes constrained to $n_1 = 250$ and $n_2 = 300$. Educated guesses for population proportions are $p_1 = 0.20$ and $p_2 = 0.15$. What is the power of this test?

Solution:

$q_1 = 1 - p_1 = 1 - 0.20 = 0.80$
$q_2 = 1 - p_2 = 1 - 0.15 = 0.85$

$$\bar{p} = \frac{n_1 p_1 + n_2 p_2}{n_1 + n_2} = \frac{(250)(0.2) + (300)(0.15)}{(250 + 300)} = 0.172727$$

$$\bar{q} = 1 - \bar{p} = 1 - 0.172727 = 0.827273$$

$$z_{1-(\alpha/2)} = z_{1-(0.05/2)} = z_{0.975} = 1.96$$

$$1 - \beta = \Phi\left(\frac{|p_1 - p_2|}{\sqrt{\frac{p_1 q_1}{n_1} + \frac{p_2 q_2}{n_2}}} - \frac{z_{1-\alpha/2} \sqrt{\bar{p}\,\bar{q}\left(\frac{1}{n_1} + \frac{1}{n_2}\right)}}{\sqrt{\frac{p_1 q_1}{n_1} + \frac{p_2 q_2}{n_2}}} \right)$$

$$= \Phi\left(\frac{|0.20 - 0.15|}{\sqrt{[(.2)(.8)/250] + [(.15)(.85)/300]}} \right.$$

$$\left. - \frac{1.96\sqrt{(0.172727)(0.827273)\left(\frac{1}{250} + \frac{1}{300}\right)}}{\sqrt{[(.2)(.8)/250] + [(.15)(.85)/300]}} \right)$$

$$= \Phi(1.532 - 1.944) = \Phi(-0.41) = 0.3409 \cong 34\%$$

Figure 17.7 contains a screenshot of results for this illustration from *WinPepi* Compare2.exe. Our hand calculations correspond to the output labeled "Power of chi-square test—not continuity corrected." Chi-square tests are equivalent to z-tests for this type of problem (Section 18.3).

[ee] Sahai, H., & Khurshid, A. (1996). Formulae and tables for the determination of sample sizes and power in clinical trials for testing differences in proportions for the two-sample design: A review. *Statistics in Medicine, 15*(1), 1–21.

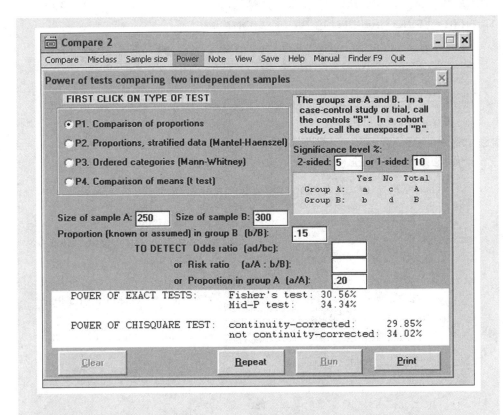

FIGURE 17.7 Output from *WinPepi*'s power program for comparing proportions. Results correspond to those of the Power illustrative example. Abramson J.H. (2011). WINPEPI updated: computer programs for epidemiologists, and their teaching potential. *Epidemiologic Perspectives & Innovations, 8*(1), 1.

Exercises

17.10 *Determinants of sample-size requirements.* List the factors that determine the sample-size requirements for a test of two proportions.

17.11 *Sample-size plan.* It is important to develop a detailed plan of study early in the development of an investigation in order to assure adequacy of the sample size. Consider a test of H_0: $p_1 = p_2$ against H_a: $p_1 \neq p_2$ at $\alpha = 0.05$ with a power that is equal to 0.80. We predict that 20% of group 1 and 30% of group 2 will experience the outcome. If we recruit an equal number of individuals in each of the groups, how many individuals do we need to study?

17.12 *Sample-size plan.* Suppose it is twice as easy to recruit nonexposed subjects in the study considered in Exercise 17.11. How many individuals will we have to study with an allocation ratio of 2 to 1?

Summary Points (Comparing Two Proportions)

1. This chapter addresses the analysis of **binary responses from two groups**. Data are computerized in the form of a binary explanatory variable (the exposure) and a binary response variable (the outcome).

2. **Estimators and parameters**: Observed proportions $\hat{p}_1$ and $\hat{p}_2$ are the point estimators of parameters p_1 and p_2.

3. The relationship between the study exposure and outcome is measured in absolute terms as **risk difference** $RD = p_1 - p_2$.

   (a) The point estimator for the risk difference $\widehat{RD} = \hat{p}_1 - \hat{p}_2$.

   (b) A $(1 - \alpha)100\%$ confidence interval for the RD is calculated with a plus-four method (similar to the procedure introduced in prior chapter for a single sample) or an exact method, as needed.

   (c) A test of $H_0: p_1 - p_2 = 0$ ("no difference in the populations") is conducted with a Normal approximation method (z-statistic) or, in small samples, with an exact procedure.

4. The relationship between the study exposure and outcome can be measured in relative terms as a **risk ratio** $RR = \dfrac{p_1}{p_2}$.

   (a) The point estimator for the risk ratio $RR = \dfrac{\hat{p}_1}{\hat{p}_2}$.

   (b) A $(1 - \alpha)100\%$ confidence interval for $RR = e^{\ln\widehat{RR} \pm z_{1 - \frac{\alpha}{2}} \cdot \sqrt{a_1^{-1} - n_1^{-1} + a_2^{-1} - n_2^{-1}}}$

   (c) A test of $H_0: RR = 1$ ("no difference in the populations") is performed with a Normal approximation method (z-statistic) or an exact procedure, as needed.

5. Confidence intervals and P-values address random sampling errors only. They do not address **systematic errors**.

6. The **three major forms of systematic errors** are confounding, information bias, and selection bias.

7. **Power and sample size**:

   (a) The sample size requirements when testing H_0: "no difference" are determined by (1) the required α-level of the test, (2) the required power level, (3) the expected proportion in group 1, (4) the expected proportion in group 2, and (5) the ratio of the two sample sizes n_2/n_1. Maximum efficiency is gained when $n_1 = n_2$. Formulas are provided in the chapter.

(b) The power of a test of H_0: "no difference" is determined by (1) the required α-level of the test, (2) n_1 and n_2, (3) the expected proportion in group 1, and (4) the expected proportion in group 2. Formulas are provided in the chapter.

Vocabulary

Allocation ratio	Prevalence ratio
Cluster	Random error
Confounding	Relative risk
Differential misclassification	Risk
Drop-out rate	Risk difference
Incidence	Risk ratio
Information bias	Selection bias
Lurking variables	Standard error of the risk difference
Nondifferential misclassification	Systematic error
Prevalence	Two-by-two cross tabulation
Prevalence difference	

Review Questions

17.1 This is the symbol the text uses to denote the observed number of successes in group i.

17.2 This symbol is used to denote the sample proportion in group i.

17.3 This symbol denotes the population proportion in group i.

17.4 Select the best response: This statistic is a direct estimate of risk.

(a) prevalence

(b) incidence proportion

(c) risk ratio

17.5 Select the best response: The term _____ is often used to refer to the explanatory variable in studies of risk.

(a) outcome

(b) exposure

(c) dependent variable

17.6 What term refers to the difference between two incidence proportions?

17.7 What term refers to the ratio of two incidence proportions?

17.8 Fill in the blank: In large samples, the sampling distribution of a risk difference is _____ distributed with a mean of $p_1 - p_2$ and standard deviation of $\sqrt{\frac{p_1 q_1}{n_1} + \frac{p_2 q_2}{n_2}}$.

17.9 In large samples, the sampling distribution of the _____ (mathematical function) of the risk ratio is Normally distributed with a mean of $\ln(RR)$ and standard deviation of $\sqrt{\frac{1}{a_1} - \frac{1}{n_1} + \frac{1}{a_2} - \frac{1}{n_2}}$.

17.10 State the null hypothesis for testing the equality of proportions in three different unique but equivalent ways.

17.11 What is wrong with the statement $H_0: \hat{p}_1 = \hat{p}_2$?

17.12 What is wrong with the following statement? A 95% confidence interval for the risk difference has a 95% chance of capturing $\hat{p}_1 - \hat{p}_2$.

17.13 Select the best response: The function of the P-values is to protect us from making too many

   (a) systematic errors

   (b) type I errors

   (c) type II errors

17.14 Which z-statistic for testing the equality of proportions provides the more conservative result, the regular z-statistic or the continuity-corrected z-statistic?

17.15 What does it mean when we say a test statistic is conservative?

17.16 What benchmark is used to determine whether a Normal-approximation (z) test is appropriate for testing two proportions?

17.17 Name two exact methods that can be used for testing proportions when the samples are small.

17.18 What is the value of the RR when the two groups being compared have the same risk of the outcome?

17.19 Which RR indicates the stronger association: 0.25 or 2?

17.20 "ln" represents what mathematical function?

17.21 List three types of systematic errors in comparative research.

17.22 Change one word in this false statement so that it becomes true: Confidence intervals protect against all errors with $(1 - \alpha)100\%$ confidence.

17.23 Nondifferential misclassification tends to bias ratio measures of association in what direction?

Exercises

17.13 *Smoking cessation trial.* Exercise 17.4 considered a randomized trial of smoking cessation in which treatment with bupropion in combination with a nicotine patch was compared to use of a nicotine patch alone. The treatment group had success in 87 of 245 participants. The control group had successes in 40 of 244 participants. Calculate the difference in proportions and a 95% confidence interval for $p_1 - p_2$. Why is this confidence interval more useful than the hypothesis test results calculated in Exercise 17.4?

17.14 *Telephone survey contact rates.* Telephone surveys typically have high rates of nonresponse. This can cause bias when the variables being studied are associated with factors that determine the response rate. For mail and home surveys, it is known that advanced-warning letters letting participants know that a survey is on its way increases overall response. A study investigated the utility of leaving messages on answering machines as a means of encouraging participation in telephone surveys. A message was left or not left at random when an answering machine picked up the first call of a telephone survey. Table 17.7 lists cross-tabulated results for ultimately making contact by phone.

TABLE 17.7 Data for Exercise 17.14. Success in making telephone contact according to whether an advanced message was left.

	Contacted	Not contacted	Households
Advanced warning	200	91	291
No advanced warning	58	42	100

Data from Xu, M., Bates, B., & Schweitzer, J. C. (1993). The impact of messages on survey participation in answering machine households. *Public Opinion Quarterly, 57*(2), 232–237, Tables 1 and 2.

(a) Descriptive statistics: Calculate contact "rates" (proportions) for each of the groups.

(b) Hypothesis test: Test the proportions for a significant difference. Show all hypothesis-testing steps.

(c) Estimation of the effect size: Calculate the difference in proportions and its 95% confidence interval. Does this effect seem large enough to be important?

17.15 *Telephone survey completion rates.* The study described in Exercise 17.14 also kept tabs of whether leaving advanced messages increased the likelihood of ultimately completing the interview. Data for this outcome are tallied in Table 17.8. Perform analyses similar to those requested in the prior exercise. To what extent did the advanced messages increase the response rate?

TABLE 17.8 Data for Exercise 17.15. Response to telephone survey according to whether an advanced message was left.

	Completed interview	Did not complete interview	Households
Advanced warning	134	157	291
No advanced warning	33	67	100

Data from Xu, M., Bates, B., & Schweitzer, J. C. (1993). The impact of messages on survey participation in answering machine households. *Public Opinion Quarterly, 57*(2), 232–237, Tables 1 and 2.

17.16 *Framingham women.* Exercise 17.9 considered 6-year coronary disease risk in Framingham men. Comparable data for Framingham women revealed 30 coronary incidents in 689 women with high serum cholesterol. Among 445 women with low serum cholesterol, there were 8 incidents. Calculate the risk difference and its 95% confidence interval.

17.17 *4S coronary mortality.* The Scandinavian simvastatin survival study (4S) was a randomized clinical trial designed to evaluate the effects of the cholesterol-lowering agent simvastatin in patients with coronary heart disease. Over 5.4 years of follow-up, the treatment group consisting of 2221 individuals experienced 111 fatal heart attacks. The placebo group of 2223 individuals experienced 189 such events.[ff] Calculate the risks in the groups and test the difference for significance. In relative terms, how much did simvastatin lower heart attack mortality?

17.18 *4S overall mortality.* The 4S study introduced in Exercise 17.17 also tallied deaths due to any cause. The treatment group experienced 182 deaths ($n_1 = 2221$) and the control group experienced 256 deaths ($n_2 = 2223$). Compare the mortality experience in the two groups in the form of a relative risk and 95% confidence interval.

17.19 *Acute otitis media.* A double-blind randomized trial compared cefaclor to amoxicillin in the treatment of acute otitis media in children. There was no difference in clinical failure rates (all but four children in each group had a good clinical response), but after 14 days, 59 (55.7%) of 106 cefaclor-treated

[ff]Scandinavian Simvastatin Survival Study Group. (1994). Randomized trial of cholesterol lowering in 4444 patients with coronary heart disease: The Scandinavian Simvastatin Survival Study (4S). *Lancet, 344*(8934), 1383–1389.

subjects had effusion-free ears compared to 40 of the 97 amoxicillin-treated subjects.[gg] Determine whether the effusion-free rates differed significantly in the groups. Show all hypothesis-testing steps.

17.20 *Drug testing student athletes.* The Supreme Court of the United States ruled in 2002 that schools could require random drug testing of students who participate in after-school activities. At that time, it was not known whether random drug testing reduced use of illicit drugs. To address this question, researchers at the Oregon Health and Science University completed a study in which student athletes at Wahtonka High School were subject to random drug testing, while student athletes at Warrenton High School were not subject to random testing. Five of the 95 student athletes at the Wahtonka school were positive for illicit drugs, while 12 of 62 student athletes at Warrenton High School were positive.[hh] Compare the experience of these two schools with methods introduced in this chapter. This question is intentionally left open to give students the opportunity to analyze the data in ways they see fit. More than one correct response is possible. Include both descriptive and inferential statistics in your response.

[gg] Mandel, E. M., Bluestone, C. D., Rockette, H. E., Blatter, M. M., Reisinger, K. S., Wucher, F. P., et al. (1982). Duration of effusion after antibiotic treatment for acute otitis media: Comparison of cefaclor and amoxicillin. *Pediatric infectious disease, 1*(5), 310–316.

[hh] Goldberg, L., Elliot, D. L., MacKinnon, D. P., Moe, E., Kuehl, K. S., Nohre, L., et al. (2003). Drug testing athletes to prevent substance abuse: Background and pilot study results of the SATURN (Student Athlete Testing Using Random Notification) study. *Journal of Adolescent Health, 32*(1), 16–25.

18 | Cross-Tabulated Counts

■ 18.1 Types of Samples

Chapter 17 compared binary responses from two groups. Data were occasionally displayed as two-by-two cross tabulations. The idea of using cross tabulations to analyze categorical outcomes is extended in this chapter to accommodate categorical variables with more than two levels.

We start by considering how data may be generated for cross tabulation. Three different sampling methods are considered. Sampling method I takes a simple random sample from the source population and later cross classifies observations. Sampling method II takes fixed numbers of individuals from groups defined by the explanatory variable. Sampling method III takes fixed numbers of individuals from groups defined by the response variable.[a]

Sampling Method I (Naturalistic): Take a simple random sample of N individuals from a population. Cross classify these observations with respect to the explanatory variable and response variable. This is a **naturalistic sample,** also called *a multinomial* or *cross-sectional sample.*

Comment: The term *cross-sectional* has several meanings in biostatistics. It can mean that the sample is a random cross-section of the population (as it does here), or it can mean that information about the explanatory variable and response variables refer to a cross-section in time (i.e., are nonlongitudinal). To avoid confusion, we will use the term *naturalistic* to refer to sampling method I and *cross-sectional* to refer to nonlongitudinal data.

---

[a] This classification scheme is by no means standard; there is no agreed-upon taxonomy for classifying sample types. See the following references for historical context: (1) Barnard, G. A. (1947). Significance testing for 2 × 2 tables. *Biometrika, 34*(1/2), 123–138. (2) Cochran, W. G. (1952). The χ^2 test of goodness of fit. *Annals of Mathematical Statistics, 23*(3), 325–327. (3) Fleiss, J. L. (1981). *Statistical Methods for Rates and Proportions* (2nd ed.). New York: John Wiley & Sons, pp. 20–24. (4) Miettinen, O. (1976). Estimability and estimation in case-referent studies. *American Journal of Epidemiology, 103*(2), 226–235.

ILLUSTRATIVE EXAMPLE

Naturalistic sample (Cytomegalovirus and coronary restenosis). Investigators prospectively studied 75 consecutive patients undergoing coronary atherectomy for symptomatic coronary artery disease. Before atherectomy was performed, blood levels of anticytomegalovirus (CMV) antibodies were measured. Forty-nine patients were seropositive for CMV, and 26 patients were seronegative. Subjects were then followed for 6 months after atherectomy to determine incidents of restenosis.[b] If we make the simplifying assumption that the sample represents a simple random sample of like patients, we can view this as a naturalistic sample of $N = 75$ in which $n_1 = 49$ are seropositive and $n_2 = 26$ are seronegatives for CMV.

Sampling Method II (Purposive Cohorts): Take a fixed number of individuals who are exposed (n_1) and not exposed (n_2) to an explanatory factor. These are **purposive cohorts samples**. Such cohorts may be constructed experimentally or nonexperimentally. The cohorts are then monitored with data ultimately cross classified according to the explanatory variable and response variable.

ILLUSTRATIVE EXAMPLE

Purposive cohort sample (Estrogen trial). We examined data from an experimental cohort that was part of the NIH-sponsored Women's Health Initiative study. This experimental cohort study involved 16,608 postmenopausal women. Roughly half of these women were assigned conjugated estrogens ($n_1 = 8506$). The remaining subjects ($n_2 = 8102$) received a placebo. Coronary heart disease, invasive breast cancer, and other adverse outcomes were monitored in the groups on an ongoing basis.[c]

Sampling Method III (Case–control): Take a predetermined number (m_1) of individuals who are "successes" with respect to the response variable; these are the *cases*. Select a predetermined number (m_2) of individuals who are "failures" with respect to

---

[b] Zhou, Y. F., Leon, M. B., Waclawiw, M. A., Popma, J. J., Yu, Z. X., Finkel, T., et al. (1996). Association between prior cytomegalovirus infection and the risk of restenosis after coronary atherectomy. *New England Journal of Medicine, 335*(9), 624–630.
[c] Writing Group for the Women's Health Initiative Investigators. (2002). Risks and benefits of estrogen plus progestin in healthy postmenopausal women: Principal results from the Women's Health Initiative randomized controlled trial. *JAMA, 288*(3), 321–333.

response variable; these are the *controls*.[d] Historical information about the explanatory variable is collected from both groups, after which data are cross classified according to the explanatory and response variables. These are **case–control samples** (also called *case–referent samples*).

Case–control sample (Baldness and myocardial infarction). To examine the relationship between male-pattern baldness and myocardial infarction, investigators studied 655 men admitted to a hospital for a first nonfatal myocardial infarction (cases) and 772 admitted to the same hospitals with noncardiac diagnoses (controls). This is a case–control sample because the investigator selected fixed numbers of subjects based on the status of the response variable (myocardial infarction). The extent of baldness was assessed in subjects using several methods.[e]

Case–control studies require distinct methods of analysis and are therefore discussed separately in Section 18.5.

■ 18.2 Describing Naturalistic and Cohort Samples

This section considers the description and exploration of naturalistic and cohort samples. Let us start with a data example.

Data (Education level and smoking). We want to learn about the relationship between education and the prevalence of smoking in a particular community. A sample of 585 adult individuals with at least a high school degree is selected from a certain high socioeconomic status neighborhood. Educational level is classified by highest degree received as follows: (1) high school graduate,

continues

---

[d] Controls in this context are more accurately termed the *reference group*, because they are not controls in the sense of a controlled trial. However, the term *controls* is standard, so we adopt its use.
[e] Lesko, S. M., Rosenberg, L., & Shapiro, S. (1993). A case-control study of baldness in relation to myocardial infarction in men. *JAMA, 269*(8), 998–1003.

continued

(2) associate degree, (3) some college, (4) undergraduate degree, or (5) graduate degree. Individuals are classified as smokers if they had smoked at least 100 cigarettes during their lifetime and currently smoke every day or nearly every day (binary response variable). The explanatory variable is education, with five categorical levels. The data may be viewed as a naturalistic sample of a particular high socioeconomic community.[f] Table 18.1 shows cross-tabulated results.

TABLE 18.1   Education level and smoking illustrative data.

	Current smoker		
Highest educational degree	**1 (Yes)**	**2 (No)**	**Total**
1 (high school graduate)	12	38	50
2 (associate degree)	18	67	85
3 (some college)	27	95	122
4 (undergraduate degree)	32	239	271
5 (graduate degree)	5	52	57
Total	94	491	585

Data are fictitious but realistic. Prevalences are based on CDC. (2005). Cigarette smoking among adults–United States, 2004. *MMWR, 54*(44), 1121–1124. Data are stored online in EDU_SMOKE.*.

Table 18.1 is an example of an ***R-by-C contingency table***. This is a five-by-two cross tabulation, with five rows and two columns. The same data could have been displayed in a two-by-five contingency table with the orientation of the explanatory and response variables reversed. This would not materially affect the results, but it would alter some of our formulas. For now, let us set up cross tabulations with the explanatory variable in the rows of the table and response variable in its columns. Table 18.2 shows how we will label table cells.

---

[f]It is often difficult to decide whether a given sample is naturalistic or purposive cohorts. This distinction depends on the whether the distribution of the explanatory factor can be viewed as a simple random sample of a source population. This, in turn, depends on how we define the source population. In this particular example, data are not an SRS of the United States but are an SRS of the particular community.

TABLE 18.2 Notation for labeling table cells.

(A) R-by-C table

Explanatory variable	Response variable					
	1	**2**	**3**	$\rightarrow$	**C**	**Total**
1	a_1	b_1	c_1	...	...	n_1
2	a_2	b_2	c_2	...	...	n_2
$\downarrow$	$\downarrow$	$\downarrow$	$\downarrow$	...	...	—
R	a_R	b_R	c_R	...	...	n_R
Total	m_1	m_2	m_3	$\rightarrow$	m_C	N

(B) R-by-two table

Explanatory variable	Response variable		Total
	+	**−**	
1	a_1	b_1	n_1
2	a_2	b_2	n_2
$\downarrow$	$\downarrow$	$\downarrow$	$\downarrow$
R	a_R	b_R	n_R
Total	m_1	m_2	N

(C) two-by-two table

Explanatory variable	Response variable		Total
	+	**−**	
+	a_1	b_1	n_1
−	a_2	b_2	n_2
Total	m_1	m_2	N

Marginal Percents

Column and row totals are listed along the margins of the R-by-C table. These are the **marginal distributions.** One may convert these to percents or display them with a bar chart. Figure 18.1 displays row and column marginal distributions for our illustrative data set. The top chart shows that we have slightly less than 100 smokers in the sample. The bottom chart reveals the predominance of college-educated individuals in the sample.

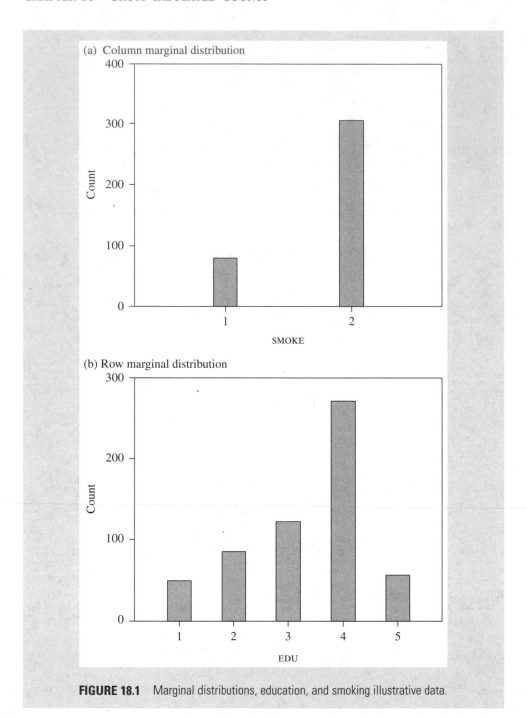

FIGURE 18.1   Marginal distributions, education, and smoking illustrative data.

Beware that marginal distributions will reflect underlying population distributions only in naturalistic samples. They will not reflect population distributions with other types of samples.

Conditional Percents

The relation between the explanatory variable and response variable is assessed by comparing relevant **conditional percents**. There are two types of conditional percents: percents conditioned on row totals (**row percents**) and percents conditioned on column totals (**column percents**). These have formulas:

$$\text{Row percents} = \frac{\text{cell counts}}{\text{row totals}} \times 100\%$$

$$\text{Column percents} = \frac{\text{cell counts}}{\text{column totals}} \times 100\%$$

When data are laid out with explanatory variable categories in rows and response variable categories in columns:

- Naturalistic samples allow for the analysis of both row and column percents.
- Purposive cohort samples allow for the analysis of row percents only.
- Case–control samples allow for the analysis of column percents only.

For naturalistic and cohort data in R-by-two cross tabulations, the row percent in column 1 will reflect the **incidence proportion** or **prevalence proportion** within each groups. These are:

$$\hat{p}_i = \frac{a_i}{n_i}$$

For example, in the illustrative data (Table 18.1), the prevalence of smoking in group 1 is $\hat{p}_1 = \frac{12}{50} = 0.240$, the prevalence in group 2 is $\hat{p}_2 = \frac{18}{85} = 0.212$, and so on. Figure 18.2 contains computer output for the illustrative data. I have highlighted the relevant row percents and placed a bar chart displaying these prevalences following the table. This analysis reveals a negative association between smoking and education, especially at the higher education levels.

Relative risk statistics as discussed in Chapter 17 may also be applied. Traditionally, the "least-exposed" group is denoted "group 1" to serve as the baseline for comparisons. If an explanatory variable cannot be placed in rank order, a baseline group is selected arbitrarily. The relative risk associated with exposure category i can now be calculated:

$$\widehat{RR}_i = \frac{\hat{p}_i}{\hat{p}_1} = \frac{a_i/n_i}{a_1/n_1}$$

EDU * SMOKE Cross tabulation

| | | | SMOKE | | |
			1	2	Total
EDU	1	Count	12	38	50
		% within EDU	24.0%	76.0%	100.0%
	2	Count	18	67	85
		% within EDU	21.2%	78.8%	100.0%
	3	Count	27	95	122
		% within EDU	22.1%	77.9%	100.0%
	4	Count	32	239	271
		% within EDU	11.8%	88.2%	100.0%
	5	Count	5	52	57
		% within EDU	8.8%	91.2%	100.0%
Total		Count	94	491	585
		% within EDU	16.1%	83.9%	100.0%

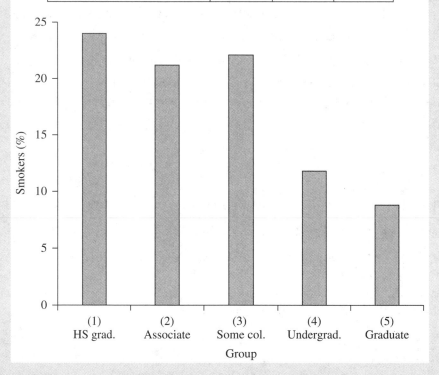

FIGURE 18.2   Prevalence of smoking by education, illustrative data. Graph produced with SPSS for Windows, Rel. 11.0.1.2001. Chicago: SPSS Inc. Reprint Courtesy of International Business Machines Corporation.

ILLUSTRATIVE EXAMPLE

Relative risks, R-by-two table. In the smoking and education illustrative data, let group 1 (high school graduates) serve as the baseline group for comparisons. Table 18.3 shows calculations of relative risks as follows: $\widehat{RR}_1 = 1.00$ (reference group), $\widehat{RR}_2 = 0.88$, $\widehat{RR}_3 = 0.92$, $\widehat{RR}_4 = 0.49$, and $\widehat{RR}_5 = 0.37$.

TABLE 18.3 Prevalences and prevalence ("risk") ratios, education, and smoking illustrative data.

	Smoker			Prevalence	"Relative risk"
Education group	1 (yes)	2 (no)	Total	$(\hat{p}_i)$	$(\widehat{RR}_i)$
1 (high school graduate)	12	38	50	0.2400^a	1.00
2 (associate degree)	18	67	85	0.2118^b	0.88^c
3 (some college)	27	95	122	0.2213	0.92
4 (undergraduate degree)	32	239	271	0.1181	0.49
5 (graduate degree)	5	52	57	0.0877	0.37

Examples of calculations:

^{a}Prevalence in group 1: $\hat{p}_1 = \dfrac{a_1}{n_1} = \dfrac{12}{50} = 0.2400$

^{b}Prevalence in group 2: $\hat{p}_2 = \dfrac{a_2}{n_2} = \dfrac{18}{85} = 0.2118$

^{c}Risk ratio in group 2: $\widehat{RR}_2 = \dfrac{\hat{p}_2}{\hat{p}_1} = \dfrac{0.2118}{0.2400} = 0.88$

Odds Ratios

The **odds ratio** is an alternative measure of effect for categorical data. The **odds** of an event is the proportion of successes divided by the proportion of failures: $\hat{o}_i = \dfrac{\hat{p}_i}{1 - \hat{p}_i}$. Using the notation in Table 18.2B, the **odds in group i** is $\hat{o}_i = \dfrac{a_i}{b_i}$. The relation between the explanatory variable and the response variable now can be quantified with the odds ratios as follows:

$$\widehat{OR}_i = \frac{\hat{o}_i}{\hat{o}_1}$$

When the odds of events are small (say, less than one to nine), odds ratios and relative risk will produce similar results. However, when events are common, the odds ratio will be more extreme (further from 1) than the comparable relative risk.

ILLUSTRATIVE EXAMPLE

Odds ratios, R-by-two table (Education and smoking). Table 18.4 illustrates calculation of the following odds ratios for the education and smoking illustrative data: $\widehat{OR}_1 = 1.00$ (reference), $\widehat{OR}_2 = 0.85$, $\widehat{OR}_3 = 0.90$, $\widehat{OR}_4 = 0.42$, and $\widehat{OR}_5 = 0.30$.

TABLE 18.4   Odds ratios, education, and smoking illustrative data.

Education group	Current smoker 1 (yes)	Current smoker 2 (no)	Odds $(\hat{o}_i)$	Odds ratio $(\widehat{OR}_i)$
1 (high school graduate)	12	38	0.3158[a]	1.00
2 (associate degree)	18	67	0.2687[b]	0.85[c]
3 (some college)	27	95	0.2842	0.90
4 (undergraduate degree)	32	239	0.1339	0.42
5 (graduate degree)	5	52	0.0962	0.30

Examples of calculations:

[a]Odds of smoking in group 1: $\hat{o}_1 = \dfrac{a_1}{b_1} = \dfrac{12}{38} = 0.3158$

[b]Odds of smoking in group 2: $\hat{o}_2 = \dfrac{a_2}{b_2} = \dfrac{18}{67} = 0.2687$

[c]Odds ration in group 2: $\widehat{OR}_2 = \dfrac{\hat{o}_2}{\hat{o}_1} = \dfrac{0.2687}{0.3158} = 0.85$

Instead of using odds to calculate $\widehat{OR}$s, you may find it easier to first rearrange the R-by-two table to form a series of two-by-two tables. Table 18.5 shows how this is done. After data are rearranged in this manner, you can use the fact that $\widehat{OR}_1 = \dfrac{\hat{o}_1}{\hat{o}_2} = \dfrac{a_1/b_1}{a_2/b_2} = \dfrac{a_1 \cdot b_2}{a_2 \cdot b_1}$ to calculate the odds ratios. Now the odds ratio is merely the **cross-product ratio** of table counts—simply multiply the diagonals and turn these products into a ratio.

Responses with More Than Two Categories

Contingency table analysis allows us to consider responses that are classified at more than two levels.

TABLE 18.5 Breaking-up the five-by-two data into separate two-by-two tables with calculation of the cross-product (odds) ratio.

Education group	Current smoker 1 (yes)	Current smoker 2 (no)
1 (high school graduate)	12	38
1 (high school graduate)	12	38

$$\widehat{OR}_1 = \frac{12 \cdot 38}{38 \cdot 12} = 1.00 \text{ (reference)}$$

Education group	Current smoker Yes	Current smoker No
2 (associate degree)	18	67
1 (high school graduate)	12	38

$$\widehat{OR}_2 = \frac{18 \cdot 38}{67 \cdot 12} = 0.85$$

Education group	Current smoker Yes	Current smoker No
3 (some college)	27	95
1 (high school graduate)	12	38

$$\widehat{OR}_3 = \frac{27 \cdot 38}{95 \cdot 12} = 0.90$$

Education group	Current smoker Yes	Current smoker No
4 (undergraduate degree)	32	239
1 (high school graduate)	12	38

$$\widehat{OR}_4 = \frac{32 \cdot 38}{239 \cdot 12} = 0.42$$

Education group	Current smoker Yes	Current smoker No
5 (graduate degree)	5	52
1 (high school graduate)	12	38

$$\widehat{OR}_5 = \frac{5 \cdot 38}{52 \cdot 12} = 0.30$$

ILLUSTRATIVE EXAMPLE

Categorical response with more than two levels (Efficacy of Echinacea). A randomized, double-blind, placebo-controlled trial evaluated the efficacy of the herbal remedy *Echinacea purpurea* in treating upper respiratory tract infections in children. Each time a subject had an upper respiratory tract infection,

continues

continued

treatment with either Echinacea or a placebo was taken for the duration of illness. The severity of the illness was rated by parents of subjects as mild, moderate, or severe. Table 18.6A shows the results. Table 18.6B shows the row percents of responses in the groups. Approximately 46% of the cases were rated "mild" in both groups. A slightly higher percentage of the treatment group experienced severe illness (14.6% vs. 10.9%), suggesting that Echinacea is ineffective in treating upper respiratory tract illness in children.

TABLE 18.6   Efficacy of Echinacea illustrative data. Parental assessments of the severity of illness.

| | **(A) Cross-tabulated counts** | | | |
| | **Parental assessment** | | | |
	Mild	**Moderate**	**Severe**	**Total**
Echinacea	153	128	48	329
Placebo	170	157	40	367
Total	323	285	88	696

| | **(B) Conditional distributions (row percents)** | | | |
| | **Parental assessment** | | | |
	Mild	**Moderate**	**Severe**	**Total**
Echinacea	46.5	38.9	14.6	100.0
Placebo	46.3	42.8	10.9	100.0
Total	46.4	40.9	12.6	100.0

Data from Taylor, J. A., Weber, W., Standish, L., Quinn, H., Goesling, J., McGann, M., et al. (2003). Efficacy and safety of echinacea in treating upper respiratory tract infections in children: A randomized controlled trial. *JAMA, 290*(21), 2824–2830.

Simple descriptions of conditional distributions are essential! They allow us to stop and think about the data before moving on to more abstract analyses.

Exercises

18.1   *YRBS, prevalence proportions.* Table 18.7 is based on results from the Youth Risk Behavior Surveillance (YRBS) system. The number of adolescents who had been in a car or other vehicle when the driver had been drinking alcohol is cross tabulated according to racial/ethnic group classifications. Calculate prevalences proportions for each of the groups. Then, determine the margin of error for each proportion according to this formula:

$$m \approx 2 \cdot \mathrm{SE}_{\hat{p}_i}$$

where $\mathrm{SE}_{\hat{p}_i} = \sqrt{\frac{\hat{p}_i \hat{q}_i}{n_i}}$

TABLE 18.7 Data for Exercises 18.1 and 18.8. Drove or rode in a car or other vehicle when driver was drinking alcohol within past 30 days by racial/ethnic group.

Group	+	−	Total
1 (non-Hisp. White)	243	1911	2154
2 (non-Hisp. Black)	25	483	508
3 (Hispanic)	55	471	526
Total	323	2865	3188

Source: Frequency counts are projections based on data in Tables 1 and 4 of Eaton, D. K., Kann, L., Kinchen, S., Ross, J., Hawkins, J., Harris, W. A., et al. (2006). Youth risk behavior surveillance—United States, 2005. *MMWR Surveillance Summaries, 55*(5), 1–108. The actual survey used multistage sampling to collect data. Counts have been adjusted to emulate the design effect of the complex sample.

18.2 *YRBS, relative risks.* Use the results from Exercise 18.1 to express the relation between driving while drinking and ethnic/racial group in the form of relative risks. Let group 1 (non-Hispanic white) serve as the reference category. Interpret the results.

18.3 *Cytomegalovirus infection and coronary restenosis.* Table 18.8 contains results from the cytomegalovirus (CMV) and restenosis study considered as an illustrative example of naturalistic samples in Section 18.1. The occurrence of coronary restenosis within 6 months of surgery is cross tabulated according to CMV status.

TABLE 18.8 Data for Exercises 18.3 and 18.9. Restenosis within 6 months of angioplasty by CMV serological status.

	Restenosed	Did not restenose	Total
CMV+	21	28	49
CMV−	2	24	26
Total	23	52	75

Data from Zhou, Y. F., Leon, M. B., Waclawiw, M. A., Popma, J. J., Yu, Z. X., Finkel, T., et al. (1996). Association between prior cytomegalovirus infection and the risk of restenosis after coronary atherectomy. *New England Journal of Medicine, 335*(9), 624–630.

(a) Calculate the prevalence of CMV in the sample as a whole.

(b) Calculate the incidence of restenosis in the sample as a whole.

(c) What percent of CMV+ patients experienced restenosis? What percent of CMV− patients restenosed? What type of conditional percents are these?

(d) Calculate the relative risk of restenosis associated with CMV.

(e) Calculate the odds ratio of restenosis associated with CMV exposure. Why is this OR larger than the RR calculated in part (d) of the exercise?

18.4 *Anger and heart disease.* A nonexperimental study looked at whether people who angered easily were more likely to develop coronary heart disease than those who angered less easily. Spielberger anger scale scores were classified into low, moderate, and high categories. Individuals were followed for up to 72 months (median follow-up time: 53 months). Table 18.9 shows cross-tabulated results for acute and fatal coronary incidents. Calculate incidence proportions for each group. Describe the relationship between angering easily and coronary heart disease.

TABLE 18.9   Data for Exercises 18.4 and 18.10. Coronary heart disease (CHD) incidents according to anger-trait category.

Anger-level	CHD+	CHD−	Total
Low	31	3079	3110
Moderate	63	4668	4731
High	18	615	633
Total	112	8362	8474

Data from Williams, J. E., Paton, C. C., Siegler, I. C., Eigenbrodt, M. L., Nieto, F. J., & Tyroler, H. A. (2000). Anger proneness predicts coronary heart disease risk: Prospective analysis from the atherosclerosis risk in communities (ARIC) study. *Circulation, 101*(17), 2034–2039.

18.5 *Response to leprosy treatment.* In 1954, the prominent statistician W. G. Cochran (1909–1980) published an important article on the analysis of cross-tabulated counts. In this article, techniques were illustrated with an example from a study on the treatment of leprosy. Table 18.10 lists the cross-tabulated results. The row variable in this table classifies patients according to their initial degree of skin damage. The column variable classifies response to therapy according to five levels of improvement. Address whether patients with high levels of skin damage progressed differently than those with low levels of skin damage by calculating the conditional percents of improvement in the two groups. Comment on these findings.

TABLE 18.10 Data for Exercise 18.5; response to treatment in leprosy patients according to level of skin damage at onset of treatment.

Skin infiltration	Marked improvement	Moderate improvement	Slight improvement	Stationary	Worse	Total
High	7	15	16	13	1	52
Low	11	27	42	53	11	144
Total	18	42	58	66	12	196

Data from Cochran, W. G. (1954). Some methods for strengthening the common chi-square tests. *Biometrics, 10*, 435.

■ 18.3 Chi-Square Test of Association

Hypotheses

A chi-square statistic is used to test the association between the row and column variables that comprise the *R*-by-*C* table. This statistic tests the independence of variables when data are from naturalistic samples and the homogeneity of proportions when data are from cohort and case–control samples. The test hypotheses are:

H_0: no association between the row and column variables in the source population

Against

H_a: the null hypothesis is false

Note that the alternative hypothesis is neither one sided nor two sided, allowing for different types of associations.

Test Statistic

Chi-square random variables are derived by summing *n* independent *squared* Standard Normal (*z*) variables[g]:

$$\chi_n^2 = z_1^2 + z_2^2 + \cdots + z_n^2$$

where *n* represents the **degrees of freedom** of the chi-square random variable. A chi-square random variable based on a single Standard Normal random variable has 1 degree of freedom. Therefore, the square root of a chi-square random variable with 1 degree of freedom is a Standard Normal random variable: $\sqrt{\chi_1^2} = z$

---

[g] The paper that introduced the chi-square test statistic was published in 1900 by the influential 20th-century statistician and scientist Karl Pearson (1857–1936). This paper is considered to be one of the foundations of modern statistics. See: Pearson, K. (1900). On the criterion that a given system of deviations from the probable in the case of a correlated system of variables is such that it can be reasonably supposed to have arisen from random sampling. *London, Edinburgh and Dublin Philosophical Magazine and Journal of Science Series 5, 50,* 157–175.

Chi-square distributions have the following characteristics.

- They start at 0.
- They are asymmetrical.
- They become more and more symmetrical as their df increases.
- They have a mean equal to their df.
- They have variance equal to $2 \times$ df.
- Chi-square distributions with 1 and 2 df have modes of 0. All other chi-square distributions have modes $=$ df $-$ 2.

Figure 18.3 depicts chi-square distributions with 1, 2, and 3 degrees of freedom.

Table E. Appendix is our chi-square table. Each row in this table corresponds to a chi-square distribution with a different df. Each column in this table contains percentile and upper-tail landmarks on the distribution. The body of the table contains "critical values" on the distributions.

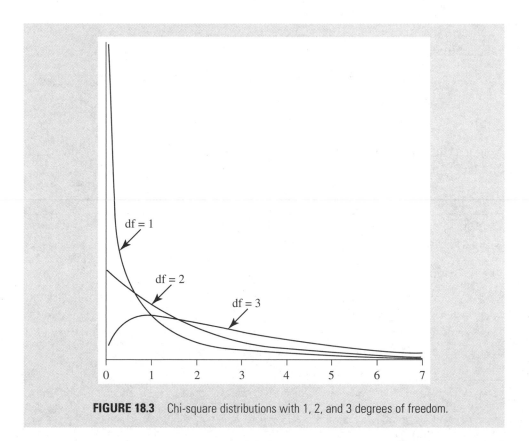

FIGURE 18.3   Chi-square distributions with 1, 2, and 3 degrees of freedom.

Figure 18.4 depicts a χ^2 distribution with 1 df. From Table E, we learn that the 95th percentile on this distribution is 3.84.

Chi-square statistics are calculated from observed and expected frequencies. **Observed frequencies** (O_i) are the data in the R-by-C table. **Expected frequencies** (E_i) are calculated according to this formula:

$$E_i = \frac{\text{row total} \times \text{column total}}{\text{table total}}$$

Table 18.11 lists expected frequencies for the education and smoking illustrative example.

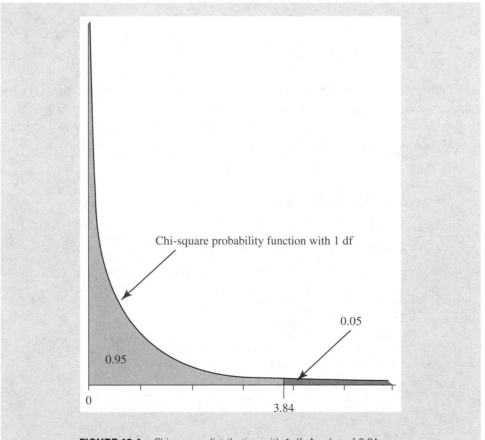

FIGURE 18.4 Chi-square distribution with 1 df. A value of 3.84 on this curve has cumulative probability 0.95 and right-tail probability 0.05.

TABLE 18.11　Expected frequencies for the education and smoking data example.

Education	Current smoker Yes	No	Total
High school graduate	8.034^{a_1}	41.966^{b_1}	50
Associate degree	13.658	71.342	85
Some college	19.603	102.397	122
Undergraduate degree	43.545	227.455	271
Graduate degree	9.159	47.841	57
Total	94	491	585

Examples of calculations:

$$^{a_1}\text{Expected} = \frac{\text{row total} \times \text{column total}}{\text{table total}} = \frac{50 \times 94}{585} = 8.034$$

$$^{b_1}\text{Expected} = \frac{\text{row total} \times \text{column total}}{\text{table total}} = \frac{50 \times 491}{585} = 41.966$$

Dissection of the Expected Frequency Formula: If there were no association between education and smoking, the proportion of smokers at each education level would be the same and the best estimate of the proportion at each education level would be the proportion of smokers for all groups combined, $\bar{p}$. For the illustrative example, $\bar{p} = \dfrac{m_1}{N} = \dfrac{94}{585} = 0.1607$. By applying this proportion to group 1, the expected frequency in cell a_1 is $n_1\bar{p} = (50)(0.1607) = 8.034$. Notice that $n_1\bar{p} = n_1 \times \dfrac{m_1}{N} = \dfrac{n_1 m_1}{N} = \dfrac{\text{row total} \times \text{column total}}{\text{table total}}$. By the same token, the proportion of nonsmokers in all groups combined is $\bar{q} = \dfrac{m_2}{N} = \dfrac{491}{585} = 0.8394$. Applying this proportion to group 1, the expected number of nonsmokers in group 1 $= n_1\bar{q} = \dfrac{n_1 m_2}{N} = \dfrac{50 \cdot 491}{585} = 41.966$. The same reasoning applies to each cell in the table.

Two different chi-square statistics are used in practice: Pearson's chi-square statistic and the continuity-corrected (Yates') chi-square statistic.

Pearson's chi-square statistic is:

$$\chi^2_{\text{stat}} = \sum_{\text{all}} \left[\frac{(O_i - E_i)^2}{E_i} \right]$$

where O_i represents the observed count in table cell i and E_i represents the expected count in cell i.

The **continuity-corrected (Yates') chi-square statistic** is:

$$\chi^2_{stat,c} = \Sigma\left[\frac{\left(|O_i - E_i| - \frac{1}{2}\right)^2}{E_i}\right]$$

These statistics have $(R - 1) \times (C - 1)$ degrees of freedom, where R represents the number of rows in the table and C represents the number of columns.

P-Value

Under the null hypothesis, the χ^2_{stat} has a χ^2 distribution with $(R - 1)(C - 1)$ df and the *P*-value corresponds to the area under the χ^2 curve to the right of χ^2_{stat}. Use Table E or a software utility[h] to find this probability. When using Table E, wedge the χ^2_{stat} between adjacent table entries to find the approximate *P*-value.

ILLUSTRATIVE EXAMPLE

Chi-square test (Education and smoking). We have established that prevalence proportions in the education and smoking illustrative data vary from 24% to 9%. Now we test the association for significance.

A. Hypotheses. H_0: no association between educational level and smoking in the source population versus H_a: "association."

B. Test statistic. Table 18.1 lists observed counts. Table 18.11 lists expected counts. The Pearson's chi-square statistic is:

$$\chi^2_{stat} = \Sigma\left(\frac{(O_i - E_i)^2}{E_i}\right) = \frac{(12 - 8.034)^2}{8.034} + \frac{(38 - 41.966)^2}{41.966} +$$

$$\frac{(18 - 13.658)^2}{13.658} + \frac{(67 - 71.342)^2}{71.342} + \frac{(27 - 19.603)^2}{19.603} +$$

$$\frac{(95 - 102.397)^2}{102.397} + \frac{(32 - 43.545)^2}{43.545} + \frac{(239 - 227.455)^2}{227.455} +$$

$$\frac{(5 - 9.159)^2}{9.159} + \frac{(52 - 47.841)^2}{47.841} = 1.958 + 0.375 + 1.380 +$$

$0.264 + 2.791 + 0.534 + 3.061 + 0.586 + 1.889 +$

$0.362 = 13.20$

continues

---

[h] Examples of software utilities that calculate chi-square probabilities include the Microsoft Excel CHIDIST function, StaTable (www.cytel.com/Products/StaTable/), and *WinPepi* WhatIs.exe.

continued

Degrees of freedom $= (R - 1) \times (C - 1) = (5 - 1) \times (2 - 1) = 4 \times 1 = 4$

C. **P-value.** The χ^2_{stat} is wedged between landmarks of 11.14 (right tail = 0.025) and 13.28 (right tail = 0.01) on the χ^2_4 distribution. Figure 18.5 depicts the placement of this test statistic, demonstrating that $P \approx 0.01$. Using a software utility, $P = 0.0103$. The association is statistically significant.

D. **Significance level.** The association is significant at $\alpha = 0.05$ and is *almost* significant at $\alpha = 0.01$.

E. **Conclusion.** The data demonstrate a negative (inverse) association between education level and the prevalence of smoking ($P = 0.010$).

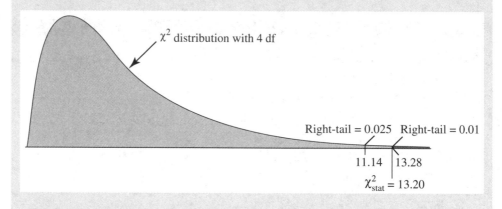

FIGURE 18.5   Sampling distribution for χ^2_{stat} from illustrative example. The *P*-value is about equal to 0.01.

Notes

1. **How the chi-square test works.** When all observed counts equal expected counts, $\chi^2_{stat} = 0$. When observed counts are far from expected, the χ^2_{stat} gets big and evidence against H_0 mounts.

2. **Significant chi-square statistics.** The 68–95–99.7 rule for the *z*-distribution allowed us to suspect significance when *z*-statistics (and *t*-statistics with more than just a few degrees of freedom) exceeded 2. Although there is no comparable rule for chi-square statistics, we do know that chi-square distributions have a mean equal to their degrees of freedom. Therefore, when the χ^2_{stat} is much larger than its df, we suspect observed values deviate significantly from expected values.

3. **Avoid chi-square statistics in small samples.** Chi-square tests are valid only in large samples.[i] The rule is that it is acceptable to use chi-square statistics as long as no table cell has an expected frequency less than 1.0 and not more than 20% of the cells have expected frequencies less than 5.0.[j] If this criterion cannot be met, an exact test (such as the one introduced in Section 17.3) must be used. *WinPepi* Compare2.exe calculates exact procedures for *R*-by-two tables.

4. **Pearson's or Yates'?** The question of whether to use Pearson's chi-square statistics or the continuity-corrected statistic has been debated for decades. There is no consensus opinion on this topic.[k]

5. **Chi-square tests do not quantify the strength of an association.** Chi-square tests of association provide insight into whether chance is a reasonable explanation for an observed association. They provide no insight into the type and strength of an association.

6. **Cells contributions.** Cell contributions to chi-square statistics are called **residuals**. Figure 18.6 shows a table listing observed counts, expected counts, and three different types of residuals.

The **unstandardized residual** for table cell *i* is the difference between observed and expected frequencies:

$$\text{Unstandardized residual}_i = O_i - E_i$$

The **standardized residual** for a cell is the unstandardized residuals divided by the square root of the expected frequency:

$$\text{Standardized residual}_i = \frac{O_i - E_i}{\sqrt{E_i}}$$

---

[i] Under the null hypothesis, observed frequencies deviate from expected frequencies according to a Standard Normal distribution. This fact is aided by the central limit theorem and operates best in large samples.

[j] Cochran, W. G. (1952). The χ^2 test of goodness of fit. *Annals of Mathematical Statistics, 23*(3), 315–345. Cochran, W. G. (1954). Some methods for strengthening the common chi-square tests. *Biometrics, 10*, 417–451.

[k] Conover, W. J. (1974). Some reasons for not using the Yates continuity correction on 2×2 contingency tables {w/Rejoinder}. *Journal of the American Statistical Association, 69*, 374–376; 382. Mantel, N. (1974). Comment and a suggestion. *Journal of the American Statistical Association,* 378–380. Miettinen, O. (1974). Comment. *Journal of the American Statistical Association, 69,* 380–382. Overall, J. E. (1990). Comment. *Statistics in Medicine, 9*, 379–382.

EDU * SMOKE Cross tabulation

			SMOKE 1	SMOKE 2	Total
EDU	1	Count	12	38	50
		Expected count	8.0	42.0	50.0
		Residual	4.0	−4.0	
		Std. residual	1.4	−0.6	
		Adjusted residual	1.6	−1.6	
	2	Count	18	67	85
		Expected count	13.7	71.3	85.0
		Residual	4.3	−4.3	
		Std. residual	1.2	−0.5	
		Adjusted residual	1.4	−1.4	
	3	Count	27	95	122
		Expected count	19.6	102.4	122.0
		Residual	7.4	−7.4	
		Std. residual	1.7	−0.7	
		Adjusted residual	2.0	−2.0	
	4	Count	32	239	271
		Expected count	43.5	227.5	271.0
		Residual	−11.5	11.5	
		Std. residual	−1.7	0.8	
		Adjusted residual	−2.6	2.6	
	5	Count	5	52	57
		Expected count	9.2	47.8	57.0
		Residual	−4.2	4.2	
		Std. residual	−1.4	0.6	
		Adjusted residual	−1.6	1.6	
Total		Count	94	491	585
		Expected count	94.0	491.0	585.0

FIGURE 18.6 Observed counts, expected counts and residuals, education and smoking illustrative example. Graph produced with SPSS for Windows, Rel. 11.0.1.2001. Chicago: SPSS Inc. Reprint Courtesy of International Business Machines Corporation.

The **adjusted standardized residual** for a cell is:

$$\text{Adjusted standardized residual}_i = \frac{O_i - E_i}{\sqrt{E_i\left(1 - \frac{n_i}{N}\right)\left(1 - \frac{m_i}{N}\right)}}$$

Adjusted standardized residuals can be interpreted as z-statistics[1] and can be converted to cell P-values. Here are the adjusted residuals for the illustrative data (computed by *WinPepi*):

ADJUSTED RESIDUALS (CELL-BY-CELL ANALYSES):			
CATEG.	A	B	P (FOR A OR B)
1	1.60	−1.60	$P = 0.110$
2	1.39	−1.39	$P = 0.165$
3	2.05	−2.05	$P = 0.040$
4	−2.61	2.61	$P = 0.009$ [9.1E−3]
5	−1.58	1.58	$P = 0.114$

7. **Equivalence of the z and chi-square tests.** Chi-square tests for two-by-two cross tabulations produce results identical to z-tests (Section 17.3). The square root of the chi-square statistic is equal to the z-statistic: $\sqrt{\chi^2_{\text{stat with 1df}}} = z_{\text{stat}}$. This is true for "uncorrected" and continuity-corrected forms of the statistics.

ILLUSTRATIVE EXAMPLE

(Estrogen trial). We considered results from the estrogen trial of the Women's Health Initiative study. The incidence of the combined morbidity index in the treatment group was 751 per 8506 (8.83%). The incidence in the control group was 623 of 8102 (7.69%).[m] The test of H_0: $p_1 = p_2$ yielded $z_{\text{stat}} = 2.66$, $P = 0.0078$ (two-sided). We propose to test the same information with a chi-square statistic.

A. **Hypotheses.** H_0: no association in source population is equivalent to H_0: $p_1 = p_2$. H_a: association in source population is equivalent to H_a: $p_1 \neq p_2$.
B. **Test statistic.** Table 18.12 lists observed and expected frequencies for the problem.

continues

---

[1] Haberman, S. J. (1973). The analysis of residuals in cross-classified tables. *Biometrics, 29,* 205–220.
[m] Writing Group for the Women's Health Initiative Investigators. (2002). Risks and benefits of estrogen plus progestin in healthy postmenopausal women: Principal results from the Women's Health Initiative randomized controlled trial. *JAMA, 288*(3), 321–333.

continued

TABLE 18.12   Observed and expected frequencies for the estrogen trial illustrative example.

Observed frequencies

	Disease +	Disease −	
Estrogen +	751	7755	8506
Estrogen −	623	7479	8102
	1374	15,234	16,608

Expected frequencies

	Disease +	Disease −	Total
Estrogen +	703.712[a]	7802.288	8506
Estrogen −	670.288	7431.712	8102
Total	1374	15,234	16,608

[a]Example of calculation: $E_{\text{cell } a_1} = (8506)(1374)/16{,}608 = 703.712$.

$$\chi^2_{\text{stat}} = \Sigma\left[\frac{(O_i - E_i)^2}{E_i}\right]$$

$$= \frac{(751 - 703.712)^2}{703.712} + \frac{(7755 - 7802.288)^2}{7802.288} +$$

$$\frac{(623 - 670.288)^2}{670.288} + \frac{(7479 - 7431.712)^2}{7431.712}$$

$$= \quad 3.178 \quad + \quad 0.287 \quad + \quad 3.336 \quad + \quad 0.301$$

$$= \quad 7.102$$

This chi-square statistic has df $= (R - 1) \times (C - 1) = (2 - 1) \times (2 - 1) = 1$.

Notice that $\sqrt{\chi^2_{\text{stat}}} = \sqrt{7.102} = 2.66 = z_{\text{stat}}$.

C. **P-value.** The χ^2_{stat} of 7.102 with 1 df converts to $P = 0.0077$, producing the same results as the z-test (except for rounding error).

D. **Significance level.** The results are statistically significant at the $\alpha = 0.01$ level but not at the $\alpha = 0.005$ level.

E. **Conclusion.** The group that received estrogen was significantly more likely to experience the combined index outcome than the placebo group (8.8% vs. 7.7%, $P = 0.0077$).

Exercises

18.6 *Chi-square landmarks.* Use Table E to find the area under the χ^2 curve with 3 degrees of freedom to the right of:

(a) 4.64

(b) 6.25

(c) 11.34

18.7 *Chi-square approximation.* Use Table E to find the approximate area under the curve to the right of 5.22 on a χ^2 distribution with 3 degrees of freedom. (Suggestion: Draw the chi-square curve showing 5.22 on the horizontal axis. Shade the area to the right of 5.22. Bracket the shaded region with critical value landmarks from Table E.)

18.8 *Drove when drinking alcohol.* Table 18.7 presented cross-tabulated results for the number of adolescents who drove or rode in a car or other vehicle when the driver was drinking alcohol. Prevalence proportions varied from 11.3% (non-Hispanic white) to 4.9% (non-Hispanic black). Test the data in Table 18.7 for significance by calculating its chi-square statistic, df, and *P*-value.

18.9 *Cytomegalovirus infection and coronary restenosis.* Exercise 18.3 found that 43% of CMV+ patients experienced arterial restenosis within 6 months of atherectomy. In contrast, 8% of CMV− patients experienced a similar outcome. Table 18.8 contains observed counts.

(a) Use a chi-square statistic to test the association for statistical significance.

(b) Repeat the test with a *z*-statistic (Section 17.3). Show the relation between the chi-square statistic and the *z*-statistic.

18.10 *Anger and heart disease.* Exercise 18.4 found a higher risk of coronary disease in individuals with higher anger levels. Table 18.9 contains the data. Test the association for statistical significance. Report the chi-square statistic, its df, and *P*-value.

■ 18.4 Test for Trend

Ordinal Explanatory Variable

In situations in which the explanatory variable is ordinal and response variable is binary, it is natural to ask whether observed proportions follow a trend. We can test an observed trend for statistical significance as follows:

A. Hypotheses. The null hypothesis (H_0) is that there is no linear trend in proportions in the source population. The two-sided alternative hypothesis (H_a) declares a linear trend. One-sided alternatives specify the direction of the trend as either upward or downward.

B. **Test statistic.** Each level of the explanatory variable is assigned an **exposure score** based on either ranks or quantitative assessments. Call this s_i. The test statistic is[n]

$$z_{\text{stat, trend}} = \frac{O - E}{\sqrt{V}}$$

where $O \equiv$ the observed sum of scores in cases $= \Sigma a_i s_i$

$E \equiv$ expected sum of scores if H_0 were true $= \dfrac{m_1}{N} \Sigma n_i s_i$

$V \equiv$ the variance of scores $= \dfrac{m_1 m_2}{N^2(N-1)} \left[N \Sigma n_i s_i^2 - \left(\Sigma n_i s_i \right)^2 \right]$

Under the null hypothesis, the test statistic has a Standard Normal distribution.

C. **P-value.** The *P*-value is derived from the *z*-statistic in the usual manner.

D. **Significance level.** The *P*-value is used to gauge the strength of evidence against the null hypothesis.

E. **Conclusion.** The test results are addressed in the context of the data and research question.

ILLUSTRATIVE EXAMPLE

Test for trend (Education and smoking). A downward trend in prevalence proportions was noted in the education and smoking illustrative data (Figure 18.2). We test whether this trend is statistically significant.

A. **Hypotheses.** H_0: no linear trend between education and smoking in the source population against H_a: linear trend between education and smoking in the source population.

B. **Test statistic.**

$O = \Sigma a_i s_i = (12)(1) + (18)(2) + \cdots + (5)(5) = 282$

$E = \dfrac{m_1}{N} \Sigma n_i s_i = \dfrac{94}{585} \left[(50)(1) + (85)(2) + \ldots + (57)(5) \right] =$
$(0.1607)(1955) = 314.14$

$V = \dfrac{m_1 m_2}{N^2(N-1)} \left[N \Sigma n_i s_i^2 - \left(\Sigma n_i s_i \right)^2 \right]$

---

[n] Mantel, N. (1963). Chi-square tests with one degree of freedom; extensions of the Mantel-Haenszel procedure. *Journal of the American Statistical Association, 58,* 690–700.

$$= \frac{(94)(491)}{585^2(585-1)}\Big[585\big((50)(1^2)+(85)(2^2)+\ldots+(57)(5^2)\big)-$$

$$\big((50)(1)+(85)(2)+\ldots+(57)(5)\big)^2\Big]=96.678$$

$$z_{stat,trend}=\frac{O-E}{\sqrt{V}}=\frac{282-314.14}{\sqrt{96.678}}=-3.27.$$

C. *P*-value. $P = 0.00108$ (via Table F).

D. Significance level. The evidence against H_0 is significant at the $\alpha = 0.002$ level and is almost significant at the $\alpha = 0.0010$ level.

E. Conclusion. The negative "dose–response" trend observed between education level and the prevalence of smoking is statistically significant ($P = 0.0011$).

Figure 18.7 is a screenshot from *WinPepi*'s Compare2.exe program showing output for this problem. The square root of the "Extended Mantel–Haenszel chi-sq." is equivalent to our $z_{stat,trend}$:

$$\sqrt{\text{Extended Mantel–Haenszel chi-sq. for trend}} = \sqrt{10.683} = 3.27 = z_{stat,trend}.$$

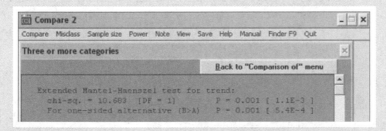

FIGURE 18.7 *WinPepi* screenshot, test of trend, education and smoking illustrative example. Abramson J.H. (2011). WINPEPI updated: computer programs for epidemiologists, and their teaching potential. *Epidemiologic Perspectives & Innovations, 8*(1), 1.

Ordinal Response Variable

The same $z_{stat,trend}$ can be applied to an ordinal response by merely transposing data from a two-by-*C* format to form an *R*-by-two format. Categorical responses now are laid out in rows.

ILLUSTRATIVE EXAMPLE

Test for trend in response, ordinal response variable (Leprosy treatment). Exercise 18.5 examined responses to treatment in 196 leprosy patients. Descriptive statistics suggested patients in the high skin-infiltration group were

continues

continued

more likely to demonstrate a favorable response (see answer key to Exercise 18.5). Table 18.13 has taken the two-by-five cross tabulation from Table 18.10 and transposed it into a five-by-two table. It also calculates odds and odds ratios, which we note become progressively smaller as we proceed down the table. We now ask whether this trend is statistically significant.

TABLE 18.13 Leprosy illustrative example test for trend; response to treatment based on level of skin damage at onset of observation.

Improvement score	Skin infiltration High	Low	Odds ($\hat{o}_i$)	Odds ratio (OR_i)
(1) marked improvement	7	11	0.6364[a]	1.00 (ref.)
(2) moderate	15	27	0.5556	0.87[b]
(3) slight	16	42	0.3810	0.60
(4) stationary	13	53	0.2453	0.39
(5) worse	1	11	0.0909	0.14

Examples of calculations:

$$^a\hat{o}_1 = \frac{a_1}{b_1} = \frac{7}{11} = 0.6364$$

$$^b\widehat{OR}_2 = \frac{\hat{o}_2}{\hat{o}_1} = \frac{15/27}{7/11} = 0.87$$

A. Hypothesis statements. H_0: no trend between amount of skin infiltration and response to leprosy treatment versus H_a: trend between amount of skin infiltration and response to leprosy treatment.

B. Test statistic.

$$O = \Sigma a_i s_i = (7)(1) + (15)(2) + \cdots + (1)(5) = 142$$

$$E = \frac{m_i}{N} \Sigma n_i s_i = \frac{52}{196} \left((18)(1) + (42)(2) + \cdots + (12)(5) \right) = 159.184$$

$$V = \frac{m_1 m_2}{N^2(N-1)} \left[N\Sigma n_i s_i^2 - \left(\Sigma n_i s_i \right)^2 \right]$$

$$\frac{(52)(144)}{196^2(196-1)} \left[196\left((18)(1^2) + (42)(2^2) + \ldots + (12)(5^2) \right) - \right.$$

$$= \left. \left((18)(1) + (42)(2) + \ldots + (12)(5) \right)^2 \right] = 44.525$$

$$z_{\text{stat,trend}} = \frac{O - E}{\sqrt{V}} = \frac{142 - 159.184}{\sqrt{44.525}} = -2.58$$

C. *P*-value. The two-sided $P = 0.00988$, providing strong evidence against the null hypothesis.
D. **Significance level.** The evidence against H_0 is significant at the $\alpha = 0.01$ level but not at the $\alpha = 0.005$ level.
E. **Conclusion.** Leprosy patients with high levels of skin infiltration were more likely to demonstrate improvement than those with low levels of skin infiltration ($P = 0.0099$).

Note that the test of trend is more sensitive than the analogous test for association. The leprosy treatment data test of trend yields $P = 0.00988$, whereas the chi-square test of association for these data (calculations not shown) finds $P = 0.14$.

Exercises

18.11 *Anger and heart disease.* Exercise 18.4 found coronary heart disease risks of 1.0%, 1.3%, and 2.8% in low, intermediate, and high anger-trait groups. Table 18.9 lists observed frequencies for the problem. Apply a test of trend to these data. Show all hypothesis-testing steps.

18.12 *Do seatbelt laws prevent serious injury?* Table 18.14 lists frequencies for severity of motor vehicle accident injuries before and after enactment of a seatbelt law. Conditional (column) percents are included in the table.

TABLE 18.14   Level of injury prior to and after enactment of seatbelt legislation.

| | Time period | | |
Injury level	After enactment	Prior to enactment	Total
No injury	1281 (92.6%)	6596 (90.5%)	7877 (90.8%)
Minimal	64 (4.6%)	400 (5.5%)	464 (5.3%)
Minor	35 (2.5%)	264 (3.6%)	299 (3.4%)
Major/fatal	4 (0.3%)	30 (0.4%)	34 (0.4%)
Total	1384 (100.0%)	7290 (100.0%)	8674 (100%)

Source: Unknown.

(a) Comment on the conditional distributions of injuries before and after enactment of the law.

(b) Calculate the odds ratios for each category of injury using "no injury" as the reference category. Then test the trend in odds ratios for significance.

(c) Use a regular chi-square test of association (not a test for trend) to determine whether there is a statistically significant association between severity of injury and enactment of the seat belt law.

(d) Which test was more sensitive—the test for trend or the "regular" test for association?

■ 18.5 Case–Control Samples

Data

This chapter began by considering ways to generate cross-tabulated counts. Three sampling methods were considered: naturalistic, purposive cohorts, and case–control. Until this point, we have addressed naturalistic and cohort samples. Now we tackle case–control samples.

Case–control sampling provides an efficient way to study rare outcomes. In its simplest form, case–control studies select a fixed number of individuals who are positive for the response (these are the cases) and a fixed-number who are negative for the response (these are the controls). This sampling method excludes the possibility of estimating incidence or prevalence directly, but still allows for quantifying the association between the explanatory variable and response variable when *cases and controls are selected in a manner that allows for nonbiased ascertainment of the exposure experience of the source population.* To appreciate this type of sampling, consider Figure 18.8. This figure depicts the longitudinal experience of five individuals. Individual 1 develops a disease at time t_1. At that time (shaded region), a control is selected *at random*

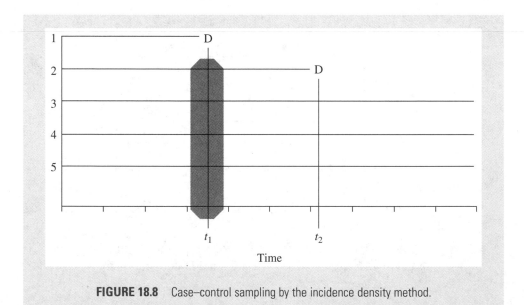

FIGURE 18.8   Case–control sampling by the incidence density method.

from the remaining individuals in the population. Notice that individual 2 is eligible for selection as a control even though they later develop the disease, at time t_2. This sampling method, known as **incidence density sampling**, allows for the direct estimation of the rate ratio associated with exposure in the underlying population.[o]

After the cases and controls are selected, information about the explanatory variable is collected and data are cross tabulated to form an R-by-C contingency table.

ILLUSTRATIVE EXAMPLE

Case–control data (Alcohol and esophageal cancer). Data for this example come from a case–control study on esophageal cancer. Cases were men diagnosed with esophageal cancer from a region in France. Controls were selected at random from electoral lists from the same geographic region. Table 18.15 cross tabulates the data for cases and controls based on daily alcohol consumption dichotomized at 80 g/day.

TABLE 18.15 Alcohol and esophageal cancer illustrative case-control data.

Alcohol grams/day	Esophageal cancer		Total
	Cases	Noncases	
≥80	96	109	205
80	104	666	770
Total	200	775	975

Data from Tuyns, A. J., Pequignot, G., & Jensen, O. M. (1977). [Esophageal cancer in Ille-et-Vilaine in relation to levels of alcohol and tobacco consumption. Risks are multiplying]. *Bulletin du Cancer, 64*(1), 45–60. Data stored online as individual records in the file BDI.* as variables ALC2 and CASE. Data set from APHA data exchange, November 1988.

Odds Ratio

The relationship between the explanatory variable and response variable is quantified with an **odds ratio**. Using the notation established in Table 18.2C, the odds of exposure in cases is $\hat{o}_1 = a_1/a_2$ and the odds of exposure in controls is $\hat{o}_2 = b_1/b_2$. The ratio of these odds forms the **odds ratio**:

$$\widehat{OR} = \frac{a_1/a_2}{b_1/b_2} = \frac{a_1 b_2}{b_1 a_2}$$

---

[o]Miettinen, O. (1976). Estimability and estimation in case-referent studies. *American Journal of Epidemiology, 103*(2), 226–235.

This odds ratio is stochastically equivalent to a rate ratio. There are several justifications for this statement. The first justification was proposed by Cornfield in 1951.[p] The more modern justification comes from Miettinen (1976).[q] Briefly, the incidence rate (incidence density) in the exposed population is a_1/T_1, where a_1 represents the number of exposed cases and T_1 represents the sum of person-time in the exposed subgroup in the population. The incidence rate in the non-exposed population is a_2/T_2. The rate ratio $= \dfrac{a_1/T_1}{a_2/T_2} = \dfrac{a_1/a_2}{T_1/T_2}$. The numerator of the final expression (a_1/a_2) is the exposure odds in the case series. The denominator (T_1/T_2) is estimated by the exposure odds (b_1/b_2) in the control series. This shows that the purpose of the control group in a case–control study is to estimate the exposed to nonexposed person-time in the source population.

ILLUSTRATIVE EXAMPLE

Odds ratio (Alcohol and esophageal cancer). The odds ratio for the case–control data in Table 18.15 is $\widehat{OR} = \dfrac{96 \cdot 666}{104 \cdot 109} = 5.64.$

This suggests that high-alcohol consumers have 5.6 times the rate of esophageal cancer compared to low-alcohol consumers.

Confidence Interval for the Odds Ratio

Let OR represent the odds ratio parameter and $\widehat{OR}$ represent the odds ratio estimate. The **confidence interval for the OR** is

$$e^{\ln\widehat{OR} \pm z_{1-\alpha/2} \cdot SE_{\ln\widehat{OR}}}$$

where e is the base on the natural logarithm (e $\approx$ 2.71828...), z is the Standard Normal deviate corresponding to the desired level of confidence (e.g., $z = 1.96$ for 95% confidence), and $SE_{\ln\widehat{OR}} = \sqrt{\dfrac{1}{a_1} + \dfrac{1}{a_2} + \dfrac{1}{b_1} + \dfrac{1}{b_2}}.$

ILLUSTRATIVE EXAMPLE

Confidence interval for the odds ratio (Alcohol and esophageal cancer). Ninety and ninety-five percent confidence intervals for the alcohol and esophageal cancer illustrative case–control data are calculated:

- We've established $\widehat{OR} = 5.640$. Therefore, $\ln\widehat{OR} = \ln(5.640) = 1.7299$ (by calculator).

[p]Cornfield, J. (1951). A method of estimating comparative rates from clinical data. Application to cancer of the lung, breast, and cervix. *Journal of the National Cancer Institute, 11*, 1269–1275.
[q]Miettinen, O. (1976). Estimability and estimation in case-referent studies. *American Journal of Epidemiology, 103*(2), 226–235.

- $SE_{\ln\widehat{OR}} = \sqrt{\dfrac{1}{a_1} + \dfrac{1}{a_2} + \dfrac{1}{b_1} + \dfrac{1}{b_2}} = \sqrt{\dfrac{1}{96} + \dfrac{1}{104} + \dfrac{1}{109} + \dfrac{1}{666}} = 0.1752.$

- The 90% confidence interval for the OR $= e^{1.7299 \pm (1.645)(0.1752)} = e^{1.7299 \pm 0.2882} = e^{1.4417,\, 2.0181} = 4.23$ to 7.52.

- The 95% confidence interval for the OR $= e^{1.7299 \pm (1.96)(0.1752)} = e^{1.7299 \pm 0.3433} = e^{1.3866,\, 2.0732} = 4.00$ to 7.95.

Hypothesis Test

Because an odds ratio of 1 indicates no association between the explanatory variable and response variable, we test H_0: OR $= 1$ versus H_a: OR $\neq 1$. When expected values exceed 5, a chi-square statistic (Section 18.3) or z-statistic (Section 17.3) may be used for the test. With smaller samples, Fisher's exact test or a mid-P procedure is required (Section 17.3).

Recall: P-values and the confidence intervals address random sampling errors only. Stay vigilant to the fact that they provide no protection against the making of systematic errors.

ILLUSTRATIVE EXAMPLE

Hypothesis test (Alcohol and esophageal cancer). A chi-square statistic will be used to test the odds ratio from the alcohol and esophageal cancer illustration.

A. Hypotheses. H_0: OR $= 1$ vs. H_a: OR $\neq 1$.
B. Test statistic. Table 18.15 contains observed frequencies for the data. Table 18.16 contains expected frequencies under the null hypothesis.

TABLE 18.16   Alcohol consumption and esophageal cancer, expected counts, two-by-two analysis.

Alcohol grams/day	Esophageal cancer Cases	Noncases	Total
≥ 80	42.051^a	162.949	205
80	157.949	612.051	770
Total	200	775	975

Example of calculation:
$^a E_{a_1} = (205)(200)/975 = 42.051$

continues

continued

Pearson's chi-square is

$$\chi^2_{stat} = \Sigma \left[\frac{(O_i - E_i)^2}{E_i} \right]$$

$$= \frac{(96 - 42.051)^2}{42.051} + \frac{(109 - 162.949)^2}{162.949} + \frac{(104 - 157.949)^2}{157.949} +$$

$$\frac{(666 - 612.051)^2}{612.051}$$

$$= 69.21 + 17.86 + 18.43 + 4.76$$
$$= 110.26$$

This test statistic has df $= (R - 1)(C - 1) = (2 - 1)(2 - 1) = 1$.

C. **P-value.** This χ^2_{stat} is "off the chart" in Table E, indicating $P < 0.0005$. ($P = 8.6 \times 10^{-26}$). This provides strong evidence against the null hypothesis.

D. **Significance level.** The evidence against H_0 is significant at the $\alpha = 0.0001$ level.

E. **Conclusion.** The association between alcohol consumption and esophageal cancer is highly significant ($P < 0.0001$).

Multiple Levels of Exposure

When the explanatory variable in a case–control study is ordinal, the odds ratios are calculated for each exposure level relative to the "least-exposed" level of exposure. A test for trend and chi-square test of association may be applied as needed.

ILLUSTRATIVE EXAMPLE

Multiple levels of exposure, case–control data. The case–control study on esophageal cancer discussed earlier included a variable in which alcohol consumption was recorded according to the following four levels:

Level 1 = 0–39 g/day
Level 2 = 40–79 g/day
Level 3 = 80–119 g/day
Level 4 = at least 120 g/day

Table 18.17 shows the cross-tabulated results.

TABLE 18.17 Case–control data of alcohol and esophageal cancer with alcohol consumption recorded according to four levels.

	Esophageal cancer	
Alcohol (g/day)	Cases	Controls
1 (0–39)	29	386
2 (40–79)	75	280
3 (80–119)	51	87
4 (120+)	45	22
Total	200	775

Data from *See* Table 18.15. Data can be downloaded in the file BD1.* as variables ALC and CASE.

- **Descriptive statistics (exposure proportions).** Table 18.18 shows the conditional distributions (column percents) of alcohol consumption in cases and controls. This reveals that cases were less likely than controls to fall into exposure category 1 and were more likely to fall into the other categories.

TABLE 18.18 Conditional distributions of exposure proportions from case–control study of alcohol and esophageal cancer with alcohol consumptions recorded according to four levels.

	Esophageal cancer	
Alcohol (g/day)	Cases	Controls
1 (0–39)	14.5%	49.8%
2 (40–79)	37.5	36.1
3 (80–119)	25.5	11.2
4 (120+)	22.5	2.8
Total	100.0%	100.0%

- **Odds ratios.** Odds ratios are calculated with the lowest exposure level (0–39 g/day) serving as the reference category. Table 18.19 shows data rearranged into four separate two-by-two tables with the following odds ratios: $\widehat{OR}_1 = 1$ (reference group), $\widehat{OR}_2 = 3.57$, $\widehat{OR}_3 = 7.80$, and $\widehat{OR}_4 = 27.23$.
- **Test of association.** The chi-square test of association addresses H_0: no association between alcohol and esophageal cancer in the source population vs.

continues

continued

H_a: "association." A chi-square test yields $\chi^2_{stat} = 158.955$, df $= 3$, $P = 3.1 \times 10^{-34}$. (Calculations not shown. You may confirm these results for practice.)

TABLE 18.19   Four levels of alcohol consumption in the illustrative case–control data with data parsed into separate two-by-two tables.

Alcohol g/day	Esophageal cancer Yes	No	
0–39	29	386	$\widehat{OR}_1 = \dfrac{29 \cdot 386}{386 \cdot 29} = 1.00$ (reference)
0–39	29	386	

Alcohol g/day	Esophageal cancer Yes	No	
40–79	75	280	$\widehat{OR}_2 = \dfrac{a_1 b_2}{b_1 a_2} = \dfrac{75 \cdot 386}{280 \cdot 29} = 3.57$
0–39	29	386	

Alcohol g/day	Esophageal cancer Yes	No	
80–119	51	87	$\widehat{OR}_3 = \dfrac{51 \cdot 386}{87 \cdot 29} = 7.80$
0–39	29	386	

Alcohol g/day	Esophageal cancer Yes	No	
120+	45	22	$\widehat{OR}_4 = \dfrac{45 \cdot 386}{22 \cdot 29} = 27.23$
0–39	29	386	

- **Test of trend.** The test of H_0: no linear relationship between alcohol consumption and esophageal cancer in the population (Section 18.4) yields $z_{stat, trend} = 12.37$, $P < 0.0005$.
- These data demonstrate a dose–response relationship between amount of alcohol consumed and esophageal cancer risk.

Exercises

18.13   *Cell phones and brain tumors, study 1.* A case–control study examined the relationship between cellular telephone use and intracranial tumors. The study (completed between 1994 and 1998) included 782 cases with various types of intracranial tumors and 799 controls with a variety of nonmalignant conditions. Subjects were considered to be "exposed" if they reported cellular telephone use for more than 100 h. The odds ratios (and 95% confidence intervals) for various tumor types were for glioma 0.9 (0.5 to 1.6), for meningioma 0.7 (0.3 to 1.7), for acoustic neuroma 1.4 (0.6 to 3.5), for all tumor types combined 1.0 (0.6 to 1.5).[r] Comment on these findings. Do results support or fail to support the theory that handheld cellular telephone use causes brain tumors?

18.14 *Cell phones and intracranial tumors, study 2.* A (different) case–control study on intracranial tumors and cell phone use completed between 1994 and 1998 used a structured questionnaire to explore the relationship between cell phone use and primary brain cancer in 469 cases and 422 controls. Compared with patients who never used handheld cellular telephones, the odds ratio associated with regular use was 0.85 (95% confidence interval 0.6 to 1.2). The OR for infrequent users was 1.0 (0.5 to 2.0). The odds ratio for frequent users was 0.7 (0.3 to 1.4). Odds ratios were less than 1.0 for all histologic categories of brain cancer except for neuroepitheliomatous cancers (odds ratio 2.1; 95% confidence interval 0.9 to 4.7).[s] Interpret these results in light of the findings from the study described in Exercise 18.13.[t]

18.15 *Doll and Hill, 1950.* An important historical case–control completed by Doll and Hill found that 647 of the 649 lung cancer cases were smokers. In contrast, 622 of 649 controls smoked.[u] Display these data as a 2-by-2 table and calculate the odds ratio associated with smoking. Include a 95% confidence interval for the OR.

18.16 *Vasectomy and prostate cancer.* Table 18.20 displays data from a case–control study on vasectomy and prostate cancer. Calculate the odds ratio; include a 95% confidence interval for the OR. (*Optional*: Calculate a *P*-value to test the association.)

---

[r] Inskip, P. D., Tarone, R. E., Hatch, E. E., Wilcosky, T. C., Shapiro, W. R., Selker, R. G., et al. (2001). Cellular-telephone use and brain tumors. *New England Journal of Medicine, 344*(2), 79–86.

[s] Muscat, J. E., Malkin, M. G., Thompson, S., Shore, R. E., Stellman, S. D., McRee, D., et al. (2000). Handheld cellular telephone use and risk of brain cancer. *JAMA, 284*(23), 3001–3007.

[t] The studies considered in Exercises 18.13 and 18.14 afford the opportunity to make two relevant points: (1) These studies looked at different types of intracranial tumors (e.g., gliomas, meningiomas, neuromas, and epitheliomas). These different tumor types behave differently and probably have different etiologies. This argues against pooling their results, at least initially; (2) without the option to assign cell phone use experimentally, other information must come into play when interpreting these statistics. One important fact is that the type of energy emitted by cell phones is nonionizing and does not cause damage to DNA chemical bonds. Combined with corroborating statistical evidence, this suggests that cell phone use is unlikely to cause brain tumors.

[u] Doll, R., & Hill, A. B. (1950). Smoking and carcinoma of the lung. *British Medical Journal, 2,* 739–748.

TABLE 18.20   Data for Exercise 18.16. Vasectomy and prostate cancer.

	Cases	Noncases
Vasectomy +	61	93
Vasectomy −	114	165

Data from Zhu, K., Stanford, J. L., Daling, J. R., McKnight, B., Stergachis, A., Brawer, M. K., et al. (1996). Vasectomy and prostate cancer: A case-control study in a health maintenance organization. *American Journal of Epidemiology, 144*(8), 717–722.

18.17   ***Baldness and myocardial infarction, self-assessed baldness.*** A case–control study was completed to learn about the relationship between male-pattern baldness and the risk of cardiovascular disease. Cases were men less than 55 years of age hospitalized for a heart attack. Cases were excluded if they had a prior history of serious heart problems. Controls were men with no history of heart disease admitted to the same hospitals for nonfatal, noncardiac problems. Table 18.21 cross tabulates results according to self-assessed baldness, with baldness graded on a scale of 1 (no baldness) to 5 (extreme baldness).

TABLE 18.21   Data for Exercise 18.17. Case–control study of baldness and myocardial infarction. Baldness was self-assessed by study subjects.

Baldness	Cases	Controls
1 (no baldness)	251	331
2	165	221
3 (moderate baldness)	195	185
4	50	34
5 (extreme baldness)	2	1
Total	663	772

Data from Table 6 in Lesko, S. M., Rosenberg, L., & Shapiro, S. (1993). A case-control study of baldness in relation to myocardial infarction in men. *JAMA, 269*(8), 998–1003.

(a) Calculate the odds ratios associated with each level of baldness. Interpret your results.

(b) Conduct a chi-square test of association. Report the chi-square statistic, its df, and *P*-value.

(c) Conduct a test for trend.

18.18 *Baldness and myocardial infarction, lurking variables.* This exercise is a continuation of Exercise 18.17. The investigators found differences in cases and controls besides those have to due to baldness. For example, the median age of cases was 47 years, while the median age of controls was 43 years.

(a) Explain why careful consideration of age differences is important when studying the relationship between baldness and heart disease.

(b) The investigators used a logistic regression model to adjust odds ratios for the following factors: age, race, religion, years of education, body mass index, use of alcohol and cigarettes, family history of myocardial infarction, history of angina, hypertension, diabetes, hypercholesterolemia, gout, exercise, personality, and number of doctor visits in the prior year. The unadjusted and multivariate adjusted odds ratios reported by the article are listed in Table 18.22. Do the multivariate adjustments materially affect your interpretation?

TABLE 18.22 Unadjusted and multivariate adjusted odds ratios for baldness and myocardial infarction study, Exercise 18.18.

Baldness level	Unadjusted OR	Multivariate adjusted $\widehat{OR}$
1	1.0 (reference)	1.0 (reference)
2	1.0	0.8
3	1.4	1.1
4	1.9	2.0
5	2.6 (very small sample)	Could not estimate

■ 18.6 Matched Pairs

Data

Section 12.1 considered the distinction between independent and paired samples in the context of quantitative responses. The same distinction applies to categorical responses. Recall that independent samples can be considered as distinct SRSs from separate populations. In contrast, paired samples match each observation in the first sample to a unique observation in the second sample. For example, we can match selections based on age, race, concurrent conditions, or any characteristics strongly associated with the response. This can help control for confounding by these matching variables, and also necessitates particular methods of analysis.

Matched-pair designs may be applied to cohort and case–control sampling methods. For **cohort** samples, the matched-paired data are displayed as follows:

TABLE 18.23   Matched-pair data table for cohort samples.

Nonexposed pair-member (sample 2)	Exposed pair-member (sample 1)	
	Disease +	Disease −
Disease +	a	b
Disease −	c	d

For **case–control** samples, the matched-pair table is:

TABLE 18.24   Matched-pair data table for case-control samples.

Control pair-member (sample 2)	Case pair-member (sample 1)	
	Exposed +	Exposed −
Exposed +	a	b
Exposed −	c	d

The cells in these tables receive counts of matched pairs, not counts of individuals. Cells a and d contain counts of **concordant pairs**; these are pairs that are the same with respect to the factor that can vary. Cells b and c contain counts of **discordant pairs**; these are pairs that differ with respect to the factor.

Odds Ratio

The **odds ratio** for the matched-pair data is provided by this formula:

$$\widehat{\text{OR}} = \frac{c}{b}$$

Notice that only the discordant pairs contribute information to this statistic.

Confidence Interval for the OR by a Normal Approximation

The confidence interval for the OR by a Normal approximation method is given by:

$$e^{\ln\widehat{\text{OR}} \pm z_{1-\alpha/2} \cdot \text{SE}_{\ln\widehat{\text{OR}}}}$$

where e is the base on the natural logarithms (e $\approx$ 2.71828...), z is a Standard Normal deviate corresponding to the desired level of confidence ($z = 1.645$ for 90% confidence, $z = 1.96$ for 95% confidence, and $z = 2.576$ for 99% confidence), and $\text{SE}_{\ln\widehat{\text{OR}}} = \sqrt{\dfrac{1}{c} + \dfrac{1}{b}}$.

ILLUSTRATIVE EXAMPLE

Matched case–control data (Diet and colonic polyps). A case–control study used matched pairs to study the relationship between adenomatous polyps of the colon and diet. Cases and controls underwent sigmoidoscopic screening. Controls were matched to cases on several factors, including date, clinic, age, and sex. There were 45 pairs in which the case but not the control reported low fruit and vegetable consumption (category c). There were 24 pairs in which the control but not the case reported low fruit and vegetable consumption (category b).[v] Based on this information:

- $\widehat{OR} = \dfrac{c}{b} = \dfrac{45}{24} = 1.88.$
- $\ln\widehat{OR} = \ln(1.875) = 0.6286$ (by calculator).
- $SE_{\ln\widehat{OR}} = \sqrt{\dfrac{1}{c} + \dfrac{1}{b}} = \sqrt{\dfrac{1}{45} + \dfrac{1}{24}} = 0.2528.$
- The 95% confidence interval for $OR = e^{0.6286 \pm (1.96)(0.2528)} = e^{0.6286 \pm 0.4959} = e^{(0.1331,\, 1.1241)} = (1.14 \text{ to } 3.08).$

This Normal-approximation–based confidence interval can be used when there are at least 10 positive-discordant pairs (matched-pair category c) and 10 negative-discordant pairs (category b) in the data set. With smaller samples, an exact method of calculating probabilities is needed.

Exact Confidence Interval

A confidence interval for the OR can be calculated by an exact method by treating the number of pairs in category c as the random number of successes of a binomial random variable based on $n = c + b$ independent trials. Let C represent the random number of successes in n matched pairs. Lowercase c represents the observed number of successes. C is distributed as a binomial random variable with parameters $n = c + b$ and p. The value of parameter p is estimated by $\hat{p} = \dfrac{c}{c + b}$. Confidence limits for binomial parameter p are derived with a plus-four method (Section 16.5) or exact method. Confidence limits for p are then transformed to odds ratio with the lower confidence limit $OR_{LCL} = \dfrac{p_{LCL}}{1 - p_{LCL}}$ and the upper confidence limit $OR_{UCL} = \dfrac{p_{UCL}}{1 - p_{UCL}}$. Alternatively, a procedure based on a relationship between the F probability distribution

[v] Witte, J. S., Longnecker, M. P., Bird, C. L., Lee, E. R., Frankl, H. D., & Haile, R. W. (1996). Relation of vegetable, fruit, and grain consumption to colorectal adenomatous polyps. *American Journal of Epidemiology, 144*(11), 1015–1025. Frequencies reported in Rothman & Greenland (1998, p. 287).

and binomial distribution may be applied as follows.[w] The **lower confidence limit of the 95% confidence limit for OR** is

$$\frac{c}{(b+1)F_{0.025,2(b+1),2c}}$$

and the **upper limit** is

$$\frac{(c+1)F_{0.025,2(c+1),2b}}{b}$$

where $F_{0.025,df_1,df_2}$ is an F random variable with df_1 and df_2 degrees of freedom and upper tail P region of 0.025. Because F tables are spotty in listing critical values for F random variables with high degrees of freedom, you might need to use a software utility to derive F values with accuracy. Alternatively, we can use *WinPepi*'s **PAIRSetc** program (Figure 18.9) to calculate exact confidence intervals directly.

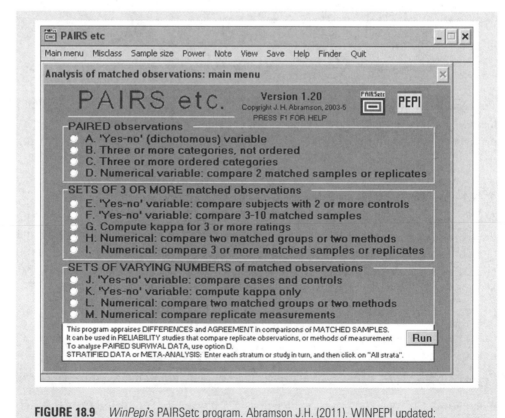

FIGURE 18.9   *WinPepi*'s PAIRSetc program. Abramson J.H. (2011). WINPEPI updated: computer programs for epidemiologists, and their teaching potential. *Epidemiologic Perspectives & Innovations, 8*(1), 1.

---

[w] Liddell, F. D. (1983). Simplified exact analysis of case-referent studies: Matched pairs; dichotomous exposure. *Journal of Epidemiology and Community Health, 37*(1), 82–84.

ILLUSTRATIVE EXAMPLE

Matched cohort data (Smoking and mortality in identical twins). When smoking was first suspected as causing premature mortality, R. A. Fisher offered the *constitutional hypothesis* as a possible explanation for the association. This hypothesis suggested that people genetically disposed to risk taking and adverse health outcomes were more likely to smoke, and hence more likely to experience premature death.[x] Years later, the constitutional hypothesis was put to a test by a study in which 22 smoking-discordant monozygotic twins were studied to see which twin first succumbed to death.[y] In this study, the smoking-twin died first in 17 of the pairs (matched-pair category *c*) and the nonsmoking twin died first in 5 instances (category *b*). Therefore, $\widehat{OR} = \dfrac{c}{b} = \dfrac{17}{5} = 3.40$.

Because there are only five negative-discordant pairs in this study, an exact method is used to calculate the confidence interval for the OR. For 95% confidence, the lower limit is $\dfrac{17}{(5 + 1)F_{0.025,12,34}} = \dfrac{17}{(5 + 1)(2.35)} = 1.21$. The upper confidence limit is $\dfrac{(17 + 1)F_{0.025,36,10}}{5} = \dfrac{(18)(3.27)}{5} = 11.77$.

Figure 18.10 is a screenshot of the data input into *WinPepi* PAIRSetc.exe (program A). (Zeros are entered for cells *a* and *d*, but these cells do not contribute to estimates.) Figure 18.11 is a screenshot of output from the program. The limits for Fisher's 95% confidence interval match those we calculated by hand.

continues

[x] Fisher did not entirely rule-out smoking as a cause of premature death. He merely put forward the constitutional hypothesis as an alternative explanation. Fisher, R. A. (1957). Dangers of cigarette-smoking. *British Medical Journal* (5034), 1518–1520. Fisher, R. A. (1958). Cancer and smoking. *Nature, 182*(4635), 596. Fisher, R. A. (1958). Lung cancer and cigarettes. *Nature, 182*(4628), 108.
[y] Kaprio, J., & Koskenvuo, M. (1989). Twins, smoking and mortality: A 12-year prospective study of smoking-discordant twin pairs. *Social Science & Medicine, 29*(9), 1083–1089.

continued

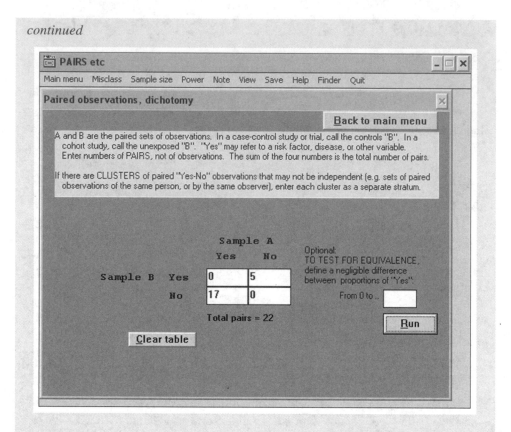

FIGURE 18.10   Screenshot of *WinPepi* data entry screen for smoking and mortality in identical twins illustrative example. Abramson J.H. (2011). WINPEPI updated: computer programs for epidemiologists, and their teaching potential. *Epidemiologic Perspectives & Innovations, 8*(1), 1.

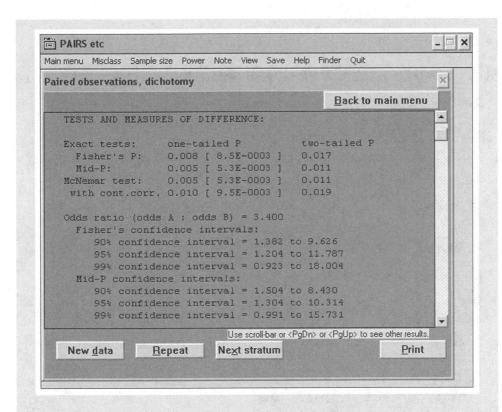

FIGURE 18.11 Screenshot of *WinPepi* output for the smoking and mortality in identical twins illustrative data. Abramson J.H. (2011). WINPEPI updated: computer programs for epidemiologists, and their teaching potential. *Epidemiologic Perspectives & Innovations, 8*(1), 1.

Hypothesis Test

Normal Approximation Method

When the number of discordant pairs ($b + c$) exceeds 5, we can use a z-statistic to test H_0: OR = 1 (no association in the population) against H_a: OR $\neq$ 1 (association in the population). The test statistic is[z]:

$$z_{\text{stat,McN}} = \sqrt{\frac{(c - b)^2}{c + b}}$$

With continuity correction, the test statistic is:

$$z_{\text{stat,McN,c}} = \sqrt{\frac{(|c - b| - 1)^2}{c + b}}$$

---

[z] This is equivalent to a McNemar test. The McNemar procedure is usually presented as a chi-square statistic with 1 df. We have taken the liberty of transforming the chi-square statistics to a z-statistic to take advantage of Table F.

Either form of the test statistic is acceptable.[aa]

ILLUSTRATIVE EXAMPLE

Hypothesis test for matched pairs, Normal approximation (Smoking and mortality in identical twins). Recall the illustrative data in which smoking-discordant monozygotic twins were studied to see which twin first succumbed to death. In 17 twin sets, the smoking twin died first. In 5 twin sets, the nonsmoking twin died first (OR = 3.4). The test of association follows.

A. **Hypotheses.** H_0: no association between smoking and earlier mortality in population (OR = 1) vs. H_a: "association" (OR ≠ 1).

B. **Test statistic.** $z_{\text{stat,McN}} = \sqrt{\dfrac{(c - b)^2}{c + b}} = \sqrt{\dfrac{(17 - 5)^2}{17 + 5}} = 2.56$

C. ***P*-value.** $P = 0.01047$ (from Table F). With continuity correction,

$$z_{\text{stat,McN,c}} = \sqrt{\dfrac{(|c - b| - 1)^2}{c + b}} = \sqrt{\dfrac{(|17 - 5| - 1)^2}{17 + 5}} = 2.35 \text{ and } P = 0.019.$$

D. **Significance level.** The evidence against H_0 is significant at the $\alpha = 0.025$ level but not at the $\alpha = 0.01$ level.

E. **Conclusion.** The twin who smoked tended to succumb to death earlier than their nonsmoking sibling ($P < 0.019$).

Exact Test

With five or fewer discordant pairs, an exact binomial procedure is used to test H_0: OR = 1. As was the case with exact confidence intervals, this is accomplished by treating the number of positive discordant pairs (matched-pair category c) as the observed number of successes for a binomial random variable based on $n = c + b$ Bernoulli trails. The observed proportion in the sample is $\hat{p} = \dfrac{c}{c + b}$. If there were no association between the exposure and disease in the population, half the discordant pairs would fall into matched-pair category c and half would fall into category b. Therefore, under the null hypothesis $p = \frac{1}{2}$ and the one-sided P-value looking for a positive association is $= \Pr(C \geq c)$ assuming $C \sim b(c + b, \frac{1}{2})$.

[aa] The continuity-corrected statistic is more popular, but here are two references that suggest we do not use the continuity-correction form of this statistic: (1) Bennett, B. M., & Underwood, R. E. (1970). On McNemar's test for the 2 × 2 table and its power function. *Biometrics, 26,* 339–343; (2) Lui, K.-K. (2001). Notes on testing equality in dichotomous data with matched pairs. *Biometrical Journal, 43*(3), 313–321.

ILLUSTRATIVE EXAMPLE

Hypothesis test for matched pairs, exact binomial test (Fictitious case–control data). Suppose a matched-pair case–control study finds three discordant pairs in which the case is exposed to a risk factor ($c = 3$) and one discordant pair in which the control is exposed ($b = 1$). We will test demonstrate a one-sided test of association for this observation.

A. **Hypotheses.** H_0: no association between the risk factor and the disease in the population (OR = 1) vs. H_a: "positive association" (OR > 1).
B. **Test statistic.** The effective sample size $n = c + b = 3 + 1 = 4$. The observed number of successes (c) = 3.
C. **P-value.** Under the null hypothesis, $p = \frac{1}{2}$ and the one-sided (Fisher's) P-value = $\Pr(C \geq 3)$ assuming $C \sim$ binomial(4, $\frac{1}{2}$). Using the binomial formula $\Pr(C = c) = {}_nC_c \cdot p^c \cdot q^{n-c}$ we calculate:

   - $\Pr(C = 3) = {}_4C_3 \cdot (\frac{1}{2})^3 \cdot (\frac{1}{2})^1 = 4 \cdot 0.125 \cdot 0.5 = 0.25$.
   - $\Pr(C = 4) = {}_4C_4 \cdot (\frac{1}{2})^4 \cdot (\frac{1}{2})^0 = 1 \cdot 0.0625 \cdot 1 = 0.0625$.
   - By Fisher's method, the one-sided P-value = $\Pr(C \geq 3) = \Pr(C = 3) + \Pr(C = 4) = 0.25 + 0.0625 = 0.3125$.
   - To get the mid-P P-value, use half of $\Pr(C = 3)$ in your calculations, so $P = \frac{1}{2}(0.25) + 0.0625 = 0.1875$.

Figure 18.12 is a screenshot of output from *WinPepi* PAIRSetc.exe replicating these results.

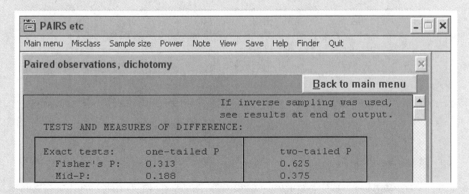

FIGURE 18.12 Screenshot from *WinPepi* PAIRSetc program showing results from exact test for matched-pair analysis. Abramson J.H. (2011). WINPEPI updated: computer programs for epidemiologists, and their teaching potential. *Epidemiologic Perspectives & Innovations, 8*(1), 1.

continues

continued

D. Significance level. The result of the one-tailed Fisher's exact test is not significant at the $\alpha = 0.30$ level.

E. Conclusion. The evidence for an association between the exposure and the outcome is weak ($P = 0.32$).

Exercises

18.19 *Diet and adenomatous polyps.* Recall the illustrative example concerning a matched case–control study of adenomatous colon polyps and diet. There were 45 matched pairs in which the case but not the control reported low fruit and vegetable consumption. There were 24 pairs in which the control but not the case reported low fruit and vegetable consumption. Test the association for significance. Show all hypothesis-testing steps.

18.20 *Estrogen and endometrial cancer.* Table 18.25 contains data from a matched case-control study on conjugated estrogen use and endometrial cancer.

TABLE 18.25   Estrogen use and endometrial cancer, case–control, matched pairs.

	Case exposed	Case not exposed	Total
Control exposed	12	7	19
Control not exposed	43	121	164
Total	55	128	183

Data from Antunes, C. M., Strolley, P. D., Rosenshein, N. B., Davies, J. L., Tonascia, J. A., Brown, C., et al. (1979). Endometrial cancer and estrogen use. Report of a large case-control study. *New England Journal of Medicine, 300*(1), 9–13 as reported on p. 137 of Abramson, J. H., & Gahlinger, P. M. (2001). *Computer Programs for Epidemiologic Analyses: PEPI v. 4.0.* Salt Lake City, UT: Sagebrush Press.

(a) Calculate the odds ratio for estrogen exposure and endometrial cancer. Interpret this result

(b) Calculate a 95% confidence interval for the OR. Interpret this result.

(c) Test the hypothesis.

18.21 *Thrombotic stroke in young women.* A 1970s case–control study on cerebrovascular disease (stroke) and oral contraceptive use in young women matched cases to controls according to neighborhood, age, sex, and race.[bb] Table 18.26 displays data from this study for thrombotic stroke.

[bb] Oral contraceptives and stroke in young women. Associated risk factors. (1975). *JAMA, 231*(7), 718–722.

TABLE 18.26 Matched case–control data on thrombotic stroke and oral contraceptive use in women between 14 and 44 years of age.

Matched Pairs	Case exposed	Case not exposed	Total
Control exposed	2	5	7
Control not exposed	44	55	99
Total	46	60	106

Data from Collaborative group for the study of stroke in young women. (1973). Oral contraception and increased risk of cerebral ischemia or thrombosis. *New England Journal of Medicine, 288*(17), 871–878.

(a) Calculate the odds ratio for these data. Briefly, interpret the results in narrative form.

(b) Suppose the match was broken and investigators analyzed the data as if derived from independent samples. Rearrange the information in Table 18.26 to show how it would appear in a two-by-two table of unmatched counts. Because the matched-pair table (Table 18.26) includes 106 pairs, your new table should total 212 individuals. Calculate the odds ratio for the data with the match broken. How does this compare to that of the (proper) matched-pair odds ratio?

Summary Points (Cross-Tabulated Counts)

1. This chapter considers categorical data that are cross tabulated to form an **R-by-C table** with the explanatory variable along the rows and response variable along the columns.

2. For naturalistic and purposive cohort samples, the **row proportions** in the cross tabulation represent group incidence proportions or prevalence proportions. (Row proportions are not relevant for case–control data.)

3. **This chi-square statistic** $X^2_{\text{stat}} = \Sigma \dfrac{(O_i - E_i)^2}{E_i}$ is used to test H_0: no association between the row variable and column variable in source population.

4. When the explanatory variable is ordinal, a **test for trend** may be applied. We rely on statistical software for this calculation.

5. **Case–control studies** select study subjects based on the status of the outcome variable in the study. This disables the calculation of incidence and prevalence but permits the estimation of relative risk through the odds ratio $\widehat{\text{OR}} = \dfrac{a_1 b_2}{b_1 a_2}$.

   (a) Cases and controls must be selected in a manner that accurately reflects the exposure experience of diseases and nondiseased individuals in the source population.

(b) The odds ratio has the same interpretation as a rate ratio, that is, it is a measure of relative effect.

(c) The $(1 - \alpha)100\%$ confidence interval for the OR $=$ $e^{\ln\widehat{OR} \pm z_{1-\frac{\alpha}{2}} \cdot \sqrt{a_1^{-1} + b_1^{-1} + a_2^{-1} + b_2^{-1}}}$.

(d) H_0: OR $= 1$ is tested with the chi-square statistic presented in item 3.

(e) When the explanatory variable is ordinal, a test for trend may be applied.

6. **Matched-pair samples** may be used to address potential confounders. This requires adaptation of computational methods.

(a) The matched-pair is the unit of analysis. A table that counts concordant and discordant pairs is constructed (Table 18.24).

(b) The odds ratio estimator $\widehat{OR} = \dfrac{c}{b}$, where c represents the number of discordant pairs with an exposed case and b represents the number of discordant pairs with an exposed control.

(c) The $(1 - \alpha)100\%$ confidence interval for the OR $= e^{\ln\widehat{OR} \pm z_{1-\frac{\alpha}{2}} \cdot \sqrt{c^{-1} + b^{-1}}}$.

(d) H_0: OR $= 1$ is tested with McNemar's test (large samples) or exact binomial procedure (small samples).

Vocabulary

Case–control samples
Column percents
Conditional percents
Continuity-corrected (Yates')
  chi-square statistic
Cross-product ratio
Degrees of freedom
Expected frequencies
Incidence density sampling

Marginal distributions
Naturalistic samples
Observed frequencies
Odds
Odds ratio
Pearson's chi-square statistic
Purposive cohort samples
R-by-C contingency table
Row percents

Review Questions

18.1 Select the best response: This type of sample selects a random cross section of the population.

(a) naturalistic sample

(b) purposive cohort

(c) case–control

18.2 Select the best response: This type of sample selects a fixed number of exposed and nonexposed individuals from the population.

(a) naturalistic sample

(b) purposive cohort

(c) case–control

18.3 Select the best response: This type of sample selects a fixed numbers of persons who have experienced the study outcome and fixed number of individuals who have not yet experienced the study outcome from the population.

(a) naturalistic sample

(b) purposive cohort

(c) case–control

18.4 What does the *R* stand for in an *R-by-C table*?

18.5 What does the *C* stand for in an *R-by-C table*?

18.6 What is a marginal distribution?

18.7 How do you calculate a marginal percent?

18.8 What is a conditional distribution?

18.9 How do you calculate a conditional row percent?

18.10 How do you calculate a conditional column percent?

18.11 This symbol represents the sample proportion in group *i*.

18.12 Select the best response: This statistic is an estimate of the risk of an outcome.

(a) prevalence

(b) incidence proportion

(c) risk ratio

18.13 This statistic represents the incidence proportion in the exposed group *minus* the incidence proportion in the nonexposed group.

18.14 This statistic represents the incidence proportion in the exposed group *divided* by the incidence proportion in the nonexposed group.

18.15 Select the best response: A risk difference of 0.56 per 1000 indicates _____ association between the exposure and the outcome.

(a) a positive

(b) no

(c) a negative

18.16 Select the best response: A risk ratio of 0.56 indicates _____ association between the study exposure and the outcome.

(a) a positive

(b) no

(c) a negative

18.17 This is the ratio of "successes" to "failures."

18.18 This is the ratio of two odds.

18.19 This is the most common statistic used to test for associations among categorical variables.

18.20 A chi-square statistic for testing for an association between the row variable and the column variable in an R-by-C table has this many degrees of freedom.

18.21 Fill in the blanks: Chi-square probability density functions are _____ (symmetrical or asymmetrical) continuous distributions that become increasingly _____ (symmetrical or asymmetrical) as their degrees of freedom increase.

18.22 Select the best response: Chi-square distributions _____ 0.

(a) are centered on

(b) start on

(c) end on

18.23 True or false? Say whether each of these statements about chi-square probability density functions is true or false.

(a) have a mean equal to 0

(b) have a mean equal to their degrees of freedom

(c) have a mode of 0

(d) have a mode of 0 when df = 1 or 2

(e) have a variance of 1

(f) have a variance of two times their degrees of freedom

18.24 Select the best response: Expected values in R-by-C table cells assume

(a) null hypothesis is true.

(b) null hypothesis is false.

(c) alternative hypothesis is true.

18.25 Fill in the blank: The expected value in table cell is equal to the (row total) × (column total) divided by _____.

18.26 State the null hypothesis for the chi-square tests of association in words.

18.27 State the null hypothesis for the Mantel test for trend in words.

18.28 Select the best response: This measure of association is used in case–control studies.

(a) risk

(b) risk ratio

(c) odds ratio

18.29 Select the best response: When the outcome is rare (incidence less than 5%), the odds ratio has the same interpretation as the

(a) risk

(b) risk ratio

(c) risk difference

18.30 Select the best response: These types of samples match individuals to one other based on factors thought to have the potential to confound the relationship being studied.

(a) independent samples

(b) paired samples

(c) case–control samples

Exercises

18.22 *Binge drinking by gender.* Wechsler and coworkers (1994) found that 19.4% of students at 4-year U.S. colleges engaged in frequent binge drinking. Table 18.27 reports the data from this study by gender.

TABLE 18.27 Data for Exercise 18.22. Frequency of binge drinking by gender. Binge drinking is defined as having five or more drinks in a row three or more times during the past 2-week period.

	Binge +	Binge −
Men	1630	5550
Women	1684	8232

Data from Wechsler, H., Davenport, A., Dowdall, G., Moeykens, B., & Castillo, S. (1994). Health and behavioral consequences of binge drinking in college. A national survey of students at 140 campuses. *JAMA, 272*(21), 1672–1677.

(a) Calculate the prevalence of binge drinking for males and for females separately.

(b) Fill in this blank: In *absolute* terms, males had a ____% greater prevalence of binge drinking than females.

(c) Fill in this blank: In *relative* terms, males had a ____% greater prevalence of binge drinking than females.

(d) Calculate a 95% confidence interval for the prevalence ratio of binge drinking for males.

18.23 ***Don't sweat the small stuff.*** Consider a study in which 40 of 320 individuals (12.5%) in the treatment group experience an outcome. In contrast, 26 of 336 (7.7%) of the individuals in the control group experience the outcome. Table 18.28 displays the data in cross-tabulated form. Calculate chi-square statistics and *P*-values for this problem using both the Pearson's and continuity-corrected (Yates') chi-square statistics. (If your instructor allows, consider using a software program such as *WinPepi*'s Compare2.exe program to calculate results.) Interpret the results of the tests. Is it reasonable to derive different conclusions from these different test statistics?

TABLE 18.28   Data for Exercise 18.23.

	Adverse event		
	Yes	No	Total
Treatment	40	280	320
Control	26	310	336
Total	66	590	656

18.24 ***University Group Diabetes Program.*** A multicentered clinical trial completed in the 1960s assessed the efficacy of various oral hypoglycemics, insulin, and diet in preventing vascular complications of diabetes.[cc] It found that 26 of 204 patients (13%) treated with the oral hypoglycemic phenformin died unexpectedly from cardiovascular events. In contrast, 2 of 64 control patients (3%) experienced a similar outcome.

(a) Is an exact procedure (e.g., Fisher's) necessary to test these data or can a chi-square test be used? Justify your response.

(b) Test the difference for significance.

18.25 ***Yates, 1934 (three-by-two).*** Table 18.29 displays results from a historically important methodological paper written by Frank Yates on the analysis of

[cc] UGDP. (1970). University Group Diabetes Program: A study of the effects of hypoglycemic agents on vascular complications in patients with adult-onset diabetes. *Diabetes, 19*(Suppl. 2).

cross-tabulated counts.[dd] Explain why a chi-square statistic should not be used to test these data. Then use WinPepi > Compare2.exe > Program F1 to calculate a Fisher's P-value for the problem. Comment on your findings.

TABLE 18.29   Data for Exercise 18.25.

	No malocclusion	Malocclusion
Breastfed	4	16
Bottlefed	1	21
Breastfed & bottlefed	3	47

Data from Yates, F. (1934). Contingency tables involving small numbers and the χ^2 test. *Supplement to the Journal of the Royal Statistical Society, 1*(2), 217–235.

18.26   ***IUDs and infertility.*** A case–control study found a history of intrauterine device use in 89 of 283 infertile women. In contrast, 640 of the 3833 controls had used IUDs.[ee] Calculate the odds ratio and its 95% confidence interval for IUD use and infertility. (Note: Data must first be put into a two-by-two table before the odds ratio is calculated.) Interpret the results.

18.27   ***Esophageal cancer and tobacco use.*** The case–control study on esophageal cancer and alcohol use that we used to illustrate case–control methods in this chapter also collected data on tobacco use as a potential risk factor. Table 18.30 shows data from the study with tobacco use dichotomized at 20 g/day. Calculate the odds ratio and 95% confidence interval for these data. Interpret the results.

TABLE 18.30   Data for Exercise 18.27.

Tobacco	Cases	Noncases
20+ g/day	64	150
0–19 g/day	136	625

Data from Tuyns, A. J., Pequignot, G., & Jensen, O. M. (1977). [Esophageal cancer in Ille-et-Vilaine in relation to levels of alcohol and tobacco consumption]. *Bulletin du Cancer, 64*(1), 45–60. Data set are stored online in the file BD1.* as variables TOB2 and CASE.

18.28   ***Brain tumors and electric blanket use.*** A case–control study assessed risks of brain tumors in children. Table 18.31 cross tabulates the data according to case status and electric blanket exposure. Calculate the odds ratio and its 95% confidence interval. Comment on the results.

[dd] Frank Yates (1902–1994)—student and colleague of R. A. Fisher; originator of the continuity-corrected chi-square statistic that bears his name.
[ee] Cramer, D. W., Schiff, I., Schoenbaum, S. C., Gibson, M., Belisle, S., Albrecht, B., et al. (1985). Tubal infertility and the intrauterine device. *New England Journal of Medicine, 312*(15), 941–947.

TABLE 18.31   Data for Exercise 18.28.

	Cases	Noncases
EI. blanket +	53	102
EI. blanket −	485	693

Data from Preston-Martin, S., Gurney, J. G., Pogoda, J. M., Holly, E. A., & Mueller, B. A. (1996). Brain tumor risk in children in relation to use of electric blankets and waterbed heaters. Results from the United States West Coast Childhood Brain Tumor Study. *American Journal of Epidemiology, 143*(11), 1116–1122. Data are available in the file BRAINTUM.*.

18.29   *Baldness and myocardial infarction, interviewer assessments.* Exercises 18.17 and 18.18 considered a case–control study of baldness and heart attacks in men using self-assessments of baldness. This same study also had the interviewers assess the extent of baldness of study subjects using a Hamilton baldness scale (Figure 18.13).

(a) Which method of ascertaining baldness do you think would be preferable in this study, the self-assessment or interviewer-based assessment?

(b) Table 18.32 cross tabulates the data according to the ascertainments made by the interviewers. Calculate the odds ratios of myocardial infarction associated with each level of baldness according to this classification.

TABLE 18.32   Data for Exercise 18.29. Baldness and myocardial infarction, case–control data. Baldness was assessed by interviewers according to the Hamilton baldness scale (Figure 18.13).

Baldness	Cases	Controls
None[a]	238	480
Frontal[b]	44	82
Mild vertex[c]	108	137
Mod vertex[d]	40	46
Severe vertex[e]	35	23
Total	465	768

[a]Hamilton baldness categories I and II.

[b]Hamilton categories IIa, III, IIIa, and IVa.

[c]Hamilton categories III and IV.

[d]Hamilton categories V and Va.

[e]Hamilton categories VI and VI.

Abramson J.H. (2011). WINPEPI updated: computer programs for epidemiologists, and their teaching potential. *Epidemiologic Perspectives & Innovations, 8*(1), 1.

(c) Compare results from this analysis to the results achieved with the self-assessed baldness in Exercise 18.17.

18.30 ***Wynder and Graham's case–control study of smoking and lung cancer.*** A historically important case–control study of smoking and lung cancer compared smoking histories of 605 lung cancer cases to 780 noncancer controls. Table 18.33 displays data from this study.

(a) Calculate the odds ratio associated with each level of smoking; comment on the findings.

(b) Conduct a test for trend in the odds ratios.

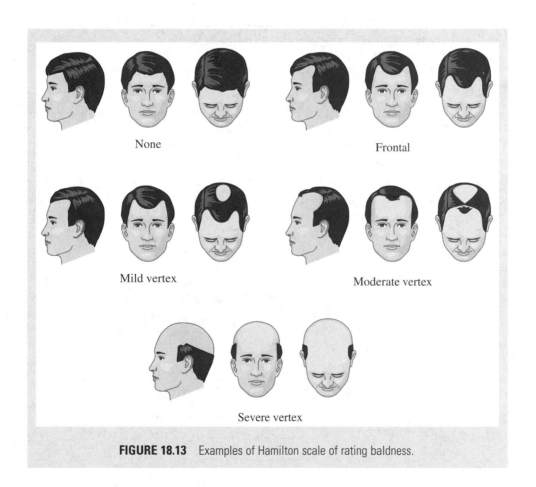

None

Frontal

Mild vertex

Moderate vertex

Severe vertex

FIGURE 18.13 Examples of Hamilton scale of rating baldness.

TABLE 18.33   Data for Exercise 18.30. Case–control data on long-term smoking habit and lung cancer.

Smoking level*	Cases	Noncases
1 Nonsmoker (<1 cigarette per day)	8	115
2 Light smoker (1–9 cigs per day)	14	82
3 Moderate (10–15 cigs per day)	61	147
4 Heavy (16–20 cigs per day)	213	274
5 Excessive (21–34 cigs per day)	186	98
6 Chain (35 of more cigs per day)	123	64
Total	605	780

*If subject smoked for less than 20 years, the amount of smoking was reduced in proportion to duration.

Data from Wynder, E. L., & Graham, E. A. (1950). Tobacco smoking as a possible etiologic factor in bronchiogenic carcinoma. *JAMA, 143 (1)*(4), 329–336.

18.31   *Practice with chi-square.* Calculate a chi-square test of association statistic and *P*-value for the case–control data in Table 18.21. Justify that it is acceptable to use a chi-square statistic in this situation even though two of the cells have small numbers.

18.32   *Hemorrhagic stroke in young women.* Exercise 18.21 addressed data from a matched-pair case–control study on oral contraceptive use and thrombotic stroke in young women. The investigators also collected information on hemorrhagic stroke. Table 18.34 lists data for this outcome.

(a)   Calculate the odds ratio for these data. Interpret the results.

(b)   Break the match and rearrange the data to form a 2-by-2 cross tabulation for unmatched data. Calculate the odds ratio ignoring the match. How does this odds ratio compare to that of the (proper) matched-pair analysis?

TABLE 18.34   Data for Exercise 18.32. Thrombotic stroke and oral contraceptive case-control study. Data are matched pairs.

	Case exposed	Case nonexposed	Total
Control exposed	5	13	18
Control nonexposed	30	107	137
Total	35	120	155

Data from Oral contraceptives and stroke in young women. Associated risk factors. (1975). *JAMA, 231*(7), 718–722.

19 | Stratified Two-by-Two Tables

■ 19.1 Preventing Confounding

Confounding is a systematic error in inference due to the influence of extraneous variables (Section 2.2). This chapter reviews concepts about confounding and introduces a method to adjust for confounding when working with binary variables.

The word *confounding* comes from the Latin term meaning "to mix together." When groups are unbalanced with respect to extraneous factors that influence the response, the effects of these extraneous factors get mixed up with the effects of the explanatory variable. This causes a distortion in the observed association, which we call *confounding*; thus, confounding is a bias due to the influence of extraneous variables. Extraneous variables that cause confounding arc called **confounders**.[a]

Confounder variables exhibit the following three properties:

1. They are associated with the explanatory variable in the study.
2. They are independent determinants of the response.
3. They are not an intermediary in the causal pathway between the explanatory variable and the response.

One way to understand confounding is to consider what would have happened in the exposed group had the exposure been absent. Of course this cannot occur in nature, so it is *counterfactual*. However, this type of thing helps clarify the nature of confounding. This type of "thought experiment" isolates the effects of the explanatory variable independent of the effect of confounders. The general idea is to make comparisons **ceteris paribus** (Latin for "with other things equal").

Several statistical techniques can be used to prevent or mitigate the effects of confounders. Methods include randomization, restriction, matching, multivariate regression, and stratification.

---

[a]For an extraneous factor to be a confounder, it must be associated with the explanatory variable and must be an independent predictor of the response.

- **Randomization** (Section 2.2) is often the most effective way to prevent confounding. It works by balancing the extraneous factors that can confound results among the comparison groups. Randomization is especially effective in large samples where the *law of large numbers* balances confounders most incisively.

- **Restriction** is a method that imposes uniformity in the study base by limiting the type of individuals who may participate in the study. By imposing restrictions, we effectively define a source population that is homogeneous with respect to the potential confounders. For example, if we limit a study to nonsmokers, smoking can no longer confound observed relationships. Restriction is applicable to both experimental and nonexperimental study designs.

- **Matching** adjusts for factors by making like-to-like comparisons. We considered matched analyses in Section 11.5 (quantitative responses) and Section 18.6 (binary responses).

- **Regression** adjusts for potential confounders through mathematical modeling. Chapter 15 used multiple linear regression for this purpose. Multiple linear regression applies to quantitative outcomes. Other types of regression models (e.g., logistic regression) apply to categorical responses.

- **Stratification** divides the data set into homogenous subgroups before pooling results. This chapter introduces stratified methods for two-by-two tables.

■ 19.2 Simpson's Paradox

Simpson's paradox[b] is an extreme form of confounding in which the observed association switches direction after considering the influence of the confounder.

ILLUSTRATIVE EXAMPLE

Simpson's paradox (Effectiveness of a treatment).[c] A trial tests a treatment at two separate clinics. Upon completion of the study, cross tabulation of results from both clinics combined is:

[b] Simpson, E. H. (1951). The interpretation of interaction in contingency tables. *Journal of the Royal Statistical Society. Series B, 13*(2), 238–241.
[c] Blyth, C. R. (1972). On Simpson's paradox and the sure thing principle. *Journal of the American Statistical Association, 67*, 364–366.

TABLE 19.1   Both clinics combined.

Group	Success	Failure	Total
1 (treatment)	1095	9005	10,100
2 (control)	5050	5950	11,000
Total	6145	14,955	21,100

The incidence proportion of success in the treatment group $\hat{p}_1 = 1095/$ $10,100 = 11\%$. The incidence proportion in the control group $\hat{p}_2 = 5050/11,000$ $= 46\%$. Therefore, the relative incidence ("relative risk"[d]) of success associated with the treatment is $11\% / 46\% = 0.24$. The treatment appears to be 76% less effective than the alternative.

Terminology and Notation: The analysis in the prior illustration is "crude" or "unadjusted" in the sense that it applies to all study subjects combined without mathematical adjustment. Let $\widehat{RR}$ (no subscript) represent the **crude relative risk**. Data will be divided into subgroups or **strata**. Statistics within strata are **strata-specific statistics**. Let $\hat{p}_{ik}$ represent the incidence proportion in treatment group i in stratum k, and let $\widehat{RR}_k$ represent the relative risk in stratum k. Let us now return to the illustrative example. Within clinic ("stratum") 1, a different picture emerges:

TABLE 19.2   Clinic 1.

Group	Success	Failure	Total
1 (treatment)	1000	9000	10,000
2 (control)	50	950	1,000
Total	1050	9950	11,000

Here, the incidence of success in the treatment group $\hat{p}_{11} = 1000/10,000 = 10\%$. The incidence in the control group $\hat{p}_{21} = 50/1000 = 5\%$. Therefore, the relative incidence of success $\widehat{RR}_1 = 10\%/5\% = 2.00$. In this clinic, the new treatment is twice as effective as the old treatment.

---

[d]We have established use of the term "relative risk" to apply to incidence proportion ratios and prevalence proportion ratios.

Data for clinic 2 are similarly revealing:

TABLE 19.3   Clinic 2.

Group	Success	Failure	Total
1 (treatment)	95	5	100
2 (control)	5000	5000	10,000
Total	5095	5005	10,100

In this clinic, $\hat{p}_{12} = 95/100 = 95\%$, $\hat{p}_{22} = 5000/10{,}000 = 50\%$, and $\widehat{RR}_2 = 1.9$. Again, the new treatment is (almost) twice as effective as the alternative. Therefore, what initially appeared to be poor performance by the treatment is reversed when we look within strata.

How do we reconcile this paradox? The answer lies in the fact that subjects at clinic 1 were much less likely to recover than the subjects at clinic 2 *and* the treatment was used primarily at clinic 1. Thus, the severity of illness (as measured by "clinic") confounded the relationship between the explanatory variable (treatment) and response variable (success). What initially appeared as a negative association reversed itself after controlling for severity of illness. This is an example of Simpson's paradox.

■ 19.3 Mantel–Haenszel Methods

Summary Measure of Effect

Confounding derives from the mixing of effects of confounding variables with those of the explanatory variable. In the previous example, the clinic was a surrogate for severity of illness. By stratifying the data into separate clinics, we achieved like-to-like comparisons and thus mitigated the effects of the confounding variable.

Our analysis could end here with data reported separately for each clinic. However, it is often desirable to have a single measure of effect that summarizes the unconfounded relationship between the explanatory variable and response variable. A **Mantel–Haenszel procedure** can be used for this purpose.

Here is the **stratum-specific notation** we need to calculate Mantel–Haenszel statistics:

TABLE 19.4   Notation, stratum k.

Group	Success	Failure	Total
Treatment	a_{1k}	b_{1k}	n_{1k}
Control	a_{2k}	b_{2k}	n_{2k}
Total	m_{1k}	m_{2k}	n_k

The **Mantel–Haenszel summary relative risk** $(\widehat{RR}_{MH})$ is a weighted average of strata-specific relative risks with weights equal to $w_k = n_{1k} n_{2k}/n_k$. It follows that

$$\widehat{RR}_{MH} = \frac{\Sigma_k w_k a_{1k}/n_{1k}}{\Sigma_k w_k a_{2k}/n_{2k}} = \frac{\Sigma_k a_{1k} n_{2k}/n_k}{\Sigma_k a_{2k} n_{1k}/n_k}$$

For the data in Table 19.2 and 19.3, the Mantel–Haenszel summary relative risk estimate is

$$\widehat{RR}_{MH} = \frac{\Sigma_k a_{1k} n_{2k}/n_k}{\Sigma_k a_{2k} n_{1k}/n_k} = \frac{1000 \cdot 1000/11{,}000 + 95 \cdot 10{,}000/10{,}100}{50 \cdot 10{,}000/11{,}000 + 5000 \cdot 100/10{,}100} = 1.95$$

Recall that the strata-specific relative risks in this data set are $\widehat{RR}_1 = 2.0$ and $\widehat{RR}_2 = 1.9$. It makes sense that the summary relative risk is between these two values.

Confidence Interval

The sampling distribution of the *natural log* of the Mantel–Haenszel summary relative risk estimator is approximately Normal with this **standard error:**[e]

$$SE_{\ln \widehat{RR}_{MH}} = \sqrt{\frac{\Sigma(m_{1k} n_{1k} n_{2k}/n_k^2 - a_{1k} a_{2k}/n_k)}{\Sigma(a_{1k} n_{2k}/n_k)(\Sigma a_{2k} n_{1k}/n_k)}}$$

Therefore, the $(1 - \alpha)100\%$ confidence for the $\ln RR_{MH}$ is

$$\ln \widehat{RR}_{MH} \pm (z_{1-\frac{\alpha}{2}})(SE_{\ln \widehat{RR}_{MH}})$$

Take the antilogs of these limits to move the confidence interval to a nonlogarithmic scale.

For the illustrative data:

- $SE_{\ln \widehat{RR}_{MH}} =$

$$\sqrt{\frac{(1050 \cdot 10{,}000 \cdot 1000/11{,}000^2 - 1000 \cdot 50/11{,}000) + (5095 \cdot 100 \cdot 10{,}000/10{,}100^2 - 95 \cdot 5000/10{,}100)}{(1000 \cdot 1000/11{,}000 + 95 \cdot 10{,}000/10{,}100)(50 \cdot 10{,}000/11{,}000 + 5000 \cdot 100/10{,}100)}}$$

  $= 0.06963.$

- The 95% confidence interval for the $\ln RR_{MH} = \ln(1.95) \pm (1.96)(0.06963) = 0.66783 \pm 0.13647 = (0.53136, 0.80430)$.

- These limits are exponentiated to derive the 95% confidence interval for the Mantel–Haenszel RR: $e^{(0.53136, \, 0.80430)} = (1.70, 2.24)$.

Thus, after adjusting for "clinic" (a surrogate for the severity of illness), we can be 95% confident that the treatment is associated with between a 1.7-fold and a 2.2-fold increase in the success rate.

---

[e] Greenland, S., & Robins, J. M. (1985). Estimation of a common effect parameter from sparse follow-up data. *Biometrics, 41*(1), 55–68.

Calculation with a Software Utility

Because hand calculations of Mantel–Haenszel statistics are tedious and prone to error, it is helpful to use statistical programs and applications to perform these calculations. Most commercial statistical programs permit Mantel–Haenszel calculations on individual-level data (i.e., data with exposure, outcome, and confounder status recorded on individuals). In addition, many free public domain applications (e.g., OpenEpi .com and WinPEPI) will perform Mantel–Haenszel calculations based on stratified two-by-two cross tabulations. Let us use *WinPEPI* to demonstrate one such process.

WinPepi calculates Mantel–Haenszel statistics with its Compare.exe program (option B).[f] After entering the data and calculating statistics for stratum 1, click the "Next stratum" button; then enter data for stratum 2. Figure 19.1 is a screenshot from the program showing the cursor arrow pointing toward the "Next stratum" button.

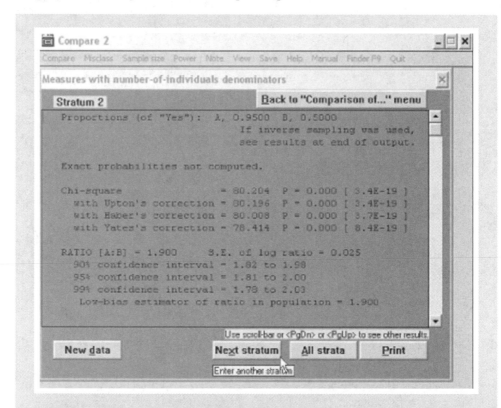

FIGURE 19.1 Screenshot of *WinPepi*'s program for stratified table analysis. The cursor is pointing to the button to add a new stratum. Click the adjacent button to calculate Mantel–Haenszel and other statistics for all strata combined. Abramson J.H. (2011). WINPEPI updated: computer programs for epidemiologists, and their teaching potential. *Epidemiologic Perspectives & Innovations, 8*(1), 1.

---

[f] Abramson, J. H. (2004). *WINPEPI* (PEPI-for-Windows): Computer programs for epidemiologists. *Epidemiologic Perspectives & Innovations, 1*(1), 6.

When data for all strata are entered, click the "All strata" button to calculate Mantel–Haenszel statistics. Figure 19.2 is a screenshot of output from the illustrative example. Mantel–Haenszel summary relative risk estimates (highlighted) match our prior calculations.

Other Summary Measures of Effect

Mantel–Haenszel procedures may also be applied to odds ratios, risk differences, and other such measures of effect. For example, the **Mantel–Haenszel summary odds ratio** is

$$\widehat{OR}_{MH} = \frac{\Sigma_k a_{1k} b_{2k}/n_k}{\Sigma_k a_{2k} b_{1k}/n_k}$$

The standard error of the natural log of this estimate is

$$SE_{\ln \widehat{OR}_{MH}} = \sqrt{\frac{\Sigma_k G_k P_k}{2\left(\Sigma_k G_k\right)^2} + \frac{\Sigma_k (G_k Q_k + H_k P_k)}{2\left(\Sigma_k G_k \Sigma_k H_k\right)} + \frac{\Sigma_k H_k Q_k}{2\left(\Sigma_k H_k\right)^2}}$$

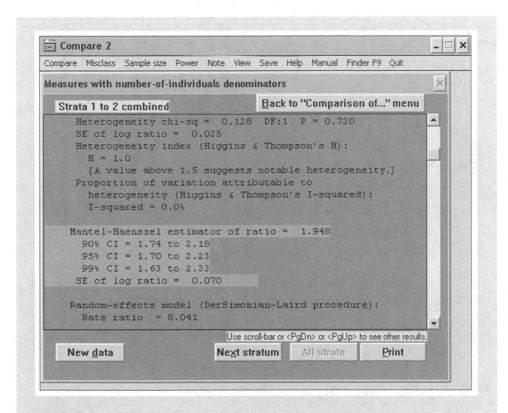

FIGURE 19.2 Screenshot of *WinPepi* output highlighting Mantel–Haenszel summary risk ratio estimates. Abramson J.H. (2011). WINPEPI updated: computer programs for epidemiologists, and their teaching potential. Epidemiologic Perspectives & Innovations, 8(1), 1.

where $G_k = a_{1k}b_{2k}/n_k$; $H_k = a_{2k}b_{1k}/n_k$; $P_k = (a_{1k} + b_{2k})/n_k$; and $Q_k = (a_{2k} + b_{1k})/n_k$.[g] These and other Mantel–Haenszel procedures are easily calculated with *WinPepi*.

Test Statistic

The null hypothesis of "no association"[h] is tested with the **Mantel–Haenszel test statistic**:

$$\chi^2_{stat,MH} = \frac{(\Sigma_k a_{1k} - \Sigma_k n_{1k}m_{1k}/n_k)^2}{\Sigma_k(n_{1k}n_{2k}m_{1k}m_{2k})/(n_k^2(n_k - 1))}$$

Under the null hypothesis, this statistic is a chi-square random variable with 1 degree of freedom.

The Mantel–Haenszel test statistic for the illustrative data is $\chi^2_{stat,MH} =$

$$\frac{\Big((1000 + 95) - (10{,}000 \cdot 1050/11{,}000 + 100 \cdot 5095/10{,}100)\Big)^2}{10{,}000 \cdot 1000 \cdot 1050 \cdot 9950/\Big(11{,}000^2(11{,}000 - 1)\Big) + 100 \cdot 10{,}000 \cdot 5095 \cdot 5005/\Big(10{,}100^2(10{,}100 - 1)\Big)} =$$

78.463 with 1 df. This converts to $P = 8.2 \times 10^{-19}$. Therefore, the association between the new treatment and "success" is highly significant.

Notes

1. A continuity correction may be applied to the Mantel–Haenszel test statistic as follows: $\chi^2_{stat,MH,c} = \dfrac{\left(|\Sigma_k a_{1k} - \Sigma_k n_{1k}m_{1k}/n_k| - \frac{1}{2}\right)^2}{\Sigma_k(n_{1k}n_{2k}m_{1k}m_{2k})/(n_k^2(n_k - 1))}$. The continuity-corrected

   Mantel–Haenszel statistic for the illustrative data is 77.594 with 1 df. $P = 1.3 \times 10^{-18}$. (Mantel chi-squares always have 1 df.)

2. Figure 19.3 is a screenshot from *WinPepi* showing the Mantel–Haenszel test statistic for the illustrative data.

3. The Mantel–Haenszel chi-square statistic has 1 df, so it can be reexpressed as

$$z_{stat,MH} = \sqrt{\frac{\left(\Sigma_k a_{1k} - \Sigma_k n_{1k}m_{1k}/n_k\right)^2}{\Sigma_k(n_{1k}n_{2k}m_{1k}m_{2k})/(n_k^2(n - 1))}}$$

---

[g] Robins, J., Breslow, N., & Greenland, S. (1986). Estimators of the Mantel–Haenszel variance consistent in both sparse data and large-strata limiting models. *Biometrics, 42*, 311–323.
[h] In testing a relative risk, H_0:$RR_{MH} = 1$; in testing an odds ratio H_0:$OR_{MH} = 1$; in testing a risk difference, H_0:$RD_{MH} = 0$; and so on.

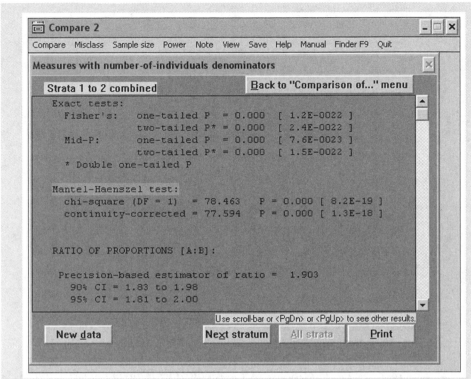

FIGURE 19.3 Screenshot of *WinPepi* output showing Mantel–Haenszel test statistics. Abramson J.H. (2011). WINPEPI updated: computer programs for epidemiologists, and their teaching potential. *Epidemiologic Perspectives & Innovations, 8*(1), 1.

Exercises

19.1 *Is participating in a follow-up survey associated with having medical aid?* A South African study compared whether participating in a 5-year follow-up interview was associated with having medical aid (a type of health insurance). Based on the setup of the problem in the original source, whether an individual was followed-up is the explanatory variable and receipt of medical aid is the response variable. The extraneous variable ("potential confounder") is race.

(a) Table 19.5A tabulates the results for all subjects combined. Calculate the prevalence of medical aid in the group that was followed up and the group that was lost to follow up. Describe the relationship between follow up and having medical aid.

(b) Subjects were subclassified by race. Table 19.5B shows the results for white participants. Describe the relationship in this stratum.

(c) Data for black participants are shown in Table 19.5C. Analyze the results for this stratum.

(d) Why did there appear to be a negative association in the crude data and no association within strata?

TABLE 19.5   Data for Exercise 19.1. Participation in follow-up survey and having medical aid (health insurance).

Table A: All subjects

	Medical Aid +	Medical Aid −	Total
Follow-up +	46	370	416
Follow-up −	195	979	1174

Table B: Whites

	Medical Aid +	Medical Aid −	Total
Follow-up +	10	2	12
Follow-up −	104	22	126

Table C: Blacks

	Medical Aid +	Medical Aid −	Total
Follow-up +	36	368	404
Follow-up −	91	957	1048

Data from Morrell, C. H. (1999). Simpson's paradox: an example from a longitudinal study in South Africa. *Journal of Statistical Education, 7*(3), http://www.amstat.org/publications/jse/secure/v7n3/datasets.morrell.cfm.

19.2 *Is participating in a follow-up survey associated with having medical aid?* Use the strata-specific data in Table 19.5B and C to calculate the Mantel–Haenszel summary prevalence proportion ratio ("relative risk") for the relationship between follow up while adjusting for race. Include a confidence interval for your estimate. (*Suggestion*: Check your calculations with *WinPepi* or another statistical utility.)

■ 19.4 Interaction

Statistical interaction[i] refers to a situation in which a statistical model does not adequately predict the joint effects of two or more explanatory factors. With stratified two-by-two tables, this corresponds to heterogeneity in the strata-specific measures of effect. When the strata-specific measures of effect are relatively homogeneous, statistical interaction is absent.

---

[i] Statistical interaction is distinct from biological interaction. Biologic interaction refers to the interdependent operation of two or more causes working together to produce a given effect. Biological interactions are always present when causes act in consort, whereas statistical interaction depends on the statistical model used to describe the data—you can have interaction in a multiplicative statistical model and no interaction in an additive model for the same factors, for instance. See Rothman, K. J., Greenland, S., & Walker, A. M. (1980). Concepts of interaction. *American Journal of Epidemiology, 112*(4), 467–470.

In the prior illustrative example, the strata-specific relative risks were similar to each other. It therefore made sense to combine these estimates to form a single summary estimate. Had the relative risks been different at the two clinics, an interaction would have existed and it would *not* have made sense to summarize the effect of the treatment with a single summary statistic.

ILLUSTRATIVE EXAMPLE

Statistical interaction (Asbestos, cigarettes, and lung cancer). This fictitious but realistic example considers odds ratios from a case-control study on asbestos and lung cancer while considering the concurrent effects of smoking. Data for all subjects (smokers and nonsmokers) are:

TABLE 19.6 Smokers and nonsmokers combined.

	Lung CA+	Lung CA−
Asbestos +	80	38
Asbestos −	15	152

$$\widehat{OR} = \frac{80 \cdot 152}{38 \cdot 15} = 21.3$$

Within smokers, however, we find a much stronger association:

TABLE 19.7 Stratum 1 (smokers).

	Lung CA+	Lung CA−
Asbestos +	75	20
Asbestos −	5	80

$$\widehat{OR}_1 = \frac{75 \cdot 80}{20 \cdot 5} = 60.00$$

For nonsmokers, cross tabulation reveals a much weaker association:

TABLE 19.8 Stratum 2 (nonsmokers).

	Lung CA+	Lung CA−
Asbestos +	5	18
Asbestos −	10	72

$$\widehat{OR}_2 = \frac{5 \cdot 72}{18 \cdot 10} = 2.00$$

Combining these heterogeneous odds ratios to form a single summary odds ratio would not adequately predict the joint effects of smoking and asbestos on lung cancer risk. Therefore, a statistical interaction is said to exist. Whenever a statistical interaction exists, Mantel–Haenszel summary statistics should *not* be calculated. Instead, strata-specific statistics should be reported.

Chi-Square Test for Statistical Interaction

Strata-specific estimates of effects will deviate from each other randomly even when no interaction is present in the population. A statistical test may be applied to help distinguish random from systematic heterogeneity of the strata-specific effect estimates.

The Heterogeneity Test Applied to Relative Risks: We test H_0: $RR_1 = RR_2 = \cdots = RR_K$ (no interaction in population) versus H_a: at least one RR_k differs (statistical interaction). An ad hoc statistic to test these claims is

$$\chi^2_{\text{stat,int}} = \Sigma_k \frac{(\ln\widehat{RR}_k - \ln\widehat{RR}_{\text{MH}})^2}{(SE_{\ln\widehat{RR}_k})^2}$$

where $\ln\widehat{RR}_k$ is the natural log of the relative risk in stratum k, $\ln\widehat{RR}_{\text{MH}}$ is the natural log of the Mantel–Haenszel summary relative risk estimate, and $SE_{\ln\widehat{RR}_k} = \sqrt{\frac{1}{a_{1k}} - \frac{1}{n_{1k}} + \frac{1}{a_{2k}} - \frac{1}{n_{2k}}}$. Under the null hypothesis, the $\chi^2_{\text{stat,int}}$ has a chi-square distribution with $K - 1$ degrees of freedom, where K represents the number of strata.

The Heterogeneity Test Applied to Odds Ratios: We test H_0: $OR_1 = OR_2 = \cdots = OR_K$ (no interaction in the population) against H_a: at least one of the OR_ks differs (interaction). The test statistic is:

$$\chi^2_{\text{stat,int}} = \Sigma_k \frac{(\ln\widehat{OR}_k - \ln\widehat{OR}_{\text{MH}})^2}{(SE_{\ln\widehat{OR}_k})^2}$$

where $\ln\widehat{OR}_k$ is the natural log of the odds ratio in stratum k, $\ln\widehat{OR}_{\text{MH}}$ is the natural log of the Mantel–Haenszel summary odds ratio, and $SE_{\ln\widehat{OR}_k} = \sqrt{\frac{1}{a_{1k}} + \frac{1}{b_{1k}} + \frac{1}{a_{2k}} + \frac{1}{b_{2k}}}$. Under the null hypothesis, this statistic has a chi-square distribution with $K - 1$ degrees of freedom, where K represents the number of strata.

ILLUSTRATIVE EXAMPLE

Test for interaction (Asbestos, cigarettes, and lung cancer). We submit the asbestos, smoking, and lung cancer data to a test of interaction.

1. **Hypotheses.** H_0: $OR_1 = OR_2$ (no interaction) against H_a: $OR_1 \neq OR_2$ (interaction).
2. **Test statistic.**
   (a) For smokers: $\ln\widehat{OR}_1 = \ln(60.00) = 4.094$ with $SE_{\ln\widehat{OR}_1} = 0.525$.
   (b) For nonsmokers: $\ln\widehat{OR}_2 = \ln(2.00) = 0.693$ with $SE_{\ln\widehat{OR}_2} = 0.608$.

(c) The Mantel–Haenszel summary odds ratio (calculated with *WinPepi*) $\widehat{OR}_{MH} = 16.20$. Therefore, $\ln\widehat{OR}_{MH} = \ln(16.20) = 2.785$.

(d) The test statistic is $\chi^2_{stat,int} = \Sigma_k \dfrac{(\ln\widehat{OR}_k - \ln\widehat{OR}_{MH})^2}{(SE_{\ln\widehat{OR}_k})^2} =$

$$\frac{(4.094 - 2.785)^2}{0.525^2} + \frac{(0.693 - 2.785)^2}{0.608^2} = 18.06 \text{ with}$$

$df = K - 1 = 2 - 1 = 1$.

3. **P-value.** The observed $\chi^2_{stat,int}$ with 1 df corresponds to $P = 2.1 \times 10^{-5}$.
4. **Significance level.** The evidence against H_0 is significant at the $\alpha = 0.0001$ level.
5. **Conclusion.** The observed strata-specific odds ratio estimates (60.0 in smokers and 2.0 in non-smokers) differ significantly ($P < 0.0001$). Thus, statistical interaction is present.

Many different chi-square statistics for interaction are used in practice. The ad hoc interaction statistics presented in this chapter are of the form:

$$\chi^2_{stat} = \Sigma_{k=1}^{k} \frac{(\text{observed measure of effect in stratum } i - \text{summary measure of effect})^2}{\text{variance of the measure of effect in strata } i}$$

This statistic is based on the principle that the sum of K standard Normal random variables follows a chi-square distribution with $K - 1$ degrees of freedom. Be aware that *WinPEPI* (and other software applications) may use different chi-square interaction statistics.[j] For example, when the illustrative data are submitted to *WinPEPI*'s interaction test, the chi-square test's statistic is 21.38 and *P*-value is 3.8×1^{-6}. This test result does not differ materially from those derived by our ad hoc formula.

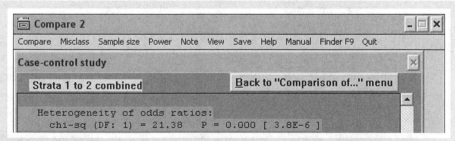

FIGURE 19.4 *WinPepi* screenshot showing interaction statistic for the illustrative example. Abramson J.H. (2011). WINPEPI updated: computer programs for epidemiologists, and their teaching potential. *Epidemiologic Perspectives & Innovations, 8*(1), 1.

---

[j] For details on how *WinPepi* calculated the interaction statistics, see formula 10.35 (p. 170) in Fleiss, J. L. (1981). *Statistical Methods for Rates and Proportions.* (Second ed.). New York: John Wiley & Sons.

Exercises

19.3 *Is participating in a follow-up survey associated with having medical aid?* Exercise 19.1 considered whether participation in a follow-up survey was associated with having medical aid. Race was a confounding variable in this analysis. Table 19.5 lists both crude and race-specific cross tabulations.

(a) Calculate the race-specific prevalence ratios ("relative risks") for medical aid. Do you think statistical interaction is present?

(b) Conduct a chi-square test of interaction for the prevalence ratios. Show all hypothesis-testing steps. Report the $\chi^2_{stat,int}$, its df, and *P*-value. Is the opinion you formed about interaction in part (a) of the exercise confirmed?

19.4 *Smoking and cervical cancer.* Table 19.9 displays data from a case–control study on smoking and invasive cervical cancer with data stratified by number of sexual partners.

TABLE 19.9   Data for Exercise 19.4. Smoking and cervical cancer stratified by number of sexual partners, case–control data.

Stratum 1 (zero to one partner)

	Cervical CA+	Cervical CA−
Smoke +	12	21
Smoke −	25	118

Stratum 2 (two or more partners)

	Cervical CA+	Cervical CA−
Smoke +	96	142
Smoke −	92	150

Data from Nischan, P., Ebeling, K., & Schindler, C. (1988). Smoking and invasive cervical cancer risk. Results from a case-control study. *American Journal of Epidemiology, 128*(1), 74–77.

(a) Calculate the odds ratios within each stratum. Comment on your findings.

(b) Test the odds ratios for interaction. Report the chi-square interaction statistic, its df, and *P*-value.

(c) Merge the data from the two strata, and then calculate the crude odds ratio. How would data have been misinterpreted had it not been stratified by the number of sexual partners?

(d) Rearrange the data to determine whether having two or more partners is a risk factor for invasive cervical cancer. Calculate the odds ratio. Consider these results in light of prior analyses in this exercise.

Summary Points (Stratified Two-by-Two Tables)

1. This chapter considers confounding and interaction by examining data that have been **stratified** to form subgroups according to potentially confounding variables.

2. **Confounding** is a systematic error in inference brought about by the influence of extraneous variables ("confounders") lurking in the background.

3. **Properties of a confounder:** (i) associated with the exposure, (ii) an independent determinant of the study outcome, and (iii) not an intermediary in the causal pathway between the explanatory factor and the outcome variable.

4. **Methods to prevent confounding** include: (i) randomization of the exposure, (ii) restricting the study to subjects or individuals that are homogenous with respect to the confounding factor, (iii) matching on the confounding factor, (iv) using regression models to adjust mathematically for the confounding factor, and (v) stratified analysis (this chapter).

5. **Simpson's paradox** is an extreme form of confounding that occurs when the confounding variable reverses the direction of the association between the explanatory variable and the study outcome.

6. **Evidence of confounding** is present when strata-specific *measures of effect* differ from the crude (overall) measure of effect. (The chapter considers the following *measures of effect*: risk ratio and odds ratio.)

7. **When confounding is present**, the association between the study exposure and the outcome can be tested with a Mantel–Haenszel test statistic, and the strength of the association can be summarized with a Mantel–Haenszel summary measure of effect.

8. **Statistical interaction** (effect measure modification and effect measure heterogeneity) occurs when strata-specific measures of effect differ significantly from each other. You can test for interaction with a chi-square test for heterogeneity.

9. **When statistical interaction is present**, separate measures of effect should be reported for each stratum.

Vocabulary

Ceteris paribus	Regression
Confounders	Restriction
Confounding	Simpson's paradox
Crude relative risk	Statistical interaction
Mantel–Haenszel summary relative risk	Strata-specific statistics
Mantel–Haenszel test statistic	Stratification
Matching	Stratum-specific notation
Randomization	

Review Questions

19.1 Select the best response: Confounding is

  (a) any error in inference.

  (b) any systematic error in inference.

  (c) any systematic error in inference due to the influence of extraneous variables.

19.2 Select the best response: The Latin term *confundere* means

  (a) to mix together.

  (b) with all things equal.

  (c) bias.

19.3 Select the best response: The Latin term *ceteris paribus* means

  (a) to mix together.

  (b) with other things equal.

  (c) bias.

19.4 List the three properties of a confounder.

19.5 List five statistical methods that are used to mitigate confounding.

19.6 What is Simpson's paradox?

19.7 What is a measure of effect?

19.8 What is a crude measure of effect?

19.9 What is a strata-specific measure of effect?

19.10 What is a Mantel–Haenszel summary measure of effect?

19.11 Select the best response: Which of the following is addressed by the Mantel–Haenszel test statistic for risk ratios?

  (a) H_0: $RR_1 = RR_2$

  (b) H_0: $RR = 1$

  (c) H_0: $RR = 0$

19.12 Select the best response: Mantel–Haenszel summary measures of effect are weighted averages of _____ measures of effect.

  (a) crude

  (b) strata-specific

  (c) adjusted

19.13 Select the best response: A synonym for "statistical interaction" is

(a) confounding.

(b) bias.

(c) effect measure modification.

19.14 Select the best response: Which of the following is associated with a statistical test for interaction?

(a) H_0: $RR_1 = RR_2$

(b) H_0: $RR = 1$

(c) H_0: $RR = 0$

19.15 Provide the null hypothesis when testing for an interaction in odds ratios from three strata.

19.16 Provide the alternative hypothesis when testing for an interaction in the odds ratios from three strata.

19.17 Why should we *not* use a Mantel–Haenszel summary measure of effect when interaction is present?

Exercises

19.5 *Sex bias in graduate school admissions?* Table 19.3 contains data on the acceptance of applications to graduate programs at the University of California Berkeley for 1973. Assuming men and women who applied for admission were equally well-qualified, one would expect an equal proportion of acceptance by gender.

(a) Table 19.10A cross-tabulates acceptance by gender for both majors combined. What proportion of male applicants were accepted? What proportion of female applicants were accepted? What is your initial impression?

(b) Table 19.10B cross-tabulates data for major 1. Did males fare better than females in this major?

(c) Table 19.10C cross tabulates data for major 2. What is the experience in this stratum?

(d) Was there sex bias in graduate school admissions? Did it favor male or female applicants?

(e) Explain why the crude analysis (part a) produced a faulty conclusion.

(f) Calculate the relative incidence ("relative risk") of acceptance by gender in major 1. Do the same for major 2. Test these relative risks for heterogeneity. Show all hypothesis-testing steps. Was an interaction present?

TABLE 19.10   Data for Exercise 19.5. Acceptance to UC Berkeley
Graduate programs by gender and major.

Table A: Applicants to both majors (combined)

	+	−	Total
Male	534	664	1198
Female	113	336	449

Table B: Applicants to major 1

	+	−	Total
Male	512	313	825
Female	89	19	108

Table C: Applicants to major 2

	+	−	Total
Male	22	351	373
Female	24	317	341

Data from Bickel, P. J., Hammel, E. A., & O'Connell, W. (1975).
Sex bias in graduate admission: Data from Berkeley. *Science, 187*,
398–404. Data reported on p. 17 of Freedman, D. A., Pisani, R.,
Purves, R., & Adhikari, A. (1991). *Statistics* (2nd ed.). New York:
W. W. Norton.

(g) Calculate the Mantel–Haenszel summary relative risk for acceptance by gender adjusting for major. In relative terms, to what extent were male applicants less likely to be accepted?

19.6 *Helicopter evacuation and survival following trauma.* Accident victims may be transported to the hospital by helicopter or, more typically, by road ambulance. Does the use of helicopter actually save lives? Table 19.11 presents a hypothetical example comparing survival rates in victims by evacuation method.

TABLE 19.11   Data for Exercise 19.6. Survival following motor
vehicle accidents by evacuation method and seriousness of accident. Data are hypothetical.

Table A: All accidents

	Died	Survived	Total
Helicopter	64	136	200
Road	260	840	1100

(continues)

(*continued*)

Table B: Serious accidents

	Died	Survived	Total
Helicopter	48	52	100
Road	60	40	100

Table C: Less-serious accidents

	Died	Survived	Total
Helicopter	16	84	100
Road	200	800	1000

Data from Oppe, S., & De Charro, F. T. (2001). The effect of medical care by a helicopter trauma team on the probability of survival and the quality of life of hospitalized victims. *Accident Analysis & Prevention, 33*(1), 132.

(a) Table 19.11A cross-tabulates the data for all accidents. Calculate the crude relative risk of death associated with helicopter evacuation.

(b) Data stratified by the seriousness of the accident are reported in Table 19.11B and 19.11C. Calculate these strata-specific relative risks.

(c) How do you explain the discrepancy between the crude results and strata-specific results?

(d) Calculate the Mantel–Haenszel adjusted summary relative risk for helicopter evaluation and death while adjusting for the seriousness of the accident.

19.7 *Infant survival.* Data on the survival of 715 infants in relation to the amount of medical care received are as follows:

Crude data (clinics 1 and 2 combined)

	Did not survive	Survived
Less care	20	373
More care	6	316

Data from Whittaker, J. (1990). *Graphical Models in Applied Multivariate Statistics*. Chichester: John Wiley & Sons; Bishop, Y. M. M., Fienberg, S. E., & Holland, P. W. (1975). *Discrete Multivariate Analysis: Theory and Practice*. Cambridge, MA: MIT Press.

(a) Determine the proportion of infants who did not survive in each group. Report the association in terms of a risk ratio. What do you conclude from this crude comparison?

(b) The experience within the two clinics that contributed to the earlier table is shown in the following tables. Address the effect of receiving less care within each clinic. What do you now conclude?

Clinic 1

	Did not survive	Survived
Less care	3	176
More care	4	293

Clinic 2

	Did not survive	Survived
Less care	17	197
More care	2	23

(c) How do you explain the apparent contradictory results of the crude analysis in part (a) and the stratified analysis in part (b)?

(d) Using a computer application such as *WinPEPI* or OpenEpi.com, test for interaction in the risk ratios. Show all hypothesis testing steps.

(e) Calculate the Mantel–Haenszel summary risk ratio associated with less care. Include a 95% confidence for the RR. Summarize your findings.

19.8 *Oral cancer.* Data from a case–control study on oral cancer and smoking stratified by alcohol consumption are shown in the following tables. Quantify the effects of smoking on oral cancer risk while controlling for smoking. Briefly, summarize your findings.

Alcohol use +

	Cases	Controls
Smoking +	225	166
Smoking −	6	12

Alcohol use −

	Cases	Controls
Smoking +	8	18
Smoking −	0	3

Data from Richardson, D. B., & Kaufman, J. S. (2009). Estimation of the relative excess risk due to interaction and associated confidence bounds. *American Journal of Epidemiology, 169*(6), 756–760; Rothman, K., & Keller, A. (1972). The effect of joint exposure to alcohol and tobacco on risk of cancer of the mouth and pharynx. *Journal of Chronic Diseases, 25*(12), 711–716.

19.9 *Herniated lumbar discs.* Data from a case–control study on participation in sports and herniated lumbar discs stratified by smoking status are shown in the following tables. Quantify the effect of "no sports participation" on the risk of disc herniation.

Stratum 1: Smokers

	Cases	Controls
No sports participation	36	28
Sports participation	138	113

Stratum 2: Smokers

	Cases	Controls
No sports participation	31	20
Sports participation	82	126

Data from Richardson, D. B., & Kaufman, J. S. (2009). Estimation of the relative excess risk due to interaction and associated confidence bounds. *American Journal of Epidemiology, 169*(6), 756–760; Mundt, D. J., Kelsey, J. L., Golden, A. L., Panjabi, M. M., Pastides, H., Berg, A. T., et al. (1993). An epidemiologic study of sports and weightlifting as possible risk factors for herniated lumbar and cervical discs. The Northeast Collaborative Group on Low Back Pain. *American Journal of Sports Medicine, 21*(6), 854–860.

A Appendix

TABLE A.   Table of 2000 Random Digits.

Line									
01	79587	19407	49825	58687	99639	82670	73457	53546	30292	75741
02	02213	54407	22917	67392	51745	53341	74452	66258	19597	38440
03	77633	43390	63003	55825	63714	40243	91576	90982	71540	04987
04	02927	39916	38879	97492	54232	26582	75594	31430	62481	48852
05	27673	08260	19904	22537	85260	03805	27138	83323	82080	65863
06	53302	10918	20917	50444	34147	78213	19541	55366	81300	98651
07	43372	88167	59836	05054	51874	59309	72740	58205	60603	55196
08	67893	05723	37080	64029	75438	13959	16442	50847	33442	99647
09	54532	47973	68704	47487	29668	31437	11068	11238	24304	15632
10	34867	89777	96947	44092	49866	94813	71694	78305	33524	30622
11	95162	43739	48362	85438	70133	18178	56655	48265	53784	36693
12	70230	91840	05955	30586	13850	24182	88039	16226	03304	28002
13	19551	63026	59709	55085	18293	50503	75710	24402	62411	97615
14	31237	82396	46680	94704	69287	24926	38249	68858	62146	00131
15	76931	95289	55809	19381	56686	37898	36275	15881	98125	55618
16	88630	59115	76942	53000	89109	61901	55927	96619	34893	97543
17	50728	87768	16193	90514	58042	64398	18491	96407	97303	93459
18	52677	87418	65211	04353	71242	43041	24940	59906	61926	36837
19	04247	38798	73286	99890	09907	17260	04619	47185	71470	98872
20	76012	83064	66743	58110	49524	51685	51815	11837	06368	68488
21	77403	60931	68951	69023	02578	08934	89067	96693	07387	94489
22	70045	45404	80652	60568	94238	08517	34838	60958	94947	98568
23	74424	09905	65366	62295	26118	87077	19265	97192	45317	67620
24	39950	05637	14388	10366	67923	29927	72973	55083	83840	45719
25	13510	32969	80172	86599	57381	52330	38380	28773	97261	75126
26	30108	01696	59451	01073	27760	86472	04865	51333	83736	52416
27	00982	91303	72173	72499	26938	78075	74684	98037	18851	11754
28	47948	47652	25224	65500	86080	47438	11404	56085	04416	22130

(continues)

TABLE A.   Table of 2000 Random Digits. (*continued*)

Line										
29	54985	64122	15648	24313	46612	28442	74549	69001	89813	61596
30	48786	26571	21652	54949	57714	05975	82721	05667	13121	31879
31	81552	66957	51926	54171	50576	41745	87903	80302	76901	18060
32	99551	02072	20173	01563	01602	63964	59429	81601	74924	87038
33	67881	88556	16382	85038	67970	31366	67243	78854	63456	16789
34	06162	40256	69688	98904	82391	82920	13214	25743	31805	82401
35	63716	64311	26224	94569	18043	26137	99795	19047	92258	95604
36	46558	56764	32508	81263	43490	29181	38375	99015	37766	52912
37	14202	65556	24283	65881	37766	54388	80069	78335	79539	60511
38	53265	52355	13913	22834	95995	24878	14148	60663	03207	95208
39	15815	72512	32388	93730	31126	11194	91331	19052	64565	87124
40	56492	44200	29678	29214	08990	01549	40625	52756	62466	96748

Random numbers generated with www.random.org/nform.html, January 2006.

B | Appendix

TABLE B: z TABLE.   Cumulative probabilities for a Standard Normal random variable.

Table entries are the area under the curve to the left of z.

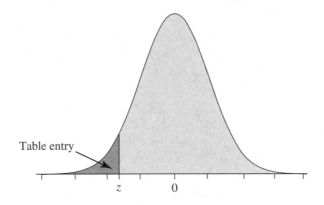

Table entry

z	**Hundredths**									
Tenths	**0.00**	**0.01**	**0.02**	**0.03**	**0.04**	**0.05**	**0.06**	**0.07**	**0.08**	**0.09**
-3.4	0.0003	0.0003	0.0003	0.0003	0.0003	0.0003	0.0003	0.0003	0.0003	0.0002
-3.3	0.0005	0.0005	0.0005	0.0004	0.0004	0.0004	0.0004	0.0004	0.0004	0.0003
-3.2	0.0007	0.0007	0.0006	0.0006	0.0006	0.0006	0.0006	0.0005	0.0005	0.0005
-3.1	0.0010	0.0009	0.0009	0.0009	0.0008	0.0008	0.0008	0.0008	0.0007	0.0007
-3.0	0.0013	0.0013	0.0013	0.0012	0.0012	0.0011	0.0011	0.0011	0.0010	0.0010
-2.9	0.0019	0.0018	0.0018	0.0017	0.0016	0.0016	0.0015	0.0015	0.0014	0.0014
-2.8	0.0026	0.0025	0.0024	0.0023	0.0023	0.0022	0.0021	0.0021	0.0020	0.0019
-2.7	0.0035	0.0034	0.0033	0.0032	0.0031	0.0030	0.0029	0.0028	0.0027	0.0026
-2.6	0.0047	0.0045	0.0044	0.0043	0.0041	0.0040	0.0039	0.0038	0.0037	0.0036
-2.5	0.0062	0.0060	0.0059	0.0057	0.0055	0.0054	0.0052	0.0051	0.0049	0.0048
-2.4	0.0082	0.0080	0.0078	0.0075	0.0073	0.0071	0.0069	0.0068	0.0066	0.0064
-2.3	0.0107	0.0104	0.0102	0.0099	0.0096	0.0094	0.0091	0.0089	0.0087	0.0084

(continues)

TABLE B: z TABLE.   Cumulative probabilities for a Standard Normal random variable. (*continued*)

z Tenths	0.00	0.01	0.02	0.03	0.04	Hundredths 0.05	0.06	0.07	0.08	0.09
−2.2	0.0139	0.0136	0.0132	0.0129	0.0125	0.0122	0.0119	0.0116	0.0113	0.0110
−2.1	0.0179	0.0174	0.0170	0.0166	0.0162	0.0158	0.0154	0.0150	0.0146	0.0143
−2.0	0.0228	0.0222	0.0217	0.0212	0.0207	0.0202	0.0197	0.0192	0.0188	0.0183
−1.9	0.0287	0.0281	0.0274	0.0268	0.0262	0.0256	0.0250	0.0244	0.0239	0.0233
−1.8	0.0359	0.0351	0.0344	0.0336	0.0329	0.0322	0.0314	0.0307	0.0301	0.0294
−1.7	0.0446	0.0436	0.0427	0.0418	0.0409	0.0401	0.0392	0.0384	0.0375	0.0367
−1.6	0.0548	0.0537	0.0526	0.0516	0.0505	0.0495	0.0485	0.0475	0.0465	0.0455
−1.5	0.0668	0.0655	0.0643	0.0630	0.0618	0.0606	0.0594	0.0582	0.0571	0.0559
−1.4	0.0808	0.0793	0.0778	0.0764	0.0749	0.0735	0.0721	0.0708	0.0694	0.0681
−1.3	0.0968	0.0951	0.0934	0.0918	0.0901	0.0885	0.0869	0.0853	0.0838	0.0823
−1.2	0.1151	0.1131	0.1112	0.1093	0.1075	0.1056	0.1038	0.1020	0.1003	0.0985
−1.1	0.1357	0.1335	0.1314	0.1292	0.1271	0.1251	0.1230	0.1210	0.1190	0.1170
−1.0	0.1587	0.1562	0.1539	0.1515	0.1492	0.1469	0.1446	0.1423	0.1401	0.1379
−0.9	0.1841	0.1814	0.1788	0.1762	0.1736	0.1711	0.1685	0.1660	0.1635	0.1611
−0.8	0.2119	0.2090	0.2061	0.2033	0.2005	0.1977	0.1949	0.1922	0.1894	0.1867
−0.7	0.2420	0.2389	0.2358	0.2327	0.2296	0.2266	0.2236	0.2206	0.2177	0.2148
−0.6	0.2743	0.2709	0.2676	0.2643	0.2611	0.2578	0.2546	0.2514	0.2483	0.2451
−0.5	0.3085	0.3050	0.3015	0.2981	0.2946	0.2912	0.2877	0.2843	0.2810	0.2776
−0.4	0.3446	0.3409	0.3372	0.3336	0.3300	0.3264	0.3228	0.3192	0.3156	0.3121
−0.3	0.3821	0.3783	0.3745	0.3707	0.3669	0.3632	0.3594	0.3557	0.3520	0.3483
−0.2	0.4207	0.4168	0.4129	0.4090	0.4052	0.4013	0.3974	0.3936	0.3897	0.3859
−0.1	0.4602	0.4562	0.4522	0.4483	0.4443	0.4404	0.4364	0.4325	0.4286	0.4247
−0.0	0.5000	0.4960	0.4920	0.4880	0.4840	0.4801	0.4761	0.4721	0.4681	0.4641

Cumulative probabilities computed with Microsoft Excel 9.0 NORMSDIST function.

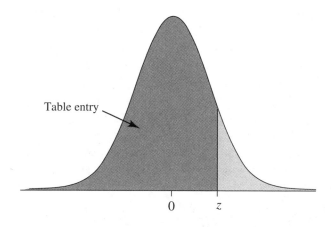

Table entry

0   z

z Tenths	**Hundredths**									
	0.00	**0.01**	**0.02**	**0.03**	**0.04**	**0.05**	**0.06**	**0.07**	**0.08**	**0.09**
0.0	0.5000	0.5040	0.5080	0.5120	0.5160	0.5199	0.5239	0.5279	0.5319	0.5359
0.1	0.5398	0.5438	0.5478	0.5517	0.5557	0.5596	0.5636	0.5675	0.5714	0.5753
0.2	0.5793	0.5832	0.5871	0.5910	0.5948	0.5987	0.6026	0.6064	0.6103	0.6141
0.3	0.6179	0.6217	0.6255	0.6293	0.6331	0.6368	0.6406	0.6443	0.6480	0.6517
0.4	0.6554	0.6591	0.6628	0.6664	0.6700	0.6736	0.6772	0.6808	0.6844	0.6879
0.5	0.6915	0.6950	0.6985	0.7019	0.7054	0.7088	0.7123	0.7157	0.7190	0.7224
0.6	0.7257	0.7291	0.7324	0.7357	0.7389	0.7422	0.7454	0.7486	0.7517	0.7549
0.7	0.7580	0.7611	0.7642	0.7673	0.7704	0.7734	0.7764	0.7794	0.7823	0.7852
0.8	0.7881	0.7910	0.7939	0.7967	0.7995	0.8023	0.8051	0.8078	0.8106	0.8133
0.9	0.8159	0.8186	0.8212	0.8238	0.8264	0.8289	0.8315	0.8340	0.8365	0.8389
1.0	0.8413	0.8438	0.8461	0.8485	0.8508	0.8531	0.8554	0.8577	0.8599	0.8621
1.1	0.8643	0.8665	0.8686	0.8708	0.8729	0.8749	0.8770	0.8790	0.8810	0.8830
1.2	0.8849	0.8869	0.8888	0.8907	0.8925	0.8944	0.8962	0.8980	0.8997	0.9015
1.3	0.9032	0.9049	0.9066	0.9082	0.9099	0.9115	0.9131	0.9147	0.9162	0.9177
1.4	0.9192	0.9207	0.9222	0.9236	0.9251	0.9265	0.9279	0.9292	0.9306	0.9319
1.5	0.9332	0.9345	0.9357	0.9370	0.9382	0.9394	0.9406	0.9418	0.9429	0.9441
1.6	0.9452	0.9463	0.9474	0.9484	0.9495	0.9505	0.9515	0.9525	0.9535	0.9545
1.7	0.9554	0.9564	0.9573	0.9582	0.9591	0.9599	0.9608	0.9616	0.9625	0.9633
1.8	0.9641	0.9649	0.9656	0.9664	0.9671	0.9678	0.9686	0.9693	0.9699	0.9706
1.9	0.9713	0.9719	0.9726	0.9732	0.9738	0.9744	0.9750	0.9756	0.9761	0.9767
2.0	0.9772	0.9778	0.9783	0.9788	0.9793	0.9798	0.9803	0.9808	0.9812	0.9817
2.1	0.9821	0.9826	0.9830	0.9834	0.9838	0.9842	0.9846	0.9850	0.9854	0.9857
2.2	0.9861	0.9864	0.9868	0.9871	0.9875	0.9878	0.9881	0.9884	0.9887	0.9890
2.3	0.9893	0.9896	0.9898	0.9901	0.9904	0.9906	0.9909	0.9911	0.9913	0.9916
2.4	0.9918	0.9920	0.9922	0.9925	0.9927	0.9929	0.9931	0.9932	0.9934	0.9936
2.5	0.9938	0.9940	0.9941	0.9943	0.9945	0.9946	0.9948	0.9949	0.9951	0.9952
2.6	0.9953	0.9955	0.9956	0.9957	0.9959	0.9960	0.9961	0.9962	0.9963	0.9964
2.7	0.9965	0.9966	0.9967	0.9968	0.9969	0.9970	0.9971	0.9972	0.9973	0.9974
2.8	0.9974	0.9975	0.9976	0.9977	0.9977	0.9978	0.9979	0.9979	0.9980	0.9981
2.9	0.9981	0.9982	0.9982	0.9983	0.9984	0.9984	0.9985	0.9985	0.9986	0.9986
3.0	0.9987	0.9987	0.9987	0.9988	0.9988	0.9989	0.9989	0.9989	0.9990	0.9990
3.1	0.9990	0.9991	0.9991	0.9991	0.9992	0.9992	0.9992	0.9992	0.9993	0.9993
3.2	0.9993	0.9993	0.9994	0.9994	0.9994	0.9994	0.9994	0.9995	0.9995	0.9995
3.3	0.9995	0.9995	0.9995	0.9996	0.9996	0.9996	0.9996	0.9996	0.9996	0.9997
3.4	0.9997	0.9997	0.9997	0.9997	0.9997	0.9997	0.9997	0.9997	0.9997	0.9998

C | Appendix

TABLE C. *t* TABLE.   Table entries are values of *t* random variables.

Cumulative probability	0.75	0.80	0.85	0.90	0.95	0.975	0.99	0.995	0.9975	0.999	0.9995
Upper tail probability	0.25	0.20	0.15	0.10	0.05	0.025	0.01	0.005	0.0025	0.001	0.0005
Probability in two tails	0.50	0.40	0.30	0.20	0.10	0.05	0.02	0.01	0.005	0.002	0.001
Degrees of freedom 1	1.000	1.376	1.963	3.078	6.314	12.71	31.82	63.66	127.3	318.3	636.6
2	0.816	1.061	1.386	1.886	2.920	4.303	6.965	9.925	14.09	22.33	31.60
3	0.765	0.978	1.250	1.638	2.353	3.182	4.541	5.841	7.453	10.21	12.92
4	0.741	0.941	1.190	1.533	2.132	2.776	3.747	4.604	5.598	7.173	8.610
5	0.727	0.920	1.156	1.476	2.015	2.571	3.365	4.032	4.773	5.893	6.869

(continues)

TABLE C. *t* TABLE. Table entries are values of *t* random variables. *(continued)*

Cumulative probability	0.75	0.80	0.85	0.90	0.95	0.975	0.99	0.995	0.9975	0.999	0.9995
Upper tail probability	0.25	0.20	0.15	0.10	0.05	0.025	0.01	0.005	0.0025	0.001	0.0005
Probability in two tails	0.50	0.40	0.30	0.20	0.10	0.05	0.02	0.01	0.005	0.002	0.001
6	0.718	0.906	1.134	1.440	1.943	2.447	3.143	3.707	4.317	5.208	5.959
7	0.711	0.896	1.119	1.415	1.895	2.365	2.998	3.499	4.029	4.785	5.408
8	0.706	0.889	1.108	1.397	1.860	2.306	2.896	3.355	3.833	4.501	5.041
9	0.703	0.883	1.100	1.383	1.833	2.262	2.821	3.250	3.690	4.297	4.781
10	0.700	0.879	1.093	1.372	1.812	2.228	2.764	3.169	3.581	4.144	4.587
11	0.697	0.876	1.088	1.363	1.796	2.201	2.718	3.106	3.497	4.025	4.437
12	0.695	0.873	1.083	1.356	1.782	2.179	2.681	3.055	3.428	3.930	4.318
13	0.694	0.870	1.079	1.350	1.771	2.160	2.650	3.012	3.372	3.852	4.221
14	0.692	0.868	1.076	1.345	1.761	2.145	2.624	2.977	3.326	3.787	4.140
15	0.691	0.866	1.074	1.341	1.753	2.131	2.602	2.947	3.286	3.733	4.073
16	0.690	0.865	1.071	1.337	1.746	2.120	2.583	2.921	3.252	3.686	4.015
17	0.689	0.863	1.069	1.333	1.740	2.110	2.567	2.898	3.222	3.646	3.965
18	0.688	0.862	1.067	1.330	1.734	2.101	2.552	2.878	3.197	3.610	3.922
19	0.688	0.861	1.066	1.328	1.729	2.093	2.539	2.861	3.174	3.579	3.883
20	0.687	0.860	1.064	1.325	1.725	2.086	2.528	2.845	3.153	3.552	3.850
21	0.686	0.859	1.063	1.323	1.721	2.080	2.518	2.831	3.135	3.527	3.819
22	0.686	0.858	1.061	1.321	1.717	2.074	2.508	2.819	3.119	3.505	3.792
23	0.685	0.858	1.060	1.319	1.714	2.069	2.500	2.807	3.104	3.485	3.768
24	0.685	0.857	1.059	1.318	1.711	2.064	2.492	2.797	3.091	3.467	3.745
25	0.684	0.856	1.058	1.316	1.708	2.060	2.485	2.787	3.078	3.450	3.725
26	0.684	0.856	1.058	1.315	1.706	2.056	2.479	2.779	3.067	3.435	3.707
27	0.684	0.855	1.057	1.314	1.703	2.052	2.473	2.771	3.057	3.421	3.690
28	0.683	0.855	1.056	1.313	1.701	2.048	2.467	2.763	3.047	3.408	3.674

Degrees of freedom

(continues)

(continued)

Cumulative probability	0.75	0.80	0.85	0.90	0.95	0.975	0.99	0.995	0.9975	0.999	0.9995
Upper tail probability	0.25	0.20	0.15	0.10	0.05	0.025	0.01	0.005	0.0025	0.001	0.0005
Probability in two tails	0.50	0.40	0.30	0.20	0.10	0.05	0.02	0.01	0.005	0.002	0.001
29	0.683	0.854	1.055	1.311	1.699	2.045	2.462	2.756	3.038	3.396	3.659
30	0.683	0.854	1.055	1.310	1.697	2.042	2.457	2.750	3.030	3.385	3.646
40	0.681	0.851	1.050	1.303	1.684	2.021	2.423	2.704	2.971	3.307	3.551
60	0.679	0.848	1.045	1.296	1.671	2.000	2.390	2.660	2.915	3.232	3.460
80	0.678	0.846	1.043	1.292	1.664	1.990	2.374	2.639	2.887	3.195	3.416
100	0.677	0.845	1.042	1.290	1.660	1.984	2.364	2.626	2.871	3.174	3.390
1000	0.675	0.842	1.037	1.282	1.646	1.962	2.330	2.581	2.813	3.098	3.300
z	0.674	0.842	1.036	1.282	1.645	1.960	2.326	2.576	2.807	3.090	3.291
Confidence level	50%	60%	70%	80%	90%	95%	98%	99%	99.5%	99.8%	99.9%

t-values computed with Microsoft Excel 9.0 TINV function.

D | Appendix

TABLE D. *F* TABLE.   Table entries are *F*-values with right-tail probability *P*.

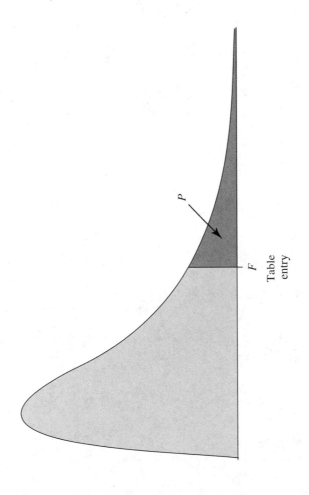

	P	1	2	3	4	5	6	7	8	12	24	1000
Degrees of freedom in numerator (df$_1$)												
1	0.100	39.86	49.50	53.59	55.83	57.24	58.20	58.91	59.44	60.71	62.00	63.30
	0.050	161.4	199.5	215.7	224.6	230.2	234.0	236.8	238.9	243.9	249.1	254.2
	0.025	647.8	799.5	864.2	899.6	921.8	937.1	948.2	956.7	976.7	997.2	1017.7
	0.010	4052	4999	5403	5625	5764	5859	5928	5981	6106	6235	6363
	0.001	405284	499999	540379	562500	576405	585937	592873	598144	610668	623497	6363011
2	0.100	8.53	9.00	9.16	9.24	9.29	9.33	9.35	9.37	9.41	9.45	9.49
	0.050	18.51	19.00	19.16	19.25	19.30	19.33	19.35	19.37	19.41	19.45	19.49
	0.025	38.51	39.00	39.17	39.25	39.30	39.33	39.36	39.37	39.41	39.46	39.50
	0.010	98.50	99.00	99.17	99.25	99.30	99.33	99.36	99.37	99.42	99.46	99.50
	0.001	998.50	999.00	999.17	999.25	999.30	999.33	999.36	999.37	999.42	999.46	999.50
3	0.100	5.54	5.46	5.39	5.34	5.31	5.28	5.27	5.25	5.22	5.18	5.13
	0.050	10.13	9.55	9.28	9.12	9.01	8.94	8.89	8.85	8.74	8.64	8.53
	0.025	17.44	16.04	15.44	15.10	14.88	14.73	14.62	14.54	14.34	14.12	13.91
	0.010	34.12	30.82	29.46	28.71	28.24	27.91	27.67	27.49	27.05	26.60	26.14
	0.001	167.03	148.50	141.11	137.10	134.58	132.85	131.58	130.62	128.32	125.93	123.53
4	0.100	4.54	4.32	4.19	4.11	4.05	4.01	3.98	3.95	3.90	3.83	3.76
	0.050	7.71	6.94	6.59	6.39	6.26	6.16	6.09	6.04	5.91	5.77	5.63
	0.025	12.22	10.65	9.98	9.60	9.36	9.20	9.07	8.98	8.75	8.51	8.26
	0.010	21.20	18.00	16.69	15.98	15.52	15.21	14.98	14.80	14.37	13.93	13.47
	0.001	74.14	61.25	56.18	53.44	51.71	50.53	49.66	49.00	47.41	45.77	44.09

Degrees of freedom in denominator (df$_2$)

df₂	P	\multicolumn{11}{c}{Degrees of freedom in numerator (df_1)}
		1	2	3	4	5	6	7	8	12	24	1000
5	0.100	4.06	3.78	3.62	3.52	3.45	3.40	3.37	3.34	3.27	3.19	3.11
	0.050	6.61	5.79	5.41	5.19	5.05	4.95	4.88	4.82	4.68	4.53	4.37
	0.025	10.01	8.43	7.76	7.39	7.15	6.98	6.85	6.76	6.52	6.28	6.02
	0.010	16.26	13.27	12.06	11.39	10.97	10.67	10.46	10.29	9.89	9.47	9.03
	0.001	47.18	37.12	33.20	31.09	29.75	28.83	28.16	27.65	26.42	25.13	23.82
6	0.100	3.78	3.46	3.29	3.18	3.11	3.05	3.01	2.98	2.90	2.82	2.72
	0.050	5.99	5.14	4.76	4.53	4.39	4.28	4.21	4.15	4.00	3.84	3.67
	0.025	8.81	7.26	6.60	6.23	5.99	5.82	5.70	5.60	5.37	5.12	4.86
	0.010	13.75	10.92	9.78	9.15	8.75	8.47	8.26	8.10	7.72	7.31	6.89
	0.001	35.51	27.00	23.70	21.92	20.80	20.03	19.46	19.03	17.99	16.90	15.77
7	0.100	3.59	3.26	3.07	2.96	2.88	2.83	2.78	2.75	2.67	2.58	2.47
	0.050	5.59	4.74	4.35	4.12	3.97	3.87	3.79	3.73	3.57	3.41	3.23
	0.025	8.07	6.54	5.89	5.52	5.29	5.12	4.99	4.90	4.67	4.41	4.15
	0.010	12.25	9.55	8.45	7.85	7.46	7.19	6.99	6.84	6.47	6.07	5.66
	0.001	29.25	21.69	18.77	17.20	16.21	15.52	15.02	14.63	13.71	12.73	11.72
8	0.100	3.46	3.11	2.92	2.81	2.73	2.67	2.62	2.59	2.50	2.40	2.30
	0.050	5.32	4.46	4.07	3.84	3.69	3.58	3.50	3.44	3.28	3.12	2.93
	0.025	7.57	6.06	5.42	5.05	4.82	4.65	4.53	4.43	4.20	3.95	3.68
	0.010	11.26	8.65	7.59	7.01	6.63	6.37	6.18	6.03	5.67	5.28	4.87
	0.001	25.41	18.49	15.83	14.39	13.48	12.86	12.40	12.05	11.19	10.30	9.36

Degrees of freedom in denominator (df_2)

(continues)

TABLE D. F TABLE.   Table entries are F-values with right-tail probability P. (continued)

df_2	P	\multicolumn{11}{c}{Degrees of freedom in numerator (df_1)}
		1	**2**	**3**	**4**	**5**	**6**	**7**	**8**	**12**	**24**	**1000**
9	0.100	3.36	3.01	2.81	2.69	2.61	2.55	2.51	2.47	2.38	2.28	2.16
	0.050	5.12	4.26	3.86	3.63	3.48	3.37	3.29	3.23	3.07	2.90	2.71
	0.025	7.21	5.71	5.08	4.72	4.48	4.32	4.20	4.10	3.87	3.61	3.34
	0.010	10.56	8.02	6.99	6.42	6.06	5.80	5.61	5.47	5.11	4.73	4.32
	0.001	22.86	16.39	13.90	12.56	11.71	11.13	10.70	10.37	9.57	8.72	7.84
10	0.100	3.29	2.92	2.73	2.61	2.52	2.46	2.41	2.38	2.28	2.18	2.06
	0.050	4.96	4.10	3.71	3.48	3.33	3.22	3.14	3.07	2.91	2.74	2.54
	0.025	6.94	5.46	4.83	4.47	4.24	4.07	3.95	3.85	3.62	3.37	3.09
	0.010	10.04	7.56	6.55	5.99	5.64	5.39	5.20	5.06	4.71	4.33	3.92
	0.001	21.04	14.91	12.55	11.28	10.48	9.93	9.52	9.20	8.45	7.64	6.78
12	0.100	3.18	2.81	2.61	2.48	2.39	2.33	2.28	2.24	2.15	2.04	1.91
	0.050	4.75	3.89	3.49	3.26	3.11	3.00	2.91	2.85	2.69	2.51	2.30
	0.025	6.55	5.10	4.47	4.12	3.89	3.73	3.61	3.51	3.28	3.02	2.73
	0.010	9.33	6.93	5.95	5.41	5.06	4.82	4.64	4.50	4.16	3.78	3.37
	0.001	18.64	12.97	10.80	9.63	8.89	8.38	8.00	7.71	7.00	6.25	5.44
14	0.100	3.10	2.73	2.52	2.39	2.31	2.24	2.19	2.15	2.05	1.94	1.80
	0.050	4.60	3.74	3.34	3.11	2.96	2.85	2.76	2.70	2.53	2.35	2.14
	0.025	6.30	4.86	4.24	3.89	3.66	3.50	3.38	3.29	3.05	2.79	2.50
	0.010	8.86	6.51	5.56	5.04	4.69	4.46	4.28	4.14	3.80	3.43	3.02
	0.001	17.14	11.78	9.73	8.62	7.92	7.44	7.08	6.80	6.13	5.41	4.62

Degrees of freedom in denominator (df_2)

df₂	P	\multicolumn{11}{c}{Degrees of freedom in numerator (df₁)}

df₂	P	1	2	3	4	5	6	7	8	12	24	1000
16	0.100	3.05	2.67	2.46	2.33	2.24	2.18	2.13	2.09	1.99	1.87	1.72
	0.050	4.49	3.63	3.24	3.01	2.85	2.74	2.66	2.59	2.42	2.24	2.02
	0.025	6.12	4.69	4.08	3.73	3.50	3.34	3.22	3.12	2.89	2.63	2.32
	0.010	8.53	6.23	5.29	4.77	4.44	4.20	4.03	3.89	3.55	3.18	2.76
	0.001	16.12	10.97	9.01	7.94	7.27	6.80	6.46	6.19	5.55	4.85	4.08
18	0.100	3.01	2.62	2.42	2.29	2.20	2.13	2.08	2.04	1.93	1.81	1.66
	0.050	4.41	3.55	3.16	2.93	2.77	2.66	2.58	2.51	2.34	2.15	1.92
	0.025	5.98	4.56	3.95	3.61	3.38	3.22	3.10	3.01	2.77	2.50	2.20
18	0.010	8.29	6.01	5.09	4.58	4.25	4.01	3.84	3.71	3.37	3.00	2.58
	0.001	15.38	10.39	8.49	7.46	6.81	6.35	6.02	5.76	5.13	4.45	3.69
20	0.100	2.97	2.59	2.38	2.25	2.16	2.09	2.04	2.00	1.89	1.77	1.61
	0.050	4.35	3.49	3.10	2.87	2.71	2.60	2.51	2.45	2.28	2.08	1.85
	0.025	5.87	4.46	3.86	3.51	3.29	3.13	3.01	2.91	2.68	2.41	2.09
	0.010	8.10	5.85	4.94	4.43	4.10	3.87	3.70	3.56	3.23	2.86	2.43
	0.001	14.82	9.95	8.10	7.10	6.46	6.02	5.69	5.44	4.82	4.15	3.40
30	0.100	2.88	2.49	2.28	2.14	2.05	1.98	1.93	1.88	1.77	1.64	1.46
	0.050	4.17	3.32	2.92	2.69	2.53	2.42	2.33	2.27	2.09	1.89	1.63
	0.025	5.57	4.18	3.59	3.25	3.03	2.87	2.75	2.65	2.41	2.14	1.80
	0.010	7.56	5.39	4.51	4.02	3.70	3.47	3.30	3.17	2.84	2.47	2.02
	0.001	13.29	8.77	7.05	6.12	5.53	5.12	4.82	4.58	4.00	3.36	2.61

Degrees of freedom in denominator (df₂)

(continues)

TABLE D. F TABLE. Table entries are F-values with right-tail probability P. (*continued*)

df₂	P	1	2	3	4	5	6	7	8	12	24	1000
50	0.100	2.81	2.41	2.20	2.06	1.97	1.90	1.84	1.80	1.68	1.54	1.33
	0.050	4.03	3.18	2.79	2.56	2.40	2.29	2.20	2.13	1.95	1.74	1.45
	0.025	5.34	3.97	3.39	3.05	2.83	2.67	2.55	2.46	2.22	1.93	1.56
	0.010	7.17	5.06	4.20	3.72	3.41	3.19	3.02	2.89	2.56	2.18	1.70
	0.001	12.22	7.96	6.34	5.46	4.90	4.51	4.22	4.00	3.44	2.82	2.05
100	0.100	2.76	2.36	2.14	2.00	1.91	1.83	1.78	1.73	1.61	1.46	1.22
	0.050	3.94	3.09	2.70	2.46	2.31	2.19	2.10	2.03	1.85	1.63	1.30
	0.025	5.18	3.83	3.25	2.92	2.70	2.54	2.42	2.32	2.08	1.78	1.36
	0.010	6.90	4.82	3.98	3.51	3.21	2.99	2.82	2.69	2.37	1.98	1.45
	0.001	11.50	7.41	5.86	5.02	4.48	4.11	3.83	3.61	3.07	2.46	1.64
1000	0.100	2.71	2.31	2.09	1.95	1.85	1.78	1.72	1.68	1.55	1.39	1.08
	0.050	3.85	3.00	2.61	2.38	2.22	2.11	2.02	1.95	1.76	1.53	1.11
	0.025	5.04	3.70	3.13	2.80	2.58	2.42	2.30	2.20	1.96	1.65	1.13
	0.010	6.66	4.63	3.80	3.34	3.04	2.82	2.66	2.53	2.20	1.81	1.16
	0.001	10.89	6.96	5.46	4.65	4.14	3.78	3.51	3.30	2.77	2.16	1.22

Degrees of freedom in numerator (df₁)

Degrees of freedom in denominator (df₂)

F-values computed with Microsoft Excel 9.0 FINV function.

E | Appendix

TABLE E. Chi-square table.

Table entries are chi-square values with right-tail probability P.

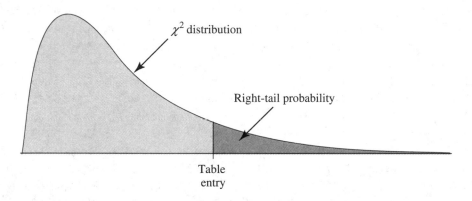

| df | \| | \| | \| | \| | \| | \| | \| | \| | \| | \| | \| |

Probability in right tail

df	0.975	0.25	0.20	0.15	0.10	0.05	0.025	0.01	0.005	0.001	0.0005
1	0.00098	1.32	1.64	2.07	2.71	3.84	5.02	6.63	7.88	10.83	12.12
2	0.051	2.77	3.22	3.79	4.61	5.99	7.38	9.21	10.60	13.82	15.20
3	0.216	4.11	4.64	5.32	6.25	7.81	9.35	11.34	12.84	16.27	17.73
4	0.48	5.39	5.99	6.74	7.78	9.49	11.14	13.28	14.86	18.47	20.00
5	0.83	6.63	7.29	8.12	9.24	11.07	12.83	15.09	16.75	20.52	22.11
6	1.24	7.84	8.56	9.45	10.64	12.59	14.45	16.81	18.55	22.46	24.10
7	1.69	9.04	9.80	10.75	12.02	14.07	16.01	18.48	20.28	24.32	26.02
8	2.18	10.22	11.03	12.03	13.36	15.51	17.53	20.09	21.95	26.12	27.87
9	2.70	11.39	12.24	13.29	14.68	16.92	19.02	21.67	23.59	27.88	29.67
10	3.25	12.55	13.44	14.53	15.99	18.31	20.48	23.21	25.19	29.59	31.42
11	3.82	13.70	14.63	15.77	17.28	19.68	21.92	24.72	26.76	31.26	33.14
12	4.40	14.85	15.81	16.99	18.55	21.03	23.34	26.22	28.30	32.91	34.82

(continues)

TABLE E.   Chi-square table. (*continued*)

df	0.975	0.25	0.20	0.15	Probability in right tail 0.10	0.05	0.025	0.01	0.005	0.001	0.0005
13	5.01	15.98	16.98	18.20	19.81	22.36	24.74	27.69	29.82	34.53	36.48
14	5.63	17.12	18.15	19.41	21.06	23.68	26.12	29.14	31.32	36.12	38.11
15	6.26	18.25	19.31	20.60	22.31	25.00	27.49	30.58	32.80	37.70	39.72
16	6.91	19.37	20.47	21.79	23.54	26.30	28.85	32.00	34.27	39.25	41.31
17	7.56	20.49	21.61	22.98	24.77	27.59	30.19	33.41	35.72	40.79	42.88
18	8.23	21.60	22.76	24.16	25.99	28.87	31.53	34.81	37.16	42.31	44.43
19	8.91	22.72	23.90	25.33	27.20	30.14	32.85	36.19	38.58	43.82	45.97
20	9.59	23.83	25.04	26.50	28.41	31.41	34.17	37.57	40.00	45.31	47.50
21	10.28	24.93	26.17	27.66	29.62	32.67	35.48	38.93	41.40	46.80	49.01
22	10.98	26.04	27.30	28.82	30.81	33.92	36.78	40.29	42.80	48.27	50.51
23	11.69	27.14	28.43	29.98	32.01	35.17	38.08	41.64	44.18	49.73	52.00
24	12.40	28.24	29.55	31.13	33.20	36.42	39.36	42.98	45.56	51.18	53.48
25	13.12	29.34	30.68	32.28	34.38	37.65	40.65	44.31	46.93	52.62	54.95
26	13.84	30.43	31.79	33.43	35.56	38.89	41.92	45.64	48.29	54.05	56.41
27	14.57	31.53	32.91	34.57	36.74	40.11	43.19	46.96	49.64	55.48	57.86
28	15.31	32.62	34.03	35.71	37.92	41.34	44.46	48.28	50.99	56.89	59.30
29	16.05	33.71	35.14	36.85	39.09	42.56	45.72	49.59	52.34	58.30	60.7
30	16.79	34.80	36.25	37.99	40.26	43.8	47.0	50.9	53.7	59.7	62.2
40	24.43	45.62	47.27	49.24	51.81	55.76	59.34	63.69	66.77	73.40	76.09
50	32.36	56.33	58.16	60.35	63.17	67.50	71.42	76.15	79.49	86.66	89.56
60	40.48	66.98	68.97	71.34	74.40	79.08	83.30	88.38	91.95	99.61	102.69
80	57.15	88.13	90.41	93.11	96.58	101.88	106.63	112.33	116.32	124.84	128.26
100	74.22	109.1	111.7	114.7	118.5	124.3	129.6	135.8	140.2	149.4	153.2

Chi-square values computed with Microsoft Excel 9.0 CHINV function.

Appendix

TABLE F. Two tails of z.   Entries in the table represent two-tailed *P*-values for absolute values of *z*-statistics.

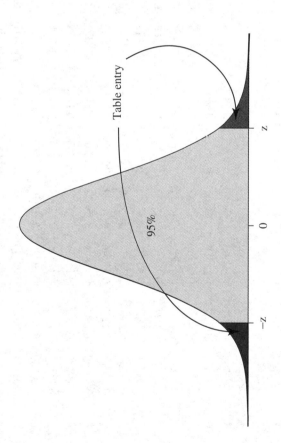

z Tenths	Hundredths									
	0.00	0.01	0.02	0.03	0.04	0.05	0.06	0.07	0.08	0.09
0.0	1.00000	0.99202	0.98404	0.97607	0.96809	0.96012	0.95216	0.94419	0.93624	0.92829
0.1	0.92034	0.91241	0.90448	0.89657	0.88866	0.88076	0.87288	0.86501	0.85715	0.84931
0.2	0.84148	0.83367	0.82587	0.81809	0.81033	0.80259	0.79486	0.78716	0.77948	0.77182
0.3	0.76418	0.75656	0.74897	0.74140	0.73386	0.72634	0.71885	0.71138	0.70395	0.69654
0.4	0.68916	0.68181	0.67449	0.66720	0.65994	0.65271	0.64552	0.63836	0.63123	0.62413
0.5	0.61708	0.61005	0.60306	0.59611	0.58920	0.58232	0.57548	0.56868	0.56191	0.55519
0.6	0.54851	0.54186	0.53526	0.52869	0.52217	0.51569	0.50925	0.50286	0.49650	0.49019
0.7	0.48393	0.47770	0.47152	0.46539	0.45930	0.45325	0.44725	0.44130	0.43539	0.42953
0.8	0.42371	0.41794	0.41222	0.40654	0.40091	0.39533	0.38979	0.38430	0.37886	0.37347
0.9	0.36812	0.36282	0.35757	0.35237	0.34722	0.34211	0.33706	0.33205	0.32709	0.32217
1.0	0.31731	0.31250	0.30773	0.30301	0.29834	0.29372	0.28914	0.28462	0.28014	0.27571
1.1	0.27133	0.26700	0.26271	0.25848	0.25429	0.25014	0.24605	0.24200	0.23800	0.23405
1.2	0.23014	0.22628	0.22246	0.21870	0.21498	0.21130	0.20767	0.20408	0.20055	0.19705
1.3	0.19360	0.19020	0.18684	0.18352	0.18025	0.17702	0.17383	0.17069	0.16759	0.16453
1.4	0.16151	0.15854	0.15561	0.15272	0.14987	0.14706	0.14429	0.14156	0.13887	0.13622
1.5	0.13361	0.13104	0.12851	0.12602	0.12356	0.12114	0.11876	0.11642	0.11411	0.11183
1.6	0.10960	0.10740	0.10523	0.10310	0.10101	0.09894	0.09691	0.09492	0.09296	0.09103
1.7	0.08913	0.08727	0.08543	0.08363	0.08186	0.08012	0.07841	0.07673	0.07508	0.07345
1.8	0.07186	0.07030	0.06876	0.06725	0.06577	0.06431	0.06289	0.06148	0.06011	0.05876
1.9	0.05743	0.05613	0.05486	0.05361	0.05238	0.05118	0.05000	0.04884	0.04770	0.04659
2.0	0.04550	0.04443	0.04338	0.04236	0.04135	0.04036	0.03940	0.03845	0.03753	0.03662
2.1	0.03573	0.03486	0.03401	0.03317	0.03235	0.03156	0.03077	0.03001	0.02926	0.02852
2.2	0.02781	0.02711	0.02642	0.02575	0.02509	0.02445	0.02382	0.02321	0.02261	0.02202
2.3	0.02145	0.02089	0.02034	0.01981	0.01928	0.01877	0.01827	0.01779	0.01731	0.01685
2.4	0.01640	0.01595	0.01552	0.01510	0.01469	0.01429	0.01389	0.01351	0.01314	0.01277

(continues)

(*continued*)

z Tenths				Hundredths						
	0.00	0.01	0.02	0.03	0.04	0.05	0.06	0.07	0.08	0.09
2.5	0.01242	0.01207	0.01174	0.01141	0.01109	0.01077	0.01047	0.01017	0.00988	0.00960
2.6	0.00932	0.00905	0.00879	0.00854	0.00829	0.00805	0.00781	0.00759	0.00736	0.00715
2.7	0.00693	0.00673	0.00653	0.00633	0.00614	0.00596	0.00578	0.00561	0.00544	0.00527
2.8	0.00511	0.00495	0.00480	0.00465	0.00451	0.00437	0.00424	0.00410	0.00398	0.00385
2.9	0.00373	0.00361	0.00350	0.00339	0.00328	0.00318	0.00308	0.00298	0.00288	0.00279
3.0	0.00270	0.00261	0.00253	0.00245	0.00237	0.00229	0.00221	0.00214	0.00207	0.00200
3.1	0.00194	0.00187	0.00181	0.00175	0.00169	0.00163	0.00158	0.00152	0.00147	0.00142
3.2	0.00137	0.00133	0.00128	0.00124	0.00120	0.00115	0.00111	0.00108	0.00104	0.00100
3.3	0.00097	0.00093	0.00090	0.00087	0.00084	0.00081	0.00078	0.00075	0.00072	0.00070
3.4	0.00067	0.00065	0.00063	0.00060	0.00058	0.00056	0.00054	0.00052	0.00050	0.00048
3.5	0.00047	0.00045	0.00043	0.00042	0.00040	0.00039	0.00037	0.00036	0.00034	0.00033
3.6	0.00032	0.00031	0.00029	0.00028	0.00027	0.00026	0.00025	0.00024	0.00023	0.00022
3.7	0.00022	0.00021	0.00020	0.00019	0.00018	0.00018	0.00017	0.00016	0.00016	0.00015
3.8	0.00014	0.00014	0.00013	0.00013	0.00012	0.00012	0.00011	0.00011	0.00010	0.00010
3.9	0.00010	0.00009	0.00009	0.00008	0.00008	0.00008	0.00007	0.00007	0.00007	0.00007

Probabilities computed with Microsoft Excel 9.0 2*(1-NORMSDIST(z_i)) function.

Answers to Odd-Numbered Exercises

■ Chapter 1: Measurement

1.1 *Value, variable, observation.*

The value of LUNGCA for the 7th observation is 18.
The value of COUNTRY for the 11th observation is Iceland.

1.3 *Value, variable, observation (cont.).*

VAR3 records the gender of subjects.

1.5 *Measurement scale.*

VAR1 – Categorical
VAR2 – Categorical
VAR3 – Categorical
VAR4 – Quantitative
VAR5 – Categorical

1.7 *Duration of hospitalization.*

(a) The following variables are categorical: SEX, AB, CULT, and SERV. The following variables are quantitative: DUR, AGE, TEMP, and WBC. There are no ordinal variables in this data set.

(b) The value of DUR for observation 4 is 11.

(c) The value of AGE for observation 24 is 43.

1.9 *Dietary histories.*

Prospective dietary logs are more accurate than retrospective recall. Memory tends to be unreliable.

1.11   *Variable types 2.*

(a) Quantitative	(f) Ordinal	(k) Categorical
(b) Quantitative	(g) Categorical	(l) Ordinal
(c) Categorical	(h) Ordinal	(m) Ordinal
(d) Quantitative	(i) Categorical	(n) Categorical
(e) Quantitative	(j) Quantitative	

1.13   *Age recorded on different measurement scales.*

Age can be recorded quantitatively in months, years, and so on. It can also be grouped into categories, for example, 1 = youngest age group and 2 = next age category.

1.15   *Binge drinking.*

AGEGRP is ordinal. BINGE2003 and BINGE2008 are quantitative.

■ Chapter 2: Types of Studies

2.1   *Sample and population.*

(a) The sample consists of the 125 individuals in the study. The source population can be viewed as either (a) all patients that attended this hospital during the study period or (b) all individuals in the hospital's geographic capturement area, that is, the population who might have been hospitalized had there been a need. It would be useful to know additional information about these source populations, for example, the time-frame of the study, the demographic make-up of the capturement area, and so on.

(b) The sample consists of the 18 diabetics in the study. The population consists of 35- to 44-year-old male diabetics. It would be useful to know other person, place, and time factors that define the source population.

2.3   *California counties.*

There are 58 counties to choose from. Starting on line 33 of Table A, the first four random numbers that apply are 18, 56, 16, and 38. These numbers identify the following counties: Lassen, Ventura, Kings, and San Francisco.

2.5   *Experimental or nonexperimental?*

(a) Nonexperimental

(b) Nonexperimental

(c) Experimental

2.7   Five-City Project. Study design schematic:

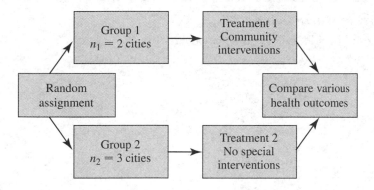

2.9   Five-City Project (cont.).
The first five digits in line 17 of Table A are 50728. Therefore, cities 5 and 2 will be designated at treatment cities.

2.11   Campus survey.
(a) The high nonresponse rate hinders our ability to make generalizations about the population; the distribution of behaviors in responders may not randomly reflect behaviors in the campus populations.
(b) The information on the questionnaires is of uncertain quality. What people say about their behavior often does not match how they behave.

2.13   Telephone directory sampling frame.
This sampling frame omits households that lack land telephones lines. It also excludes those with unlisted numbers and double counts those with more than one phone line.

2.15   Four-naughts.
Four zeros in a row would occur 1 in every 10,000 four-tuples.

2.17   Employee counseling.
(a) Depending on how one structures the research question, the population can be defined as either the 1000 employees who used the service or all employees who might potentially use the service.
(b) The sample consists of the 25 employees who completed and returned their questionnaire.
(c) Since only one in four potential respondents returned a completed questionnaire, there is the potential for *nonresponse bias* (a form of selection bias in which nonresponders differ systematically from responders).
(d) This would still not be an SRS. This would be a systematic sample.

■ Chapter 3: Frequency Distributions

3.1 *Poverty in eastern states, 2000.*

Percent of people living below the poverty line by state.

```
 7│346
 8│1128
 9│889
10│01225
11│19
12│156
13│23
14│67
15│58
×1
```

- Shape: The distribution has a mild positive skew. There are no apparent outliers.
- Location: The median (underlined) has a depth of $(26 + 1) / 2 = 13.5$ and a value of 10.2. Michigan and Massachusetts are the "median states."
- Spread: Values vary from 7.3% to 15.8%.

3.3 *Leaves on a common stem.*

(a) Comparison A. Groups have the same central locations; group 1 has greater variability (spread).

```
Group 1│ │Group 2
      0│1│
       │2│
      0│3│0
       │4│0
      0│5│0
       │6│0
      0│7│0
       │8│
      0│9│
     ×10
```

(b) Comparison B. Group 1 has larger values on average; groups have the same variability.

```
Group 1|  |Group 2
_ _ _ _ _ _ _ _ .
      |1|
      |2|
      |3|0
      |4|0
    0|5|0
    0|6|0
    0|7|0
    0|8|
    0|9|
    ×10
```

(c) Comparison C. Group 1 has a lower average and greater spread.

```
Group 1|  |Group 2
_ _ _ _ _ _ _ _ .
    0|1|
      |2|
    0|3|
      |4|
    0|5|0
      |6|0
    0|7|0
      |8|0
    0|9|0
    ×10
```

3.5 *Hospital stay duration.*

  (a) Frequency table

Duration (days)	Frequency	Relative Frequency (%)	Cumulative Frequency (%)
0–4	5	20.0	20
5–9	12	48.0	68
10–14	6	24.0	92
15–19	1	4.0	96
20–24	0	0.0	96
25–29	0	0.0	96
30–34	1	4.0	100
Total	25	100	–

  (b) Five of 25 (20.0%) were less than 5 days.
  (c) Ninety-two percent were less than 15 days (see Cumulative Frequency column).
  (d) Two of 25 (8%) were at least 15 days in length.

3.7   *Body weight expressed as a percentage of ideal.*

```
08|8
09|59
10|0147
11|444679
12|0145
13|
14|
15|2
×10
```

- Shape: The distribution has a negative skew and high outlier.
- Location: The median is 114 (underlined).
- Spread: Data range from 88 to 152.

3.9   *Seizures following bacterial meningitis.*

```
0|0004
1|22
2|44
3|16
4|2
5|5
6|
7|
8|
9|6
×10 (months)
```

Data have a pronounced positive skew and a high outlier (shape). The median is 24 (location). Induction times are highly variable, ranging from 0.1 to 96 months (spread).

3.11   *U.S. Hispanic population.*

Here is the stemplot with single stem values:

```
0|0001111111122222222333344444555667777 8899
1|035679
2|5
3|22
4|2
×10
```

Here is the stemplot with split stem values:

```
0|00011111112222222233334444
0|5556677778899
1|03
1|5679
2|
2|5
3|22
3|
4|2
×10
```

The stemplot with split stem values does a better job showing the distribution's shape. Notice the positive skew with possible outlier values: 42 (New Mexico), 32 (California), 32 (Texas), and 25 (Arizona). The median (about 4) is underscored.

3.13 *Air samples.*

Here's a stemplot with a regular stem:

```
Site 1|  |  Site 2
      842|2|
      862|3|2346689
        2|4|0
         |5|
        8|6|
       (×10)
```

Here's a stemplot with split stem values:

```
Site 1|  |Site 2
       42|2|
        8|2|
        2|3|234
       86|3|6689
        2|4|0
         |4|
         |5|
         |5|
         |6|
        8|6|
       (×10)
```

Interpretation:

- The sites have similar central locations.
- Site 1 exhibits greater variability.
- A high outlier value is apparent at site 1.

3.15   *Practicing docs.*

(a)   2004 data. The distribution has a positive skew and several potential outliers (labeled on plot). The median has a depth of $(51 + 1)/2 = 26$ and an *approximate* value of 22. (Median is approximate because it is based on counting the truncated leaves on the stemplot.) Data range from *about* 15 to 64.

```
1F|5
1S|66677
1.|888999
2*|000000111111
2T|2222233333333
2F|44455
2S|6
2.|8
3*|011
3T|233
3F|
3S|7              (MA)
3.|
4*|
4T|
4F|
4S|
4.|
5*|
5T|
5F|
5S|
5.|
6*|
6T|
6F|4              (DC)
 ×10
```

Here is how SPSS plots the data:

```
Frequency          Stem & Leaf
   1.00                1.5
   5.00                1.66677
   6.00                1.888999
  12.00                2.000000111111
  13.00                2.2222233333333
   5.00                2.44455
   1.00                2.6
   1.00                2.8
   7.00 Extremes      (>=31)


Stem width:         10.0
Each leaf:          1 case(s)
```

Notice that SPSS includes frequency counts to the left of the stem. The stem exhibits a quintuple split but does not use our convention of using "*, T, F, S, and ." to keep track of stem values. Seven observations are identified as EXTREMES (values of 31 or greater). STEM WIDTH 10.0 refers to the stem multiplier (×10).

(b) 1975 data. The distribution has a positive skew and several potential outliers. The median has a depth of (51 + 1)/2 = 26 and an approximate value of 11. Values range from 7.7 to 34.6.

```
0S|77
0.|8888899999
1*|0000000000111111
1T|22233333333
1F|44455
1S|6677
1.|8
2*|0    (NY)
2T|
2F|
2S|
2.|
3*|
3T|
3F|4    (DC)
3S|
3.|
  ×10
```

(c) 1975 and 2004 data in back-to-back stemplot form. The back-to-back placement of the plots facilitates comparisons. Note that there is little overlap of the distributions.

```
        1975 data  |  | 2004 data
 - - - - - - - - - - - - -  -
                77|0S|
         9999988888|0.|
1111110000000000|1*|
        33333333222|1T|
              55444|1F|5
               7766|1S|66677
                  8|1.|888999
                  0|2*|000000111111
                   |2T|2222233333333
                   |2F|44455
                   |2S|6
                   |2.|8
                   |3*|011
                   |3T|233
                4|3F|
                   |3S|7
                   |3.|
                   |4*|
                   |4T|
                   |4F|
                   |4S|
                   |4.|
                   |5*|
                   |5T|
                   |5F|
                   |5S|
                   |5.|
                   |6*|
                   |6T|
                   |6F|4
                ×10
```

3.17 *Cancer treatment.*

```
0|0012238
1|0
2|1
3|
4|
5|1
6|
7|0
×100
```

The distribution has a strong positive skew with at least two potential outliers (510 and 700). The median has a depth of $(11 + 1)/2 = 6$ and a value of *approximately* 30 (underlined); the actual median is 34 (precision lost when data points were truncated to draw the plot). Data range from 0 to 700.

■ Chapter 4: Summary Statistics

4.1 *Gravitational center.*

Locations of means are shown with a ^.

(a) $\bar{x} = 3$

```
    X                       X
    1----2----3----4----5
              ^
```

(b) $\bar{x} = 3.7$

```
                        X
    X                   X
    1----2----3----4----5
              ^
```

(c) $\bar{x} = 3$

```
              XXX
    1----2----3----4----5
              ^
```

(d) $\bar{x} = 3.7$

```
              X
          X       X
    1----2----3----4----5
              ^
```

4.3 *More visualization.*

The calculated means are (a) 9.9, (b) 159, and (c) 2.8. How good were your visual estimates?

4.5 *Outside?*

Five-point summary: 88, 101, 114, 120, 152

$IQR = 120 - 101 = 19$

$Fence_{Upper} = 120 + (1.5)(19) = 148.5$

The value 152 is outside the upper fence.

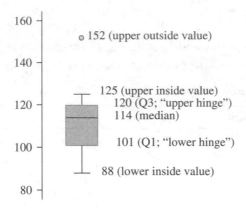

4.7   Spread.

No arithmetics is required to see that batch A has the greatest variability.

4.9   Standard deviation (and variance) via technology.

For site 1, $s = 14.56$ and $s^2 = 211.93$.

For site 2, $s = 2.88$ and $s^2 = 8.286$.

4.11   Units of measure changes numeric values of the standard deviation.

The distributions are identical and are merely expressed in different units. The standard deviations are $s_1 = 1$ year, $s_2 = 12$ months, and $s_3 = 365$ days, respectively.

4.13   Which statistics?

Data set (a) is fairly symmetrical. Therefore, use of the mean and standard deviation are recommended.

Data set (b) is bimodal and has an outlier. The median and IQR should be used to describe this distribution.

Data set (c) has a mild negative skew. It would be prudent to calculate both the mean and median to see whether they differed. If they differed in a meaningful way, you should report the median and IQR.

4.15   Leaves on stems.

(a) Comparison A

Group 1: $\bar{x} = 50$ and $s = 31.6$.

Group 2: $\bar{x} = 50$ and $s = 15.8$.

How statistics relate to what we see: Same central locations; greater spread is group 1.

(b) Comparison B

Group 1: $\bar{x} = 70$ and $s = 15.8$.

Group 2: $\bar{x} = 50$ and $s = 15.8$.

How statistics relate to what we see: Higher central location in group 1; equal group spreads.

(c) Comparison C

Group 1: $\bar{x} = 50$ and $s = 31.6$.

Group 2: $\bar{x} = 70$ and $s = 15.8$.

How statistics relate to what we see: Smaller central location and greater spread in group 1.

4.17 *Health insurance by state.*

(a) Mean = 14.28, median = 13.8. This suggests the distribution has a positive skew or high outlier.

(b) Five-point summary: 8.5, 11.2, 13.8, 16.75, 25.1

(c) IQR = 16.75 − 11.2 = 5.55

Fence$_{\text{Upper}}$ = 16.75 + (1.5)(5.55) = 25.075. There is an outside value on top. This value is 25.1.

Fence$_{\text{Lower}}$ = 11.2 − (1.5)(5.55) = 2.875. There are no outside values on the bottom.

4.19 *What would you report?*

The stemplot is mound shaped and symmetrical. (Symmetry is confirmed by the fact that both $\bar{x}$ and the median are equal to 5.5.) Therefore, the mean and standard deviation should do a good job summarizing this distribution.

```
0*|0
0T|3
0F|445
0S|77
0.|88
 ×1
```

4.21 *Practicing docs (side-by-side boxplots).*

Notes for 1975 data: 5-point summary: 7.7, 10.05, 11.4, 13.85, and 34.6. IQR = 13.85 − 10.05 = 3.8. Fence$_L$ = 10.05 − (1.5) · (3.8) = 4.35; there are no lower outside values. Fence$_U$ = 13.85 + (1.5) · (3.8) = 19.55; there are two upper outside values: 20.2 and 34.6. The upper inside value is 18.3.

Notes for 2004 data: 5-point summary: 15.6, 20.05, 22.2, 24.05, and 64.2. IQR = 24.05 − 20.05 = 4. Fence$_L$ = 20.05 − (1.5) · (4) = 14.05; therefore, there are no lower outside values. Fence$_U$ = 24.05 + (1.5) · (4) = 30.05;

therefore, there are seven upper outside values: 30.7, 31.3, 31.9, 33.0, 33.8, 37.4, and 64.2. The upper inside value is 28.1.

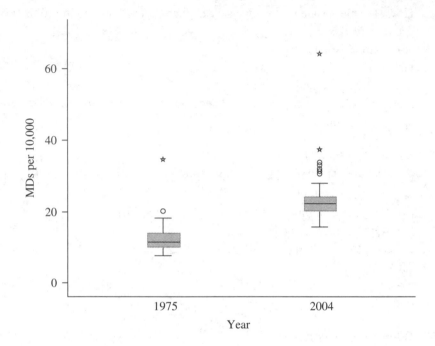

- *Shapes*: Both distributions have upper outside values indicating positive skews. The tail of the 2004 distribution appears to be more prominent than the tail of the 1975 distribution.

- *Locations*: In 2004, the states demonstrated a much higher number of practicing medical doctors per capita, so much so that there is no overlap in the boxes and little overlap in the whiskers.

- *Spreads*: Their IQRs appear to be similar. However, their range of values is greater in 2004.

4.23   *Melanoma treatment.*

The response variable is cell doubling time in days (DOUBLING) and the explanatory variable is the manner in which the cells were cultured (indicated by the variable COHORT). These side-by-side boxplots indicate much longer doubling times in cohort 1 (extended *ex vivo* culturing of cells) and less variability in doubling times in cohort 2 (short *ex vivo* culturing).

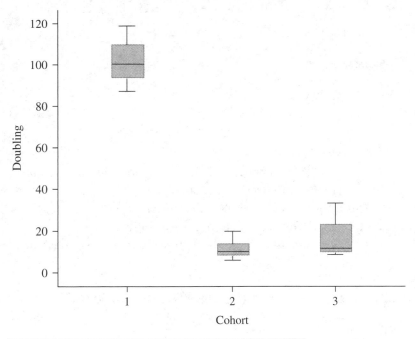

	N	Mean	Standard deviation
Cohort 1	3	10.20	1.61
Cohort 2	11	1.17	0.46
Cohort 3	4	1.63	1.12

Note: The cell doubling times are multiplied by a factor 10 for the plot.

■ Chapter 5: Probability Concepts

5.1 *Explaining probability.*

We cannot say for certain whether this particular patient will survive, but we can say that in a large number of patients with identical characteristics, 60% will survive at least five years and 40% will not.

5.3 *February birthdays.*

(a) February 28 occurs four times every 4 years. Three of the years have 365 days, and every fourth year (leap year) has 366 days. Therefore,

$$\Pr(\text{Feb 28}) = \frac{4}{365 + 365 + 365 + 366} = 0.0027378.$$

(b) February 29 occurs once every 4 years. Therefore, $\Pr(\text{Feb 29}) =$

$$\frac{1}{365 + 365 + 365 + 366} = 0.0006845.$$

(c) (February 28 or 29) occur five times every four years. Therefore, $\Pr($Feb 28 or Feb 29$) = \dfrac{5}{365 + 365 + 365 + 366} = 0.003422313.$

5.5. $N = 26.$

(a) 1 in 26 = 0.0385

(b) 1 in 26 = 0.0385

(c) 0

5.7 Expressions of probability.

(a) This event seldom happens. It has a very low chance of occurring.   5%

(b) This event is infrequent. It is unlikely.   20%

(c) This happens as often as not; chances are even.   50%

(d) This is very frequent. It has high probability.   80%

(e) This event almost always occurs. It has a very high probability.   95%

5.9 Lottery.

(a) $\mu = \Sigma\, x_i \cdot \Pr(X = x_i) = (0 \cdot 0.999999982) + (10{,}000{,}000 \cdot 0.000000018) = 0 + 0.18 = 0.18$

(b) 19 cents

5.11 Uniform (0, 1) pdf.

(a) $\Pr(X \le 0.8) = 0.8$

(b) $\Pr(X \le 0.2) = 0.2$

(c) $\Pr(0.2 \le X \le 0.8) = 0.8 - 0.2 = 0.6$

5.13 The sum of two uniform (0,1) random variables.

(a) $\Pr(X \le 1) = $ half the area under the curve $= 0.5$

Also note that the area is that of a right angle triangle with height $= 1$ and base $= 1$. Area $= \frac{1}{2} \times h \times b = \frac{1}{2} \times 1 \times 1 = 0.5$.

(b) $\Pr(X \le 0.5) = \frac{1}{2} \times h \times b = \frac{1}{2} \times 0.5 \times 0.5 = 0.125$

(c) $\Pr(0.5 \le X \le 1.5) = 1 - \Pr(X \le 0.5) - \Pr(X \ge 1.5) = 1 - 0.125 - 0.125 = 0.75$.

Explanation:

- $\Pr(X \le 0.5) = 0.125$ as explained in part (b).

- By symmetry, $\Pr(X \ge 1.5)$ is also equal to 0.125.

- The area under the curve sums to exactly 1 (property 2 of probabilities). Subtract $\Pr(X \le 0.5)$ and $\Pr(X \ge 1.5)$ from 1 to get $\Pr(0.5 \le X \le 1.5)$.

5.15   Uniform distribution of highway accidents.

(a) Pr(in the first mile) = shaded area = height × base = $1 \times \frac{1}{5} = \frac{1}{5}$ or 0.20.

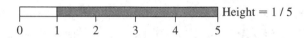

(b) Pr(not in the first mile) = 1 − Pr(in the first mile) = 1 − 0.20 = 0.80

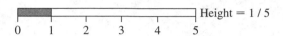

(c) Pr(between miles 2.5 and 4) = $1.5 \times \frac{1}{5} = 0.30$.

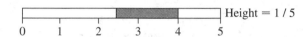

(d) Pr(in first mile OR between 2.5 and 4 miles) = Pr(in the first mile) + Pr(between 2.5 and 4 miles) = 0.20 + 0.30 − 0.50.

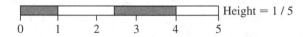

5.17   Bound for Glory (variance).

$\sigma^2 = \Sigma \, (x_i - \mu)^2 \cdot \Pr(X = x_i) = [(0 - 0.1667)^2 \cdot 59/60] + [(10 - 0.1667)^2 \cdot 1/60] = 0.0273 + 1.6116 = 1.6389$ (units are dollars²).

5.19   The sum of two uniform (0,1) random variables (areas under the curve).

(a) $\Pr(X < 1) = \frac{1}{2} \, hb = \frac{1}{2} \cdot 1 \cdot 1 = 0.50$:

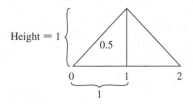

(b) Note that $\Pr(X > 1.5) = \frac{1}{2}\,hb = \frac{1}{2} \cdot \frac{1}{2} \cdot \frac{1}{2} = \frac{1}{8}$ (right tail in this figure):

By the law of complements, $\Pr(X \le 1.5) = 1 - \Pr(X > 1.5) = 1 - \frac{1}{8} = \frac{7}{8}$, as shown as the area under the curve to the right of 1.5 in the above figure.

(d) $\Pr(1 < X < 1.5) = \Pr(X < 1.5) - \Pr(X < 1) = \frac{7}{8} - \frac{1}{2} = \frac{3}{8}$.

■ Chapter 6: Binomial Probability Distributions

6.1 Tay-Sachs.
(a) Yes, this is a binomial random variable because it is based on $n = 3$ independent Bernoulli trials, each with probability of success $p = 0.25$.

(b) This is not a binomial because the number of trials n is not fixed.

6.3 Tay-Sachs inheritance.
Based on Mendelian genetics, there is a one-in-four chance that both carriers will contribute the Tay-Sachs allele during conception. Let X represent the number of Tay-Sachs affected children out of 3. $X \sim b(3, 0.25)$. Therefore:

$\Pr(X = 0) = {}_nC_x \cdot p^x \cdot (1 - p)^{n-x} = {}_3C_0 \cdot 0.25^0 \cdot (1 - 0.25)^3 = 1 \cdot 1 \cdot 0.4219 = 0.4219$

$\Pr(X = 1) = {}_3C_1 \cdot 0.25^1 \cdot (1 - 0.25)^2 = 3 \cdot 0.25 \cdot 0.5625 = 0.4219$

$\Pr(X = 2) = {}_3C_2 \cdot 0.25^2 \cdot (1 - 0.25)^1 = 3 \cdot 0.0625 \cdot 0.75 = 0.1406$

$\Pr(X = 3) = {}_3C_3 \cdot 0.25^3 \cdot (1 - 0.25)^0 = 1 \cdot 0.0156 \cdot 1 = 0.0156$

6.5 Telephone survey.
Given: $X \sim b(8, 0.15)$

$\Pr(X = 2) = {}_8C_2 \cdot 0.15^2 \cdot 0.85^6 = 28 \cdot 0.0225 \cdot 0.3771 = 0.2376$.

6.7 Tay-Sachs.
$\mu = np = (3)(0.25) = 0.75$

$\sigma^2 = npq = (3)(0.25)(0.75) = 0.5625$

6.9   *Prevalence 76.8%.*

(a) $\mu = (5)(0.768) = 3.84$

(b) $\mu = (10)(0.768) = 7.68$

(c) $\Pr(X \geq 9) = \Pr(X = 9) + \Pr(X = 10) = 0.2156 + 0.0714 = 0.2870$

6.11   *Prevalence 10%.*

(a) $X \sim b(15, 0.10)$

(b) $\Pr(X = 0) = 0.2059$

(c) $\Pr(X = 1) = 0.3432$

(d) $\Pr(X \leq 1) = \Pr(X = 0) + \Pr(X = 1) = 0.2059 + 0.3432 = 0.5491$

(e) $\Pr(X \geq 2) = 1 - \Pr(X \leq 1) = 1 - 0.5491 = 0.4509$

6.13   *Linda's omelets.*

Let X represent the number of contaminated eggs in the three-egg omelet. Given: $X \sim b(n = 3, p = 0.16667)$. Calculate: $\Pr(X = 0) = {}_3C_0 \cdot 0.16667^0 \cdot (1 - 0.16667)^3 = 1 \cdot 1 \cdot 0.5787 = 0.5787$. Therefore, the probability of "at least one" $\Pr(X \geq 1) = 1 - \Pr(X = 0) = 1 - 0.5787 = 0.4213$.

6.15   *Decayed teeth.*

$${}_{20}C_2 = \frac{20!}{2! \cdot 18!} = \frac{20 \cdot 19 \cdot 18!}{(2 \cdot 1) \cdot 18!} = 190.$$ Therefore, there are 190 possible combinations.

6.17   *Human papillomavirus.*

(a) The pmf for $X \sim b(4, 0.20)$:

x	0	1	2	3	4
$\Pr(X = x)$	0.4096	0.4096	0.1536	0.0256	0.0016

(b) $\Pr(X \geq 1) = 1 - \Pr(X = 0) = 1 - 0.4096 = 0.5904$

■ Chapter 7: Normal Probability Distributions

7.1   *Heights of 10-year-olds.*

Let X represent heights of 10-year-old males in centimeters. Given: $X \sim N(138, 7)$. Visually, this probability density function is represented as a bell-shaped curve centered on a μ value of 138 with inflection points at $\mu \pm \sigma = 138 \pm 7$, as depicted in the following drawing. According to the "68 part" of the 68–95–99.7 rule, 68% of the values from this distribution will fall in this range. The remaining 32% fall outside this range, with equal numbers at either extreme: 16% fall below 131 and 16% fall above 145.

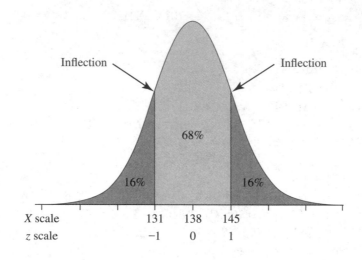

7.3   Visualizing the distribution of gestational length.

$X \sim N(39, 2)$

The curve appears as Figure 7.13 on page 160.

7.5   Heights of 10-year-old boys.

$X \sim N(138, 7)$

(a) What proportion of the population is less than 150 cm tall?
The four-step solution is:

1. State: We want to determine $Pr(X < 150)$

2. Standardize: $z = \dfrac{150 - 138}{7} = 1.71$

3. Sketch :

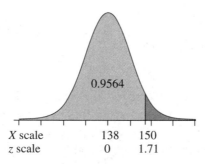

4. Use Table B: $Pr(z < 1.71) = 0.9564$

(b) What proportion of the population is less than 140 cm tall?

    1. State: $\Pr(X < 140)$

    2. Standardize: $z = \dfrac{140 - 138}{7} = 0.29$

    3. Sketch:

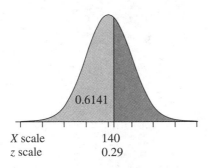

X scale   140
z scale   0.29

    4. Use Appendix Table B: $\Pr(z < 0.29) = 0.6141$

(c) What proportion is between 150 and 140 cm?

Make use of the fact demonstrated in Figure 7.12 (page 159): $\Pr(140 \le X \le 150) = \Pr(X \le 150) - \Pr(X \le 140) = 0.9564 - 0.6141 = 0.3423$

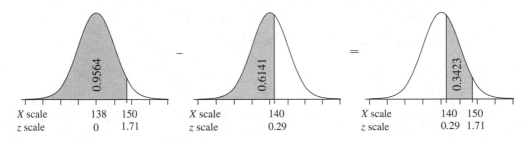

7.7 *45th percentile on a Standard Normal curve.*

    $z_{0.45} = -0.13$

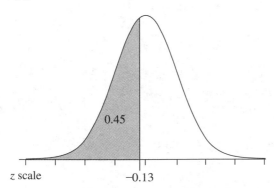

7.9   *Middle 50% of WAISs.*

1. State: $X \sim N(100, 15)$. We want to find the range for the middle 50% of values, that is, from the 25th percentile to the 75th percentile.

2. Table B: The z-scores for these percentiles are $z_{0.25} = -0.67$ and $z_{0.75} = 0.67$

3. Sketch:

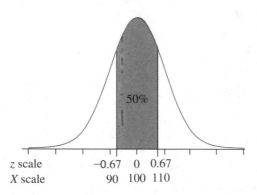

4. Unstandardize using the formula $x = \mu + (z_p)(\sigma)$

The 25th percentile is $x = 100 + (-0.67)(15) = 100 - 10.05 = 89.95 \approx 90$.
The 75th percentile is $x = 100 + (0.67)(15) = 100 + 10.05 = 110.05 \approx 110$.
The range 90 to 110 captures the middle 50% of values.

7.11   *Death row inmate.*

Again, the four-step procedure is used:

1. State: $X \sim N(100, 15)$. We want to determine $\Pr(X < 51)$

2. Standardize: $z = \dfrac{51 - 100}{15} = -3.27$

3. Sketch: not shown

4. Table B: $\Pr(z \le -3.27) = 0.0005$ (about 0.05%)

7.13   *Alzheimer brains.*

1. State: $X \sim N(1077, 106)$. We want to determine $\Pr(X \ge 1250)$

2. Standardize: $z = \dfrac{1250 - 1077}{106} = 1.63$

3. Sketch: not shown

4. Use Table B: $\Pr(z \ge 1.63) = 1 - 0.9484 = 0.0516$

7.15   *z percentiles.*

(a) $z_{0.10} = -1.28$ (The 10th percentile on a Standard Normal curve is 1.28.)

(b) $z_{0.35} = -0.39$

(c) $z_{0.74} = 0.64$

(d) $z_{0.85} = 1.04$

(e) $z_{0.999} = 3.09$

7.17   *Gestation less than 32 weeks.*

- *State the problem.* Let X represent normal gestational length from conception to birth in weeks: $X \sim N(39, 2)$. This question asks "What percentage of gestational lengths is less than 32 weeks?" In notation, $\Pr(X < 32) = ?$

- *Standardize the value.* The z-score corresponding to 32 weeks is $z = (32 - 39)/2 = -3.50$.

- *Sketch the curve and shade the probability area.* The drawing is not shown in this key. However, one can imagine the standardized value of -3.50 located in the far left tail of the Standard Normal curve. The AUC to the left of this point is very tiny.

- *Use Appendix Table B to look up the probability.* Table B does not include the cumulative probability for -3.50. However, it does include the fact that $\Pr(z \leq -3.49) = 0.0002$. Our standardized value is very close to this point, so we can assume $\Pr(z \leq -3.50) \approx 0.0002$. Using the StaTable probability application, $\Pr(z \leq -3.50) = 0.000233$. Therefore, 0.02% of gestations are less than 32 weeks in length.

7.19   *A six-foot seven-inch tall man.*

Let X represent male height in inches. We are given $X \sim N(70, 3)$. The probability of seeing a man who is 70″ tall or taller is $\Pr(X \geq 70) = \Pr(z \geq (70 - 70)/3) = \Pr(z \geq 0) = 0.5$. Therefore, half the men in the population will be taller than 5′10″. In contrast, the probability of seeing a man who is 79″ tall or taller is $\Pr(X \geq 79) = \Pr(z \geq (79 - 70)/3) = \Pr(z \geq 3) = 1 - \Pr(z < 3) = 1 - 0.9987 = 0.0013$. Therefore, only 0.13% of men are taller than 6′7″. It follows that the 6′7″ man is a rarity, while the 5′10″ man is common. That is why 6′7″ seems so much taller than 5′10″.

7.21 $|z| \geq 2.56.$

$\Pr(z \geq 2.56) = 0.0052$ and $\Pr(z \leq -2.56) = 0.0052$. Since these two events are disjoint (in separate tails of the pdf), we can add their probabilities to get the probability of their union: $\Pr(z \geq 2.56 \text{ or } z \leq -2.56) = \Pr(z \geq -2.56) + \Pr(z \leq -2.56) = 0.0052 + 0.0052 = 0.0104$.

7.23   MCATs.

Let X represent scores on the biological section of the MCATs. We are given the fact that $X \sim N(9.2, 2.2)$. The probability of a score of 10.8 or greater is $\Pr(X \geq 10.8) = \Pr(z \geq (10.8 - 9.2)/2.2) = \Pr(z \geq 0.73) = 0.2327$. Therefore, approximately 23% of those taking the exam will get a score of 10.8 or better.

■ Chapter 8: Introduction to Statistical Inference

8.1   Breast cancer survival.

Whether a value is a parameter or a statistic often depends on how the research question is stated. The current research question seems to address survival of breast cancer cases in general. Therefore, these 1225 cases must be considered a sample of the larger population of breast cancer cases and all the highlighted calculations represent sample statistics.

8.3   Parameter or statistic?

(a) Statistic

(b) The number 12% is a parameter. The number 8% is a statistic.

(c) All are statistics. (These statistics will be used to infer costs at online and community pharmacies, respectively.)

8.5   Survey of health problems.

(a) True. The standard deviation of $\bar{x} = \dfrac{1.65}{\sqrt{500}} = 0.0740$.

(b) False. It is not reasonable to assume that the number of health problems per person is Normal. Most people will have 0 or 1 health problem and a small number will have 2, 3, 4, or more problems; the distribution is likely to have a positive skew.

(c) True. Because n is large, we can count on the central limit theorem to make the sampling distribution of $\bar{x}$ fairly Normal.

8.7   Repeated lab measurements.

(a) The standard deviation of the mean of four measurements $= \dfrac{\sigma}{\sqrt{n}} = \dfrac{1}{\sqrt{4}} = 0.5$.

(b) Assuming the measurements are unbiased, the average is more likely to reflect the true value the measurement than is a single measurement.

(c) $\mathrm{SE}_{\bar{x}} = \dfrac{\sigma}{\sqrt{n}}$. Solving for n, $n = \left(\dfrac{\sigma}{\mathrm{SE}_{\bar{x}}}\right)^2$. To achieve a standard error of 0.2, use $n = \left(\dfrac{1}{0.2}\right)^2 = 25$.

8.9 Pediatric asthma survey, n = 50.

$npq = (50)(0.05)(0.95) = 2.375$. Therefore, the sample is too small to apply the Normal approximation.

The binomial function $X \sim b(50, 0.05)$ can be applied to random variable X.

8.11 Pediatric asthma survey, n = 250.

$npq = (250)(0.05)(0.95) = 11.875$. Therefore, the Normal approximation to the binomial can be applied. Note that $\mu = 250 \cdot 0.05 = 12.5$ and $\sigma = \sqrt{250 \cdot 0.05 \cdot 0.95} = 3.446$, that is, $X \sim N(12.5, 3.446)$.

1. State: We want to determine $\Pr(X \geq 25)$

2. Standardize: $z = \dfrac{25 - 12.5}{3.446} = 3.63$

3. Sketch:

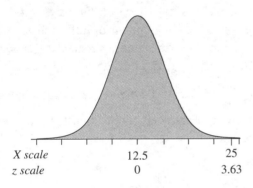

| X scale | 12.5 | 25 |
| z scale | 0 | 3.63 |

4. Use Table B: $\Pr(z \geq 3.63) \approx 0.000$

8.13 Fill in the blanks.

(a) binomial

(b) p

(c) np

(d) $\sqrt{npq}$

(e) $\sqrt{\dfrac{pq}{n}}$

8.15 Patient preference.

We start by assuming $X \sim b(10, 0.5)$.

$\Pr(X \geq 7)$

$= \Pr(X = 7) + \Pr(X = 8) + \Pr(X = 9) + \Pr(X = 10)$

$= 0.117188 + 0.043945 + 0.009766 + 0.000977$

$= 0.1719$

■ Chapter 9: Basics of Hypothesis Testing

9.1 *Misconceived hypotheses.*
   (a) The null and alternative hypotheses must be set up so that either H_0 or H_a is true. Here, it is possible for neither to be true.
   (b) Hypotheses must address the parameter (for example, μ), not the statistic (for example, $\bar{x}$).
   (c) Same problem as we identified in (b). These hypothesis statements address sample statistic $\hat{p}$. They should address population parameter p.

9.3 *Patient satisfaction.*
   (a) $\text{SE}_{\bar{x}} = \dfrac{7.5}{\sqrt{36}} = 1.25$
   (b) The sketch of $\bar{x} \sim N(50, 1.25)$ is not shown in this key. Note that the curve should be centered on $\mu = 50$ with points of inflection at 48.75 and 51.25. The standard deviation landmarks starting 2 standard errors below the mean are at 47.5, 48.75, 50.0, 51.25, and 52.5.
   (c) Notice that $z_{\text{stat}} = \dfrac{48.8 - 50}{1.25} = -0.96$. Therefore, 48.8 is a little less than 1 standard deviation below μ_0. This would not be unusual and would *not* provide strong evidence against H_0.

9.5 *P from z.*
   One-sided $P = \Pr(z \le -2.45) = 0.0071$ (from Table B)
   Two-sided P-value $2 \times 0.0071 = 0.0142$

9.7 *Patient satisfaction (sample mean of 48.8).*
   (a) H_a: $\mu < 50$
   (b) $z_{\text{stat}} = \dfrac{48.8 - 50}{1.25} = -0.96$
   (c) $P = \Pr(z < -0.96) = 0.1685$. Interpretation: If H_0 were correct, results this extreme or more extreme would occur about 17% of the time (that is, would not be that unusual). Thus, the sample mean of 48.8 is not significantly different from a population mean of 50.

9.9 *LDL and fiber.*
   The P-value lets you know that the observed difference (or one more extreme) could occur 1 in 100 observations if there was no reduction in the population. Because this is unlikely, the results are considered to be statistically significant.

9.11 **Gestational length, African American women, hypothesis test.**

    A. Hypotheses. $H_0: \mu = 39$ versus $H_a: \mu \neq 39$

    B. Test statistic. $\mathrm{SE}_{\bar{x}} = \dfrac{2}{\sqrt{22}} = 0.4264$ and $z_{\mathrm{stat}} = \dfrac{38.5 - 39}{0.4264} = -1.17$

    C. *P*-value. One-sided $P = \Pr(z \leq -1.17) = 0.1210$ (from Table B) and two-sided $P = 2 \times 0.1210 = 0.2420$. The evidence against H_0 is nonsignificant by usual conventions.

    D. Significance level. The results are not significant at $\alpha = 0.10$ (retain H_0).

    E. Conclusion. The mean gestational length in this sample of African-American women (38.5 weeks) is not significantly different from the expected population mean of 39 weeks ($P = 0.24$).

9.13 **Gestational length, African American women, sample size.**

$$n = \frac{\sigma^2 (z_{1-\beta} + z_{1-\alpha/2})^2}{(\mu_0 - \mu_a)^2} = \frac{2^2 (1.28 + 1.96)^2}{(39 - 38.5)^2} = 167.96. \text{ Round this up to 168.}$$

9.15 **Female administrators.**

    (a) Conditions for the test: Data represent an SRS of female executives. The distribution of $\bar{x}$ is approximately Normal.

    (b) First note that $\mathrm{SE}_{\bar{x}} = \dfrac{10{,}000}{\sqrt{20}} = 2236.$ Under the null hypothesis $\bar{x} \sim N(85{,}100, 2236)$. The sketch of $\bar{x} \sim N(85{,}100, 2236)$ is shown below.

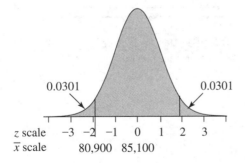

    (c) $z_{\mathrm{stat}} = \dfrac{80{,}900 - 85{,}100}{2236} = -1.88$

    One-sided $P = \Pr(z \leq -1.88) = 0.0301$

    Two-sided $P = 2 \times 0.0301 = 0.0602$

(d) Explanations for the observed difference:
1) Chance ($P = 0.0602$)
2) Selection bias: The sample is not an SRS of female executives.
3) Confounding: An extraneous variable is lurking in the background. For example, these executives may be younger and/or less experienced on average than the population.
4) Gender bias.

9.17 University men.

The four-step procedure is used to solve the problem:

A. Hypotheses. H_0: $\mu = 69$ versus H_a: $\mu \neq 69$

B. Test statistic. $z_{stat} = 1.52$

C. *P*-value. One-sided $P = \Pr(z \geq 1.52) = 0.0643$. Therefore, the two-sided $P = 2 \times 0.0643 = 0.1286$. This is considered to be nonsignificant by usual conventions.

D. Significance level. Results are not significant at $\alpha = 0.10$ (retain H_0).

E. Conclusion. This group is not taller than average ($P = 0.13$).

9.19 The criminal justice analogy.

	TRUTH	
DECISION OF JURY	Did not do crime	Did crime
Not guilty	(a)	(c)
Guilty	(b)	(d)

Declaring an innocent person guilty (b) is analogous to a type I error. Declaring a criminal not guilty (c) is analogous to a type II error. In both hypothesis testing and in the criminal justice system, it is important to avoid a type I error.

9.21 Lab reagent, power analysis.

Conditions: $\alpha = 0.05$ (two-sided), $n = 6$, $\mu_0 = 5$ weeks, $\mu_a = 4.75$ weeks, $\sigma = 0.2$ weeks. Based on these conditions, $1 - \beta = \phi\left(-z_{1-\frac{\alpha}{2}} + \frac{|\mu_0 - \mu_a| \sqrt{n}}{\sigma}\right)$

$= \phi\left(-1.96 + \frac{|5 - 4.75| \sqrt{6}}{0.2}\right) = \phi(1.10) = 0.8643.$

■ Chapter 10: Basics of Confidence Intervals

10.1 *Misinterpreting a confidence interval.*

The pharmacist is incorrect. The confidence interval applies to the population mean μ; it does *not* apply to the distribution of individual observations.

10.3 *Newborn weight.*

(a) $6.1 \pm (1.96)\dfrac{2}{\sqrt{81}} = 6.1 \pm 0.43 = (5.7 \text{ to } 6.5)$

(b) $7.0 \pm (1.96)\dfrac{2}{\sqrt{36}} = 7.0 \pm 0.653 = (6.3 \text{ to } 7.7)$

(c) $5.8 \pm (1.96)\dfrac{2}{\sqrt{9}} = 5.8 \pm 1.31 = (4.5 \text{ to } 7.1)$

10.5 *SIDS.*

The 95% confidence interval for $\mu = 2998 \pm (1.96)\dfrac{800}{\sqrt{49}} = 2998 \pm 224 = (2774 \text{ to } 3222)$.

Interpret your results. Based on this sample, we have 95% confidence the population mean μ is between 2774 and 3222.

10.7 *Hemoglobin.*

(a) $n = \left(1.96 \cdot \dfrac{1.2}{0.5}\right)^2 = 22.1 \rightarrow$ round up to 23

(b) $n = \left(2.58 \cdot \dfrac{1.2}{0.5}\right)^2 = 38.3 \rightarrow$ round up to 39

10.9 *P-value and confidence interval.*

The 95% confidence interval for μ will exclude 0 because the mean is significantly different from 0 at $\alpha = 0.05$. However, the 99% confidence interval will capture 0 because the mean is not significantly different from 0 at $\alpha = 0.01$.

10.11 *Antigen titer.*

$\bar{x} = 7.4033$

$SE_{\bar{x}} = \dfrac{0.070}{\sqrt{3}} = 0.04041$

95% confidence interval for $\mu = 7.4033 \pm (1.96)(0.0404) = 7.4033 \pm 0.0792 = (7.3241 \text{ to } 7.4825)$

10.13 *Reverse engineering the confidence interval.*

(a) Because the sample mean is the center of the confidence interval, $\bar{x} = (6.5 + 5.7) / 2 = 6.1$.

(b) The margin of error is half the confidence interval length: $m = \frac{1}{2} \cdot (6.5 - 5.7) = 0.4$.

(c) For 95% confidence, $m = 1.96 \times SE$. Therefore, $SE = m/1.96 = 0.4/1.96 = 0.204$.

(d) 99% confidence interval for $\mu = 6.1 \pm (2.576)(0.204) = 6.1 \pm 0.53 = (5.57 \text{ to } 6.63)$ pounds.

(e) Yes, the sample mean is significantly different from 7.2 pounds at $\alpha = 0.01$ because it excludes 7.2 with 99% confidence.

10.15 *True or false?*

(a) False; 5 is the margin of error.

(b) False; 13 is the point estimate.

(c) True.

10.17 *Lab reagent, 90% confidence interval for true concentration.*

90% confidence interval for $\mu = \bar{x} \pm z_1 - \frac{0.05}{2} \cdot \frac{\sigma}{\sqrt{n}} = 4.9883 \pm 1.645\frac{0.2}{\sqrt{6}} = 4.9883 \pm 0.1343 = (4.854 \text{ to } 5.123)$. We conclude with 90% confidence that the true concentration of the solution is between 4.854 and 5.123 (mg/dL).

■ Chapter 11: Inference About a Mean

11.1 *Blood pressure.*

(a) $SE_{\bar{x}} = \frac{10.3}{\sqrt{35}} = 1.74$

(b) $SE_{\bar{x}} = \frac{s}{\sqrt{n}}$ Therefore, $n = \left(\frac{s}{SE_{\bar{x}}}\right)^2$ and $n = \left(\frac{10.3}{1}\right)^2 \approx 106.09$. Round up to the next integer to ensure the stated level of precision. Therefore, resolve to use 107 observations.

11.3 *Sketch a curve.*

The middle 95% of the curve is defined by $t_{22, 0.975} = 2.074$ and $t_{22, 0.025} = -2.074$.

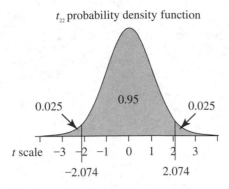

t_{22} probability density function

11.5 Probabilities not in Table C.

$\Pr(T_8 > 2.65)$ is between 0.01 and 0.025 (VIA TABLE C). Using a computer program, $\Pr(T_8 > 2.65) = 0.015$.

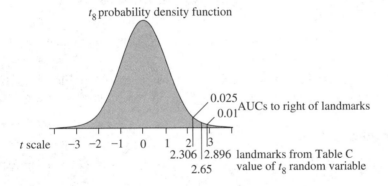

t_8 probability density function

11.7 Software utility programs.

(a) $\Pr(T_8 \geq 2.65) = 0.0150$

(b) $\Pr(T_8 \geq 2.98) = 0.0088$

(c) $\Pr(T_{19} \leq 2.98) = 0.9962$

11.9 Critical values for a t-statistic.

First note that df $= 21 - 1 = 20$. For a one-sided test we get a P-value less than 0.05 when the t_{stat} is either less than -1.725 or more than 1.725, that is, $|t_{stat}| \geq 1.725$. For a two-tailed test, we get a P-value less than 0.05 when the t_{stat} is either less than -2.086 or more than 2.086, that is, $|t_{stat}| \geq 2.086$.

11.11 *Menstrual cycle length.*

A. Hypotheses. H_0: μ = 29.5 days versus H_a: μ ≠ 29.5 days

B. Test statistic.

$n = 9$

$\bar{x} = 27.78$

$\sigma = 2.906$

$$SE_{\bar{x}} = \frac{s}{\sqrt{n}} = \frac{2.906}{\sqrt{9}} = 0.9687$$

$$t_{stat} = \frac{\bar{x}_d - \mu_0}{SE_{\bar{x}_d}} = \frac{27.78 - 29.5}{0.9687} = -1.78$$

df = 9 − 1 = 8

C. *P*-value. Use Table C to determine that $t_{8,0.90}$ = 1.40 (right tail = 0.10) and $t_{8,0.95}$ = 1.86 (right tail = 0.05). Therefore, the one-sided *P*-value is between 0.05 and 0.10 and the two-tailed *P*-value is 0.10 < *P* < 0.20. Using a utility program, *P* = 0.11.

D. Significance level. The evidence against H_0 is not significant at α = 0.10 (retain H_0).

E. Conclusion. The sample mean (27.78 days) is not significantly different from the hypothesized value of 29.5 days (*P* = 0.11).

11.13 *Menstrual cycle length.*

(a) $SE_{\bar{x}} = \dfrac{2.906}{\sqrt{9}}$ = 0.9687, df = 9 − 1 = 8, $t_{8, 0.975}$ = 2.306. Therefore, the 95% CI for μ = 27.78 ± (2.306)(0.9687) = 27.78 ± 2.23 = (25.55 to 30.01) days.

(b) The confidence interval includes 28.5. It also includes 30. Therefore, the sample mean is not significantly different from either 28.5 or 30 at α = 0.05.

11.15 *Water fluoridation.*

(a) DELTA values:

AFTER	BEFORE	DELTA
49.2	18.2	31.0
30.0	21.9	8.1
16.0	5.2	10.8
47.8	20.4	27.4
3.4	2.8	0.6
16.8	21.0	−4.2

AFTER	BEFORE	DELTA
10.7	11.3	−0.6
5.7	6.1	−0.4
23.0	25.0	−2.0
17.0	13.0	4.0
79.0	76.0	3.0
66.0	59.0	7.0
46.8	25.6	21.2
84.9	50.4	34.5
65.2	41.2	24.0
52.0	21.0	31.0

Stemplot of DELTA values:

```
−0|0024
 0|03478
 1|0
 2|147
 3|114
 ×10 (change in cavity-free rate per 100 children)
```

Interpretation:

- Data spread from −4 to 34.
- The median is *about* 7.5 (taken from the stemplot).
- The shape of the distribution is difficult to assess because of the small sample size. There are no prominent outliers.

(b) All but 4 of the 16 cities (25%) showed improvement.

(c) $n = 16$

$\bar{x}_d = 12.21$

$s_d = 13.62$

$SE = \dfrac{13.62}{\sqrt{16}} = 3.405$

$df = 16 - 1 = 15$

$t_{15, 0.975} = 2.131$

95% CI for $\mu_d = 12.21 \pm (2.131)(3.405) = 12.21 \pm 7.26 = (4.95$ to $19.47)$ additional cavity-free children per 100.

11.17 *Large t-statistic.*

When the sample contains more than just a few observations, the associated t-statistic will have more than a few df and will look very much like a Standard Normal z-distribution. Based on the 68–95–99.7 rule, we would almost never get a test statistic that is 6.6 standard deviations away from the 0. Therefore, we can say that the *t*-test statistic is in the far right-hand tail of the sampling distribution and the *P*-value will be very, very small (e.g., less than 0.01).

11.19   Vector control in an African village.

(a) $SE = \dfrac{39.82}{\sqrt{100}} = 3.982$

$t_{100-1,1-(0.05/2)} = t_{99,0.975} = 1.984$ (via computer applet). If a computer program is not available, you can make use of the fact that program $t_{99,0.975} \approx t_{100,0.975} = 1.984$

95% CI for $\mu = 249 \pm (1.984)(3.982) = 249 \pm 7.9 = (241.1$ to $256.9)$ square feet.

(b) It would not be correct to make this statement. The confidence interval applies to population mean μ, not to individual observations.

11.21   Boy height.

$SE = \dfrac{3.1}{\sqrt{26}} = 0.608$

$df = 26 - 1 = 25$

$t_{25,0.975} = 2.060$

95% CI for $\mu = 63.8 \pm (2.060)(0.608) = 63.8 \pm 1.3 = (62.5$ to $65.1)$ inches.

11.23   Faux pas.

(a) Data

VISIT1	VISIT2	DELTA
5	4	−1
13	11	−2
17	12	−5
3	3	0
20	14	−6
18	14	−4
8	10	2
15	9	−6

(b) $\bar{x}_1 = 12.38$, $s_1 = 6.32$, $\bar{x}_2 = 9.63$, $s_2 = 4.17$, $\bar{x}_d = -2.75$, $n = 8$.

(c) The stemplot is:

```
−0|566
−0|124
 0|02
 ×10
```

- Six of the eight students (75%) showed a decline in the number of *faux pas*.
- Data range from -6 to 2 (spread).
- The median is about -1.5 (location).
- The distribution is mound shaped with no apparent outliers, and shows no dramatic departures from Normality.

(d) Yes, a *t*-procedure can be used because the data are mound-shaped and there are no major departures from Normality.

(e) Hypothesis test.

A. Hypotheses. H_0: $\mu_d = 0$ versus H_a: $\mu_d \neq 0$

B. Test statistic. $n = 8$, $\bar{x}_d = -2.75$, $s_d = 2.964$, $SE = \dfrac{2.964}{\sqrt{8}} = 1.048$

$$t_{stat} = \frac{-2.75 - 0}{1.048} = -2.62 \text{ with df} = 8 - 1 = 7.$$

C. *P*-value. $P = 0.034$ (via applet). Using Table C, $0.025 < P < 0.05$. These *P*-value provide good evidence against H_0.

D. Significance level. The difference is significant at $\alpha = 0.05$ but not at $\alpha = 0.01$.

E. Conclusion. There was a significant reduction in the number of *faux pas* after the intervention ($P = 0.034$).

11.25 *Beware $\alpha = 0.05$.*
(a) $P = 0.0464$. Yes, the test is statistically significant at $\alpha = 0.05$.
(b) $P = 0.0514$. No, the test is not statistically significant at $\alpha = 0.05$ ($P > \alpha$).
(c) It is *not* reasonable to derive different conclusions because the observed mean changes are identical and the *P*-values are nearly identical, both providing fairly strong evidence against the null hypothesis.

11.27 *Benign prostatic hyperplasia, maximum flow.*

A. Hypotheses. H_0: $\mu_d = 0$ versus H_a: $\mu_d \neq 0$

B. Test statistic. $n = 10$, sample mean difference $= 3$, $s_d = 4.6188$, $t_{stat} = 2.054$ with df $= 10 - 1 = 9$

C. *P*-value. $P = 0.072$. The evidence against H_0 is marginally significant.

D. Significance level. The evidence against the null hypothesis is significant at $\alpha = 0.10$ but not at $\alpha = 0.05$.

E. Conclusion. Maximum urine flow increased by an average of 2.91 units. By standard conventions, this result is deemed marginally significant ($P = 0.072$).

11.29   *Therapeutic touch.*

$$SE = \frac{1.74}{\sqrt{15}} = 0.4493$$

$t_{14, 0.975} = 2.145$

95% confidence interval for $\mu = 4.67 \pm (2.145)(0.4493) = 4.67 \pm 0.96 =$ (3.71 to 5.63).

This result is compatible with random "5 out of 10" guessing.

■ Chapter 12: Comparing Independent Means

12.1   *Sampling designs.*

   (a) Independent samples

   (b) Paired samples

   (c) Single sample

12.3   *Facetious data.*

   (a) The mean in group 1 is 98. The mean in group 2 is 108. The mean difference is $98 - 108 = -10$. This mean difference is based on independent samples.

ID	GROUP	BEFORE	AFTER	DELTA
1	1	100	104	4
2	1	88	93	5
3	1	106	109	3
4	2	116	117	1
5	2	102	104	2
6	2	106	106	0

   (b) The mean change in group 1 is equal to 4. This mean difference is based on paired samples.

   (c) The mean change in group 2 is equal to 1. This mean difference is based on paired samples.

   (d) Group 1 had a greater mean change by $4 - 1 = 3$ units. This is an independent comparison.

12.5   *Air samples.*

(a) Stemplots

```
Site 1|  |Site 2
_ _ _ _  _ _ _ _
   842|2|
   862|3|2346689
     2|4|0
      |5|
     8|6|
     ×10
```

Discussion:

* Site 1 has a high outlier.

* The distributions have similar central locations.

* Site 1 has greater variability.

(b) Means and standard deviations

	n	**Mean**	**Std. Dev.**
1	8	36.25	14.56
2	8	36.00	2.88

(c) The summary statistics confirm that the distributions have similar central locations and that site 1 has much greater variability.

12.7   *Air samples.*

(a) $\text{SE}_{\bar{x}_1 - \bar{x}_2} = \sqrt{\dfrac{14.56^2}{8} + \dfrac{2.88^2}{8}} = 5.247$

(b) $\text{df}_{\text{conserv}} = 7$; $t_{7,0.975} = 2.365$
The 95% CI for $\mu_1 - \mu_2 = (36.25 - 36.00) \pm (2.365)(5.247) = 0.25 \pm 12.41 = (-12.16$ to $12.66)$ $\mu\text{g/m}^3$

(c) $t_{\text{stat}} = \dfrac{36.25 - 36.00}{5.247} = 0.05$. The one-side P-value (by Appendix Table C) is greater 0.25. Therefore, the two-tailed $P > 0.50$. Using a software utility (www.cytel.com/Products/StaTable/), $P = 0.96$. There is no significant difference in means.

12.9   *Sample size calculation.*

$n = (2)(0.67^2)(1.28 + 1.96)^2 / 0.25^2 = 150.80$. Resolve to study 151 individuals in each group.

12.11 *Testing a test kit.*

    (a) One sample

    (b) Paired samples

12.13 *Risk taking behavior in boys and girls.*

The boxplots is:

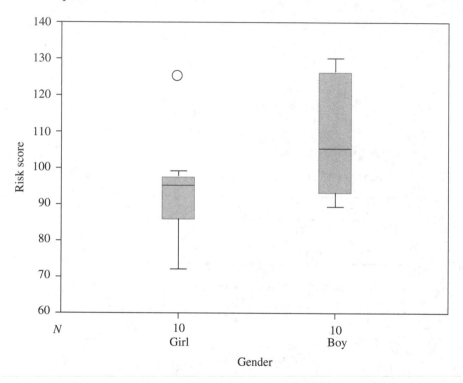

The girls have lower scores on average and less variability. There is one outlier in the girls group.

12.15 *Scrapie treatment, delay of death.*

    A. Hypotheses. H_0: $\mu_1 = \mu_2$ versus H_a: $\mu_1 \neq \mu_2$

    B. Test statistic. $\mathrm{SE} = \sqrt{\mathrm{SE}_1^2 + \mathrm{SE}_2^2} = \sqrt{5.6^2 + 1.9^2} = 5.91$ and $t_{\mathrm{stat}} = (116 - 88.5)/5.91 = 4.65$. Since $n_1 = 10$ and $n_2 = 10$, $\mathrm{df}_{\mathrm{conserv}} = 10 - 1 = 9$.

    C. *P*-value. The two-sided *P*-value is between 0.001 and 0.002. Using a computer applet the two-sided *P*-value = 0.0012.

    D. Significance level. The evidence against the null hypothesis is significant at $\alpha = 0.01$.

E. Conclusion. The treated hamsters survived significantly longer than the control hamsters (mean survival 116 vs. 88.5 days, $P = 0.0012$).

12.17 Bone density in newborns.

   (a) The infants of smoking mothers had slightly higher bone density on average compared to those of the nonsmoking mothers (0.098 compared to 0.095 g/cm^3).

$$SE_{\bar{x}_1 - \bar{x}_2} = \sqrt{\frac{0.026^2}{77} + \frac{0.025^2}{161}} = 0.003558. \quad df_{conserv} = \text{the lesser of}$$

   $(n_1 - 1)$ or $(n_2 - 1)$ = $77 - 1 = 76$. Since this df is not in Appendix Table C, use the next smallest df (60) to derive $t_{60,1-(0.05/2)} = t_{60,0.975} = 2.000$. The 95% confidence for $\mu_1 - \mu_2 = (0.098 - 0.095) \pm (2.000)(0.003558) = 0.003 \pm 0.007 = (-0.004$ to $0.010)$ g/cm^3.

   (b) In testing $H_0: \mu_1 - \mu_2 = 0$, the value of the mean difference under the null hypothesis is 0. Since the 95% confidence interval for $\mu_1 - \mu_2$ includes a value of 0, you would retain H_0 at $\alpha = 0.05$ and conclude no significant difference in the mean bone densities of newborns from smoking and nonsmoking mothers.

12.19 Efficacy of echinacea, severity of symptoms.

   A. Hypotheses. $H_0: \mu_1 - \mu_2 = 0$ versus $H_a: \mu_1 - \mu_2 \neq 0$.

   B. Test statistic. $t_{stat} = \dfrac{6.0 - 6.1}{\sqrt{\dfrac{2.3^2}{337} + \dfrac{2.4^2}{370}}} = \dfrac{-0.1}{0.1768} = -0.57;$

   $df_{conserv} = 337 - 1 = 336$.

   C. P-value. Use the row for 100 df in Table C to derive a conservative estimate of the P-value estimate. Thus, $P > 0.50$. (A statistical applet derived $P = 0.5691$.)

   D. Significance level. The evidence against the null hypothesis is not statistically significant at any reasonable level for α.

   E. Conclusion. There was no significant difference in the mean severity of symptoms in the echinacea treatment group and the control group ($P = 0.57$).

12.21 Calcium supplementation and blood pressure, exploration.
   Here are the side-by-side boxplots:

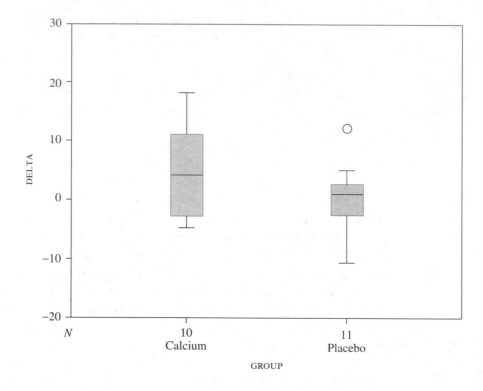

The plot shows that the calcium treated group had a greater average decline in blood pressure. They also exhibited greater variability. There is a high outside value in the placebo group.

12.23 *Delay in discharge.*

A. Hypotheses. $H_0: \mu_1 = \mu_2$ versus $H_a: \mu_1 \neq \mu_2$

B. Statistics calculated with SPSS > Analyze > Compare means > Independent Samples T

Facility	n	Mean	Std. Deviation
A	12	14.08	3.777
B	12	11.00	3.330

$t_{stat} = 2.121$ with $df_{Welch} = 21.7$

C. $P = 0.046$

D. Significance level. The observed difference is significant at $\alpha = 0.05$ but not at $\alpha = 0.04$.

E. Conclusion. The mean delay at facility A was significantly greater than that at facility B (14.1 vs. 11.0 days, $P = 0.046$).

12.25   *Time spent sitting or walking.*

   A. Hypotheses. This exercise seeks to answer whether lean and obese people differ in the average time they spend sitting. Under the null hypothesis, the population means are equal, so we test $H_0: \mu_1 = \mu_2$ versus $H_a: \mu_1 \neq \mu_2$.

   B. Test statistics. Statistics calculated with SPSS > Analyze > Compare means > Independent Samples T. Output shown.

Group Statistics

	N	Mean	Std. Devation	Std. Error Mean
1	10	407.450	104.1248	32.9271
2	10	571.180	65.8646	20.8282

Independent Samples Test

	Levene's Test for Equality of Variances		*t*-test for Equality of Means						95% Confidence Interval of the Difference	
	F	Sig.	t	df	Sig. (2-tailed)	Mean Difference	Std. Error Difference	Lower	Upper	
Equal variances assumed	3.207	0.090	−4.202	18	0.001	−163.7300	38.9617	−245.5854	−81.8746	
Equal variances not assumed			−4.202	15.208	0.001	−163.7300	38.9617	−246.6759	−80.7841	

$t_{stat} = -4.201$ with $df_{Welch} = 15.2$ (unequal variance *t*-procedure).

   C. $P = 0.001$.

   D. Significance level. The observed difference is significant at $\alpha = 0.001$.

   E. Conclusion. The lean individuals spent significantly less time sitting per day than the obese individuals (407 vs. 571 min, $P = 0.001$).

■ Chapter 13: Comparing Several Means (One-Way Analysis of Variance)

13.1   *Birth weight, exploration.*

   (a) This study is nonexperimental ("observational") because the investigator did *not* assign the explanatory factor (smoking) to study participants.

   (b) Outline of study design:

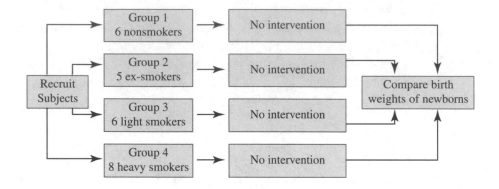

(c) Interpretation of Figure 13.5.

- Location: Nonsmokers and ex-smokers have higher average birth weights than smokers.

- Spread: It is difficult to evaluate variability in such small samples. However, hinge-lengths (IRQs) seem comparable.

- Shape: The samples are too small (six to eight in each group) to make definitive statements about their shapes; there is an outside value in group 2.

13.3   Smoking and birth weight, ANOVA.

A. Hypotheses. H_0: $\mu_1 = \mu_2 = \mu_3 = \mu_4$ versus H_a: at least two of the μ_is differ

B. Test statistic. ANOVA table (below). $F_{stat} = 4.096/1.255 = 3.26$ with 3 and 21 df.

	Sum of Squares	df	Mean Square
Between groups	12.289	3	4.096
Within groups	26.364	21	1.255
Total	38.653	24	

C. *P*-value. $P < 0.05$ by Appendix Table D (use the 3 and 20 df column). $P = 0.042$ by a software utility.

D. Significance level. The results are significant at $\alpha = 0.05$ but are not significant at $\alpha = 0.04$.

E. Conclusion. Mean birth weights differed significantly according to the smoking status of the mothers ($P = 0.042$). Infants of nonsmoking mothers demonstrated the highest average birth weight (7.9 pounds), while the mothers who smoked at least half a pack per day demonstrated the lowest (6.3 pounds).

13.5   Smoking and birth weight, post hoc comparisons.

(a) Least squared difference (LSD) tests

Null hypothesis	$\bar{x}_i - \bar{x}_j$	SE	*P*-value (two-sided)
$H_0: \mu_1 - \mu_2 = 0$	0.0857	0.67848	0.901
$H_0: \mu_1 - \mu_3 = 0$	1.1191	0.64690	0.098
$H_0: \mu_1 - \mu_4 = 0$	1.6129	0.60512	0.014
$H_0: \mu_2 - \mu_3 = 0$	1.0334	0.67848	0.143
$H_0: \mu_2 - \mu_4 = 0$	1.5272	0.63876	0.026
$H_0: \mu_3 - \mu_4 = 0$	0.4938	0.60512	0.424

Calculated by SPSS (Rel. 11.0.1. 2001. Chicago: SPSS Inc.)

Summary in concise narrative form: Differences between groups 1 and 4 and between groups 2 and 4 are statistically significant. The difference between groups 1 and 3 is marginally significant.

(b) Bonferroni's method tests

Null hypothesis	$\bar{x}_i - \bar{x}_j$	SE	*P*-value (two-sided)
$H_0: \mu_1 - \mu_2 = 0$	0.0857	0.67848	1.000
$H_0: \mu_1 - \mu_3 = 0$	1.1191	0.64690	0.590
$H_0: \mu_1 - \mu_4 = 0$	1.6129	0.60512	0.087
$H_0: \mu_2 - \mu_3 = 0$	1.0334	0.67848	0.856
$H_0: \mu_2 - \mu_4 = 0$	1.5272	0.63876	0.158
$H_0: \mu_3 - \mu_4 = 0$	0.4938	0.60512	1.000

Calculated with SPSS (Rel. 11.0.1. 2001. Chicago: SPSS Inc.)

Summary in concise narrative form: The difference between groups 1 and 4 is marginally significant. All other differences are not statistically significant.

(c) Post hoc confidence intervals incorporating Bonferroni's correction

Parameter	$\bar{x}_i - \bar{x}_j$	SE	95% confidence interval	
			LCL	UCL
$\mu_1 - \mu_2$	0.0857	0.67848	−1.8901	2.0615
$\mu_1 - \mu_3$	1.1191	0.64690	−0.7647	3.0030
$\mu_1 - \mu_4$	1.6129	0.60512	−0.1492	3.3751
$\mu_2 - \mu_3$	1.0334	0.67848	−0.9424	3.0092
$\mu_2 - \mu_4$	1.5272	0.63876	−0.3329	3.3873
$\mu_3 - \mu_4$	0.4938	0.60512	−1.2684	2.2559

Calculated with SPSS (Rel. 11.0.1. 2001. Chicago: SPSS Inc.)

13.7   Smoking and birth weight. Kruskal–Wallis test.

A. Hypotheses. H_0: birth weight distributions in the four populations are the same versus H_a: the populations differs

B. Test statistic. Chi-square = 7.305 with 3 df (Calculated with SPSS, Rel. 11.0.1. 2001. Chicago: SPSS Inc.)

C. P-value. $P = 0.063$ ("marginal significance")

D. Significance level. The results are significant at $\alpha = 0.10$ (reject H_0) but are not significant at $\alpha = 0.05$ (retain H_0).

E. Conclusion. The birth weight distributions differ, with evidence rising to marginal significance ($P = 0.063$).

13.9.   Antipyretic trial.

(a) Here is the plot:

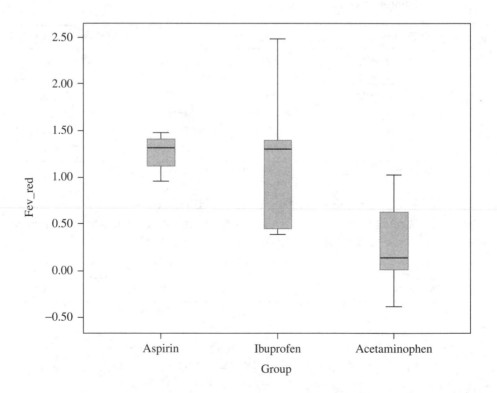

This plot shows that aspirin and ibuprofen are associated with greater average fever reduction than acetaminophen. Results with ibuprofen are more variable (larger IQR) than with aspirin.

(b) Means and standard deviations (calculated with SPSS, Rel. 11.0.1. 2001. Chicago: SPSS Inc.)

fev_red Group	Mean	N	Std. Deviation
Aspirin	1.2600	4	0.22346
Ibuprofen	1.2020	5	0.85444
Acetimenophen	0.2533	6	0.49657
Total	0.8380	15	0.74301

(c) Hypothesis test

A. Hypotheses. H_0: $\mu_1 = \mu_2 = \mu_3$ versus H_a: at least two of the μ_is differ

B. Test statistic. Calculated with SPSS, Rel. 11.0.1. 2001. Chicago: SPSS Inc.

fev_red	Sum of Squares	df	Mean Square	F	Sig.
Between groups	3.426	2	1.713	4.777	0.030
Within groups	4.303	12	0.359		
Total	7.729	14			

C. *P*-value. $P = 0.030$, showing the differences to be statistically significant.

D. Significance level. The results are significant at $\alpha = 0.05$ (reject H_0) but are not significant at $\alpha = 0.01$ (retain H_0).

E. Conclusion. There is significant difference in the mean reduction in body temperature according to analgesic type ($P = 0.030$). Aspirin reduced fever by an average of 1.26°F, ibuprofen reduced fever by an average of 1.20°F, and acetaminophen reduced fever by an average of 0.25°F.

(d) Here are the post hoc comparisons via the LSD method:

(I) group	(J) group	Mean Difference (I-J)	Std. Error	Sig.	95% Confidence interval Lower Bound	Upper Bound
aspirin	ibuprofen	0.0580	0.40170	0.888	−0.8172	0.9332
aspirin	acetimenophen	1.0067*	0.38654	0.023	0.1645	1.8489
ibuprofen	acetimenophen	0.9487*	0.36260	0.023	0.1586	1.7387

*The mean difference is significant at the 0.05 level.

The aspirin group and acetaminophen group differ significantly ($P = 0.023$), as do the ibuprofen group and acetaminophen group (also $P = 0.023$). The aspirin group and ibuprofen group means do not differ significantly ($P = 0.888$).

■ Chapter 14: Correlation and Regression

14.1 *Bicycle helmet use.*

(a) Here is the scatterplot:

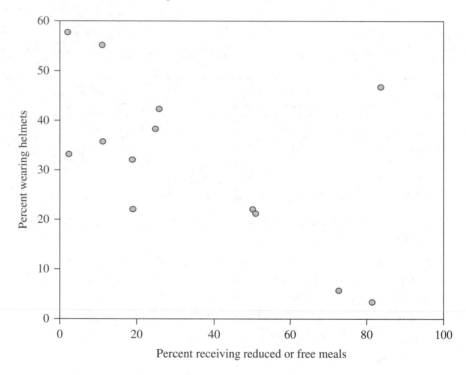

A negative linear relationship is evident. There is a possible outlier in the upper right-hand quadrant (observation 13, Los Arboles).

(b) $r = -0.581$ ($n = 13$)

(c) *Discuss what this (the potential outlier) means in plain terms...* This observation had high helmet use and low socioeconomic status.

(d) $r = -0.849$ ($n = 12$)

The correlation went from moderate strength to strong after removing the outlier, indicating a much better fit of data points to the negative trend line.

(e) Hypothesis test

A. Hypotheses. H_0: $\rho = 0$ versus H_a: $\rho \neq 0$

B. Test statistic. $t_{\text{stat}} = -5.08$ with 10 df

C. *P*-value. $P = 0.00048$ (strong evidence against the null hypothesis)

D. Significance level. The evidence against H_0 is significant at $\alpha = 0.01$ (reject H_0).

E. Conclusion. The observed negative association between the receipt of free or reduced-fee lunches at school and the prevalence of bicycle helmet use is statistically significant ($r = -0.85$, $n = 12$, $P = 0.00048$).

14.3 *Bicycle helmet use, n = 12.*

(a) Regression coefficients

$a = 47.490$

$b = -0.539$

Interpretation of *b*: For each additional "percent children receiving reduced-fee or free meals," the model predicts a 0.5% decline in bicycle helmet use.

Interpretation of *a*: This is where the regression line would cross the *Y* axis, that is, where $x = 0$.

(b) The 95% CI for $\beta = (-0.775$ to $-0.303)$.

(c) The slope is significant at $\alpha = 0.05$ because the 95% confidence interval does not include 0.

(d) Here is the stemplot of residuals:

```
−1│35
−0│0245
 0│1148
 1│13
 ×10
```

There are no major departures from Normality.

14.5 *Anscombe's quartet.*

I would use correlation or linear regression to analyze data set I because there appears to be a positive linear trend. I would not use correlation or regression to analyze the other data sets because these relations cannot be accurately described with a single straight line.

14.7 *Domestic water and dental cavities, range restriction.*

(a) The range below 1 ppm fluoride demonstrates a fairly straight relationship. The least squares regression line for this range has these coefficients:

$a = 780.34$

$b = -528.07$

Therefore, regression line for this range is $\hat{y} = 780.34 + (-528.07)X$.

The slope in this range predicts a decline of 528 caries per ppm of fluoride (equivalently, 52.8 fewer caries per 0.1 ppm fluoride).

(b) $r^2 = 0.856$. The fit of this model is not as good as the ln–ln model calculated in Exercise 14.6, in which $r^2 = 0.947$.

(c) Opinions on this matter will differ. I prefer this model. Although its fit is not as good as the ln–ln model (Exercise 14.6), this model is (a) easier to interpret and (b) addresses a useful biological range. The major declines in caries occur in the 0 to 0.8 range. Since higher levels have only modest benefits, and other sources suggest toxicity with high fluoride, it seems reasonable to restrict the analysis to this biologically relevant range.

14.9 *True or false.*

(a), (c), (e), and (g) are false. The others are true.

14.11 $(\bar{x}, \bar{y})$ *is always on the least squares regression line.*

When $X = \bar{x}$, $\hat{y} = a + bX = (\bar{y} - b\bar{x}) + b\bar{x} = \bar{y}$. Therefore, $(\bar{x}, \bar{y})$ is always on the least squares regression line.

14.13 *Nonexercise activity thermogenesis (NEAT).*

(a) The scatterplot reveals a linear negative association between NEAT and FATGAIN. There are no apparent outliers. The relationship appears to be moderate in strength. (See prior comments about the difficulty judging correlational strength by eye alone.)

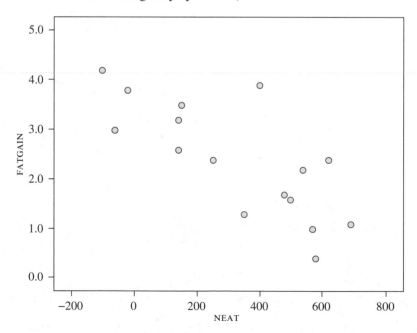

The least square regression line is $\hat{y} = 3.50 + -0.0034X$. This linear relationship is statistically significant ($P = 0.00061$). Each calorie unit of NEAT predicts 0.0034 fewer kilograms of fat gained. Equivalently, each 100 calories of NEAT predicts 0.34 fewer kilograms of fat gained.

(b) $\hat{y} = 3.50 + -0.0034 \cdot 600 = 3.50 + -2.04 = 1.46$ (kg).

(c) The regression line is not shown in this key.

(d) **Observation 1 (x_1, y_1) = (−100, 4.2).** The predicted value for this observation $\hat{y}_1 = a + bx_1 = 3.50 + (-0.0034 \cdot -100) = 3.84$ and residual$_1 = y_1 - \hat{y}_1 = 4.2 - 3.84 = 0.36$.

Observation 2 (x_2, y_2) = (−60, 3.0). The predicted value for this observation is $\hat{y}_2 = a + bx_2 = 3.50 + (-0.0034 \cdot -60) = 3.70$ and residual$_2 = y_2 - \hat{y}_2 = 3.0 - 3.70 = -0.70$.

Observation 3 (x_3, y_3) = (−20, 3.8). The predicted value for this observation is $\hat{y}_3 = 3.50 + (-0.0034 \cdot -20) = 3.57$ and residual$_3 = y_3 - \hat{y}_3 = 3.8 - 3.57 = 0.23$.

The graph showing the residuals is not shown in this key.

14.15 *Gorilla ebola.*

(a) DISTANCE is the independent variable. ONSET is the dependent variable.

(b) The scatterplot reveals a positive linear association with no apparent outliers.

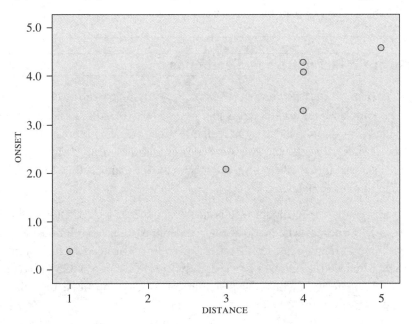

(c) $r = 0.962$, demonstrating that the correlation is extremely strong.

Correlations

		DISTANCE	ONSET
DISTANCE	Pearson Correlation	1.000	0.962
	Sig. (two-tailed)		0.002
	N	6.000	6
ONSET	Pearson Correlation	0.962	1.000
	Sig. (two-tailed)	0.002	
	N	6	6.000

(d) $r^2 = 0.962^2 = 0.93$

(e) The least square regression model is ONSET $= -8.09 + 11.26 \cdot$ DISTANCE. Results calculated with SPSS v 20.0.0 are shown in the following table.

Coefficients[a]

Model		Unstandardized Coefficients		Standardized Coefficients	t	Sig.	95% Confidence Interval for B	
		B	Std. Error	β			Lower Bound	Upper Bound
1	(Constant)	−8.088	5.917		−1.367	0.243	−24.516	8.341
	DISTANCE	11.263	1.591	0.962	7.080	0.002	6.846	15.680

[a]Dependent Variable: ONSET

The slope predicts that it takes, on average, 11.3 days for the outbreak to move from one band of gorillas to the next.

■ Chapter 15: Multiple Linear Regression

15.1 *The relation between* FEV *and* SEX *in the illustrative data set.*

The simple regression model is FEV $= 2.451 + (0.361)($SEX$)$. Because SEX is coded $0 =$ female and $1 =$ male, the intercept (2.451) represents mean FEV for female subjects and the slope represents the mean difference between females and males. Therefore, the mean FEV for males is 0.361 (L/sec) higher than that for females.

Adding AGE to the model results in FEV $= 0.281 + (0.323)($SEX$) + (0.220)$ (AGE). To address whether AGE confounded the observed relationship between SEX and FEV, we consider the effect of adding AGE to the regression model, which reduced the slope of SEX slightly, from 0.361 to 0.323 (11% decrease in relative terms). Thus, the potential for AGE to confound the relationship between SEX and FEV is minimal.

■ Chapter 16: Inference About a Proportion

16.1 AIDS-related risk factor.

(a) The population to which inference will be made is U.S. adult heterosexuals at the time of the survey. The parameter of interest is the proportion of individuals with multiple sexual partners. The sample proportion $\hat{p} = 170/2673 = 0.063598 \approx 0.064$ or 6.4%.

(b) Examples of selection biases that may be pertinent: (1) Specific high-risk groups (for example, homeless, intravenous drug users) may not have permanent telephone lines and may be underrepresented in the sample. (2) Nonresponse should be scrutinized.

(c) Without a specific validation study to address this issue, it is difficult to determine the accuracy of responses. However, we may hypothesize that data could have understimated the true prevalence is respondents under-reported sexual behavior out of fear of embarrassment or reprisal.

16.3 AIDS-related risk factor.

(a) Sampling distributions: Let X represent the number of individuals who are positive for the attribute. Random variable X has a binomial distribution with $n = 2673$ and parameter p. The value of p is unknown. Using the notation established in this chapter, $X \sim b(n = 2673, p = \text{unknown})$.

(b) A Normal approximation can be used if p is not too small. Suppose, for example, $p = 0.01$. Then $npq = (2673)(0.01)(0.99) = 26.5$. Since this exceeds 5, the sampling distribution of X is $X \sim N(\mu = 2673 \cdot p, \sigma = \sqrt{2673 \cdot p \cdot q})$. In contrast, if P is very small (say, 0.0001), then a normal approximation can not be used because $npq = 2673 \cdot 0.0001 \cdot 0.9999^0 = 0.267$.

16.5 AIDS-related risk factor.

A. Hypotheses. $H_0: p = 0.075$ versus $H_a: p \neq 0.075$

B. Test statistic. We start by checking the *npq* rule: $np_0q_0 = 2673(0.075)(0.925) = 185$. Therefore, the sample is large enough to use the z-test.

$$SE_{\hat{p}} = \sqrt{\frac{0.075 \cdot 0.925}{2673}} = 0.005095$$

$$z_{stat} = \frac{0.063598 - 0.075}{0.005095} = -2.24$$

C. *P*-value. $P = 0.025$ via Appendix Table F (good evidence against H_0)

D. Significance level. The evidence against H_0 is significant at $\alpha = 0.05$ (reject H_0) but is not significant at $\alpha = 0.01$ (retain H_0).

E. Conclusion. The current prevalence of 6.4% is significantly less than the historical level of 7.5% ($P = 0.025$).

16.7   *Patient preference, Fisher's method.*

A. Hypotheses. H_0: $p = 0.50$ versus H_a: $p > 0.50$

B. Test statistic. An exact binomial test is used because of the small sample size. Under H_0, $X \sim b(8, 0.5)$. The observed number of success in the sample is $x = 7$.

C. *P*-value. P (one-sided) $= \Pr(X = 7) + \Pr(X = 8) = 0.0313 + 0.0039 = 0.0352$. This provided good evidence against the null hypothesis.

D. Significance level. The evidence against H_0 is significant at $\alpha = 0.05$ (reject H_0) but is not significant at $\alpha = 0.01$ (retain H_0).

E. Conclusion. The data provide reliable evidence that more than half the patient population favors procedure A ($P = 0.0352$).

16.9   *AIDS-related risk factor.*

$\tilde{n} = 2677$

$\widetilde{p} = \dfrac{172}{2677} = 0.0643$

$\widetilde{q} = 1 - \widetilde{p} = 1 - 0.0643 = 0.9357$

$SE_{\widetilde{p}} = \sqrt{\dfrac{(0.0643)(0.9357)}{2677}} = 0.004740$

The 95% confidence interval for, population prevalence $p = 0.0643 \pm (1.96)(0.004740) = 0.0643 \pm 0.0093 = (0.055$ to $0.074)$ or (5.5% to 7.4%).

16.11   *Patient preference.*

The 95% confidence interval for p by Fisher's method is 0.473 to 0.997.
The 95% confidence interval for p by the Mid-P method is 0.520 to 0.994.
Confidence intervals calculated with *WinPepi* describe.exe version 1.5.1.

16.13   *Cerebral tumors and cell phone use.*

A. Hypotheses. H_0: $p = 1/2$ against H_a: $p \neq 1/2$

B. Test statistic. $SE = \sqrt{\dfrac{0.5 \cdot 0.5}{41}} = 0.078087$

$\hat{p} = \dfrac{26}{41} = 0.6341$

$$z_{stat} = \frac{0.6341 - 0.5}{0.078087} = 1.72$$

C. *P*-value. $P = 0.085$ via Table F. Results suggest that evidence against H_0 is marginally significant (by usual conventions).

D. Significance level. The results are significant at $\alpha = 0.10$ (reject H_0) but not significant at $\alpha = 0.05$ (retain H_0).

E. Conclusion. Data provide some evidence that the tumors occurred more frequently on the side of the head as cellular phone use ($P = 0.085$).

Additional notes

- With continuity correction, $z_{stat,c} = 1.56$ and $P = 0.12$
- The Fisher's test derives $P = 0.117$
- The published article (Muscat et al., 2000) reported $P = 0.06$ based on a one-sided goodness of fit test with continuity correction (Muscat 2006, personal communication). This corresponds perfectly with our two-sided continuity corrected *z*-test.

16.15   *Insulation workers.*

A. Hypotheses. H_0: $p = 0.0259$ versus H_a: $p \neq 0.0259$

B. Test statistic. First check whether the Normal approximation can be used: $np_0q_0 = (556)(0.02590)(1 - 0.0259) = 14.0$. Therefore, the Normal (*z*-statistic) method is OK.

$$SE_{\widehat{p}} = \sqrt{\frac{0.02590 \cdot (1 - 0.02590)}{556}} = 0.00674$$

$z_{stat} = (0.0468 - 0.0259) / (0.00674) = 3.10$

C. *P*-value. $P = 0.0020$ ("highly significant" by usual conventions).

D. Significance level. Data provide significant evidence against H_0 at $\alpha = 0.01$.

E. Conclusion. The incidence of cancer deaths in these insulation workers (4.68%) is significantly greater than the expected incidence of 2.59% ($P = 0.0020$).

16.17   *Kidney cancer survival.*

A. Hypotheses. H_0: $p = 0.2$ versus H_a: $p \neq 0.2$

B. Test statistic. The *z*-test can be used because $np_0q_0 = (40)(0.2)(0.8) = 6.4$. The observed proportion $\hat{p} = 16/40 = 0.4$

$$\text{The } z_{\text{stat}} = \frac{0.4 - 0.2}{\sqrt{\dfrac{0.2 \times 0.8}{40}}} = 3.16$$

C. *P*-value. $P = 0.00158$ by Table F, providing highly significant evidence against H_0.

D. Significance level. The evidence against H_0 is significant at $\alpha = 0.01$ (reject H_0).

E. Conclusion. There has been a significant improvement in survival ($P = 0.0016$).

Note: Using the exact Mid-P procedure (calculated with *WinPepi* describe.exe 1.5.1) the two-sided $P = 0.0039$.

16.19 *Sample-size requirement.*

Conditions: 95% confidence; $p^* = 0.50$ (since no educated guess for p is available); desired margin of error $m = 0.06$.

Calculation: $n = \dfrac{z_{1-\frac{\alpha}{2}}^2 \cdot p^* \cdot q^*}{m^2} = \dfrac{1.96^2 \cdot 0.5 \cdot 0.5}{0.06^2} = 266.8$. Therefore, resolve to study 267 individuals.

16.21 *Alternative medicine.*

95% CI for population prevalence $p = \widetilde{p} \pm z_{1-\frac{\alpha}{2}} \cdot \sqrt{\dfrac{\widetilde{p}\,\widetilde{q}}{\widetilde{n}}} = \dfrac{662}{1504} \pm 1.960 \cdot$

$\sqrt{\dfrac{\frac{662}{1504} \cdot \left(1 - \frac{662}{1504}\right)}{1504}} = 0.4402 \pm 0.0251 = (0.4151, 0.4653)$. We conclude with 95% confidence that between 41.5% and 46.5% of the population would use alternative medicine if traditional medical care failed to produce the desired results.

Note: The plus-four confidence interval method is unnecessary with this large sample size—a straight Normal approximation would have sufficed. However, there is no harm in using the plus-four method.

16.23 *Perinatal growth failure.*

A. Hypotheses. H_0: $p = 0.025$ versus H_a: $p > 0.025$

B. Test statistic. $x = 8$

C. *P*-value. $P = \Pr(X \geq 8 \,|\, X \sim b(33, 0.025)) = 6.5 \times 10^{-7}$ ("highly significant").

D. Significance level. Data provide significant evidence against H_0 at extremely low α levels.

E. Conclusion. Infants with perinatal grown failure syndrome have a higher incidence of very-low intelligence scores at age 8 compared to the general population ($P < 0.0001$).

16.25 *Incidence of improvement.*

Incidence $\hat{p} = \dfrac{20}{75} = 0.2667$ (about 27%)

Confidence interval for p by the plus-four method:

$\tilde{n} = 79$

$\tilde{x} = 22$

$\tilde{p} = 0.2785$

$SE_{\tilde{p}} = \sqrt{\dfrac{(0.2785)(0.7215)}{79}} = 0.0504$

95% confidence interval for $p = 0.2785 \pm (1.96)(0.0504) = 0.2785 \pm 0.0988 = 0.1797$ to 0.3773 or about 18% to 38%.

We can conclude with 95% confidence that between 18% and 38% of this population shows spontaneous improvement within a month.

16.27 *Familial history of breast cancer, sample size requirements.*

(a) $n = \left(\dfrac{z_{1-\alpha/2}\sqrt{p_0 q_0} + z_{1-\beta}\sqrt{p_1 q_1}}{p_1 - p_0} \right)^2 =$

$\left(\dfrac{1.960\sqrt{0.03 \cdot (1-0.03)} + 1.28\sqrt{0.05 \cdot (1-0.05)}}{0.05 - 0.03} \right)^2 =$

$\left(\dfrac{0.33435 + 0.27897}{0.02} \right)^2 = 940.4.$ Therefore, resolve to study 941 individuals.

(b) $n = \left(\dfrac{1.960\sqrt{0.03 \cdot (1-0.03)} + 1.28\sqrt{0.06 \cdot (1-0.06)}}{0.06 - 0.03} \right)^2 =$

$\left(\dfrac{0.33435 + 0.30398}{0.03} \right)^2 = 452.7.$ Therefore, resolve to study 453 individuals.

(c) The larger expected differences diminished the sample size requirement of the study.

(d) $n = \left(\dfrac{2.576\sqrt{0.03 \cdot (1 - 0.03)} + 1.28\sqrt{0.05 \cdot (1 - 0.05)}}{0.05 - 0.03} \right)^2 =$

$\left(\dfrac{0.439433 + 0.27897}{0.02} \right)^2 = 1290.3.$ Therefore, resolve to study 1291 individuals.

(e) The lower α level increased the sample size requirement of the study.

16.29 *Freshman binge drinking.*

(a) This is a large sample, so we could go directly to the large sample formula. However, there is no harm in using the plus-four method. Thus, $\widetilde{p} = \dfrac{1802 + 2}{5266 + 4} = 0.3423, \widetilde{q} = 1 - 0.3423 = 0.6577,$ and $\widetilde{n} = (5266 + 4) = 5270,$ and the 95% CI for $p = \widetilde{p} \pm z_{1-\frac{\alpha}{2}} \cdot \sqrt{\dfrac{\widetilde{p}\,\widetilde{q}}{\widetilde{n}}} = 0.3423 \pm 1.96$ $\cdot \sqrt{\dfrac{0.3423 \cdot 0.6577}{5270}} = 0.3424 \pm 0.0128 = (0.3296, 0.3552).$ We can now state with 95% confidence that the population prevalence is between 33.0% and 35.5%.

(b) The 99% CI for $p = \widetilde{p} \pm z_{1-\frac{\alpha}{2}} \cdot \sqrt{\dfrac{\widetilde{p}\,\widetilde{q}}{\widetilde{n}}} = 0.3423 \pm 2.576 \cdot$ $\sqrt{\dfrac{0.3423 \cdot 0.6577}{5270}} = 0.3424 \pm 0.0168 = (0.3256, 0.3592).$ We can now state with 99% confidence that the population prevalence is between 32.6% and 36.0%.

(c) Yes. Both the 95% confidence interval and 99% confidence interval for p exclude a population proportion of 20%. This is equivalent to saying that the evidence against H_0: $p = 0.20$ is reliable at both $\alpha = 0.05$ and 0.01 levels of statistical significance.

Chapter 17: Comparing Two Proportions

17.1 *Prevalence of cigarette use among two ethnic groups.*

(a) The sampling distribution of $\hat{p}_1 - \hat{p}_2$ will be approximately Normal with mean $\mu = p_1 - p_2 = 0.40 - 0.12 = 0.28$ and standard deviation $s_{\hat{p}_1 - \hat{p}_2} = \sqrt{\dfrac{p_1 q_1}{n_1} + \dfrac{p_2 q_2}{n_2}} = \sqrt{\dfrac{(0.4)(0.6)}{1000} + \dfrac{(0.12)(0.88)}{1000}} = 0.01859.$ In symbols, $\hat{p}_1 - \hat{p}_2 \sim N(0.28, 0.01859)$

(b) $\Pr(\hat{p}_1 - \hat{p}_2 \leq 0.26) = \Pr(z \leq [(0.26 - 0.28)/0.01859]) = \Pr(z \leq -1.08)$ $= 0.1401$ (from Table B)

(c) $\Pr(\hat{p}_1 - \hat{p}_2 \leq 0) = \Pr(z \leq [(0.00 - 0.28) / 0.01859]) = \Pr(z \leq -15.06)$
$= 0.0000$

17.3 *Cytomegalovirus and coronary restenosis.*

(a) Risk in the CMV+ group $\hat{p}_1 = \dfrac{21}{49} = 0.4286$

Risk in the CMV− group $\hat{p}_2 = \dfrac{2}{26} = 0.0769$

Risk difference $\hat{p}_1 - \hat{p}_2 = 0.4286 - 0.0769 = 0.3517$

(b) 95% confidence by the plus-four method:

$\hat{p}_1 = \dfrac{22}{51} = 0.4314$

$\hat{p}_2 = \dfrac{3}{28} = 0.1071$

SE = 0.0907

95% confidence interval for $p_1 - p_2 = (0.4314 - 0.1071) \pm (1.96)$
$(0.0907) = 0.3243 \pm 0.1778 = (0.1465, 0.5021)$.

We are 95% confident that CMV increases the risk of restenosis by between 14.7% and 50.2%.

(c) Here are results calculated by *WinPepi* Compare2.exe (version 1.38).
DIFFERENCE (A minus B) = 0.324 SE = 0.091

Large-sample method (Fleiss), continuity-corrected:
90% CI = 0.147 to 0.501
95% CI = 0.119 to 0.530
99% CI = 0.063 to 0.586
Wilson's score method:

Not continuity-corrected (Newcombe's method 10):
90% CI = 0.153 to 0.455
95% CI = 0.117 to 0.477
99% CI = 0.044 to 0.518

Continuity-corrected (Newcombe's method 11):
90% CI = 0.131 to 0.469
95% CI = 0.094 to 0.490
99% CI = 0.022 to 0.529

The 95% confidence interval by the Wilson score method, not continuity-corrected (bold face) most closely corresponds with our plus-four method.

17.5 *Joseph Lister and anti septic surgery.*
$\hat{p}_1 = 0.457143$
$\hat{p}_2 = 0.1500$
Hypothesis test
   A. Hypotheses. $H_0: p_1 = p_2$ versus $H_a: p_1 \neq p_2$
   B. Test statistic. $z_{stat} = 2.91$
   C. *P*-value. $P = 0.0036$, providing highly significant evidence against the null hypothesis.
   D. Significance level. The evidence against H_0 is significant at $\alpha = 0.005$ (reject H_0).
   E. Conclusion. Adoption of aseptic surgical techniques decreased post-operative mortality from 45.7% to 15.0% ($P = 0.0036$).

17.7 *Induction of labor and meconium staining.*
   (a) Estimates
       $\hat{p}_1 = 1/111 = 0.0090$
       $\hat{p}_2 = 13/117 = 0.1111$

   (b) The following table of expected values shows that all expected values exceed 5. Therefore, an exact procedure is unnecessary.

Expected values, Exercise 17.7.

	Successes	**Failures**	**Total**
Exposed Group 1	$\dfrac{14 \times 111}{228} = 6.82$	$\dfrac{111 \times 214}{228} = 104.18$	111
Nonexposed Group 2	$\dfrac{14 \times 117}{228} = 7.18$	$\dfrac{117 \times 214}{228} = 109.82$	117
Total	14	214	228

   (c) Hypothesis test using *z*-procedure
       A. Hypothesis statements. $H_0: p_1 = p_2$ versus $H_a: p_1 \neq p_2$
       B. Test statistic. $\bar{p} = 14/228 = 0.0614$

$$z_{stat} = \frac{\hat{p}_1 - \hat{p}_2}{\sqrt{\bar{p}\bar{q}\left(\dfrac{1}{n_1} + \dfrac{1}{n_2}\right)}} = \frac{0.0090 - 0.1111}{\sqrt{0.0614 \cdot (1 - 0.0614)\left(\dfrac{1}{111} + \dfrac{1}{117}\right)}}$$

$$= \frac{-0.1021}{0.0318} = -3.21.$$

    C. $P = 0.00133$ (via Appendix Table F).

    D. Significance level. The evidence against the null hypothesis is significant at the $\alpha = 0.002$ level but not at the $\alpha = 0.001$ level.

    E. Conclusion. Induction of labor significantly lowered incidence of meconium staining ($P = 0.0013$).

(d) The P-value by Fisher's test is 0.0014, which is not materially different from the P-value derived by the z-test.

17.9 *Framingham Heart Study.*

The incidence proportion in the high cholesterol group $\hat{p}_1 = \dfrac{51}{424} = 0.12028$.

The incidence proportion in the low cholesterol group $\hat{p}_2 = \dfrac{16}{454} = 0.03524$.

$\widehat{RR} = \dfrac{0.12028}{0.03524} = 3.4130$.

Note that $\ln\left(\widehat{RR}\right) = \ln(3.4130) = 1.2276$ and

$SE_{\ln\widehat{RR}} = \sqrt{(51^{-1} - 424^{-1} + 16^{-1} - 454^{-1})} = 0.27847$. The 95% CI for lnRR $= 1.2276 \pm (1.960)(0.27847) = 1.2276 \pm 0.54580 = 0.6818$ to 1.77340. Therefore, the 95% confidence interval for the RR $= e^{(0.6818,\ 1.77340)} = (1.98$ to $5.89)$. Interpretation: We can be 95% confident that the RR in the source population is between 1.98 and 5.89 (i.e., two to six times the risk of coronary heart disease in high cholesterol group compared to the low cholesterol group).

17.11 *Sample-size plan.*

Assumptions: $\alpha = 0.05$, $p_1 = 0.20$, $p_2 = 0.30$, average risk $\bar{p} = 0.25$, $1 - \beta = 0.80$, and $n_1 = n_2 = n$.

$$n = \left(\frac{z_{1-\frac{\alpha}{2}}\sqrt{2\bar{p}\bar{q}} + z_{1-\beta}\sqrt{p_1 q_1 + p_2 q_2}}{|p_1 - p_2|}\right)^2$$

$$= \left(\frac{1.96\sqrt{2 \cdot 0.25 \cdot (1 - 0.25)} + 0.84\sqrt{0.2 \cdot (1 - 0.2) + 0.3 \cdot (1 - 0.3)}}{|0.2 - 0.3|}\right)^2$$

$= 17.112^2 = 292.8$. Therefore, resolve to study 293 individuals in each group.

WinPEPI estimated a sample size requirement of 294 per group. The discrepancy between the hand calculated estimate of 293 and WinPEPI's estimate of 294 is inconsequential and is due to rounding error (e.g., using $z_{0.80} = 0.84$ instead of $z_{0.80} = 0.842$).

The earlier calculation assumes no continuity correction. Incorporation of a continuity correction factor results in $n' = \frac{n}{4}\left(1 + \sqrt{1 + \frac{4}{n|p_1 - p_2|}}\right)^2 =$

$$\frac{292.8}{4}\left(1 + \sqrt{1 + \frac{4}{292.8|0.30 - 0.20|}}\right)^2 = 312.5 \rightarrow 313 \text{ per group.}$$

17.13 *Smoking cessation trial.*

$\hat{p}_1 = 87/245 = 0.3551$

$\hat{p}_2 = 40/244 = 0.1639$

$\hat{p}_1 - \hat{p}_2 = 0.1912$

Confidence interval by the plus-four method:

$\tilde{p}_1 = 0.3563$

$\tilde{q}_1 = 0.6437$

$\tilde{n}_1 = 247$

$\tilde{p}_2 = 0.1667$

$\tilde{q}_2 = 0.8333$

$\tilde{n}_2 = 246$

SE = 0.03864

95% CI for $p_1 - p_2 = (0.3563 - 0.1667) \pm (1.96)(0.03864) = 0.1896 \pm 0.0757 = (0.1139, 0.2653)$.

The confidence interval is more useful than the hypothesis test because it quantifies the effect of the intervention. The hypothesis test merely addressed whether there was any effect.

17.15 *Telephone survey completion rates.*

(a) Descriptive statistics. Proportion that completed the survey in the group that received advanced warning $\hat{p}_1 = 134/291 = 0.4605$. Proportion in the group that did not receive advanced warning $\hat{p}_2 = 33/100 = 0.3300$.

(b) Hypothesis test.

A. $H_0: p_1 = p_2$ versus $H_a: p_1 \neq p_2$.

B. Note: $\bar{p} = (134 + 33)/(291 + 100) = 0.4271$

$$z_{stat} = \frac{\hat{p}_1 - \hat{p}_2}{\sqrt{\bar{p}\bar{q}\left(\frac{1}{n_1} + \frac{1}{n_2}\right)}} = \frac{0.4605 - 0.330}{\sqrt{0.4271 \cdot (1 - 0.4271)\left(\frac{1}{291} + \frac{1}{100}\right)}}$$

$$= \frac{0.1305}{0.05734} = 2.28.$$

C. $P = 0.023$.

D. The observed difference is significant at $\alpha = 0.025$ but not at $\alpha = 0.02$.

E. The advanced warning letter improved interview completion rates from 33.0% to 46.0% ($P = 0.023$).

(c) Estimation of effect size. The point estimate of the difference in proportions $\hat{p}_1 - \hat{p}_2 = 0.46048 - 0.3300 = 0.13048$. Thus, advanced warning improved the response rate by 13.0% (in absolute terms).

$$\tilde{p}_1 = \frac{134 + 2}{291 + 4} = 0.46102 \qquad \tilde{p}_2 = \frac{33 + 2}{100 + 4} = 0.33654$$

$$\mathrm{SE}_{\tilde{p}_1 - \tilde{p}_2} = \sqrt{\frac{0.46102(1 - 0.46102)}{295} + \frac{0.33654(1 - 0.33654)}{104}}$$

$$= 0.05467$$

95% CI for $p_1 - p_2 = (\tilde{p}_1 - \tilde{p}_2) \pm z_{1-\frac{\alpha}{2}} \cdot \mathrm{SE}_{\tilde{p}_1 - \tilde{p}_2} = (0.46102 - 0.33654) \pm (1.960)(0.05467) = 0.1245 \pm 0.1072 = (0.0173, 0.2317) = (1.7\%, 23.2\%)$. We can be 95% confident that the effect of the warning letter is to improve the response rate from 1.7% to 23.2%. A larger study is needed to derive a more precise estimate of the effect.

17.17 *4S coronary mortality.*
$\hat{p}_1 = 111/2221 = 0.049977$
$\hat{p}_2 = 189/2223 = 0.085020$

A. Hypotheses. H_0: $p_1 = p_2$ versus H_a: $p_1 \neq p_2$

B. Test statistic. $z_{\mathrm{stat}} = 4.66$

C. *P*-value. P (two-tailed) $= 0.0000032$

D. Significance level. The evidence against the null hypothesis is significant at $\alpha = 0.01$.

E. Conclusion. The simvastatin treatment group demonstrated significantly lower fatal heart attacks risk compared to the placebo group (5.0% vs. 8.5%, $P = 3.2 \times 10^{-6}$).

In relative terms, how much did simvastatin lower heart attach risk? The easiest way to address this question is to first calculate the relative risk:

$$\widehat{RR} = \frac{0.049977}{0.085020} = 0.59.$$

The relative reduction in risk $= 1 - 0.59 = 0.41$, or 41%.

17.19   Acute otitis media.

   A. Hypotheses. H_0: $p_1 = p_2$ versus H_a: $p_1 \neq p_2$

   B. Test statistic. $z_{stat} = 2.054$

   C. P-value. $P = 0.040$

   D. Significance level. The evidence against the null hypothesis is significant at $\alpha = 0.05$.

   E. Conclusion. Clearance of the effusions from ear infections after 14 days was significantly better with cefaclor than with amoxicillin (55.7% vs. 41.2%, $P = 0.040$).

   Notes

   • With continuity correction, $z_{stat, c} = 1.913$ and $P = 0.056$.

   • Take care in interpreting results, and do not extrapolate beyond the conditions of the test. The current test applies only to the 14-day follow-up point. In the published article (Mandel et al., 1982), the improvement rates equalized by 42 days (68.9% in the cefaclor group and 67.5% in the amoxicillin group). This and other studies suggest no difference in long-term failure rates with cefaclor and amoxicillin. For clinical recommendations, see AHRQ (2001). *Number 15. Management of Acute Otitis Media.* Retrieved August 30, 2006, from http://www.ncbi.nlm.nih.gov/books /bv.fcgi?rid=hstat1.chapter.21026.

■ Chapter 18: Cross-Tabulated Counts

18.1   YRBS prevalence proportions.
   $\hat{p}_1 = 243/2154 = 0.113$; SE $= 0.00682$; $m = 0.014$
   $\hat{p}_2 = 25/508 = 0.049$; SE $= 0.00958$; $m = 0.019$
   $\hat{p}_3 = 55/526 = 0.105$; SE $= 0.01337$; $m = 0.027$

18.3   Cytomegalovirus infection and coronary restenosis.

   (a) Prevalence of CMV in both groups combined $= 49/75 = 0.653 = 65.3\%$

   (b) Incidence of restenosis overall $= 23/75 = 0.307 = 30.7\%$

   (c) The proportion of the CMV+ group experiencing restenosis $\hat{p}_1 = 21/49 = 0.4286 = 42.9\%$.

   The proportion of the CMV− group experiencing restenosis $\hat{p}_2 = 2/26 = 0.0769 = 7.7\%$.

   These are row percents.

(d) $\widehat{RR} = \dfrac{0.4286}{0.0769} = 5.57$

(e) $\widehat{OR} = \dfrac{21/28}{2/24} = 9.00$

This OR is larger than the RR because the outcome is common.

18.5 Response to leprosy treatment.

Here are the relevant row percentages:

Skin infiltration	Marked improvement	Marked improvement	Slight improvement	Stationary	Worse	Total
High	13.5%*	28.8%	30.8%	25%	1.9%	100%
Low	7.6%	18.8%	29.2%	36.8%	7.6%	100%

*Example of calculation: $7/52 \times 100\% = 13.5\%$.

These distributions show that patients with high skin damage were more likely to show improvement than those with low skin infiltration.

18.7 Chi-square approximation.

The area under the curve is between the chi-square landmarks of 4.64 (right-tail 0.20) and 5.32 (right-tail 0.15). Therefore, $0.15 < P < 0.20$. The precise area computed with a software utility is 0.1564.

18.9 Cytomegalovirus infection and coronary restenosis.

(a) $\chi^2_{stat} = 9.879$, df $= 1$, $P = 0.0017$
    With continuity-correction, $\chi^2_{stat,c} = 8.294$, df $= 1$, $P = 0.0040$

(b) $z_{stat} = 3.14$, $P = 0.0017$. Note that $z^2_{stat} = 3.14^2 = 9.86 \approx \chi^2_{stat}$
    With continuity-correction, $z^2_{stat,c} = 2.88^2 = 8.30 = \chi^2_{stat,c}$, $P = 0.0040$

18.11 Anger and heart disease.

A. Hypotheses. H_0: "no trend between anger-trait and hard coronary events in the source population" versus H_a: "trend in population"

B. Test statistic. $z_{stat,\ trend} = 3.16$

C. P-value. $P = 0.0016$

D. Significance level. The evidence against H_0 is significant at $\alpha = 0.005$ but at $\alpha = 0.001$.

E. Conclusion. The positive trend between the anger trait and incidence of coronary heart disease is statistically significant ($P = 0.0016$).

18.13   *Cell phones and brain tumors, study 1.*
Results *fail* to support an association between cell phone use and brain tumors. The odds ratios for glioma and meningioma show small negative associations. The odds ratio for acoustic neuroma shows a small positive association. All confidence interval are consistent with no association. The point estimate for all tumor types combined is 1.0 indicating no association between recent cell phone use and intracranial tumors in general. All confidence interval are consistent with population odds ratios of 1 (no association).

18.15   *Doll and Hill, 1950.*

Smoke	Cases	Noncases	Total
+	647	622	1269
−	2	27	29
Total	649	649	1298

$$\widehat{OR} = \frac{647 \cdot 27}{622 \cdot 2} = 14.04$$

This suggests that the smokers had 14 times the risk of nonsmokers.
Determining the confidence interval for the OR:
$\ln(14.04) = 2.6419$
$SE = sqrt(647^{-1} + 622^{-1} + 2^{-1} + 27^{-1}) = 0.7350$
The 95% CI for OR $= e^{(2.6419 \pm 1.96 \times 0.7350)} = e^{(1.2013,\, 4.0825)} = 3.3$ to 59.3

18.17   *Baldness and myocardial infarction, self-assessed baldness.*
(a) $\widehat{OR}_1 = 1.00$ (reference)
$\widehat{OR}_2 = 0.98$
$\widehat{OR}_3 = 1.39$
$\widehat{OR}_4 = 1.94$
$\widehat{OR}_5 = 2.64$
This reveals a positive trend in odds ratios after baldness level 2.
(b) $\chi^2_{stat} = 14.570$, df $= 4$, $P = 0.0057$. The association is highly significant.
(c) $z_{stat,trend} = 3.39$, $P = 0.00070$. The trend is highly significant.

18.19   *Diet and adenomatous polyps.*

A. Hypotheses. H_0: no association between fruit and vegetable consumption and colon polyps in the population versus H_a: H_0 false.
B. Test statistic. $z_{McN} = sqrt[(45 - 24)^2 / (45 + 24)] = 2.528$
C. *P*-value. $P = 0.011$

D. Significance level. The evidence against H_0 is significant at $\alpha = 0.05$ and is *almost* significant at $\alpha = 0.01$.

E. Conclusion. The association between low fruit and vegetable consumption and the recurrence of colon polyps is statistically significant ($P = 0.011$).

Additional notes: With continuity-correction, $z_{McN, c} = \text{sqrt}[(|45 - 24| - 1)^2 / (45 + 24)] = 2.408$ and $P = 0.016$.

18.21   *Thrombotic stoke in young women.*

(a) $\widehat{OR} = \dfrac{44}{5} = 8.8$. Interpretation: Oral contraceptive use was associated with an almost nine-fold increase in the risk of thrombotic stroke.

(b) Here are the data with the match broken:

Match broken	Case	Control	Total
Exposed	46	7	53
Non-exposed	60	99	159
Total	106	106	212

The odds ratio with the match broken is $= (46)(99) / (7)(60) = 10.8$, overestimating the more appropriate matched-pair odds ratio of 8.8.

18.23   *Don't sweat the small stuff.*

Without continuity-correction: $\chi^2_{stat} = 4.107$, df $= 1$, $P = 0.043$

With continuity-correction: $\chi^2_{stat, c} = 3.598$, df $= 1$, $P = 0.058$

It would not be reasonable to derive different conclusions because the actual data has not changed. Both Pearson's test ($P = 0.043$) and Yates' test ($P = 0.058$) provide reasonably reliable evidence against the null hypothesis. Therefore, the treatment group experienced the outcome significantly more often than the control group (12.5% vs. 7.7%).

18.25   *Yates, 1934 (three-by-two).*

Here are the expected values:

	Normal teeth	Malocclusion
Breast fed	1.739	18.261
Bottle fed	1.913	20.087
Breast & bottle feed	4.348	45.652

You should not use a chi-square test in this situation because three table cells have expected values that are less than 5. *WinPepi* Compare2.exe (version 1.38) calculates a Fisher's P of 0.1503. Therefore, the evidence against H_0 is not significant. The conclusion is that the prevalence of malocclusion did not differ significantly according to whether the infant was breast fed or bottle fed ($P = 0.15$).

18.27   *Esophageal cancer and tobacco use.*

$\widehat{OR} = \dfrac{64 \cdot 625}{150 \cdot 136} = 1.96$. 95% confidence interval for the OR = (1.37, 2.81). This confidence interval was calculated with WinPEPI > Compare2 > A Proportions or odds > Cornfield's confidence interval for the odds ratio.

Interpretation: The point estimate suggests a doubling in the risk of esophageal cancer risk with tobacco use at the reported level. The 95% confidence interval suggests data are consistent with population odds ratios between 1.37 and 2.81.

18.29   *Baldness and myocardial infarction, interviewer assessments.*

(a) The interviewer assessments are likely to be more consistent and objective than the self-assessments.

(b) Baldness levels were classified as 1 = none, 2 = frontal, 3 = mild vertex, 4 = moderate vertex, and 5 = severe vertex according to interviewer assessments using the Hamilton baldness scale. Odds ratio estimates are as follows: $\widehat{OR}_1 = 1.00$ (referent); $\widehat{OR}_2 = \dfrac{44 \cdot 480}{82 \cdot 238} = 1.08$; $\widehat{OR}_3 = \dfrac{108 \cdot 480}{137 \cdot 238}$ $= 1.59$; $\widehat{OR}_4 = \dfrac{40 \cdot 480}{46 \cdot 238} = 1.75$; $\widehat{OR}_5 = \dfrac{35 \cdot 480}{23 \cdot 238} = 3.07$.

(c) This table below compares the results of the two analyses:

Baldness level	Self-assessed baldness	Interviewer-assessed baldness
1 (no baldness)	1.0 (reference)	1.0 (reference)
2	1.0	1.1
3 (moderate baldness)	1.4	1.6
4	1.9	1.8
5 (severe baldness)	2.6 (small sample)	3.1 (small sample)

Similar positive trends are observed between baldness level and myocardial infraction risk.

18.31 Practice with chi-square.

Observed

Baldness	Cases	Controls	Total
1 (none)	251	331	582
2	165	221	386
3	195	185	380
4	50	34	84
5 (extreme)	2	1	3
Total	663	772	1435

Expected

Baldness	Cases	Controls	Total
1 (none)	268.896	313.104	582
2	178.340	207.660	386
3	175.568	204.432	380
4	38.810	45.190	84
5 (extreme)	1.386	1.614	3
Total	663.000	772.000	1435

Use of the chi-square test is justified because only 20% of the table cells have expected frequencies less than 5.

$(O - E)^2 / E$

Baldness	Cases	Controls
1 (none)	1.191	1.023
2	0.998	0.857
3	2.151	1.847
4	3.227	2.771
5 (extreme)	0.272	0.234

$\chi^2_{stat} = 1.191 + 1.023 + 0.998 + 0.857 + 2.151 + 1.847 + 3.227 + 2.771 + 0.272 + 0.234 = 14.571$

$df = (5 - 1)(2 - 1) = 4$

$P = 0.0057$ (highly significant evidence against H_0)

■ Chapter 19: Stratified two-by-two Tables

19.1 *Is participating in a follow-up survey associated with having medical aid?*

(a) Among the 416 children who were followed-up, 46 (11.1%) had medical aid. In contrast, 195 of 1174 (16.6%) who were not followed-up had medical aid. Therefore, there is a negative association between follow-up and having medical aid.

(b) Among the white participants, 10 of 12 (83%) who were followed-up had medical aid. In total, 104 of the 126 white subjects who were not followed-up (83%) had medical aid. Therefore, there is no association between follow-up and having medical aid in this stratum.

(c) Nine percent of both groups had medical aid. Therefore, there is no association between follow-up and having medical aid in this stratum.

(d) Race is associated with follow-up and medical aid coverage. Therefore, race confounded the crude analysis in part (a).

19.3 *Is participating in a follow-up survey associated with having medical aid?*

(a) Race-specific prevalence ratios:

$$\widehat{RR}_1 = 1.010$$
$$\widehat{RR}_2 = 1.026$$

These strata-specific relative risks are homogeneous, so interaction is absent.

(b) Hypothesis test

A. Hypotheses. H_0: $RR_1 = RR_2$ (no interaction) versus H_a: $RR_1 \neq RR_2$ (interaction)

B. Recall that $\widehat{RR}_{MH} = 1.022$ (calculated with WinPEPI > Compare2 > A. Proportions > "Stratified tables"). Calculation of our ad hoc interaction statistic is as follows:

$$\chi^2_{\text{stat int}} = \Sigma \frac{\left(\ln\widehat{RR}_k - \ln\widehat{RR}_{MH} \right)^2}{\left(SE_{\ln\widehat{RR}_k} \right)^2} = \frac{\left(\ln(1.010) - \ln(1.022) \right)^2}{0.135^2}$$

$$+ \frac{\left(\ln(1.026) - \ln(1.022) \right)^2}{0.188^2}$$

$$= 0.0081, \text{df} = k - 1 = 2 - 1 = 1.$$

C. $P = 0.93$.

D. Significance level. The evidence against the null hypothesis is not at all significant.

E. Conclusion. There is no significant interaction in the relative risks ($\widehat{RR}_{whites}$ = 1.01, $\widehat{RR}_{blacks}$ = 1.03, P = 0.93).

Comment: The test for interaction derived by WinPEPI produces chi-square = 0.005, 1 df, P = 0.94 (i.e., nearly identical results). However, instead of using the Mantel–Haenszel relative risk estimate in its equation, WinPEPI uses an "inverse variance" pooled estimate of relative risk as its baseline summary measure.

19.5 *Sex bias in graduate school admissions?*
(a) Crude analysis
$\hat{p}_{males}$ = 534/1198 = 0.4457 or about 45%
$\hat{p}_{females}$ = 113/449 = 0.2517 or about 25%
$$\widehat{RR} = \frac{0.4457}{0.2517} = 1.77$$
Overall, a higher percentage of male applications were accepted.
(b) Applicants for major 1
$\hat{p}_{males}$ = 512/825 = 0.6206 or about 62%
$\hat{p}_{females}$ = 89/108 = 0.8241 or about 82%
In major 1, a higher percentage of female applications were accepted.
(c) Applicants for major 2
$\hat{p}_{males}$ = 22/373 = 0.0590 or about 6%
$\hat{p}_{females}$ = 24/341 = 0.070 or about 7%
Major 2 accepted approximately the same percentage of male and female applicants.
(d) There does *not* seem to be gender bias in favor of males in this graduate school's admissions practices. A higher percentage of female applicants were accepted to major 1 (82% vs. 62%). In major 2, the acceptance rates were about the same for females and males (7% and 6%, respectively).
(e) The initial analysis was confounded because males tended to apply to major 1. Major 1 had a high acceptance rate (601 of 933 = 64%), while major 2 had a low acceptance rate (46 of 714 = 6%).
(f) Relative incidence of acceptance for males by major
$\widehat{RR}_1$ = 0.753
$\widehat{RR}_2$ = 0.838
Test for interaction
A. Hypotheses. H_0: $RR_1 = RR_2$ (no interaction) versus H_a: $RR_1 \neq RR_2$ (interaction)
B. Test statistic. $\chi^2_{stat,int}$ = 0.136 with df = 1. Calculated with *WinPEPI* > Compare 2 > A. Proportions.

C. *P*-value. $P = 0.713$

D. Significance level. The evidence against the null hypothesis is not significant at any reasonable level of α.

E. Conclusion. The "RR" for males in major 1 was 0.75. In major 2, the RR was 0.84. These RRs do not differ significantly ($P = 0.71$). Therefore, interaction in the ratio measures of effect is absent.

Comment: It is possible to have no interaction in the incidence *ratios* while still having an interaction in the incidence ratio *differences*: interactions are measure of effect specific. For example, there is no significant interaction in incidence ratios for the current data ($P = 0.71$). However, a test for interaction in the incidence *difference* conducted with WinPEPI proved to be significant ($P = 1.5 \times 10^{-5}$). Thus, the ratio effect measure statistical model adequately predicted the joint effects of gender and major, while the difference effect measure model did not.

(g) $\widehat{RR}_{MH} = 0.77$. After adjusting for major, male applicants were 23% less likely to be accepted than female applicants. The 95% confidence interval for the RR is (0.68, 0.86).

19.7   *Infant survival.*

(a) Incidence in the group with less care $\hat{p}_1 = 20/393 = 0.0509 = 5.1\%$. Incidence in the group with more care $\hat{p}_2 = 6/322 = 0.01863 = 1.9\%$. $\widehat{RR} = \dfrac{20/393}{6/322} = 2.73$. "Less care" was associated with almost three times the risk of not surviving.

(b) In clinic 1, the mortality rates were 1.7% (less care) and 1.3% (more care), respectively, for a relative risk of 1.24. In clinic 2, the mortality rates were 7.9% and 8.0%, respectively, for a relative risk of 0.99. Thus, there is almost no association between survival and amount of care received within the clinics. It is also worth noting that clinical 2 has much higher mortality in absolute terms, possibly because it treats a much sicker population.

(c) The crude association was confounded by "clinic," with "clinic" perhaps representing a surrogate measure for severity of the underlying condition.

(d) Test for interaction in the RRs.

A. H_0: $RR_1 = RR_2$ versus H_a: $RR_1 \neq RR_2$.

B. Chi-square interaction statistic $= 0.047$, df $= 1$ (derived using WinPEPI > Compare2 > precision-based heterogeneity statistic for ratio measures).

C. $P = 0.83$.

D. The evidence against H_0 is not significant.

E. The risk ratio in the two clinics (1.24 and 0.99, respectively) do not differ significantly ($P = 0.83$). Therefore, evidence of interaction in risk ratios is absent.

(e) Mantel–Haenszel summary risk ratio = 1.11 (95% CI: 0.40 to 3.07) calculated with WinPEPI > Compare2 > <u>A</u> proportions. This indicates that there is no significant association between mortality and amount of care received after adjusting for the effect of "clinic."

19.9 *Herniated lumbar discs.*

$$\widehat{OR}_1 = \frac{36 \cdot 113}{28 \cdot 138} = 1.05; \widehat{OR}_2 = \frac{31 \cdot 126}{20 \cdot 82} = 2.38.$$

In testing, H_0: $OR_1 = OR_2$, $\chi^2_{\text{interaction}} = 3.69$, df = 1, $P = 0.055$ (via WinPEPI's heterogeneity of odds ratio procedure). Therefore, the odds ratios in the two strata (1.05 and 2.38, respectively) appear to be heterogeneous ($P = 0.055$); there is evidence of significant interaction in odds ratios.

Because interaction is present, we report separate odds ratios for the strata. There appears to be little effect of "no sports participation" among smokers (odds ratio = 1.05). In contrast, "no sports participation" seems to increase the risk among nonsmokers (odds ratio = 2.38).

Index

Note: Page numbers followed by *f*, *t*, and *n* indicate material in figures, tables, and footnotes respectively.